Springer INdAM Series

Volume 33

Springer INdAM Series

This series will publish textbooks, multi-authors books, thesis and monographs in English language resulting from workshops, conferences, courses, schools, seminars, doctoral thesis, and research activities carried out at INDAM - Istituto Nazionale di Alta Matematica, http://www.altamatematica.it/en. The books in the series will discuss recent results and analyze new trends in mathematics and its applications.
THE SERIES IS INDEXED IN SCOPUS

More information about this series at http://www.springer.com/series/10283

Serena Dipierro
Editor

Contemporary Research in Elliptic PDEs and Related Topics

Editor
Serena Dipierro
Dept. of Mathematics and Statistics
University of Western Australia
Crawley, Perth, WA, Australia

ISSN 2281-518X ISSN 2281-5198 (electronic)
Springer INdAM Series
ISBN 978-3-030-18923-5 ISBN 978-3-030-18921-1 (eBook)
https://doi.org/10.1007/978-3-030-18921-1

This Springer imprint is published by the registered company Springer Nature Switzerland AG.
The registered company address is: Gewerbestrasse 11, 6330 Cham, Switzerland

Preface

This book comprises a series of contributions based primarily on lectures given during the Intensive Period "Contemporary Research in Elliptic PDEs and Related Topics", which was hosted by the Department of Mathematics of the University of Bari during Spring 2017 and sponsored by the Istituto Nazionale di Alta Matematica. It also includes a small number of contributions from a complementary Specialized Conference at which research leaders and talented young researchers introduced their work, providing deep insights into problems, results, and methodologies.

The topics covered in the book reflect the variety of themes considered during the lectures. They include nonlocal equations, elliptic equations and systems, fully nonlinear equations, nonlinear parabolic equations, overdetermined boundary value problems, maximum principles, geometric analysis, control theory, mean field games, and biomathematics. Given the very challenging and complex nature of the problems addressed, many of them require a truly interdisciplinary approach in order to produce major breakthroughs in terms of both theory and applications.

The authors include world-leading experts who have contributed substantially to advances in contemporary research. All have made great efforts to present their work in a way which is simultaneously exhaustive and clearly accessible to PhD students, early career researchers and professional researchers. Accordingly, the contributions collected in this volume will serve as an excellent introduction to a variety of fundamental topics of contemporary investigation and trigger novel and high-quality research.

The INdAM Intensive Period was exceptionally effective in promoting new scientific interactions between leading experts and emerging scholars and in training a new generation of extremely promising young mathematicians. Participants displayed a thirst for knowledge and contagious enthusiasm, while the speakers showed great professionalism and exceptional communication skills. The latter are both very evident in the contributions contained in this book.

We once again thank INdAM for making all this possible.

Crawley, Perth, WA, Australia Serena Dipierro

Contents

About the Editor

Serena Dipierro received her PhD from SISSA (Trieste) in 2012, and she is currently an Associate Professor at the University of Western Australia's Department of Mathematics and Statistics. Her research interests include partial differential equations, free boundary problems, nonlinear analysis, nonlocal equations, and calculus of variations.

Getting Acquainted with the Fractional Laplacian

Nicola Abatangelo and Enrico Valdinoci

Abstract These are the handouts of an undergraduate minicourse at the Università di Bari (see Fig. 1), in the context of the 2017 INdAM Intensive Period "Contemporary Research in elliptic PDEs and related topics". Without any intention to serve as a throughout epitome to the subject, we hope that these notes can be of some help for a very initial introduction to a fascinating field of classical and modern research.

Keywords Fractional calculus · Functional Analysis · Applications

2010 Mathematics Subject Classification 35R11, 34A08, 60G22

1 The Laplace Operator

The operator mostly studied in partial differential equations is likely the so-called Laplacian, given by

$$-\Delta u(x) := -\sum_{j=1}^{n} \frac{\partial^2 u}{\partial x_j^2}(x) = \lim_{r \searrow 0} \frac{\text{const}}{r^{n+2}} \int_{B_r(x)} \big(u(x) - u(y)\big)\, dy = -\,\text{const} \int_{\partial B_1} D^2 u(x)\,\theta \cdot \theta \, d\theta \tag{1.1}$$

N. Abatangelo
Département de mathématique, Université Libre de Bruxelles, Ixelles, Boulevard du Triomphe, Belgium
e-mail: nicola.abatangelo@ulb.ac.be

E. Valdinoci (✉)
Department of Mathematics and Statistics, University of Western Australia, Crawley, WA, Australia
e-mail: enrico.valdinoci@uwa.edu.au

S. Dipierro (ed.), *Contemporary Research in Elliptic PDEs and Related Topics*, Springer INdAM Series 33, https://doi.org/10.1007/978-3-030-18921-1_1

Fig. 1 Working hard (and profitably) in Bari

Of course, one may wonder why mathematicians have a strong preference for such kind of operators—say, why not studying

$$\frac{\partial^7 u}{\partial x_1^7}(x) - \frac{\partial^8 u}{\partial x_2^8}(x) + \frac{\partial^9 u}{\partial x_3^9}(x) - \frac{\partial^{10} u}{\partial x_1 \partial x_2^4 \partial x_3^5}(x) \qquad ?$$

Since historical traditions, scientific legacies or impositions from above by education systems would not be enough to justify such a strong interest in only one operator (plus all its modifications), it may be worth to point out a simple geometric property enjoyed by the Laplacian (and not by many other operators). Namely, Eq. (1.1) somehow reveals that the fact that a function is harmonic (i.e., that its Laplace operator vanishes in some region) is deeply related to the action of "comparing with the surrounding values and reverting to the averaged values in the neighborhood".

To wit, the idea behind the integral representation of the Laplacian in formula (1.1) is that the Laplacian tries to model an "elastic" reaction: the vanishing of such operator should try to "revert the value of a function at some point to the values nearby", or, in other words, from a "political" perspective, the Laplacian is a very "democratic" operator, which aims at levelling out differences in order to make things as uniform as possible. In mathematical terms, one looks at the difference between the values of a given function u and its average in a small ball of radius r, namely

$$\delta_r(x) := u(x) - \fint_{B_r(x)} u(y)\,dy = \fint_{B_r(x)} \big(u(x) - u(y)\big)\,dy.$$

In the smooth setting, a second order Taylor expansion of u and a cancellation in the integral due to odd symmetry show that $\mathfrak{d}_r$ is quadratic in r, hence, in order to detect the "elastic", or "democratic", effect of the model at small scale, one has to divide by r^2 and take the limit as $r \searrow 0$. This is exactly the procedure that we followed in formula (1.1).

Other classical approaches to integral representations of elliptic operators come in view of potential theory and inversion operators, see e.g. [96].

This tendency to revert to the surrounding mean suggests that harmonic equations, or in general equations driven by operators "similar to the Laplacian", possess some kind of rigidity or regularity properties that prevents the solutions to oscillate too much (of course, detecting and establishing these properties is a marvelous, and technically extremely demanding, success of modern mathematics, and we do not indulge in this set of notes on this topic of great beauty and outmost importance, and we refer, e.g. to the classical books [62, 71–73]).

Interestingly, the Laplacian operator, in the perspective of (1.1), is the infinitesimal limit of integral operators. In the forthcoming sections, we will discuss some other integral operators, which recover the Laplacian in an appropriate limit, and which share the same property of averaging the values of the function. Differently from what happens in (1.1), such averaging procedure will not be necessarily confined to a small neighborhood of a given point, but will rather tend to comprise all the possible values of a certain function, by possibly "weighting more" the close-by points and "less" the contributions coming from far.

2 Some Fractional Operators

We describe here the basics of some different fractional[1] operators. The fractional exponent will be denoted by $s \in (0,1)$. For more exhaustive discussions and comparisons see e.g. [24, 49, 81, 82, 84, 91, 104, 107, 108]. For simplicity, we do not treat here the case of fractional operators of order higher than 1 (see e.g. [3–5, 50]).

2.1 The Fractional Laplacian

A very popular nonlocal operator is given by the fractional Laplacian

$$(-\Delta)^s u(x) := \text{P.V.} \int_{\mathbb{R}^n} \frac{u(x)-u(y)}{|x-y|^{n+2s}}\, dy. \tag{2.1}$$

[1]The notion (or, better to say, several possible notions) of fractional derivatives attracted the attention of many distinguished mathematicians, such as Leibniz, Bernoulli, Euler, Fourier, Abel, Liouville, Riemann, Hadamard and Riesz, among the others. A very interesting historical outline is given in pages xxvii–xxxvi of [104].

Here above, the notation "P.V." stands for "in the Principal Value sense", that is

$$(-\Delta)^s u(x) := \lim_{\varepsilon \searrow 0} \int_{\mathbb{R}^n \setminus B_\varepsilon(x)} \frac{u(x) - u(y)}{|x - y|^{n+2s}}\, dy.$$

The definition in (2.1) differs from others available in the literature since a normalizing factor has been omitted for the sake of simplicity: this multiplicative constant is only important in the limits as $s \nearrow 1$ and $s \searrow 0$, but plays no essential role for a fixed fractional parameter $s \in (0, 1)$.

The operator in (2.1) can be also conveniently written in the form

$$-(-\Delta)^s u(x) = \frac{1}{2} \int_{\mathbb{R}^n} \frac{u(x+y) + u(x-y) - 2u(x)}{|y|^{n+2s}}\, dy. \tag{2.2}$$

The expression in (2.2) reveals that the fractional Laplacian is a sort of second order difference operator, weighted by a measure supported in the whole of $\mathbb{R}^n$ and with a polynomial decay, namely

$$-(-\Delta)^s u(x) = \frac{1}{2} \int_{\mathbb{R}^n} \delta_u(x, y)\, d\mu(y),$$

$$\text{where} \quad \delta_u(x, y) := u(x+y) + u(x-y) - 2u(x) \quad \text{and} \quad d\mu(y) := \frac{dy}{|y|^{n+2s}}. \tag{2.3}$$

Of course, one can give a pointwise meaning of (2.1) and (2.2) if u is sufficiently smooth and with a controlled growth at infinity (and, in fact, it is possible to set up a suitable notion of fractional Laplacian also for functions that grow polynomially at infinity, see [59]). Besides, it is possible to provide a functional framework to define such operator in the weak sense (see e.g. [106]) and a viscosity solution approach is often extremely appropriate to construct general regularity theories (see e.g. [31]).

We refer to [49] for a gentle introduction to the fractional Laplacian.

From the point of view of the Fourier Transform, denoted, as usual, by $\widehat{\cdot}$ or by $\mathcal{F}$ (depending on the typographical convenience), an instructive computation (see e.g. Proposition 3.3 in [49]) shows that

$$\widehat{(-\Delta)^s u}(\xi) = c\, |\xi|^{2s}\, \widehat{u}(\xi),$$

for some $c > 0$. An appropriate choice of the normalization constant in (2.1) (also in dependence of n and s) allows us to take $c = 1$, and we will take this normalization for the sake of simplicity (and with the slight abuse of notation of dropping constants here and there). With this choice, the fractional Laplacian in Fourier space is simply the multiplication by the symbol $|\xi|^{2s}$, consistently with the fact that the classical Laplacian corresponds to the multiplication by $|\xi|^2$. In particular, the fractional

Laplacian recovers[2] the classical Laplacian as $s \nearrow 1$. In addition, it satisfies the semigroup property, for any $s, s' \in (0,1)$ with $s + s' \leqslant 1$,

$$\mathscr{F}(-\Delta)^s(-\Delta)^{s'}u = |\xi|^{2s}\,\mathscr{F}((-\Delta)^{s'}u) = |\xi|^{2s}\,|\xi|^{2s'}\,\widehat{u} = |\xi|^{2(s+s')}\,\widehat{u} = \mathscr{F}(-\Delta)^{s+s'}u,$$

that is

$$(-\Delta)^s(-\Delta)^{s'}u = (-\Delta)^{s'}(-\Delta)^s u = (-\Delta)^{s+s'}u. \tag{2.4}$$

As a special case of (2.4), when $s = s' = 1/2$, we have that the square root of the Laplacian applied twice produces the classical Laplacian, namely

$$\left((-\Delta)^{1/2}\right)^2 = -\Delta. \tag{2.5}$$

This observation gives that if $U : \mathbb{R}^n \times [0, +\infty) \to \mathbb{R}$ is the harmonic extension[3] of $u : \mathbb{R}^n \to \mathbb{R}$, i.e. if

$$\begin{cases} \Delta U = 0 & \text{in } \mathbb{R}^n \times [0, +\infty), \\ U(x, 0) = u(x) & \text{for any } x \in \mathbb{R}^n, \end{cases} \tag{2.6}$$

then

$$-\,\partial_y U(x, 0) = (-\Delta)^{1/2}u(x). \tag{2.7}$$

See Appendix A for a confirmation of this. In a sense, formula (2.7) is a particular case of a general approach which reduces the fractional Laplacian to a local operator which is set in a halfspace with an additional dimension and may be of singular or degenerate type, see [30].

As a rather approximative "general nonsense", we may say that the fractional Laplacian shares some common feature with the classical Laplacian. In particular,

[2]We think that it is quite remarkable that the operator obtained by the inverse Fourier Transform of $|\xi|^2\,\widehat{u}$, the classical Laplacian, reduces to a local operator. This is not true for the inverse Fourier Transform of $|\xi|^{2s}\,\widehat{u}$. In this spirit, it is interesting to remark that the fact that the classical Laplacian is a local operator is not immediate from its definition in Fourier space, since computing Fourier Transforms is always a nonlocal operation.

[3]Some care has to be used with extension methods, since the solution of (2.6) is not unique (if U solves (2.6), then so does $U(x, y) + cy$ for any $c \in \mathbb{R}$). The "right" solution of (2.6) that one has to take into account is the one with "decay at infinity", or belonging to an "energy space", or obtained by convolution with a Poisson-type kernel. See e.g. [24] for details.

Also, the extension method in (2.6) and (2.7) can be related to an engineering application of the fractional Laplacian motivated by the displacement of elastic membranes on thin (i.e. codimension one) obstacles, see [28]. The intuition for such application can be grasped from Figs. 7, 10, and 12. These pictures can be also useful to develop some intuition about extension methods for fractional operators and boundary reaction-diffusion equations.

both the classical and the fractional Laplacian are invariant under translations and rotations. Moreover, a control on the size of the fractional Laplacian of a function translates, in view of (2.3), into a control of the oscillation of the function (though in a rather "global" fashion): this "democratic" tendency of the operator of "averaging out" any unevenness in the values of a function is indeed typical of "elliptic" operators—and the classical Laplacian is the prototype example in this class of operators, while the fractional Laplacian is perhaps the most natural fractional counterpart.

To make this counterpart more clear, we will say that a function u is s-harmonic in a set Ω if $(-\Delta)^s u = 0$ at any point of Ω (for simplicity, we take this notion in the "strong" sense, but equivalently one could look at distributional definitions, see e.g. Theorem 3.12 in [18]).

For example, constant functions in $\mathbb{R}^n$ are s-harmonic in the whole space for any $s \in (0, 1)$, as both (2.1) and (2.2) imply.

Another similarity between classical and fractional Laplace equations is given by the fact that notions like those of fundamental solutions, Green functions and Poisson kernels are also well-posed in the fractional case and somehow similar formulas hold true, see e.g. Definitions 1.7 and 1.8, and Theorems 2.3, 2.10, 3.1 and 3.2 in [22] (and related formulas hold true also for higher-order fractional operators, see [3–5, 50]).

In addition, space inversions such as the Kelvin Transform also possess invariant properties in the fractional framework, see e.g. [19] (see also Lemma 2.2 and Corollary 2.3 in [63], and in addition Proposition A.1 on page 300 in [97] for a short proof). Moreover, fractional Liouville-type results hold under various assumptions, see e.g. [64] and [59].

Another interesting link between classical and fractional operators is given by subordination formulas which permit to reconstruct fractional operators from the heat flow of classical operators, such as

$$(-\Delta)^s u = -\frac{s}{\Gamma(1-s)} \int_0^{+\infty} t^{-1-s}\Big(e^{t\Delta} - 1\Big)u\,dt,$$

see [11].

In spite of all these similarities, many important structural differences between the classical and the fractional Laplacian arise. Let us list some of them.

Difference 2.1 (Locality Versus Nonlocality) The classical Laplacian of u at a point x only depends on the values of u in $B_r(x)$, for any $r > 0$.

This is not true for the fractional Laplacian. For instance, if $u \in C_0^\infty(B_2, [0, 1])$ with $u = 1$ in B_1, we have that, for any $x \in \mathbb{R}^n \setminus B_4$,

$$-(-\Delta)^s u(x) = \text{P.V.} \int_{\mathbb{R}^n} \frac{u(y) - u(x)}{|x-y|^{n+2s}}\,dy = \int_{B_2} \frac{u(y)}{|x-y|^{n+2s}}\,dy \geqslant \int_{B_1} \frac{dy}{\big(|x|+1\big)^{n+2s}} \geqslant \frac{\text{const}}{|x|^{n+2s}} \tag{2.8}$$

while of course $\Delta u(x) = 0$ in this setting.

It is worth remarking that the estimate in (2.8) is somewhat optimal. Indeed, if u belongs to the Schwartz space (or space of rapidly decreasing functions)

$$\mathcal{S} := \left\{ u \in C^\infty(\mathbb{R}^n) \text{ s.t. } \sup_{x \in \mathbb{R}^n} |x|^\alpha \left| D^\beta u(x) \right| < +\infty \text{ for all } \alpha, \beta \in \mathbb{N}^n \right\}, \tag{2.9}$$

we have that, for large $|x|$,

$$\left| (-\Delta)^s u(x) \right| \leqslant \frac{\text{const}}{|x|^{n+2s}}. \tag{2.10}$$

See Appendix B for the proof of this fact.

Difference 2.2 (Summability Assumptions) The pointwise computation of the classical Laplacian on a function u does not require integrability properties on u. Conversely, formula (2.1) for u can make sense only when

$$\int_{\mathbb{R}^n} \frac{|u(y)|}{1 + |y|^{n+2s}} \, dy \; < \; +\infty$$

which can be read as a local integrability complemented by a growth condition at infinity. This feature, which could look harmless at a first glance, can result problematic when looking for singular solutions to nonlinear problems (as, for example, in [1, 66] where there is an unavoidable integrability obstruction on a bounded domain) or in "blow-up" type arguments (as mentioned in [59], where the authors propose a way to outflank this restriction).

Difference 2.3 (Computation Along Coordinate Directions) The classical Laplacian of u at the origin only depends on the values that u attains along the coordinate directions (or, up to a rotation, along a set of n orthogonal directions).

This is not true for the fractional Laplacian. As an example, let $u \in C_0^\infty(B_2(4e_1 + 4e_2), [0, 1])$, with $u = 1$ in $B_1(4e_1 + 4e_2)$. Let also R_j be the straight line in the jth coordinate direction, that is

$$R_j := \{te_j, \; t \in \mathbb{R}\},$$

see Fig. 2. Then

$$R_j \cap B_2(4e_1 + 4e_2) = \varnothing$$

for each $j \in \{1, \ldots, n\}$, and so $u(te_j) = 0$ for all $t \in \mathbb{R}$ and $j \in \{1, \ldots, n\}$. This gives that $\Delta u(0) = 0$.

On the other hand,

$$\int_{\mathbb{R}^n} \frac{u(y) - u(0)}{|0 - y|^{n+2s}} \, dy = \int_{\mathbb{R}^n} \frac{u(y)}{|y|^{n+2s}} \, dy \geqslant \int_{B_1(4e_1+4e_2)} \frac{dy}{|y|^{n+2s}} > 0,$$

which says that $(-\Delta)^s u(0) \neq 0$.

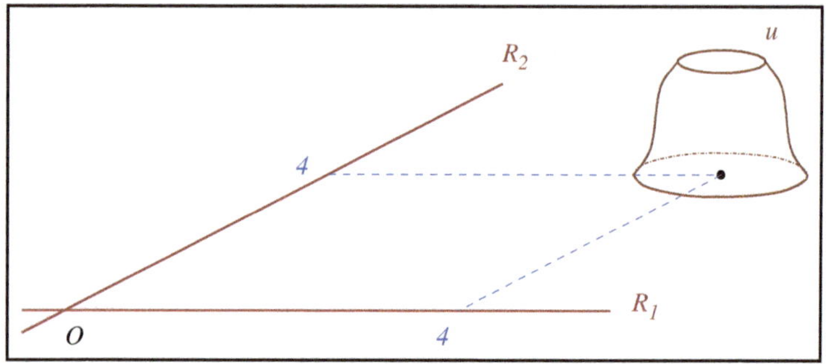

Fig. 2 Coordinate directions not meeting a bump function

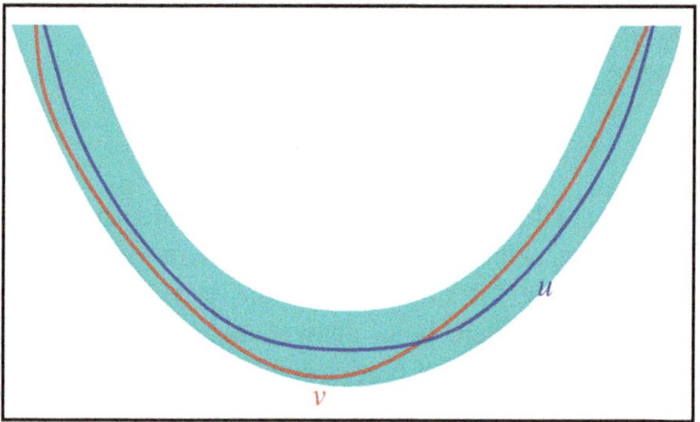

Fig. 3 A function v which is "close to u"

Difference 2.4 (Harmonic Versus s-Harmonic Functions) If $\Delta u(0) = 1$, $\|u - v\|_{C^2(B_1)} \leqslant \varepsilon$ and $\varepsilon > 0$ is sufficiently small (see Fig. 3) then $\Delta v(0) \geqslant 1 - \text{const}\,\varepsilon > 0$, and in particular $\Delta v(0) \neq 0$.

Quite surprisingly, this is not true for the fractional Laplacian. More generally, in this case, as proved in [55], for any $\varepsilon > 0$ and any (bounded, smooth) function $\bar{u}$, we can find v_ε such that

$$\begin{cases} \quad \|\bar{u} - v_\varepsilon\|_{C^2(B_1)} \leqslant \varepsilon \\ \text{and } (-\Delta)^s v_\varepsilon = 0 \text{ in } B_1. \end{cases} \tag{2.11}$$

A proof of this fact in dimension 1 for the sake of simplicity is given in [112] (the original paper [55] presents a complete proof in any dimension). See also [70, 99, 100] for different approaches to approximation methods in fractional settings which lead to new proofs, and very refined and quantitative statements.

We also mention that the phenomenon described in (2.11) (which can be summarized in the evocative statement that *all functions are locally s-harmonic (up to a small error)*) is very general, and it applies to other nonlocal operators, also independently from their possibly "elliptic" structure (for instance all functions are locally s-caloric, or s-hyperbolic, etc.). In this spirit, for completeness, in Sect. 5 we will establish the density of fractional caloric functions in one space variable, namely of the fact that for any $\varepsilon > 0$ and any (bounded, smooth) function $\bar{u} = \bar{u}(x,t)$, we can find $v_\varepsilon = v_\varepsilon(x,t)$ such that

$$\begin{cases} \|\bar{u} - v_\varepsilon\|_{C^2((-1,1)\times(-1,1))} \leqslant \varepsilon \\ \text{and } \partial_t v_\varepsilon + (-\Delta)^s v_\varepsilon = 0 \text{ for any } x \in (-1,1) \text{ and any } t \in (-1,1). \end{cases} \tag{2.12}$$

We also refer to [58] for a general approach and a series of general results on this type of approximation problems with solutions of operators which are the superposition of classical differential operators with fractional Laplacians. Furthermore, similar results hold true for other nonlocal operators with memory, see [23]. See in addition [36, 37, 79] for related results on higher order fractional operators.

Difference 2.5 (Harnack Inequality) The classical Harnack Inequality says that if u is harmonic in B_1 and $u \geqslant 0$ in B_1 then

$$\inf_{B_{1/2}} u \geqslant \text{const} \sup_{B_{1/2}} u,$$

for a suitable universal constant, only depending on the dimension.

The same result is not true for s-harmonic functions. To construct an easy counterexample, let $\bar{u}(x) = |x|^2$ and, for a small $\varepsilon > 0$, let v_ε be as in (2.11). Notice that, if $x \in B_1 \setminus B_{1/4}$

$$v_\varepsilon(x) \geqslant \bar{u}(x) - \|\bar{u} - v_\varepsilon\|_{L^\infty(B_1)} \geqslant \frac{1}{16} - \varepsilon > \frac{1}{32} \tag{2.13}$$

if ε is small enough, while

$$v_\varepsilon(0) \leqslant \bar{u}(0) + \|\bar{u} - v_\varepsilon\|_{L^\infty(B_1)} \leqslant 0 + \varepsilon < \frac{1}{32}.$$

These observations imply that $v_\varepsilon(0) < v_\varepsilon(x)$ for all $x \in B_1 \setminus B_{1/4}$ and therefore the infimum of v_ε in B_1 is taken at some point $\bar{x}$ in the closure of $B_{1/4}$. Then, we define

$$u_\varepsilon(x) := v_\varepsilon(x) - \inf_{B_1} v_\varepsilon = v_\varepsilon(x) - v_\varepsilon(\bar{x}).$$

Notice that u_ε is s-harmonic in B_1, since so is v_ε, and $u_\varepsilon \geqslant 0$ in B_1. Also, u_ε is strictly positive in $B_1 \setminus B_{1/4}$. On the other hand, since $\bar{x} \in B_{1/2}$

$$\inf_{B_{1/2}} u_\varepsilon = u_\varepsilon(\bar{x}) = 0,$$

which implies that u_ε cannot satisfy a Harnack Inequality as the one in (2.13).

In any case, it must be said that suitable Harnack Inequalities are valid also in the fractional case, under suitable “global” assumptions on the solution: for instance, the Harnack Inequality holds true for solutions that are positive in the whole of $\mathbb{R}^n$ rather than in a given ball. We refer to [75, 76] for a comprehensive discussion on this topic and for recent developments.

Difference 2.6 (Growth from the Boundary) Roughly speaking, solutions of Laplace equations have “linear (i.e. Lipschitz) growth from the boundary”, while solutions of fractional Laplace equations have only Hölder growth from the boundary. To understand this phenomenon, we point out that if u is continuous in the closure of B_1, with $\Delta u = f$ in B_1 and $u = 0$ on ∂B_1, then

$$|u(x)| \leqslant \text{const}\,(1 - |x|) \sup_{B_1} |f|. \tag{2.14}$$

Notice that the term $(1 - |x|)$ represents the distance of the point $x \in B_1$ from ∂B_1. See e.g. Appendix C for a proof of (2.14).

The case of fractional equations is very different. A first example which may be useful to keep in mind is that the function

$$\begin{aligned} &\mathbb{R}^n \ni x \mapsto (x_n)_+^s \\ &\text{is } s\text{-harmonic in the halfspace } \{x_n > 0\}. \end{aligned} \tag{2.15}$$

For an elementary proof of this fact, see e.g. Section 2.4 in [24]. Remarkably, the function in (2.15) is only Hölder continuous with Hölder exponent s near the origin.

Another interesting example is given by the function

$$\mathbb{R} \ni x \mapsto u_{1/2}(x) := (1 - |x|^2)_+^{1/2}, \tag{2.16}$$

which satisfies

$$(-\Delta)^{1/2} u_{1/2} = \text{const} \quad \text{in } (-1, 1). \tag{2.17}$$

A proof of (2.17) based on extension methods and complex analysis is given in Appendix D.

The identity in (2.17) is in fact a special case of a more general formula, according to which the function

$$\mathbb{R}^n \ni x \mapsto u_s(x) := (1 - |x|^2)_+^s \tag{2.18}$$

satisfies

$$(-\Delta)^s u_s = \text{const} \quad \text{in } B_1. \tag{2.19}$$

For this formula, and in fact even more general ones, see [61]. See also [69] for a probabilistic approach.

Interestingly, (2.15) can be obtained from (2.19) by a blow-up at a point on the zero level set.

Notice also that

$$\lim_{|x|\nearrow 1} \frac{|u_s(x)|}{1-|x|} = \lim_{|x|\nearrow 1} \frac{(1-|x|^2)_+^s}{1-|x|} = \lim_{|x|\nearrow 1} \frac{1}{(1-|x|)^{1-s}} = +\infty,$$

therefore, differently from the classical case, u_s does not satisfy an estimate like that in (2.14).

It is also interesting to observe that the function u_s is related to the function x_+^s via space inversion (namely, a Kelvin transform) and integration, and indeed one can also deduce (2.19) from (2.15): this fact was nicely remarked to us by Xavier Ros-Oton and Joaquim Serra, and the simple but instructive proof is sketched in Appendix E.

Difference 2.7 (Global (Up to the Boundary) Regularity) Roughly speaking, solutions of Laplace equations are "smooth up to the boundary", while solutions of fractional Laplace equations are not better than Hölder continuous at the boundary. To understand this phenomenon, we point out that if u is continuous in the closure of B_1,

$$\begin{cases} \Delta u = f \;\text{ in } B_1, \\ u = 0 \;\text{ on } \partial B_1, \end{cases} \tag{2.20}$$

then

$$\sup_{x \in B_1} |\nabla u(x)| \leqslant \text{const} \sup_{B_1} |f|. \tag{2.21}$$

See e.g. Appendix F for a proof of this fact.

The case of fractional equations is very different since the function u_s in (2.18) is only Hölder continuous (with Hölder exponent s) in B_1, hence the global Lipschitz estimate in (2.21) does not hold in this case. This phenomenon can be seen as a counterpart of the one discussed in Difference 2.6. The boundary regularity for fractional Laplace problems is discussed in details in [97].

Difference 2.8 (Explosive Solutions) Solutions of classical Laplace equations cannot attain infinite values in the whole of the boundary. For instance, if u is harmonic in B_1, then

$$\overline{\lim_{\rho \nearrow 1}} \inf_{\partial B_\rho} u \leqslant \text{const}\, u(0). \tag{2.22}$$

Indeed, by the Mean Value Property for harmonic functions, for any $\rho \in (0,1)$,

$$u(0) = \frac{\text{const}}{\rho^{n-1}} \int_{\partial B_\rho} u(x)\, d\mathcal{H}_x^{n-1} \geqslant \inf_{\partial B_\rho} u,$$

from which (2.22) plainly follows (another proof follows by using the Maximum Principle instead of the Mean Value Property). On the contrary, and quite remarkably, solutions of fractional Laplace equations may "explode" at the boundary and (2.22) can be violated by s-harmonic functions in B_1 which vanish outside B_1.

For example, for

$$\mathbb{R} \ni x \mapsto u_{-1/2}(x) := \begin{cases} (1-|x|^2)^{-1/2} & \text{if } x \in (-1,1), \\ 0 & \text{otherwise,} \end{cases} \tag{2.23}$$

one has

$$(-\Delta)^{1/2} u_{-1/2} = 0 \quad \text{in } (-1,1), \tag{2.24}$$

and, of course, (2.22) is violated by $u_{-1/2}$. The claim in (2.24) can be proven starting from (2.17) and by suitably differentiating both sides of the equation: the details of this computation can be found in Appendix G. For completeness, we also give in Appendix H another proof of (2.24) based on complex variable and extension methods.

A geometric interpretation of (2.24) is depicted in Fig. 4 where a point $x \in (-1,1)$ is selected and the graph of $u_{-1/2}$ above the value $u_{-1/2}(x)$ is drawn with a "dashed curve" (while a "solid curve" represents the graph of $u_{-1/2}$ below the

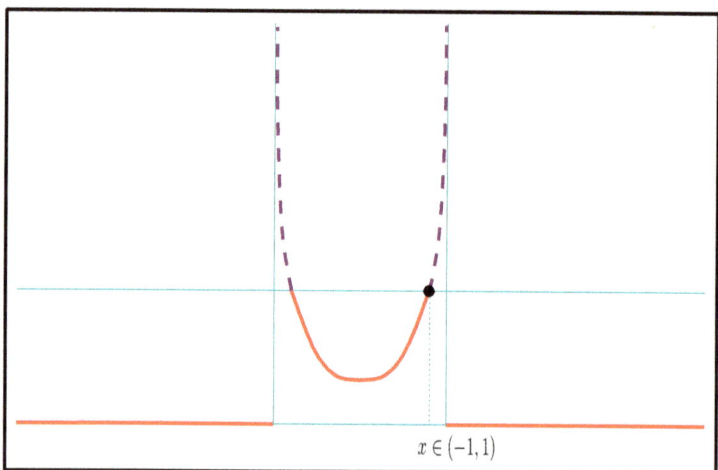

Fig. 4 The function $u_{-1/2}$ and the cancellation occurring in (2.24)

value $u_{-1/2}(x)$): then, when computing the fractional Laplacian at x, the values coming from the dashed curve, compared with $u_{-1/2}(x)$, provide an opposite sign with respect to the values coming from the solid curve. The "miracle" occurring in (2.24) is that these two contributions with opposite sign perfectly compensate and cancel each other, for any $x \in (-1, 1)$.

More generally, in every smooth bounded domain $\Omega \subset \mathbb{R}^n$ it is possible to build s-harmonic functions exploding at $\partial\Omega$ at the same rate as $\mathrm{dist}(\cdot, \partial\Omega)^{s-1}$. A phenomenon of this sort was spotted in [66], and see [1] for the explicit explosion rate. See [1] also for a justification of the boundary behavior, as well as the study of Dirichlet problems prescribing a singular boundary trace.

Concerning this feature of explosive solutions at the boundary, it is interesting to point out a simple analogy with the classical Laplacian. Indeed, in view of (2.15), if $s \in (0, 1)$ and we take the function $\mathbb{R} \ni x \mapsto x_+^s$, we know that it is s-harmonic in $(0, +\infty)$ and it vanishes on the boundary (namely, the origin), and these features have a clear classical analogue for $s = 1$. Then, since for all $s \in (0, 1]$ the derivative of x_+^s is x_+^{s-1}, up to multiplicative constants, we have that the latter is s-harmonic in $(0, +\infty)$ and it blows-up at the origin when $s \in (0, 1)$ (conversely, when $s = 1$ one can do the same computations but the resulting function is simply the characteristic function of $(0, +\infty)$ so no explosive effect arises).

Similar computations can be done in the unit ball instead of $(0, +\infty)$, and one simply gets functions that are bounded up to the boundary when $s = 1$, or explosive when $s \in (0, 1)$ (further details in Appendices G and H).

Difference 2.9 (Decay at Infinity) The Gaussian $e^{-|x|^2}$ reproduces the classical heat kernel. That is, the solution of the heat equation with initial datum concentrated at the origin, when considered at time $t = 1/4$, produces the Gaussian (of course, the choice $t = 1/4$ is only for convenience, any time t can be reduced to unit time by scaling the equation).

The fast decay prescribed by the Gaussian is special for the classical case and the fractional case exhibits power law decays at infinity. More precisely, let us consider the heat equation with initial datum concentrated at the origin, that is

$$\begin{cases} \partial_t u(x,t) = -(-\Delta)^s u(x,t) \ \text{ for } (x,t) \in \mathbb{R}^n \times (0,+\infty), \\ \qquad\qquad\qquad\qquad u(x,0) = \delta_0, \end{cases} \tag{2.25}$$

and set

$$\mathscr{G}_s(x) = u(x, 1). \tag{2.26}$$

By taking the Fourier Transform of (2.25) in the x variable (and possibly neglecting normalization constants) one finds that

$$\begin{cases} \partial_t \hat{u} = -|\xi|^{2s} \hat{u} \ \text{ in } \mathbb{R}^n \times (0,+\infty), \\ \qquad\qquad u(\xi, 0) = 1, \end{cases}$$

hence

$$\hat{u} = e^{-|\xi|^{2s}t}, \tag{2.27}$$

and consequently

$$\mathscr{G}_s(x) = \mathscr{F}^{-1}(e^{-|\xi|^{2s}}), \tag{2.28}$$

being $\mathscr{F}^{-1}$ the anti-Fourier Transform of the Fourier Transform $\mathscr{F}$. When $s = 1$, and neglecting the normalizing constants, the expression in (2.28) reduces to the Gaussian (since the Gaussian is the Fourier Transform of itself). On the other hand, as far as we know, there is no simple explicit representation of the fractional heat kernel in (2.28), except in the "miraculous" case $s = 1/2$, in which (2.28) provides the explicit representation

$$\mathscr{G}_{1/2}(x) = \frac{\text{const}}{\left(1 + |x|^2\right)^{\frac{n+1}{2}}}. \tag{2.29}$$

See Appendix I for a proof of (2.29) using Fourier methods and Appendix J for a proof based on extension methods.

We stress that, differently from the classical case, the heat kernel $\mathscr{G}_{1/2}$ decays only with a power law. This is in fact a general feature of the fractional case, since, for any $s \in (0, 1)$, it holds that

$$\lim_{|x|\to+\infty} |x|^{n+2s}\mathscr{G}_s(x) = \text{const} \tag{2.30}$$

and, for $|x| \geqslant 1$ and $s \in (0, 1)$, the heat kernel $\mathscr{G}_s(x)$ is bounded from below and from above by $\frac{\text{const}}{|x|^{n+2s}}$.

We refer to [78] for a detailed discussion on the fractional heat kernel. See also [13] for more information on the fractional heat equation. For precise asymptotics on fractional heat kernels, see [15, 17, 47, 95].

The decay of the heat kernel is also related to the associated distribution in probability theory: as we will see in Sect. 4.2, the heat kernel represents the probability density of finding a particle at a given point after a unit of time; the motion of such particle is driven by a random walk in the classical case and by a random process with long jumps in the fractional case and, as a counterpart, the fractional probability distribution exhibits a "long tail", in contrast with the rapidly decreasing classical one.

Another situation in which the classical case provides exponentially fast decaying solutions while the fractional case exhibits polynomial tails is given by the Allen-Cahn equation (see e.g. Section 1.1 in [65] for a simple description of this equation also in view of phase coexistence models). For concreteness, one can consider the one-dimensional equation

$$\begin{cases} (-\Delta)^s u = u - u^3 \ \text{ in } \mathbb{R}, \\ \dot{u} > 0, \\ u(0) = 0, \\ \lim_{t \to \pm\infty} u(t) = \pm 1. \end{cases} \tag{2.31}$$

For $s = 1$, the system in (2.31) reduces to the pendulum-like system

$$\begin{cases} -\ddot{u} = u - u^3 \ \text{ in } \mathbb{R}, \\ \dot{u} > 0, \\ u(0) = 0, \\ \lim_{t \to \pm\infty} u(t) = \pm 1. \end{cases} \tag{2.32}$$

The solution of (2.32) is explicit and it has the form

$$u(t) := \tanh \frac{t}{\sqrt{2}}, \tag{2.33}$$

as one can easily check. Also, by inspection, we see that such solution satisfies

$$\begin{aligned} &|u(t) - 1| \leqslant \text{const} \exp(-\text{const}\ t) \quad \text{for any } t \geqslant 1 \\ \text{and } &|u(t) + 1| \leqslant \text{const} \exp(-\text{const}\ |t|) \quad \text{for any } t \leqslant -1. \end{aligned} \tag{2.34}$$

Conversely, to the best of our knowledge, the solution of (2.31) has no simple explicit expression. Also, remarkably, the solution of (2.31) decays to the equilibria ± 1 only polynomially fast. Namely, as proved in Theorem 2 of [92], we have that the solution of (2.31) satisfies

$$\begin{aligned} &|u(t) - 1| \leqslant \frac{\text{const}}{t^{2s}} \quad \text{for any } t \geqslant 1 \\ \text{and } &|u(t) + 1| \leqslant \frac{\text{const}}{|t|^{2s}} \quad \text{for any } t \leqslant -1, \end{aligned} \tag{2.35}$$

and the estimates in (2.35) are optimal, namely it also holds that

$$\begin{aligned} &|u(t) - 1| \geqslant \frac{\text{const}}{t^{2s}} \quad \text{for any } t \geqslant 1 \\ \text{and } &|u(t) + 1| \geqslant \frac{\text{const}}{|t|^{2s}} \quad \text{for any } t \leqslant -1. \end{aligned} \tag{2.36}$$

See Appendix K for a proof of (2.36). In particular, (2.36) says that solutions of fractional Allen-Cahn equations such as the one in (2.31) do not satisfy the exponential decay in (2.34) which is fulfilled in the classical case.

The estimate in (2.36) can be confirmed by looking at the solution of the very similar equation

$$\begin{cases} (-\Delta)^s u = \dfrac{1}{\pi}\sin(\pi u) \ \text{ in } \mathbb{R}, \\ \dot{u} > 0, \\ u(0) = 0, \\ \lim\limits_{t\to\pm\infty} u(t) = \pm 1. \end{cases} \tag{2.37}$$

Though a simple expression of the solution of (2.37) is not available in general, the "miraculous" case $s = 1/2$ possesses an explicit solution, given by

$$u(t) := \frac{2}{\pi}\arctan t. \tag{2.38}$$

That (2.38) is a solution of (2.37) when $s = 1/2$ is proved in Appendix L. Another proof of this fact using (2.29) is given in Appendix M.

The reader should not be misled by the similar typographic forms of (2.33) and (2.38), which represent two very different behaviors at infinity: indeed

$$\lim_{t\to+\infty} t\left(1 - \frac{2}{\pi}\arctan t\right) = \frac{2}{\pi},$$

and the function in (2.38) satisfies the slow decay in (2.36) (with $s = 1/2$) and not the exponentially fast one in (2.34).

Equations like the one in (2.31) naturally arise, for instance, in long-range phase coexistence models and in models arising in atom dislocation in crystals, see e.g. [52, 110].

A similar slow decay also occurs in the study of fractional Schrödinger operators, see e.g. [38] and Lemma C.1 in [68]. For instance, the solution of

$$(-\Delta)^s\Gamma + \Gamma = \delta_0 \quad \text{in } \delta_0 \tag{2.39}$$

satisfies, for any $|x| \geqslant 1$,

$$\Gamma(x) \simeq \frac{\text{const}}{|x|^{n+2s}}.$$

A heuristic motivation for a bound of this type can be "guessed" from (2.39) by thinking that, for large $|x|$, the function Γ should decay more or less like $(-\Delta)^s\Gamma$, which has "typically" the power law decay described in (2.10).

If one wishes to keep arguing in this heuristic way, also the decays in (2.30) and (2.36) may be seen as coming from an interplay between the right and the left side of the equation, in the light of the decay of the fractional Laplace operator discussed in (2.10). For instance, to heuristically justify (2.30), one may think that the solution of the fractional heat equation which starts from a Dirac's Delta, after a unit of time (or an "infinitesimal unit" of time, if one prefers) has produced some bump, whose fractional Laplacian, in view of (2.10), may decay at infinity like $\frac{1}{|x|^{n+2s}}$. Since the time derivative of the solution has to be equal to that, the solution itself, in this unit of time, gets "pushed up" by an amount like $\frac{1}{|x|^{n+2s}}$ with respect to the initial datum, thus justifying (2.30).

A similar justification for (2.36) may seem more tricky, since the decay in (2.36) is only of the type $\frac{1}{|t|^{2s}}$ instead of $\frac{1}{|t|^{1+2s}}$, as the analysis in (2.10) would suggest. But to understand the problem, it is useful to consider the derivative of the solution $v := \dot{u}$ and deduce from (2.31) that

$$(-\Delta)^s v = (-\Delta)^s \dot{u} = \dot{u} - 3u^2\dot{u} = (1 - 3u^2)v. \tag{2.40}$$

That is, for large $|t|$, the term $1 - 3u^2$ gets close to $1 - 3 = -2$ and so the profile at infinity may locally resemble the one driven by the equation $(-\Delta)^s v = -2v$. In this range, v has to balance its fractional Laplacian, which is expected to decay like $\frac{1}{|t|^{1+2s}}$, in view of (2.10). Then, since u is the primitive of v, one may expect that its behavior at infinity is related to the primitive of $\frac{1}{|t|^{1+2s}}$, and so to $\frac{1}{|t|^{2s}}$, which is indeed the correct answer given by (2.36).

We are not attempting here to make these heuristic considerations rigorous, but perhaps these kinds of comments may be useful in understanding why the behavior of nonlocal equations is different from that of classical equations and to give at least a partial justification of the delicate quantitative aspects involved in a rigorous quantitative analysis (in any case, ideas like these are rigorously exploited for instance in Appendix K).

See also [21] for decay estimates of ground states of a nonlinear nonlocal problem.

We also mention that other very interesting differences in the decay of solutions arise in the study of different models for fractional porous medium equations, see e.g. [33, 34, 48].

Difference 2.10 (Finiteness Versus Infiniteness of the Mean Squared Displacement) The mean squared displacement is a useful notion to measure the "speed of a diffusion process", or more precisely the portion of the space that gets "invaded" at a given time by the spreading of the diffusive quantity which is concentrated at a point source at the initial time. In a formula, if $u(x, t)$ is the fundamental solution of the diffusion equation related to the diffusion operator $\mathfrak{L}$, namely

$$\begin{cases} \partial_t u = \mathfrak{L} u & \text{for any } x \in \mathbb{R}^n \text{ and } t > 0, \\ u(\cdot, 0) = \delta_0(\cdot), \end{cases} \tag{2.41}$$

being δ_0 the Dirac's Delta, one can define the mean squared displacement relative to the diffusion process $\mathfrak{L}$ as the "second moment" of u in the space variables, that is

$$\mathrm{MSD}_{\mathfrak{L}}(t) := \int_{\mathbb{R}^n} |x|^2 u(x,t)\, dx. \tag{2.42}$$

For the classical heat equation, by Fourier Transform one sees that, when $\mathfrak{L} = \Delta$, the fundamental solution of (2.41) is given by the classical heat kernel

$$u(x,t) = \frac{1}{(4\pi t)^{n/2}} e^{-\frac{|x|^2}{4t}},$$

and therefore[4] in such case, the substitution $y := \frac{x}{2\sqrt{t}}$ gives that

$$\mathrm{MSD}_{\Delta}(t) = \int_{\mathbb{R}^n} \frac{|x|^2}{(4\pi t)^{n/2}} e^{-\frac{|x|^2}{4t}}\, dx = \int_{\mathbb{R}^n} \frac{4t|y|^2}{\pi^{n/2}} e^{-|y|^2}\, dy = Ct, \tag{2.43}$$

for some $C > 0$. This says that the mean squared displacement of the classical heat equation is finite, and linear in the time variable.

On the other hand, in the fractional case in which $\mathfrak{L} = -(-\Delta)^s$, by (2.27) the fractional heat kernel is endowed with the scaling property

$$u(x,t) = \frac{1}{t^{\frac{n}{2s}}} \mathscr{G}_s\left(\frac{x}{t^{\frac{1}{2s}}}\right),$$

with $\mathscr{G}_s$ being as in (2.25) and (2.26). Consequently, in this case, the substitution $y := \frac{x}{t^{\frac{1}{2s}}}$ gives that

$$\mathrm{MSD}_{-(-\Delta)^s}(t) = \int_{\mathbb{R}^n} |x|^2 \frac{1}{t^{\frac{n}{2s}}} \mathscr{G}_s\left(\frac{x}{t^{\frac{1}{2s}}}\right) dx = t^{\frac{1}{s}} \int_{\mathbb{R}^n} |y|^2 \mathscr{G}_s(y)\, dy. \tag{2.44}$$

Now, from (2.30), we know that

$$\int_{\mathbb{R}^n} |y|^2 \mathscr{G}_s(y)\, dy = +\infty$$

and therefore we infer from (2.44) that

$$\mathrm{MSD}_{-(-\Delta)^s}(t) = +\infty. \tag{2.45}$$

[4] See Appendix A in [103] for a very nice explanation of the dimensional analysis and for a throughout discussion of its role in detecting fundamental solutions.

This computation shows that, when $s \in (0, 1)$, the diffusion process induced by $-(-\Delta)^s$ does not possess a finite mean squared displacement, in contrast with the classical case in (2.43).

Other important differences between the classical and fractional cases arise in the study of nonlocal minimal surfaces and in related fields: just to list a few features, differently than in the classical case, nonlocal minimal surfaces typically "stick" at the boundary, see [25, 53, 56], the gradient bounds of nonlocal minimal graphs are different than in the classical case, see [26], nonlocal catenoids grow linearly and nonlocal stable cones arise in lower dimension, see [45, 46], stable surfaces of vanishing nonlocal mean curvature possess uniform perimeter bounds, see Corollary 1.8 in [42], the nonlocal mean curvature flow develops singularity also in the plane, see [41], its fattening phenomena are different, see [40], and the self-shrinking solutions are also different, see [39], and genuinely nonlocal phase transitions present stronger rigidity properties than in the classical case, see e.g. Theorem 1.2 in [60] and [67]. Furthermore, from the probabilistic viewpoint, recurrence and transiency in long-jump stochastic processes are different from the case of classical random walks, see e.g. [6] and the references therein.

We would like to conclude this list of differences with one similarity, which seems to be not very well-known. There is indeed a "nonlocal representation" for the classical Laplacian in terms of a singular kernel. It reads as

$$-\Delta u(x) = \text{const} \int_{\mathbb{R}^n} \frac{u(x+2y) + u(x-2y) - 4u(x+y) - 4u(x-y) + 6u(x)}{|y|^{n+2}}\, dy. \tag{2.46}$$

This one is somehow very close to (2.2) with one important modification: the difference operator in the numerator of the integrand has been increased in order, in such a way that it is able to compensate the singularity of the kernel in 0. We include in Appendix N a computation proving (2.46) when u is $C^{2,\alpha}$ around x. For a complete proof, involving Fourier transform techniques and providing the explicit value of the constant, we refer to [3].

2.2 *The Regional (or Censored) Fractional Laplacian*

A variant of the fractional Laplacian in (2.1) consists in restricting the domain of integration to a subset of $\mathbb{R}^n$. In this direction, an interesting operator is defined by the following singular integral:

$$(-\Delta)^s_\Omega u(x) := \text{P.V.} \int_\Omega \frac{u(x) - u(y)}{|x-y|^{n+2s}}\, dy. \tag{2.47}$$

We remark that when $\Omega := \mathbb{R}^n$ the regional fractional Laplacian in (2.47) boils down to the standard fractional Laplacian in (2.1).

In spite of the apparent similarity, the regional fractional Laplacian and the fractional Laplacian are structurally two different operators. For instance, concerning Difference 2.4, we mention that solutions of regional fractional Laplace equations do not possess the same rich structure of those of fractional Laplace equations, and indeed

$$\begin{aligned}&\text{it is not true that for any } \varepsilon > 0 \text{ and any (bounded, smooth) function } \bar{u},\\&\text{we can find } v_\varepsilon \text{ such that}\\&\begin{cases} \|\bar{u} - v_\varepsilon\|_{C^2(B_1)} \leqslant \varepsilon \\ \text{and } (-\Delta)^s_\Omega v_\varepsilon = 0 \text{ in } B_1. \end{cases}\end{aligned} \tag{2.48}$$

A proof of this observation will be given in Appendix O.

Interestingly, the regional fractional Laplacian turns out to be useful also in a possible setting of Neumann-type conditions in the nonlocal case, as presented[5] in [54]. Related to this, we mention that it is possible to obtain a regional-type operator starting from the classical Laplacian coupled with Neumann boundary conditions (details about it will be given in formula (2.52) below).

[5]Some colleagues pointed out to us that the use of R and r in some steps of formula (5.5) of [54] are inadequate. We take this opportunity to amend such a flaw, presenting a short proof of (5.5) of [54]. Given $\varepsilon > 0$, we notice that

$$I_1 := \iint_{\substack{\Omega\times(\mathbb{R}^n\setminus\Omega)\\ \{|x-y|\geqslant\varepsilon\}}} \frac{|u(x)-u(y)|^2}{|x-y|^{n+2s}}\,dx\,dy \leqslant \iint_{\Omega\times(\mathbb{R}^n\setminus B_\varepsilon)} \frac{4\,\|u\|^2_{L^\infty(\mathbb{R}^n)}\,dx\,d\zeta}{|\zeta|^{n+2s}} \leqslant \frac{\text{const}}{s\,\varepsilon^{2s}},$$

where the constants are also allowed to depend on Ω and u. Furthermore, if we define Ω_ε to be the set of all the points in Ω with distance less than ε from $\partial\Omega$, the regularity of $\partial\Omega$ implies that the measure of Ω_ε is bounded by $\text{const}\,\varepsilon$, and therefore

$$\begin{aligned} I_2 &:= \iint_{\substack{\Omega\times(\mathbb{R}^n\setminus\Omega)\\ \{|x-y|<\varepsilon\}}} \frac{|u(x)-u(y)|^2}{|x-y|^{n+2s}}\,dx\,dy \leqslant \iint_{\Omega_\varepsilon\times B_\varepsilon(x)} \frac{4\,\|u\|^2_{C^1(\mathbb{R}^n)}\,|x-y|^2\,dx\,dy}{|x-y|^{n+2s}} \\ &\leqslant \int_{B_\varepsilon} \frac{\text{const}\,\varepsilon\,d\zeta}{|\zeta|^{n+2s-2}} \leqslant \frac{\text{const}\,\varepsilon^{3-2s}}{1-s}. \end{aligned}$$

These observations imply that

$$\lim_{s\nearrow 1}(1-s)\iint_{\Omega\times(\mathbb{R}^n\setminus\Omega)} \frac{|u(x)-u(y)|^2}{|x-y|^{n+2s}}\,dx\,dy \leqslant \lim_{s\nearrow 1}(1-s)\left(\frac{\text{const}}{s\,\varepsilon^{2s}} + \frac{\text{const}\,\varepsilon^{3-2s}}{1-s}\right) = \text{const}\,\varepsilon.$$

Taking ε as small as we wish, we obtain formula (5.5) in [54].

2.3 The Spectral Fractional Laplacian

Another natural fractional operator arises in taking fractional powers of the eigenvalues. For this, we write

$$u(x,t) = \sum_{k=0}^{+\infty} u_k(t)\,\phi_k(x), \tag{2.49}$$

where ϕ_k is the eigenfunction corresponding to the kth eigenvalue of the Dirichlet Laplacian, namely

$$\begin{cases} -\Delta\phi_k = \lambda_k\phi_k \text{ in } \Omega \\ \phi_k \in H_0^1(\Omega). \end{cases}$$

with $0 < \lambda_0 < \lambda_1 \leqslant \lambda_2 \leqslant \dots$. We normalize the sequence ϕ_k to make it an orthonormal basis of $L^2(\Omega)$ (see e.g. page 335 in [62]). In this setting, we define

$$(-\Delta)^s_{D,\Omega}u(x) := \sum_{k=0}^{+\infty} \lambda_k^s\, u_k(t)\,\phi_k(x). \tag{2.50}$$

We refer to [109] for extension methods for this type of operator. Furthermore, other types of fractional operators can be defined in terms of different boundary conditions: for instance, a spectral decomposition with respect to the eigenfunctions of the Laplacians with Neumann boundary data naturally leads to an operator $(-\Delta)^s_{N,\Omega}$ (and such operator also have applications in biology, see e.g. [90] and [57]).

It is also interesting to observe that the spectral fractional Laplacian with Neumann boundary conditions can also be written in terms of a regional operator with a singular kernel. Namely, given an open and bounded set $\Omega \subset \mathbb{R}^n$, denoting by $\Delta_{N,\Omega}$ the Laplacian operator coupled with Neumann boundary conditions on $\partial\Omega$, we let $\{(\mu_j,\psi_j)\}_{j\in\mathbb{N}}$ the pairs made up of eigenvalues and eigenfunctions of $-\Delta_{N,\Omega}$, that is

$$\begin{cases} -\Delta\psi_j = \mu_j\psi_j \text{ in } \Omega \\ \partial_\nu\psi_j = 0 \text{ on } \partial\Omega \\ \psi_j \in H^1(\Omega). \end{cases}$$

with $0 = \mu_0 < \mu_1 \leqslant \mu_2 \leqslant \mu_3 \leqslant \dots$.

We define the following operator by making use of a spectral decomposition

$$(-\Delta)^s_{N,\Omega} := \sum_{j=0}^{+\infty} \mu_j^s \hat{u}_j\,\psi_j, \qquad \hat{u}_j = \int_\Omega u\psi_j,\ u \in C_0^\infty(\Omega). \tag{2.51}$$

Comparing with (2.50), we can consider $(-\Delta)^s_{N,\Omega}$ a spectral fractional Laplacian with respect to classical Neumann data. In this setting, the operator $(-\Delta)^s_{N,\Omega}$ is also an integrodifferential operator of regional type, in the sense that one can write

$$(-\Delta)^s_{N,\Omega}u(x) = \text{P.V.} \int_\Omega \big(u(x) - u(y)\big)\, K(x, y)\, dy, \tag{2.52}$$

for a kernel $K(x, y)$ which is comparable to $\frac{1}{|x-y|^{n+2s}}$. We refer to Appendix P for a proof of this.

Interestingly, the fractional Laplacian and the spectral fractional Laplacian coincide, up to a constant, for periodic functions, or functions defined on the flat torus, namely

$$\text{if } u(x + k) = u(x) \text{ for any } x \in \mathbb{R}^n \text{ and } k \in \mathbb{Z}^n, \text{ then } (-\Delta)^s_{D,\Omega}u(x) = \text{const}\,(-\Delta)^s u(x). \tag{2.53}$$

See e.g. Appendix Q for a proof of this fact.

On the other hand, striking differences between the fractional Laplacian and the spectral fractional Laplacian hold true, see e.g. [91, 107].

Interestingly, it is not true that all functions are s-harmonic with respect to the spectral fractional Laplacian, up to a small error, that is

$$\begin{aligned} &\text{it is not true that for any } \varepsilon > 0 \text{ and any (bounded, smooth) function } \bar{u}, \\ &\text{we can find } v_\varepsilon \text{ such that} \\ &\begin{cases} \|\bar{u} - v_\varepsilon\|_{C^2(B_1)} \leqslant \varepsilon \\ \text{and } (-\Delta)^s_{D,\Omega}v_\varepsilon = 0 \text{ in } B_1. \end{cases} \end{aligned} \tag{2.54}$$

A proof of this will be given in Appendix R. The reader can easily compare (2.54) with the setting for the fractional Laplacian discussed in Difference 2.4.

Remarkably, in spite of these differences, the spectral fractional Laplacian can also be written as an integrodifferential operator of the form

$$\text{P.V.} \int_\Omega \big(u(x) - u(y)\big)\, K(x, y)\, dy + \beta(x)u(x), \tag{2.55}$$

for a suitable kernel K and potential β, see Lemma 38 in [2] or Lemma 10.1 in [20]. This can be proved with analogous computations to those performed in the case of the regional fractional Laplacian in the previous paragraph.

2.4 Fractional Time Derivatives

The operators described in Sections in 2.1, 2.2, and 2.3 are often used in the mathematical description of anomalous types of diffusion (i.e. diffusive processes which produce important differences with respect to the classical heat equation, as we will discuss in Sect. 4): the main role of such nonlocal operators is usually to produce a different behavior of the diffusion process with respect to the space variables.

Other types of anomalous diffusions arise from non-standard behaviors with respect to the time variable. These aspects are often the mathematical counterpart of memory effects. As a prototype example, we recall the notion of Caputo fractional derivative, which, for any $t > 0$ (and up to normalizing factors that we omit for simplicity) is given by

$$\partial^s_{C,t} u(t) := \int_0^t \frac{\dot{u}(\tau)}{(t-\tau)^s}\, d\tau. \tag{2.56}$$

We point out that, for regular enough functions u,

$$\begin{aligned}
\partial^s_{C,t} u(t) &= \int_0^t \frac{\dot{u}(\tau)}{(t-\tau)^s}\, d\tau \\
&= \int_0^t \left(\frac{d}{d\tau} \frac{\big(u(\tau)-u(t)\big)}{(t-\tau)^s} - s\, \frac{\big(u(\tau)-u(t)\big)}{(t-\tau)^{1+s}} \right) d\tau \\
&= \frac{u(t)-u(0)}{t^s} - \lim_{\tau\to t} \frac{u(t)-u(\tau)}{(t-\tau)^s} - s \int_0^t \frac{\big(u(\tau)-u(t)\big)}{(t-\tau)^{1+s}}\, d\tau \qquad (2.57) \\
&= \frac{u(t)-u(0)}{t^s} - \dot{u}(t) \lim_{\tau\to t} (t-\tau)^{1-s} - s \int_0^t \frac{u(\tau)-u(t)}{(t-\tau)^{1+s}}\, d\tau \\
&= \frac{u(t)-u(0)}{t^s} + s \int_0^t \frac{u(t)-u(\tau)}{(t-\tau)^{1+s}}\, d\tau.
\end{aligned}$$

Though in principle this expression takes into account only the values of $u(t)$ for $t \geqslant 0$, hence u does not need to be defined for negative times, as pointed out e.g. in Section 2 of [7], it may be also convenient to constantly extend u in $(-\infty, 0)$. Hence, we take the convention for which $u(t) = u(0)$ for any $t \leqslant 0$. With this extension, one has that, for any $t > 0$,

$$s \int_{-\infty}^0 \frac{u(t)-u(\tau)}{(t-\tau)^{1+s}}\, d\tau = s \int_{-\infty}^0 \frac{u(t)-u(0)}{(t-\tau)^{1+s}}\, d\tau = \frac{u(t)-u(0)}{t^s}.$$

Hence, one can write (2.57) as

$$\partial^s_{C,t} u(t) = s \int_{-\infty}^{t} \frac{u(t) - u(\tau)}{(t-\tau)^{1+s}}\, d\tau. \tag{2.58}$$

This type of formulas also relates the Caputo derivative to the so-called Marchaud derivative, see e.g. [104].

In the literature, one can also consider higher order Caputo derivatives, see e.g. [85, 89] and the references therein.

Also, it is useful to consider the Caputo derivative in light of the (unilateral) Laplace Transform (see e.g. Chapter 2.8 in [94], and [86])

$$\mathscr{L} u(\omega) := \int_0^{+\infty} u(t)\, e^{-\omega t}\, dt. \tag{2.59}$$

With this notation, up to dimensional constants, one can write (for a smooth function with exponential control at infinity) that

$$\mathscr{L}(\partial^s_{C,t} u)(\omega) = \omega^s \mathscr{L} u(\omega) - \omega^{s-1} u(0), \tag{2.60}$$

see Appendix S for a proof.

In this way, one can also link equations driven by the Caputo derivative to the so-called Volterra integral equations: namely one can invert the expression $\partial^s_{C,t} u = f$ by

$$u(t) = u(0) + C \int_0^t \frac{f(\tau)}{(t-\tau)^{1-s}}\, d\tau, \tag{2.61}$$

for some normalization constant $C > 0$, see Appendix S for a proof.

It is also worth mentioning that the Caputo derivative of order s of a power gives, up to normalizing constants, the "power minus s": more precisely, by (2.56) and using the substitution $\vartheta := \tau/t$, we see that, for any $r > 0$,

$$\partial^s_{C,t}(t^r) = r \int_0^t \frac{\tau^{r-1}}{(t-\tau)^s}\, d\tau = r\, t^{r-s} \int_0^1 \frac{\vartheta^{r-1}}{(1-\vartheta)^s}\, d\vartheta = C t^{r-s},$$

for some $C > 0$.

Moreover, in relation to the comments on page 17, we have that

> the mean squared displacement related to the diffusion operator
>
> $$\begin{cases} \partial^s_{C,t} u = \Delta u & \text{for any } x \in \mathbb{R}^n \text{ and } t > 0, \\ u(\cdot, 0) = \delta_0(\cdot), \end{cases} \tag{2.62}$$
>
> is finite and proportional to t^s.

See Appendix U for a proof of this.

The Caputo derivatives describes a process "with memory", in the sense that it "remembers the past", though "old events count less than recent ones". We sketch a memory effect of Caputo type in Appendix V.

Due to its memory effect, operators related to Caputo derivatives have found several applications in which the basic parameters of a physical system change in time, in view of the evolution of the system itself: for instance, in studying flows in porous media, when time goes, the fluid may either "obstruct" the holes of the medium, thus slowing down the diffusion, or "clean" the holes, thus making the diffusion faster, and the Caputo derivative may be a convenient approach to describe such modification in time of the diffusion coefficient, see [35].

Other applications of Caputo derivatives occur in biology and neurosciences, since the network of neurons exhibit time-fractional diffusion, also in view of their highly ramified structure, see e.g. [51] and the references therein.

We also refer to [24, 113, 115] and to the references therein for further discussions on different types of anomalous diffusions.

3 A More General Point of View: The "Master Equation"

The operators discussed in Sects. 2.1, 2.2, 2.3, and 2.4 can be framed into a more general setting, that is that of the "master equation", see e.g. [32].

Master equations describe the evolution of a quantity in terms of averages in space and time of the quantity itself. For concreteness one can consider a quantity $u = u(x, t)$ and describe its evolution by an equation of the kind

$$c\partial_t u(x,t) = Lu(x,t) + f\big(x, t, u(x,t)\big)$$

for some $c \in \mathbb{R}$ and a forcing term f, and the operator L has the integral form

$$Lu(x,t) := \iint_{\mathbb{R}^n \times (0,+\infty)} \big(u(x,t) - u(x-y, t-\tau)\big)\, \mathscr{K}(x,t,y,\tau)\, d\mu(x,\tau), \tag{3.1}$$

for a suitable measure μ (with the integral possibly taken in the principal value sense, which is omitted here for simplicity; also one can consider even more general operators by taking actions different than translations and more general ambient spaces).

Though the form of such operator is very general, one can also consider simplifying structural assumptions. For instance, one can take μ to be the space-time Lebesgue measure over $\mathbb{R}^n \times (0, +\infty)$, namely

$$d\mu(x,\tau) = dx\, d\tau.$$

Another common simplifying assumption is to assume that the kernel is induced by an uncorrelated effect of the space and time variables, with the product structure

$$\mathscr{K}(x,t,y,\tau) = \mathscr{K}_{\text{space}}(x,y)\,\mathscr{K}_{\text{time}}(t,\tau).$$

The fractional Laplacian of Sect. 2.1 is a particular case of this setting (for functions depending on the space variable), with the choice, up to normalizing constants,

$$\mathscr{K}_{\text{space}}(x,y) := \frac{1}{|y|^{n+2s}}.$$

More generally, for $\Omega \subseteq \mathbb{R}^n$, the regional fractional Laplacian in Sect. 2.2 comes from the choice

$$\mathscr{K}_{\text{space}}(x,y) := \frac{\chi_\Omega(x-y)}{|y|^{n+2s}}.$$

Finally, in view of (2.58), for time-dependent functions, the choice

$$\mathscr{K}_{\text{time}}(t,\tau) := \frac{\chi_{(-\infty,t)}(\tau)}{|\tau|^{1+s}}.$$

produces the Caputo derivative discussed in Sect. 2.4.

We recall that one of the fundamental structural differences in partial differential equations consists in the distinction between operators "in divergence form", such as

$$-\sum_{i,j=1}^{n} \frac{\partial}{\partial x_i}\left(a_{ij}(x)\frac{\partial u}{\partial x_j}(x)\right) \tag{3.2}$$

and those "in non-divergence form", such as

$$-\sum_{i,j=1}^{n} a_{ij}(x)\frac{\partial^2 u}{\partial x_i \partial x_j}(x). \tag{3.3}$$

This structural difference can also be recovered from the master equation. Indeed, if we consider a (say, for the sake of concreteness, strictly positive, bounded and smooth) matrix function $M : \mathbb{R}^n \to \text{Mat}\,(n \times n)$, we can take into account the master spatial operator induced by the kernel

$$\mathscr{K}_{\text{space}}(x,y) := \frac{1-s}{|M(x-y,y)\,y|^{n+2s}}, \tag{3.4}$$

that is, in the notation of (3.1),

$$(1-s)\int_{\mathbb{R}^n}\frac{u(x)-u(x-y)}{|M(x-y,y)\,y|^{n+2s}}\,dy. \tag{3.5}$$

Then, up to a normalizing constant, if

$$M(x-y,y)=M(x,-y), \tag{3.6}$$

then

$$\begin{aligned}&\text{the limit as } s\nearrow 1 \text{ of the operator in (3.5)}\\ &\text{recovers the classical divergence form operator in (3.2),}\\ &\text{with } a_{ij}(x):=\text{const}\int_{S^{n-1}}\frac{\omega_i\,\omega_j}{|M(x,0)\,\omega|^{n+2}}\,d\mathscr{H}^{n-1}_\omega.\end{aligned} \tag{3.7}$$

A proof of this will be given in Appendix W.

It is interesting to observe that condition (3.6) says that, if we set $z:=x-y$, then

$$M(z,x-z)=M(x,z-x) \tag{3.8}$$

and so the kernel in (3.4) is invariant by exchanging x and z. This invariance naturally leads to a (possibly formal) energy functional of the form

$$\frac{1-s}{2}\iint_{\mathbb{R}^n\times\mathbb{R}^n}\frac{\big(u(x)-u(z)\big)^2}{|M(z,x-z)\,(x-z)|^{n+2s}}\,dx\,dz. \tag{3.9}$$

We point out that condition (3.8) translates, roughly speaking, into the fact that the energy density in (3.9) "charges the variable x as much as the variable z".

The study of the energy functional in (3.9) also drives to a natural quasilinear generalization, in which the fractional energy takes the form

$$\int_{\mathbb{R}^n}\frac{\Phi\big(u(x)-u(z)\big)}{|M(z,x-z)\,(x-z)|^{n+2s}}\,dx\,dz,$$

for a suitable Φ, see e.g. [80, 114] and the references therein for further details on quasilinear nonlocal operators. See also [113] and the references therein for other type of nonlinear fractional equations.

Another case of interest (see e.g. [14]) is the one in which one considers the master equation driven by the spatial kernel

$$\mathscr{K}_{\text{space}}(x,y):=\frac{1-s}{|M(x,y)\,y|^{n+2s}}\,dy,$$

that is, in the notation of (3.1),

$$(1-s)\int_{\mathbb{R}^n}\frac{u(x)-u(x-y)}{|M(x,y)\,y|^{n+2s}}\,dy. \tag{3.10}$$

Then, up to a normalizing constant, if

$$M(x,y)=M(x,-y), \tag{3.11}$$

then

$$\begin{aligned}&\text{the limit as } s\nearrow 1 \text{ of the operator in (3.10)}\\ &\text{recovers the classical non-divergence form operator in (3.3),}\\ &\text{with } a_{ij}(x):=\text{const}\int_{S^{n-1}}\frac{\omega_i\omega_j}{|M(x,0)\,\omega|^{n+2}}\,d\mathcal{H}^{n-1}_\omega\end{aligned} \tag{3.12}$$

A proof of this will be given in Appendix X.

We recall that nonlocal linear operators in non-divergence form can also be useful in the definition of fully nonlinear nonlocal operators, by taking appropriate infima and suprema of combinations of linear operators, see e.g. [83] and the references therein for further discussions about this topic (which is also related to stochastic games).

We also remark that understanding the role of the affine transformations of the spaces on suitable nonlocal operators (as done for instance in (3.10) and (3.10)) often permits a deeper analysis of the problem in nonlinear settings too, see e.g. the very elegant way in which a fractional Monge-Ampère equation is introduced in [29] by considering the infimum of fractional linear operators corresponding to all affine transformations of determinant one of a given multiple of the fractional Laplacian.

As a general comment, we also think that an interesting consequence of the considerations given in this section is that classical, local equations can also be seen as a limit approximation of more general master equations.

We mention that there are also many other interesting kernels, both in space and time, which can be taken into account in integral equations. Though we focused here mostly on the case of singular kernels, there are several important problems that focus on "nice" (e.g. integrable) kernels, see e.g. [8, 43, 88] and the references therein.

As a technical comment let us point out that, in a sense, the nice kernels may have computational advantages, but may provide loss of compactness and loss of regularity issues: roughly speaking, convolutions with smooth kernel are always smooth, thus any smoothness information on a convolved function gives little information on the smoothness of the original function—viceversa, if the convolution of an "object" with a singular kernel is smooth, then it means that the original object has a "good order of vanishing at the origin". When the original object is built by the difference of a function and its translation, such vanishing implies some control of the oscillation of the function, hence opening a door towards a regularity result.

4 Probabilistic Motivations

We provide here some elementary, and somewhat heuristic, motivations for the operators described in Sect. 2 in view of probability and statistics applications. The treatment of this section is mostly colloquial and not to be taken at a strictly rigorous level (in particular, all functions are taken to be smooth, some uniformity problems are neglected, convergence is taken for granted, etc.). See e.g. [74] for rigorous explanations linking pseudo-differential operators and Markov/Lévy processes. See also [9, 12, 16, 101, 111] for other perspectives and links between probability and fractional calculus and [77] for a complete survey on jump processes and their connection to nonlocal operators.

The probabilistic approach to study nonlocal effects and the analysis of distributions with polynomial tails are also some of the cornerstones of the application of mathematical theories to finance, see e.g. [87, 93], and models with jump process for prices have been proposed in [44].

4.1 The Heat Equation and the Classical Laplacian

The prototype of parabolic equations is the heat equation

$$\partial_t u(x,t) = c\,\Delta u(x,t) \tag{4.1}$$

for some $c > 0$. The solution u may represent, for instance, a temperature, and the foundation of (4.1) lies on two basic assumptions:

- the variation of u in a given region $U \subset \mathbb{R}^n$ is due to the flow of some quantity $v : \mathbb{R}^n \to \mathbb{R}^n$ through U,
- v is produced by the local variation of u.

The first ansatz can be written as

$$\partial_t \int_U u(y,t)\,dy = \int_{\partial U} v(y,t)\cdot \nu(y)\,d\mathscr{H}^{n-1}_y, \tag{4.2}$$

where ν denotes the exterior normal vector of U and $\mathscr{H}^{n-1}$ is the standard $(n-1)$-dimensional surface Hausdorff measure.

The second ansatz can be written as $v = c\nabla u$, which combined with (4.2) and the Divergence Theorem gives that

$$\partial_t \int_U u(y,t)\,dy = c\int_{\partial U} \nabla u(y,t)\cdot \nu(y)\,d\mathcal{H}_y^{n-1} = c\int_U \operatorname{div}\left(\nabla u(y,t)\right)dy = c\int_U \Delta u(y,t)\,dy.$$

Since U is arbitrary, this gives (4.1).

Let us recall a probabilistic interpretation of (4.1). The idea is that (4.1) follows by taking suitable limits of a discrete "random walk". For this, we take a small space scale $h > 0$ and a time step

$$\tau = h^2. \tag{4.3}$$

We consider the random motion of a particle in the lattice $h\mathbb{Z}^n$, as follows. At each time step, the particle can move in any coordinate direction with equal probability. That is, a particle located at $h\bar{k} \in h\mathbb{Z}^n$ at time t is moved to one of the $2n$ points $h\bar{k}\pm he_1, \ldots, h\bar{k}\pm he_n$ with equal probability (here, as usual, e_j denotes the jth element of the standard Euclidean basis of $\mathbb{R}^n$).

We now look at the expectation to find the particle at a point $x \in h\mathbb{Z}^n$ at time $t \in \tau\mathbb{N}$. For this, we denote by $u(x,t)$ the probability density of such expectation. That is, the probability for the particle of lying in the spatial region $B_r(x)$ at time t is, for small r, comparable with

$$\int_{B_r(x)} u(y,t)\,dy.$$

Then, the probability of finding a particle at the point $x \in h\mathbb{Z}^n$ at time $t+\tau$ is the sum of the probabilities of finding the particle at a closest neighborhood of x at time t, times the probability of jumping from this site to x. That is,

$$u(x,t+\tau) = \frac{1}{2n}\sum_{j=1}^{n}\Big(u(x+he_j) + u(x-he_j)\Big). \tag{4.4}$$

Also,

$$\begin{aligned}
&u(x+he_j) + u(x-he_j) - 2u(x,t)\\
&= \left(u(x,t) + h\nabla u(x,t)\cdot e_j + \frac{h^2\,D^2u(x,t)\,e_j\cdot e_j}{2}\right)\\
&\qquad + \left(u(x,t) - h\nabla u(x,t)\cdot e_j + \frac{h^2\,D^2u(x,t)\,e_j\cdot e_j}{2}\right) - 2u(x,t) + O(h^3)\\
&= h^2\,\partial^2_{x_j} u(x,t) + O(h^3).
\end{aligned}$$

Thus, subtracting $u(x,t)$ to both sides in (4.4), dividing by τ, recalling (4.3), and taking the limit (and neglecting any possible regularity issue), we formally find that

$$\begin{aligned}\partial_t u(x,t) &= \lim_{\tau\searrow 0}\frac{u(x,t+\tau)-u(x,t)}{\tau}\\ &= \lim_{h\searrow 0}\frac{1}{2n}\sum_{j=1}^{n}\frac{u(x+he_j)+u(x-he_j)-2u(x,t)}{h^2}\\ &= \lim_{h\searrow 0}\frac{1}{2n}\sum_{j=1}^{n}\partial^2_{x_j}u(x,t)+O(h)\\ &= \frac{1}{2n}\Delta u(x,t),\end{aligned}$$

which is (4.1).

4.2 *The Fractional Laplacian and the Regional Fractional Laplacian*

Now we consider an open set $\Omega\subseteq\mathbb{R}^n$ and a discrete random process in $h\mathbb{Z}^n$ which can be roughly speaking described in this way. The space parameter $h>0$ is linked to the time step

$$\tau := h^{2s}. \tag{4.5}$$

A particle starts its journey from a given point $h\bar{k}\in\Omega$ of the lattice $h\mathbb{Z}^n$ and, at each time step τ, it can reach any other point of the lattice hk, with $k\neq\bar{k}$, with probability

$$P_h(\bar{k},k) := \frac{\chi_\Omega(h\bar{k})\,\chi_\Omega(hk)}{C\,|k-\bar{k}|^{n+2s}}, \tag{4.6}$$

then the process continues following the same law. Notice that the above probability density does not allow the process to leave the domain Ω, since P_h vanishes in the complement of Ω (in jargon, this process is called "censored").

In (4.6), the constant $C>0$ is needed to normalize to total probability and is defined by

$$C := \sum_{k\in\mathbb{Z}^n\setminus\{0\}}\frac{1}{|k|^{n+2s}}.$$

We let

$$c_h(\bar{k}) := \sum_{k\in\mathbb{Z}^n\setminus\{0\}} P_h(\bar{k},k) = \sum_{k\in\mathbb{Z}^n\setminus\{0\}} P_h(k,\bar{k}) \tag{4.7}$$

and

$$p_k(\bar{k}) := 1 - c_h(\bar{k}).$$

Notice that, for any $\bar{k} \in \mathbb{Z}^n$, it holds that

$$c_h(\bar{k}) \leqslant \sum_{k\in\mathbb{Z}^n\setminus\{\bar{k}\}} \frac{1}{C\,|k-\bar{k}|^{n+2s}} = 1, \tag{4.8}$$

hence, for a fixed $h > 0$ and $\bar{k} \in \mathbb{Z}^n$, this aggregate probability does not equal to 1: this means that there is a remaining probability $p_h(\bar{k}) \geqslant 0$ for which the particle does not move (in principle, such probability is small when so is h, but, for a bounded domain Ω, it is not negligible with respect to the time step, hence it must be taken into account in the analysis of the process in the general setting that we present here).

We define $u(x,t)$ to be the probability density for the particle to lie at the point $x \in \Omega \cap (h\mathbb{Z}^n)$ at time $t \in \tau\mathbb{N}$. We show that, for small space and time scale, the function u is well described by the evolution of the nonlocal heat equation

$$\partial_t u(x,t) = -c\,(-\Delta)^s_\Omega u(x,t) \qquad \text{in } \Omega, \tag{4.9}$$

for some normalization constant $c > 0$. To check this, up to a translation, we suppose that $x = 0 \in \Omega$ and we set $c_h := c_h(0)$ and $p_h := p_h(0)$. We observe that the probability of being at 0 at time $t+\tau$ is the sum of the probabilities of being somewhere else, say at $hk \in h\mathbb{Z}^n$, at time t, times the probability of jumping from hk to the origin, plus the probability of staying put: that is

$$\begin{aligned} u(0,t+\tau) &= \sum_{k\in\mathbb{Z}^n\setminus\{0\}} u(hk,t)\,P_h(k,0) + u(0,t)\,p_h \\ &= \sum_{k\in\mathbb{Z}^n\setminus\{0\}} u(hk,t)\,P_h(k,0) + (1-c_h)\,u(0,t). \end{aligned}$$

Thus, recalling (4.7),

$$\begin{aligned} u(0,t+\tau)-u(0,t) &= \sum_{k\in\mathbb{Z}^n\setminus\{0\}} u(hk,t)\,P_h(k,0)-c_h\,u(0,t) \\ &= \sum_{k\in\mathbb{Z}^n\setminus\{0\}} \Big(u(hk,t)-u(0,t)\Big)\,P_h(k,0) \\ &= \sum_{k\in\mathbb{Z}^n\setminus\{0\}} \Big(u(hk,t)-u(0,t)\Big)\,\frac{\chi_\Omega(hk)}{C\,|k|^{n+2s}} \\ &= \frac{h^{n+2s}}{C}\sum_{k\in\mathbb{Z}^n\setminus\{0\}} \Big(u(hk,t)-u(0,t)\Big)\,\frac{\chi_\Omega(hk)}{|hk|^{n+2s}}. \end{aligned}$$

So, we divide by τ and, in view of (4.5), we find that

$$C\,\frac{u(0,t+\tau)-u(0,t)}{\tau}=h^n\sum_{k\in\mathbb{Z}^n\setminus\{0\}} \Big(u(hk,t)-u(0,t)\Big)\,\frac{\chi_\Omega(hk)}{|hk|^{n+2s}}.$$

We write this identity changing k to $-k$ and we sum up: in this way, we obtain that

$$\begin{aligned} &2C\,\frac{u(0,t+\tau)-u(0,t)}{\tau} \\ &=h^n\sum_{k\in\mathbb{Z}^n\setminus\{0\}} \frac{\big(u(hk,t)-u(0,t)\big)\,\chi_\Omega(hk)+\big(u(-hk,t)-u(0,t)\big)\,\chi_\Omega(-hk)}{|hk|^{n+2s}}. \end{aligned} \tag{4.10}$$

Now, for small y, if u is smooth enough,

$$\begin{aligned} &\Big|\big(u(y,t)-u(0,t)\big)\,\chi_\Omega(y)+\big(u(-y,t)-u(0,t)\big)\,\chi_\Omega(-y)\Big| \\ &=\Big|\big(u(y,t)-u(0,t)\big)+\big(u(-y,t)-u(0,t)\big)\Big| \\ &=\Big|\big(\nabla u(0,t)y+O(|y|^2)\big)+\big(-\nabla u(0,t)y+O(|y|^2)\big)\Big| \\ &=O(|y|^2) \end{aligned}$$

and therefore, if we write

$$\varphi(y):=\frac{\big(u(y,t)-u(0,t)\big)\,\chi_\Omega(y)+\big(u(-y,t)-u(0,t)\big)\,\chi_\Omega(-y)}{|y|^{n+2s}},$$

we (formally) have that

$$\varphi(y) = O(|y|^{2-n-2s}) \tag{4.11}$$

for small $|y|$.

Now, we fix $\delta > 0$ and use the Riemann sum representation of an integral to write (for a bounded Riemann integrable function $\varphi : \mathbb{R}^n \setminus B_\delta \to \mathbb{R}$),

$$\int_{\mathbb{R}^n \setminus B_\delta} \varphi(y)\,dy = \lim_{h \searrow 0} h^n \sum_{k \in \mathbb{Z}^n} \varphi(hk)\,\chi_{\mathbb{R}^n \setminus B_\delta}(hk) = \lim_{h \searrow 0} h^n \sum_{\substack{k \in \mathbb{Z}^n \\ k \neq 0}} \varphi(hk)\,\chi_{\mathbb{R}^n \setminus B_\delta}(hk). \tag{4.12}$$

If, in addition, (4.11) is satisfied, one has that, for small δ,

$$\int_{B_\delta} \varphi(y)\,dy = O(\delta^{2-2s}).$$

From this and (4.12) we have that

$$\begin{aligned} \int_{\mathbb{R}^n} \varphi(y)\,dy &= O(\delta^{2-2s}) + \lim_{h \searrow 0} h^n \sum_{\substack{k \in \mathbb{Z}^n \\ k \neq 0}} \varphi(hk)\,\chi_{\mathbb{R}^n \setminus B_\delta}(hk) \\ &= O(\delta^{2-2s}) + \lim_{h \searrow 0} h^n \sum_{\substack{k \in \mathbb{Z}^n \\ k \neq 0}} \varphi(hk) + h^n \sum_{\substack{k \in \mathbb{Z}^n \\ k \neq 0}} \varphi(hk)\big(\chi_{\mathbb{R}^n \setminus B_\delta}(hk) - 1\big). \end{aligned} \tag{4.13}$$

Also, in view of (4.11),

$$\begin{aligned} &\left| h^n \sum_{\substack{k \in \mathbb{Z}^n \\ k \neq 0}} \varphi(hk)\big(\chi_{\mathbb{R}^n \setminus B_\delta}(hk) - 1\big) \right| = \left| h^n \sum_{\substack{k \in \mathbb{Z}^n \\ 0 < h|k| < \delta}} \varphi(hk) \right| \\ &\leqslant \operatorname{const} h^n \sum_{\substack{k \in \mathbb{Z}^n \\ 0 < |k| < \delta/h}} |hk|^{2-n-2s} = \operatorname{const} h^{2-2s} \sum_{\substack{k \in \mathbb{Z}^n \\ 0 < |k| < \delta/h}} \frac{|k|^{1-s}}{|k|^{n+s-1}} \\ &\leqslant \operatorname{const} h^{2-2s} \left(\frac{\delta}{h}\right)^{1-s} \sum_{\substack{k \in \mathbb{Z}^n \\ 1 \leqslant |k| < \delta/h}} \frac{1}{|k|^{n+s-1}} \\ &\leqslant \operatorname{const} h^{2-2s} \left(\frac{\delta}{h}\right)^{1-s} \left(\frac{\delta}{h}\right)^{1-s} = \operatorname{const} \delta^{2-2s}. \end{aligned}$$

Hence, (4.13) boils down to

$$\int_{\mathbb{R}^n} \varphi(y)\,dy = O(\delta^{2-2s}) + \lim_{h\searrow 0} h^n \sum_{\substack{k\in\mathbb{Z}^n\\ k\neq 0}} \varphi(hk)$$

and so, taking δ arbitrarily small,

$$\int_{\mathbb{R}^n} \varphi(y)\,dy = \lim_{h\searrow 0} h^n \sum_{\substack{k\in\mathbb{Z}^n\\ k\neq 0}} \varphi(hk).$$

Therefore, recalling (4.10),

$$\begin{aligned}
2C\partial_t u(0,t) &= \lim_{h\searrow 0} 2C\,\frac{u(0,t+\tau)-u(0,t)}{\tau}\\
&= \lim_{h\searrow 0} h^n \sum_{k\in\mathbb{Z}^n\setminus\{0\}} \frac{\big(u(hk,t)-u(0,t)\big)\,\chi_\Omega(hk)+\big(u(-hk,t)-u(0,t)\big)\,\chi_\Omega(-hk)}{|hk|^{n+2s}}\\
&= \lim_{h\searrow 0} h^n \sum_{\substack{k\in\mathbb{Z}^n\\ k\neq 0}} \varphi(hk)\\
&= \int_{\mathbb{R}^n} \varphi(y)\,dy\\
&= \int_{\mathbb{R}^n} \frac{\big(u(y,t)-u(0,t)\big)\,\chi_\Omega(y)+\big(u(-y,t)-u(0,t)\big)\,\chi_\Omega(-y)}{|y|^{n+2s}}\\
&= -2(-\Delta)^s_\Omega u(x,0).
\end{aligned}$$

This confirms (4.9).

As a final comment, in view of these calculations and those of Sect. 4.1, we may compare the classical random walk, which leads to the classical heat equation, and the long-jump random walk which leads to the nonlocal heat equation and relate such jumps to an "infinitely fast" diffusion, in the light of the computations of the associated mean squared displacements (recall (2.43) and (2.45)).

4.3 *The Spectral Fractional Laplacian*

Now, we briefly discuss a heuristic motivation for the fractional heat equation run by the spectral fractional Laplacian, that is

$$\partial_t u(x,t) = -c\,(-\Delta)^s_{D,\Omega} u(x,t) \qquad \text{in } \Omega, \tag{4.14}$$

for some normalization constant $c > 0$. To this end, we consider a bounded and smooth set $\Omega \subset \mathbb{R}^n$ and we define a random motion of a "distribution of particles" in Ω. For any $x \in \Omega$ and $t \geqslant 0$, the function $u(x,t)$ denotes the "number of particles" present at the point x at the time t. No particles lie outside Ω and we write u as a suitable superposition of eigenfunctions $\{\phi_k\}_{k\geqslant 1}$ of the Laplacian with Dirichlet boundary data (this is a reasonable assumption, given that such eigenfunctions provide a basis of $L^2(\Omega)$, see e.g. page 335 in [62]). In this way, we write

$$u(x,t) = \sum_{k=1}^{+\infty} u_k(t)\,\phi_k(x).$$

Namely, in the notation in (2.49), the evolution of the particle distribution u is defined on each spectral component u_k and it is taken to follow a "classical" random walk, but the space/time scale is supposed to depend on k as well: namely, spectral components relative to high frequencies will move slower than the ones relative to low frequencies (namely, the time step is taken to be longer if the frequency is higher).

More precisely, for any $k \in \mathbb{N}$, we suppose that each of the u_k particles of the kth spectral component undergo a classical random walk in a lattice $h_k\mathbb{Z}^d$, as described in Sect. 4.1, but with time step

$$\tau_k := \lambda_k^{1-s}\, h_k^2. \tag{4.15}$$

We suppose that h_k and τ_k are "small space and time increments". Namely, after a time step τ_k, each of these $u_k(t)\,\phi_k(x)$ particles will move, with equal probability $\frac{1}{2n}$, to one of the points $x \pm h_k e_1, \ldots, x \pm h_k e_n$ (for simplicity, we are imaging here u_k to be positive; the case of negative u_k represents a "lack of particles", which is supposed to diffuse with the same law). Hence, the number of particles at time $t+\tau_k$ which correspond to the kth frequency of the spectrum and which lie at the point $x \in \Omega$ is equal to the sum of the number of the particles at time t which lie somewhere else times the probability of jumping to x in this time step, that is, in formula,

$$u_k(t+\tau_k)\,\phi_k(x) = \frac{1}{2n}\sum_{j=1}^{n} u_k(t)\,\Big(\phi_k(x+h_k e_j) + \phi_k(x-h_k e_j)\Big). \tag{4.16}$$

Moreover,

$$\begin{aligned}
&\phi_k(x + h_k e_j) + \phi_k(x - h_k e_j) - 2\phi_k(x) \\
&= \left(\phi_k(x) + h_k \nabla\phi_k(x) \cdot e_j + \frac{h_k^2\, D^2\phi_k(x)e_j \cdot e_j}{2}\right) \\
&\quad + \left(\phi_k(x) - h_k \nabla\phi_k(x) \cdot e_j + \frac{h_k^2\, D^2\phi_k(x)e_j \cdot e_j}{2}\right) - 2\phi_k(x) + O(h_k^3) \\
&= h_k^2\, \partial^2_{x_j}\phi_k(x) + O(h_k^3).
\end{aligned}$$

Consequently, from this and (4.16),

$$\begin{aligned}
\Big(u_k(t+\tau_k) - u_k(t)\Big)\,\phi_k(x) &= \frac{1}{2n}\sum_{j=1}^{n} u_k(t)\Big(\phi_k(x + h_k e_j) + \phi_k(x - h_k e_j) - 2\phi_k(x)\Big) \\
&= \frac{h_k^2}{2n}\, u_k(t)\, \Delta\phi_k(x) + O(h_k^3) \\
&= -\frac{\lambda_k\, h_k^2}{2n}\, u_k(t)\,\phi_k(x) + O(h_k^3).
\end{aligned}$$

Hence, with a formal computation, dividing by τ_k, using (4.15) and sending h_k, $\tau_k \searrow 0$ (for a fixed k), we obtain

$$\begin{aligned}
\partial_t u_k(t) &= \lim_{\tau_k \searrow 0} \frac{\big(u_k(t+\tau_k) - u_k(t)\big)\,\phi_k(x)}{\tau_k} \\
&= \lim_{h_k \searrow 0} -\frac{\lambda_k^s}{2n}\, u_k(t)\,\phi_k(x) + O(h_k) = -\frac{\lambda_k^s}{2n}\, u_k(t)\,\phi_k(x).
\end{aligned}$$

Hence, from (2.49) (and neglecting converge issues in k), we have

$$\partial_t u(x,t) = \sum_{k=0}^{+\infty} \partial_t u_k(t)\,\phi_k(x) = -\sum_{k=0}^{+\infty} \frac{\lambda_k^s}{2n}\, u_k(t)\,\phi_k(x),$$

that is (4.14).

4.4 *Fractional Time Derivatives*

We consider a model in which a bunch of people is supposed to move along the real line (say, starting at the origin) with some given velocity f, which depends on time. We consider the case in which the environment surrounding the moving people is "tricky", and some of them risk to get stuck for some time, and they are able to "exit the trap" only by overcoming their past velocity. Concretely, we fix a function $\varphi : [0, +\infty) \to [0, +\infty)$ with

$$C_\varphi := \sum_{k=1}^{+\infty} \varphi(k) < +\infty. \tag{4.17}$$

Then we define

$$p_k := \frac{\varphi(k)}{C_\varphi}$$

and we notice that

$$\sum_{k=1}^{+\infty} p_k = \frac{1}{C_\varphi} \sum_{k=1}^{+\infty} \varphi(k) = 1.$$

Then, we denote by $u(t)$ the position of the "generic person" at time t, with $u(0) = 0$. We suppose that some people, say a proportion p_1 of the total population, move with the prescribed velocity for a unit of time, after which their velocity is the difference between the prescribed velocity at that time and the one at the preceding time with respect to the time unit. In formulas, this says that there is a proportion p_1 of the total people who travels with velocity

$$\dot{u}_1(t) := \begin{cases} f(t) & \text{if } t \in [0, 1], \\ f(t) - f(t-1) & \text{if } t > 1. \end{cases}$$

After integrating, we thus obtain that there is a proportion p_1 of the total people whose position is described by the function

$$\begin{aligned} u_1(t) &= \begin{cases} \displaystyle\int_0^t f(\vartheta)\, d\vartheta & \text{if } t \in [0, 1], \\ \displaystyle\int_0^1 f(\vartheta)\, d\vartheta + \int_1^t \big(f(\vartheta) - f(\vartheta - 1)\big)\, d\vartheta & \text{if } t > 1, \end{cases} \\ &= \begin{cases} \displaystyle\int_0^t f(\vartheta)\, d\vartheta & \text{if } t \in [0, 1], \\ \displaystyle\int_0^t f(\vartheta)\, d\vartheta - \int_1^t f(\vartheta - 1)\, d\vartheta & \text{if } t > 1, \end{cases} \\ &= \begin{cases} \displaystyle\int_0^t f(\vartheta)\, d\vartheta & \text{if } t \in [0, 1], \\ \displaystyle\int_0^t f(\vartheta)\, d\vartheta - \int_0^{t-1} f(\vartheta)\, d\vartheta & \text{if } t > 1, \end{cases} \\ &= \int_{(t-1)_+}^t f(\vartheta)\, d\vartheta. \end{aligned}$$

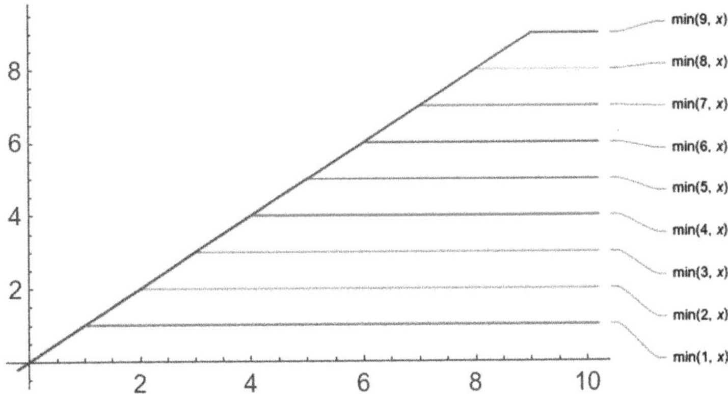

Fig. 5 The motions u_k described in Sect. 4.4 when the velocity field f is constant

For instance, if f is constant, then the position u_1 grows linearly for a unit of time and then remains put (this would correspond to consider "stopping times" in the motion, see Fig. 5).

Similarly, a proportion p_2 of the total population evolves with prescribed velocity f for two units of time, after which its velocity becomes the difference between the prescribed velocity at that time and the one at the preceding time with respect to two time units, namely

$$\dot{u}_2(t) := \begin{cases} f(t) & \text{if } t \in [0,2], \\ f(t) - f(t-2) & \text{if } t > 2. \end{cases}$$

In this case, an integration gives that there is a proportion p_2 of the total people whose position is described by the function

$$\begin{aligned} u_2(t) &= \begin{cases} \displaystyle\int_0^t f(\vartheta)\,d\vartheta & \text{if } t \in [0,2], \\ \displaystyle\int_0^2 f(\vartheta)\,d\vartheta + \int_2^t \big(f(\vartheta) - f(\vartheta-2)\big)\,d\vartheta & \text{if } t > 2, \end{cases} \\ &= \begin{cases} \displaystyle\int_0^t f(\vartheta)\,d\vartheta & \text{if } t \in [0,2], \\ \displaystyle\int_0^t f(\vartheta)\,d\vartheta - \int_2^t f(\vartheta-2)\,d\vartheta & \text{if } t > 2. \end{cases} \end{aligned}$$

$$
\begin{aligned}
&= \begin{cases} \displaystyle\int_0^t f(\vartheta)\,d\vartheta & \text{if } t \in [0,2], \\ \displaystyle\int_0^t f(\vartheta)\,d\vartheta - \int_0^{t-2} f(\vartheta)\,d\vartheta & \text{if } t > 2. \end{cases} \\
&= \int_{(t-2)_+}^t f(\vartheta)\,d\vartheta.
\end{aligned}
$$

Repeating this argument, we suppose that for each $k \in \mathbb{N}$ we have a proportion p_k of the people that move initially with the prescribed velocity f, but, after k units of time, get their velocity changed into the difference of the actual velocity field and that of k units of time before (which is indeed a "memory effect"). In this way, we have that a proportion p_k of the total population moves with law of motion given by

$$
u_k(t) = \int_{(t-k)_+}^t f(\vartheta)\,d\vartheta.
$$

The average position of the moving population is then given by

$$
u(t) := \sum_{k=1}^{+\infty} p_k\, u_k(t) = \frac{1}{C_\varphi} \sum_{k=1}^{+\infty} \varphi(k) \int_{(t-k)_+}^t f(\vartheta)\,d\vartheta. \tag{4.18}
$$

We now specialize the computation above for the case

$$
\varphi(x) := x^{s-2},
$$

with $s \in (0,1)$. Notice that the quantity in (4.17) is finite in this case, and we can denote it simply by C_s. In addition, we will consider long time asymptotics in t and introduce a small time increment h which is inversely proportional to t, namely

$$
h := \frac{1}{t}.
$$

In this way, recalling that the motion was supposed to start at the origin (i.e., $u(0) = 0$) and using the substitution $\eta := \vartheta/t$, we can write (4.18) as

$$
\begin{aligned}
u(t) - u(0) &= \frac{1}{C_s} \sum_{k=1}^{+\infty} k^{s-2} \int_{(t-k)_+}^t f(\vartheta)\,d\vartheta \\
&= \frac{t^s\, h}{C_s} \sum_{k=1}^{+\infty} (hk)^{s-2} \int_{(1-kh)_+}^1 f(t\eta)\,d\eta \qquad (4.19) \\
&\simeq \frac{t^s}{C_s} \int_0^{+\infty} \left[\lambda^{s-2} \int_{(1-\lambda)_+}^1 f(t\eta)\,d\eta \right] d\lambda,
\end{aligned}
$$

where we have recognized a Riemann sum in the last line.

We also point out that the conditions

$$\lambda \in (0, +\infty) \text{ and } 0 < \xi < \min\{1, \lambda\}$$

are equivalent to

$$0 < \xi < 1 \text{ and } \lambda \in (\xi, +\infty),$$

and, furthermore,

$$1 - (1 - \lambda)_+ = 1 - \max\{0,\ 1 - \lambda\} = \min\{1 - 0,\ 1 - (1 - \lambda)\} = \min\{1, \lambda\}.$$

Therefore we use the substitution $\xi := 1 - \eta$ and we exchange the order of integrations, to deduce from (4.19) that

$$\begin{aligned} u(t) - u(0) &= \frac{t^s}{C_s} \int_0^{+\infty} \left[\int_0^{\min\{1,\lambda\}} \lambda^{s-2} f(t - t\xi)\, d\xi \right] d\lambda \\ &= \frac{t^s}{C_s} \int_0^1 \left[\int_\xi^{+\infty} \lambda^{s-2} f(t - t\xi)\, d\lambda \right] d\xi \\ &= \frac{t^s}{C_s\,(1 - s)} \int_0^1 \xi^{s-1} f(t - t\xi)\, d\xi. \end{aligned}$$

The substitution $\tau := t\xi$ then gives

$$u(t) - u(0) = \frac{1}{C_s\,(1 - s)} \int_0^t \tau^{s-1} f(t - \tau)\, d\tau,$$

which, comparing with (2.61) and possibly redefining constants, gives that $\partial^s_{C,t} u = f$.

Of course, one can also take into account the case in which the velocity field f is induced by a classical diffusion in space, i.e. $f = \Delta u$, and in this case one obtains the time fractional diffusive equation $\partial^s_{C,t} u = \Delta u$.

4.5 *Fractional Time Diffusion Arising from Heterogeneous Media*

A very interesting phenomenon to observe is that the geometry of the diffusion medium can naturally transform classical diffusion into an anomalous one. This feature can be very well understood by an elegant model, introduced in [10] (see also [105] and the references therein for an exhaustive account of the research in this direction) consisting in random walks on a "comb", that we briefly reproduce

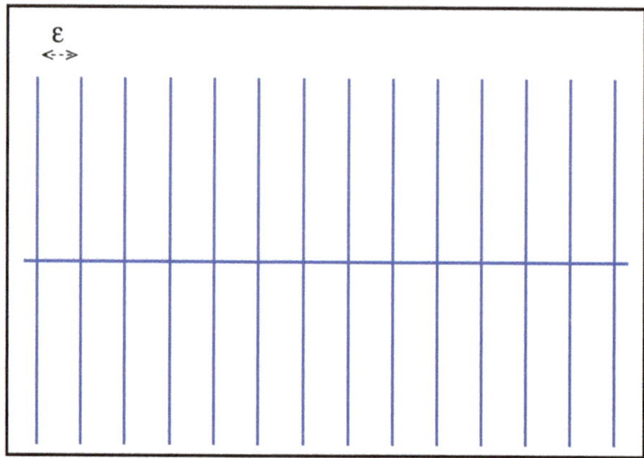

Fig. 6 The comb $\mathscr{C}_\varepsilon$

here for the facility of the reader. Given $\varepsilon > 0$, the comb may be considered as a transmission medium that is the union of a "backbone" $\mathscr{B} := \mathbb{R} \times \{0\}$ with the "fingers" $\mathscr{P}_k := \{\varepsilon k\} \times \mathbb{R}$, namely

$$\mathscr{C}_\varepsilon := \mathscr{B} \cup \left(\bigcup_{k \in \mathbb{Z}} \mathscr{P}_k \right),$$

see Fig. 6.

We suppose that a particle experiences a random walk on the comb, starting at the origin, with some given horizontal and vertical speeds. In the limit, this random walk can be modeled by the diffusive equation along the comb $\mathscr{C}_\varepsilon$

$$\begin{cases} u_t = d_1 \delta_0(y)\, u_{xx} + d_2\, \varepsilon \sum_{k \in \mathbb{Z}} \delta_0(\varepsilon k) u_{yy}, \\ u(x, y, 0) = \delta_0(x)\, \delta_0(y), \end{cases} \tag{4.20}$$

with $d_1, d_2 > 0$. The case $d_1 = d_2$ corresponds to equal horizontal and vertical speeds of the random walk (and this case is already quite interesting). Also, in the limit as $\varepsilon \searrow 0$, we can consider the Riemann sum approximation

$$\int_{\mathbb{R}} f(y)\, dy \simeq \varepsilon \sum_{k \in \mathbb{Z}} f(\varepsilon k),$$

and $\mathscr{C}_\varepsilon$ tends to cover the whole of $\mathbb{R}^2$ when ε gets small. Accordingly, at least at a formal level, as the fingers of the comb become thicker and thicker, we can think that

$$1 = \int_{\mathbb{R}} \delta_0(y)\, dy \simeq \varepsilon \sum_{k \in \mathbb{Z}} \delta_0(\varepsilon k),$$

and reduce (4.20) to the diffusive equation in $\mathbb{R}^2$ given by

$$\begin{cases} u_t = d_1 \delta_0(y)\, u_{xx} + d_2 u_{yy}, \\ u(x, y, 0) = \delta_0(x)\, \delta_0(y). \end{cases} \tag{4.21}$$

The very interesting feature of (4.21) is that it naturally induces a fractional time diffusion along the backbone. The quantity that experiences this fractional diffusion is the total diffusive mass at a point of the backbone. Namely, one sets

$$U(x, t) := \int_{\mathbb{R}} u(x, y, t)\, dy, \tag{4.22}$$

and we claim that

$$\partial^{1/2}_{C,t} U(x, t) = \frac{d_1}{2\sqrt{d_2}} \Delta U(x, t) \qquad \text{for all } (x, t) \in \mathbb{R} \times (0, +\infty). \tag{4.23}$$

Equation (4.23) reveals the very relevant phenomenon that a diffusion governed by the Caputo derivative may naturally arise from classical diffusion, only in view of the particular geometry of the domain.

To check (4.23), we first point out that

$$\hat{U}(\xi, 0) := \int_{\mathbb{R}} \hat{u}(\xi, y, 0)\, dy = \int_{\mathbb{R}} \delta_0(y)\, dy = 1. \tag{4.24}$$

Then, we observe that, if $a, b \in \mathbb{C}$, and

$$\begin{aligned} & g(y) := b\, e^{-a|y|} && \text{for any } y \in \mathbb{R}, \\ & \text{then we have that} && g''(y) = a^2 g(y) - 2ab\, \delta_0(y). \end{aligned} \tag{4.25}$$

To check this let $\varphi \in C_0^\infty(\mathbb{R})$. Then, integrating twice by parts,

$$\begin{aligned} & \frac{1}{b} \int_{\mathbb{R}} \big(g(y)\varphi''(y) - a^2 g(y)\varphi(y)\big)\, dy \\ & = \int_0^{+\infty} e^{-ay} \varphi''(y)\, dy + \int_{-\infty}^0 e^{ay} \varphi''(y)\, dy - a^2 \int_{\mathbb{R}} e^{-a|y|} \varphi(y)\, dy \\ & = a \int_0^{+\infty} e^{-ay} \varphi'(y)\, dy - a \int_{-\infty}^0 e^{ay} \varphi'(y)\, dy - a^2 \int_{\mathbb{R}} e^{-a|y|} \varphi(y)\, dy \\ & = -2a\varphi(0) + a^2 \int_0^{+\infty} e^{-ay} \varphi(y)\, dy + a^2 \int_{-\infty}^0 e^{ay} \varphi'(y)\, dy - a^2 \int_{\mathbb{R}} e^{-a|y|} \varphi(y)\, dy \\ & = -2a\varphi(0), \end{aligned}$$

thus proving (4.25).

We also remark that, in the notation of (4.25), we have that $\delta_0(y)g(y) = \delta_0(y)g(0) = b\delta_0(y)$, and so, for every $c \in \mathbb{R}$,

$$g''(y) = a^2 g(y) - b(2a+c)\delta_0(y) + c\delta_0(y)g(y). \tag{4.26}$$

Now, taking the Fourier Transform of (4.21) in the variable x, using the notation $\hat{u}(\xi, y, t)$ for the Fourier Transform of $u(x, y, t)$, and possibly neglecting normalization constants, we get

$$\begin{cases} \hat{u}_t = -d_1|\xi|^2\delta_0(y)\,\hat{u} + d_2\hat{u}_{yy}, \\ \qquad \hat{u}(\xi, y, 0) = \delta_0(y). \end{cases} \tag{4.27}$$

Now, we take the Laplace Transform of (4.27) in the variable t, using the notation $w(\xi, y, \omega)$ for the Laplace Transform of $\hat{u}(\xi, y, t)$, namely $w(\xi, y, \omega) := \mathscr{L}\hat{u}(\xi, y, \omega)$. In this way, recalling that

$$\mathscr{L}(\dot{f}) = \omega\mathscr{L}f(\omega) - f(0),$$

and therefore

$$\mathscr{L}(\hat{u}_t)(\xi, y, \omega) = \omega\mathscr{L}\hat{u}(\xi, y, \omega) - \hat{u}(\xi, y, 0) = \omega w(\xi, y, \omega) - \delta_0(y),$$

we deduce from (4.27) that

$$\omega w - \delta_0(y) = -d_1|\xi|^2\delta_0(y)\, w + d_2 w_{yy}. \tag{4.28}$$

That is, setting

$$a(\omega) := \left(\frac{\omega}{d_2}\right)^{1/2}, \qquad b(\xi, \omega) := \frac{1}{(4d_2\omega)^{1/2} + d_1\,|\xi|^2} \qquad \text{and} \qquad c(\xi) := \frac{d_1\,|\xi|^2}{d_2},$$

we see that

$$b(2a+c) = \frac{2\left(\dfrac{\omega}{d_2}\right)^{1/2} + \dfrac{d_1\,|\xi|^2}{d_2}}{(4d_2\omega)^{1/2} + d_1\,|\xi|^2} = \frac{1}{d_2},$$

and hence we can write (4.28) as

$$w_{yy} = \frac{\omega}{d_2}\, w - \frac{1}{d_2}\delta_0(y) + \frac{d_1|\xi|^2}{d_2}\delta_0(y)\, w = a^2 w - b(2a+c)\delta_0(y) + c\delta_0(y)w.$$

In light of (4.26), we know that this equation is solved by taking $w = g$, that is

$$\mathscr{L}\hat{u}(\xi, y, \omega) = w(\xi, y, \omega) = b(\xi, \omega)e^{-a(\omega)|y|}.$$

As a consequence, by (4.22),

$$\mathscr{L}\hat{U}(\xi,t)=\int_{\mathbb{R}}\mathscr{L}\hat{u}(\xi,y,\tau)\,dy=\int_{\mathbb{R}}b(\xi,\omega)e^{-a(\omega)|y|}\,dy=\frac{2b(\xi,\omega)}{a}.$$

This and (4.24) give that

$$\Big((4d_2\omega)^{1/2}+d_1\,|\xi|^2\Big)\,\mathscr{L}\hat{U}(\xi,t)=\frac{\mathscr{L}\hat{U}(\xi,t)}{b(\xi,\omega)}=\frac{2}{a}=\left(\frac{4d_2}{\omega}\right)^{1/2}=2\sqrt{d_2}\,\omega^{-1/2}\hat{U}(\xi,0),$$

that is

$$\omega^{1/2}\mathscr{L}\hat{U}(\xi,t)-\omega^{-1/2}\hat{U}(\xi,0)=-\frac{d_1}{2\sqrt{d_2}}\,|\xi|^2\,\mathscr{L}\hat{U}(\xi,t).$$

Transforming back and recalling (2.60), we obtain (4.23), as desired.

5 All Functions Are Locally s-Caloric (Up to a Small Error): Proof of (2.12)

We let $(x,t)\in\mathbb{R}\times\mathbb{R}$ and consider the operator $\mathfrak{L}:=\partial_t+(-\Delta)^s_x$. One defines

$$\mathscr{V}:=\big\{h:\mathbb{R}\times\mathbb{R}\to\mathbb{R}\text{ s.t. }\mathfrak{L}h=0\text{ in some neighborhood of the origin in }\mathbb{R}^2\big\},$$

and for any $J\in\mathbb{N}$, we define

$$\mathscr{V}_J:=\left\{\big(\partial^\alpha h(0,0)\big)_{\substack{\alpha=(\alpha_x,\alpha_t)\in\mathbb{N}\times\mathbb{N}\\ \alpha_x+\alpha_t\in[0,J]}}\text{ with }h\in\mathscr{V}\right\}.$$

Notice that $\mathscr{V}_J$ is a linear subspace of $\mathbb{R}^{N+1}$, for some $N\in\mathbb{N}$. The core of the proof is to establish the maximal span condition

$$\mathscr{V}_J=\mathbb{R}^{N+1}. \tag{5.1}$$

To this end, we argue for a contradiction and we suppose that $\mathscr{V}_J$ is a linear subspace strictly smaller than $\mathbb{R}^{N+1}$: hence, there exists

$$\nu=(\nu_0,\ldots,\nu_N)\in S^N \tag{5.2}$$

such that

$$\mathscr{V}_J \subseteq \Big\{ X = (X_0, \dots, X_N) \in \mathbb{R}^{J+1} \text{ s.t. } \nu \cdot X = 0 \Big\}. \tag{5.3}$$

One considers ϕ to be the first eigenfunctions of $(-\Delta)^s$ in $(-1, 1)$ with Dirichlet data, normalized to have unit norm in $L^2(\mathbb{R})$. Accordingly,

$$\begin{cases} (-\Delta)^s \phi(x) = \lambda\, \phi(x) & \text{for any } x \in (-1, 1), \\ \phi(x) = 0 & \text{for any } x \text{ outside } (-1, 1), \end{cases}$$

for some $\lambda > 0$.

In view of the boundary properties discussed in Difference 2.6, one can prove that

$$\partial^\ell \phi(-1 + \delta) = \text{const}\, \delta^{s-\ell}(1 + o(1)), \tag{5.4}$$

with $o(1)$ infinitesimal as $\delta \searrow 0$. So, fixed $\varepsilon, \tau > 0$, we define

$$h_{\varepsilon,\tau}(x, t) := e^{-\tau t}\, \phi\left(-1 + \varepsilon + \frac{\tau^{\frac{1}{2s}}\, x}{\lambda^{\frac{1}{2s}}}\right).$$

This function is smooth for any x in a small neighborhood of the origin and any $t \in \mathbb{R}$, and, in this domain,

$$\begin{aligned} \mathfrak{L} h_{\varepsilon,\tau}(x, t) &= \partial_t \left(e^{-\tau t}\, \phi\left(-1 + \varepsilon + \frac{\tau^{\frac{1}{2s}}\, x}{\lambda^{\frac{1}{2s}}}\right)\right) + (-\Delta)^s_x \left(e^{-\tau t}\, \phi\left(-1 + \frac{\tau^{\frac{1}{2s}}\, x}{\lambda^{\frac{1}{2s}}}\right)\right) \\ &= -\tau\, e^{-\tau t}\, \phi\left(-1 + \varepsilon + \frac{\tau^{\frac{1}{2s}}\, x}{\lambda^{\frac{1}{2s}}}\right) + \frac{\tau\, e^{-\tau t}}{\lambda} (-\Delta)^s \phi\left(-1 + \varepsilon + \frac{\tau^{\frac{1}{2s}}\, x}{\lambda^{\frac{1}{2s}}}\right) \\ &= -\tau\, e^{-\tau t}\, \phi\left(-1 + \varepsilon + \frac{\tau^{\frac{1}{2s}}\, x}{\lambda^{\frac{1}{2s}}}\right) + \tau\, e^{-\tau t}\, \phi\left(-1 + \varepsilon + \frac{\tau^{\frac{1}{2s}}\, x}{\lambda^{\frac{1}{2s}}}\right) \\ &= 0. \end{aligned}$$

This says that $h_{\varepsilon,\tau} \in \mathscr{V}$ and therefore

$$\big(\partial^\alpha h_{\varepsilon,\tau} 0, 0)\big)_{\substack{\alpha = (\alpha_x, \alpha_t) \in \mathbb{N} \times \mathbb{N} \\ \alpha_x + \alpha_t \in [0, J]}} \in \mathscr{V}_J.$$

This, together with (5.3), implies that, for any fixed and positive τ and y,

$$
\begin{aligned}
0 &= \sum_{\substack{\alpha=(\alpha_x,\alpha_t)\in\mathbb{N}\times\mathbb{N}\\ \alpha_x+\alpha_t\in[0,J]}} \nu_\alpha\, \partial^\alpha h_{\varepsilon,\tau}(0,0) = \sum_{\substack{(\alpha_x,\alpha_t)\in\mathbb{N}\times\mathbb{N}\\ \alpha_x+\alpha_t\in[0,J]}} \nu_{(\alpha_x,\alpha_t)}\, \partial_t^{\alpha_t}\partial_x^{\alpha_x} h_{\varepsilon,\tau}(0,0)\\
&= \sum_{\substack{(\alpha_x,\alpha_t)\in\mathbb{N}\times\mathbb{N}\\ \alpha_x+\alpha_t\in[0,J]}} \nu_{(\alpha_x,\alpha_t)}(-\tau)^{\alpha_t} \left(\frac{\tau^{\frac{1}{2s}}}{\lambda^{\frac{1}{2s}}}\right)^{\alpha_x} e^{-\tau t}\, \partial^{\alpha_x}\phi\left(-1+\varepsilon+\frac{\tau^{\frac{1}{2s}}x}{\lambda^{\frac{1}{2s}}}\right)\Bigg|_{(x,t)=(0,0)}\\
&= \sum_{\substack{(\alpha_x,\alpha_t)\in\mathbb{N}\times\mathbb{N}\\ \alpha_x+\alpha_t\in[0,J]}} \nu_{(\alpha_x,\alpha_t)} \frac{(-1)^{\alpha_t}}{\lambda^{\frac{\alpha_x}{2s}}}\, \tau^{\alpha_t+\frac{\alpha_x}{2s}}\, \partial^{\alpha_x}\phi(-1+\varepsilon).
\end{aligned}
$$

Hence, fixed $\tau > 0$, this identity and (5.4) yield that

$$
0 = \sum_{\substack{(\alpha_x,\alpha_t)\in\mathbb{N}\times\mathbb{N}\\ \alpha_x+\alpha_t\in[0,J]}} \nu_{(\alpha_x,\alpha_t)} \frac{(-1)^{\alpha_t}}{\lambda^{\frac{\alpha_x}{2s}}}\, \tau^{\alpha_t+\frac{\alpha_x}{2s}}\, \varepsilon^{s-\alpha_x}(1+o(1)), \tag{5.5}
$$

with $o(1)$ infinitesimal as $\varepsilon \searrow 0$.

We now take $\bar{\alpha}_x$ be the largest integer α_x for which there exists an integer α_t such that $\bar{\alpha}_x+\alpha_t \in [0, J]$ and $\nu_{(\bar{\alpha}_x,\alpha_t)} \neq 0$. Notice that this definition is well-posed, since not all the $\nu_{(\alpha_x,\alpha_t)}$ can vanish, due to (5.2). Then, (5.5) becomes

$$
0 = \sum_{\substack{(\alpha_x,\alpha_t)\in\mathbb{N}\times\mathbb{N}\\ \alpha_x+\alpha_t\in[0,J]\\ \alpha_x\leqslant\bar{\alpha}_x}} \nu_{(\alpha_x,\alpha_t)} \frac{(-1)^{\alpha_t}}{\lambda^{\frac{\alpha_x}{2s}}}\, \tau^{\alpha_t+\frac{\alpha_x}{2s}}\, \varepsilon^{s-\alpha_x}(1+o(1)), \tag{5.6}
$$

since the other coefficients vanish by definition of $\bar{\alpha}_x$.

Thus, we multiply (5.6) by $\varepsilon^{\bar{\alpha}_x-s}\tau^{-\frac{\bar{\alpha}_x}{2s}}$ and we take the limit as $\varepsilon \searrow 0$: in this way, we obtain that

$$
\begin{aligned}
0 &= \lim_{\varepsilon\searrow 0} \sum_{\substack{(\alpha_x,\alpha_t)\in\mathbb{N}\times\mathbb{N}\\ \alpha_x+\alpha_t\in[0,J]\\ \alpha_x\leqslant\bar{\alpha}_x}} \nu_{(\alpha_x,\alpha_t)} \frac{(-1)^{\alpha_t}}{\lambda^{\frac{\alpha_x}{2s}}}\, \tau^{\alpha_t+\frac{\alpha_x}{2s}-\frac{\bar{\alpha}_x}{2s}}\, \varepsilon^{\bar{\alpha}_x-\alpha_x}(1+o(1))\\
&= \sum_{\substack{\alpha_t\in\mathbb{N}\\ \bar{\alpha}_x+\alpha_t\in[0,J]}} \nu_{(\bar{\alpha}_x,\alpha_t)} \frac{(-1)^{\alpha_t}}{\lambda^{\frac{\bar{\alpha}_x}{2s}}}\, \tau^{\alpha_t}.
\end{aligned}
$$

Since this is valid for any $\tau > 0$, by the Identity Principle for Polynomials we obtain that

$$
\nu_{(\bar{\alpha}_x,\alpha_t)} \frac{(-1)^{\alpha_t}}{\lambda^{\frac{\bar{\alpha}_x}{2s}}} = 0,
$$

and thus $\nu_{(\bar{\alpha}_x,\alpha_t)} = 0$, for any integer α_t for which $\bar{\alpha}_x + \alpha_t \in [0, J]$. But this is in contradiction with the definition of $\bar{\alpha}_x$ and so we have completed the proof of (5.1).

From this maximal span property, the proof of (2.12) follows by scaling (arguing as done, for instance, in [112]).

Acknowledgements It is a great pleasure to thank the Università degli Studi di Bari for its very warm hospitality and the Istituto Nazionale di Alta Matematica for the strong financial and administrative support which made this minicourse possible. And of course special thanks go to all the participants, for their patience in attending the course, their competence, empathy and contagious enthusiasm. This work was supported by INdAM and ARC Discovery Project N.E.W. Nonlocal Equations at Work.

⟨⟨The longest appendix measured 26cm (10.24in) when it was removed from 72-year-old Safranco August (Croatia) during an autopsy at the Ljudevit Jurak University Department of Pathology, Zagreb, Croatia, on 26 August 2006.⟩⟩

(Source: http://www.guinnessworldrecords.com/world-records/largest-appendix-removed)

Appendix A: Confirmation of (2.7)

We write Δ_x to denote the Laplacian in the coordinates $x \in \mathbb{R}^n$. In this way, the total Laplacian in the variables $(x, y) \in \mathbb{R}^n \times (0, +\infty)$ can be written as

$$\Delta = \Delta_x + \partial_y^2. \tag{A.1}$$

Given a (smooth and bounded, in the light of footnote 3 on page 5) $u : \mathbb{R}^n \to \mathbb{R}$, we take $U := E_u$ be (smooth and bounded) as in (2.6).

We also consider the operator

$$Lu(x) := -\partial_y E_u(x, 0) \tag{A.2}$$

and we take $V(x, y) := -\partial_y U(x, y)$. Notice that $\Delta V = -\partial_y \Delta U = 0$ in $\mathbb{R}^n \times (0, +\infty)$ and $V(x, 0) = Lu(x)$ for any $x \in \mathbb{R}^n$. In this sense, V is the harmonic extension of Lu and so we can write $V = E_{Lu}$ and so, in the notation of (A.2), and recalling (2.6) and (A.1), we have

$$\begin{aligned} L(Lu)(x) &= -\partial_y E_{Lu}(x,0) = -\partial_y V(x,0) = \partial_y^2 U(x,0) \\ &= \Delta U(x,0) - \Delta_x U(x,0) = -\Delta_x U(x,0) = -\Delta u(x). \end{aligned}$$

This gives that $L^2 = -\Delta$, which is consistent with $L = (-\Delta)^{1/2}$, thanks to (2.5).

Appendix B: Proof of (2.10)

Let $u \in \mathcal{S}$. By (2.9), we can write

$$\sup_{x\in\mathbb{R}^n} (1+|x|^n)\,|u(x)| + \sup_{x\in\mathbb{R}^n} (1+|x|^{n+2}) \left|D^2u(x)\right| \leqslant \text{const}\,. \tag{B.1}$$

Fixed $x \in \mathbb{R}^n$ (with $|x|$ to be taken large), recalling the notation in (2.3), we consider the map $y \mapsto \delta_u(x,y)$ and we observe that

$$\begin{aligned} &\delta_u(x,0) = 0, \\ &\nabla_y \delta_u(x,y) = \nabla u(x+y) - \nabla u(x-y), \\ \text{and}\quad &D_y^2\delta_u(x,y) = D^2u(x+y) + D^2u(x-y). \end{aligned}$$

Hence, if $|Y| \leqslant |x|/2$ we have that $|x \pm Y| \geqslant |x| - |Y| \geqslant |x|/2$, and thus

$$\left|D_y^2\delta_u(x,Y)\right| \leqslant 2 \sup_{|\zeta|\geqslant|x|/2} \left|D^2u(\zeta)\right| \leqslant 2 \sup_{|\zeta|\geqslant|x|/2} \frac{(2|\zeta|)^{n+2}\left|D^2u(\zeta)\right|}{|x|^{n+2}} \leqslant \frac{\text{const}}{|x|^{n+2}},$$

thanks to (B.1).

Therefore, a second order Taylor expansion of δ_u in the variable y gives that, if $|y| \leqslant |x|/2$,

$$\begin{aligned} \left|\delta_u(x,y)\right| &\leqslant \sup_{|Y|\leqslant|x|/2} \left|\delta_u(x,0) + \nabla\delta_u(x,0)\cdot y + \frac{D^2\delta_u(x,Y)\,y\cdot y}{2}\right| \\ &= \sup_{|Y|\leqslant|x|/2} \left|\frac{D^2\delta_u(x,Y)\,y\cdot y}{2}\right| \leqslant \frac{\text{const}\,|y|^2}{|x|^{n+2}}. \end{aligned}$$

Consequently,

$$\left|\int_{B_{|x|/2}} \frac{\delta_u(x,y)}{|y|^{n+2s}}\,dy\right| \leqslant \frac{\text{const}}{|x|^{n+2}} \int_{B_{|x|/2}} \frac{|y|^2}{|y|^{n+2s}}\,dy \leqslant \frac{\text{const}\,|x|^{2-2s}}{|x|^{n+2}} = \frac{\text{const}}{|x|^{n+2s}}. \tag{B.2}$$

Moreover, by (B.1),

$$
\begin{aligned}
&\left|\int_{\mathbb{R}^n\setminus B_{|x|/2}} \frac{\delta_u(x,y)}{|y|^{n+2s}}\,dy\right| \\
&\leqslant \int_{\mathbb{R}^n\setminus B_{|x|/2}} \frac{|u(x+y)|}{|y|^{n+2s}}\,dy + \int_{\mathbb{R}^n\setminus B_{|x|/2}} \frac{|u(x-y)|}{|y|^{n+2s}}\,dy + 2\int_{\mathbb{R}^n\setminus B_{|x|/2}} \frac{|u(x)|}{|y|^{n+2s}}\,dy \\
&\leqslant \int_{\mathbb{R}^n\setminus B_{|x|/2}} \frac{|u(x+y)|}{(|x|/2)^{n+2s}}\,dy + \int_{\mathbb{R}^n\setminus B_{|x|/2}} \frac{|u(x-y)|}{(|x|/2)^{n+2s}}\,dy + \frac{\text{const}\,|u(x)|}{|x|^{2s}} \\
&\leqslant \frac{\text{const}}{|x|^{n+2s}}\int_{\mathbb{R}^n} |u(\zeta)|\,d\zeta + \frac{\text{const}\,|u(x)|}{|x|^{2s}} \\
&\leqslant \frac{\text{const}}{|x|^{n+2s}}.
\end{aligned}
$$

This and (B.2), recalling (2.3), establish (2.10).

Appendix C: Proof of (2.14)

Let $M := \frac{1}{2n}\left(1+\sup_{B_1}|f|\right)$ and $v(x) := M(1-|x|^2) - u(x)$. Notice that $v = 0$ along ∂B_1 and

$$
\Delta v = -2nM - \Delta u \leqslant -M - f \leqslant -M + \sup_{B_1}|f| \leqslant 0
$$

in B_1. Consequently, $v \geqslant 0$ in B_1, which gives that $u(x) \leqslant M(1-|x|^2)$.

Arguing similarly, by looking at $\tilde{v}(x) := M(1-|x|^2) + u(x)$, one sees that $-u(x) \leqslant M(1-|x|^2)$. Accordingly, we have that

$$
|u(x)| \leqslant M(1-|x|^2) \leqslant M(1+|x|)(1-|x|) \leqslant 2M(1-|x|).
$$

This proves (2.14).

Appendix D: Proof of (2.17)

The idea of the proof is described in Fig. 7. The trace of the function in Fig. 7 is exactly the function $u_{1/2}$ in (2.16). The function plotted in Fig. 7 is the harmonic extension of $u_{1/2}$ in the halfplane (like an elastic membrane pinned at the halfcircumference along the trace). Our objective is to show that the normal derivative of

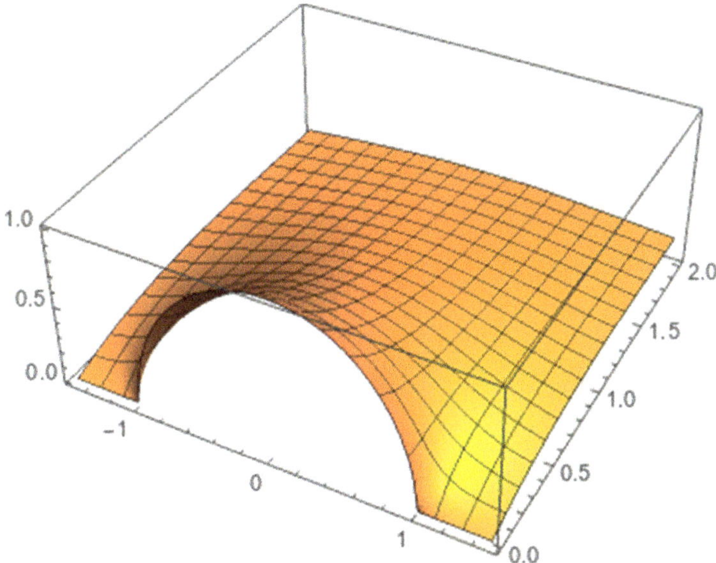

Fig. 7 Harmonic extension in the halfplane of the function $\mathbb{R} \ni x \mapsto (1-x^2)_+^{1/2}$

such extended function along the trace is constant, and so we can make use of the extension method in (2.6) and (2.7) to obtain (2.17).

In further detail, we use complex coordinates, identifying $(x, y) \in \mathbb{R} \times (0, +\infty)$ with $z := x + iy \in \mathbb{C}$ with $\Im(z) > 0$. Also, as customary, we define the principal square root in the cut complex plane

$$\mathbb{C}_\star := \{z = re^{i\varphi} \text{ with } r > 0 \text{ and } -\pi < \varphi < \pi\}$$

by defining, for any $z = re^{i\varphi} \in \mathbb{C}_\star$,

$$\surd(z) := \sqrt{r}\, e^{i\varphi/2}, \tag{D.1}$$

see Fig. 8 (for typographical convenience, we distinguish between the complex and the real square root, by using the symbols $\surd(\cdot)$ and $\sqrt{\cdot}$ respectively).

The principal square root function is defined using the nonpositive real axis as a "branch cut" and

$$\big(\surd(z)\big)^2 = r\, e^{i\varphi} = z. \tag{D.2}$$

Moreover,

$$\text{the function } \surd \text{ is holomorphic in } \mathbb{C}_\star \tag{D.3}$$

$$\text{and } \partial_z \surd(z) = \frac{1}{2\surd(z)}. \tag{D.4}$$

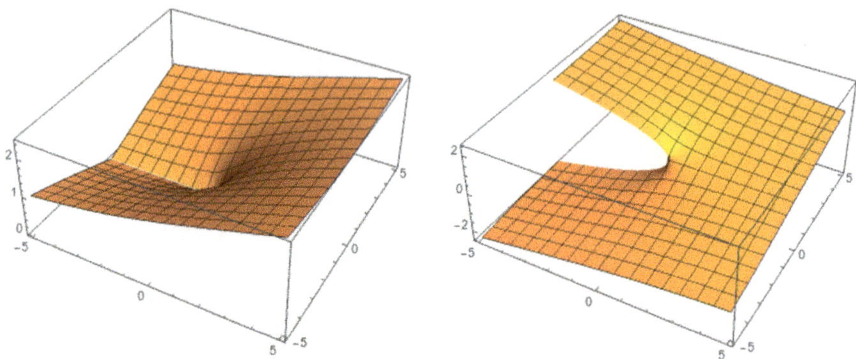

Fig. 8 Real and imaginary part of the complex principal square root

To check these facts, we take $z \in \mathbb{C}_\star$: since $\mathbb{C}_\star$ is open, we have that $z + w \in \mathbb{C}_\star$ for any $w \in \mathbb{C} \setminus \{0\}$ with small module. Consequently, by (D.2), we obtain that

$$\begin{aligned} w = (z+w) - z &= \big(\surd(z+w)\big)^2 - \big(\surd(z)\big)^2 \\ &= \Big(\surd(z+w) + \surd(z)\Big)\Big(\surd(z+w) - \surd(z)\Big). \end{aligned}$$

Dividing by w and taking the limit, we thus find that

$$\begin{aligned} 1 &= \lim_{w \to 0} \Big(\surd(z+w) + \surd(z)\Big) \frac{\surd(z+w) - \surd(z)}{w} \\ &= 2\surd(z) \lim_{w \to 0} \frac{\surd(z+w) - \surd(z)}{w} \end{aligned} \tag{D.5}$$

Since $\mathbb{C}_\star \subseteq \mathbb{C} \setminus \{0\}$, we have that $z \neq 0$, and thus $\surd(z) \neq 0$. As a result, we can divide (D.5) by $2\surd(z)$ and conclude that

$$\lim_{w \to 0} \frac{\surd(z+w) - \surd(z)}{w} = \frac{1}{2\surd(z)},$$

which establishes, at the same time, both (D.3) and (D.4), as desired.

We also remark that

$$\text{if } z \in \mathbb{C} \text{ with } \Im(z) > 0, \text{ then } 1 - z^2 \in \mathbb{C}_\star. \tag{D.6}$$

To check this, if $z = x + iy$ with $y > 0$, we observe that

$$1 - z^2 = 1 - (x + iy)^2 = 1 - x^2 + y^2 - 2ixy. \tag{D.7}$$

Hence, if $1 - z^2$ lies on the real axis, we have that $xy = 0$, and so $x = 0$. Then, the real part of $1 - z^2$ in this case is equal to $1 + y^2$ which is strictly positive. This proves (D.6).

Thanks to (D.6), for any $z \in \mathbb{C}$ with $\Im(z) > 0$ we can define the function $\sqrt{(1 - z^2)}$. From (D.7), we can write

$$\begin{aligned} 1 - z^2 &= r(x, y)\, e^{i\varphi(x,y)}, \\ \text{where } r(x, y) &= \big((1 - x^2 + y^2)^2 + 4x^2y^2\big)^{1/2}, \\ r(x, y) \cos\varphi(x, y) &= 1 - x^2 + y^2 \\ \text{and } r(x, y) \sin\varphi(x, y) &= 2xy. \end{aligned}$$

Notice that

$$\lim_{y \searrow 0} r(x, y) = \big((1 - x^2)^2\big)^{1/2} = |1 - x^2|.$$

As a consequence,

$$\begin{aligned} &|1 - x^2| \lim_{y \searrow 0} \cos\varphi(x, y) = \lim_{y \searrow 0} r(x, y) \cos\varphi(x, y) = \lim_{y \searrow 0} (1 - x^2 + y^2) = 1 - x^2 \\ \text{and} \quad &|1 - x^2| \lim_{y \searrow 0} \sin\varphi(x, y) = \lim_{y \searrow 0} r(x, y) \sin\varphi(x, y) = \lim_{y \searrow 0} 2xy = 0. \end{aligned}$$

This says that, if $x^2 > 1$ then

$$\begin{aligned} &\lim_{y \searrow 0} \cos\varphi(x, y) = -1 \\ \text{and} \quad &\lim_{y \searrow 0} \sin\varphi(x, y) = 0, \end{aligned}$$

while if $x^2 < 1$ then

$$\begin{aligned} &\lim_{y \searrow 0} \cos\varphi(x, y) = 1 \\ \text{and} \quad &\lim_{y \searrow 0} \sin\varphi(x, y) = 0. \end{aligned}$$

On this account, we deduce that

$$\lim_{y \searrow 0} \varphi(x, y) = \begin{cases} \pi & \text{if } x^2 > 1, \\ 0 & \text{if } x^2 < 1 \end{cases} \tag{D.8}$$

and therefore, recalling (D.1),

$$\lim_{y\searrow 0}\sqrt{(1-z^2)} = \lim_{y\searrow 0}\sqrt{r(x,y)}\,e^{i\varphi(x,y)/2} = \begin{cases} \sqrt{|1-x^2|}\,e^{i\pi/2} & \text{if } x^2>1,\\ \sqrt{|1-x^2|}\,e^{i0} & \text{if } x^2<1,\\ 0 & \text{if } x^2=1 \end{cases}$$
$$= \begin{cases} i\sqrt{|1-x^2|} & \text{if } x^2>1,\\ \sqrt{|1-x^2|} & \text{if } x^2<1,\\ 0 & \text{if } x^2=1. \end{cases} \tag{D.9}$$

This implies that

$$\lim_{y\searrow 0}\Re\Big(\sqrt{(1-z^2)}\Big) = \begin{cases} 0 & \text{if } x^2\geqslant 1,\\ \sqrt{|1-x^2|} & \text{if } x^2<1 \end{cases} \tag{D.10}$$
$$= (1-x^2)_+^{1/2}.$$

Now we define

$$z=x+iy\mapsto \Re\Big(\sqrt{(1-z^2)}+iz\Big) =: U_{1/2}(x,y).$$

The function $U_{1/2}$ is the harmonic extension of $u_{1/2}$ in the halfplane, as plotted in Fig. 7. Indeed, from (D.10),

$$\lim_{y\searrow 0}U_{1/2}(x,y) = \lim_{y\searrow 0}\Re\Big(\sqrt{(1-z^2)}+ix-y\Big) = (1-x^2)_+^{1/2} = u_{1/2}(x).$$

Furthermore, from (D.3), we have that $U_{1/2}$ is the real part of a holomorphic function in the halfplane and so it is harmonic.

These considerations give that $U_{1/2}$ solves the harmonic extension problem in (2.6), hence, in the light of (2.7),

$$(-\Delta)^{1/2}u_{1/2}(x) = \lim_{y\searrow 0} -\partial_y U_{1/2}(x,y) = \lim_{y\searrow 0} -\Re\Big(\partial_y\sqrt{(1-z^2)}+i\partial_y z\Big)$$
$$= \lim_{y\searrow 0} -\Re\Big(\partial_y\sqrt{(1-z^2)}-1\Big) = 1-\lim_{y\searrow 0}\Re\Big(\partial_y\sqrt{(1-z^2)}\Big). \tag{D.11}$$

Now, recalling (D.4), we see that, for any $x\in(-1,1)$ and small $y>0$,

$$\partial_y\sqrt{(1-z^2)} = \partial_z\sqrt{(1-z^2)}\,\partial_y z = \frac{1}{2\sqrt{(1-z^2)}}\,\partial_z(1-z^2)\,\partial_y(x+iy) = -\frac{iz}{\sqrt{(1-z^2)}}. \tag{D.12}$$

We stress that the latter denominator does not vanish when $x \in (-1, 1)$ and $y > 0$ is small. So, using that $\Re(ZW) = \Re Z \Re W - \Im Z \Im W$ for any $Z, W \in \mathbb{C}$, we obtain that

$$\begin{aligned} y &= \Re\big(-i(x+iy)\big) = \Re(-iz) = \Re\Big(\surd(1-z^2)\, \partial_y \surd(1-z^2)\Big) \\ &= \Re\Big(\surd(1-z^2)\Big)\, \Re\Big(\partial_y \surd(1-z^2)\Big) - \Im\Big(\surd(1-z^2)\Big)\, \Im\Big(\partial_y \surd(1-z^2)\Big). \end{aligned} \tag{D.13}$$

From (D.9), for any $x \in (-1, 1)$ we have that

$$\lim_{y \searrow 0} \Im\Big(\surd(1-z^2)\Big) = \Im\Big(\sqrt{|1-x^2|}\Big) = 0.$$

This and the fact that $\partial_y \surd(1-z^2)$ is bounded (in view of (D.12)) give that, for any $x \in (-1, 1)$,

$$\lim_{y \searrow 0} \Im\Big(\surd(1-z^2)\Big)\, \Im\Big(\partial_y \surd(1-z^2)\Big) = 0.$$

This, (D.9) and (D.13) imply that, for any $x \in (-1, 1)$,

$$\begin{aligned} 0 = \lim_{y \searrow 0} y &= \lim_{y \searrow 0} \Re\Big(\surd(1-z^2)\Big)\, \Re\Big(\partial_y \surd(1-z^2)\Big) - \Im\Big(\surd(1-z^2)\Big)\, \Im\Big(\partial_y \surd(1-z^2)\Big) \\ &= \Re\Big(\sqrt{|1-x^2|}\Big) \lim_{y \searrow 0} \Re\Big(\partial_y \surd(1-z^2)\Big) + 0 \\ &= \sqrt{|1-x^2|} \lim_{y \searrow 0} \Re\Big(\partial_y \surd(1-z^2)\Big) \end{aligned}$$

and therefore

$$\lim_{y \searrow 0} \Re\Big(\partial_y \surd(1-z^2)\Big) = 0. \tag{D.14}$$

Plugging this information into (D.11), we conclude the proof of (2.17), as desired.

Appendix E: Deducing (2.19) from (2.15) Using a Space Inversion

From (2.15), up to a translation, we know that

$$\text{the function } \mathbb{R} \ni x \mapsto v_s(x) := (x-1)_+^s \text{ is } s\text{-harmonic in } (1, +\infty). \tag{E.1}$$

We let w_s be the space inversion of v_s induced by the Kelvin transform in the fractional setting, namely

$$w_s(x) := |x|^{2s-1}\, v_s\left(\frac{x}{|x|^2}\right) = |x|^{2s-1}\left(\frac{x}{|x|^2}-1\right)_+^s = \begin{cases} x^{s-1}(1-x)^s & \text{if } x \in (0,1), \\ 0 & \text{otherwise}. \end{cases}$$

By (E.1), see Corollary 2.3 in [63], it follows that $w_s(x)$ is s-harmonic in $(0, 1)$. Consequently, the function

$$w_s^\star(x) := w_s(1-x) = \begin{cases} x^s(1-x)^{s-1} & \text{if } x \in (0,1), \\ 0 & \text{otherwise}. \end{cases}$$

is also s-harmonic in $(0, 1)$. We thereby conclude that the function

$$W_s^\star(x) := w_s(x) - w_s^\star(x) = \begin{cases} x^{s-1}(1-x)^s - x^s(1-x)^{s-1} & \text{if } x \in (0,1), \\ 0 & \text{otherwise}. \end{cases}$$

is also s-harmonic in $(0, 1)$. See Fig. 9 for a picture of w_s and $W_s^\star$ when $s = 1/2$. Let now

$$U_s(x) := x_+^s(1-x)_+^s = \begin{cases} x^s(1-x)^s & \text{if } x \in (0,1), \\ 0 & \text{otherwise}. \end{cases}$$

and notice that U_s is the primitive of $sW_s^\star$. Since the latter function is s-harmonic in $(0, 1)$, after an integration we thereby deduce that $(-\Delta)^s U_s = \text{const}$ in $(0, 1)$. This and the fact that

$$U_s\left(\frac{x+1}{2}\right) = 2^{-s}\, u_s(x)$$

imply (2.19).

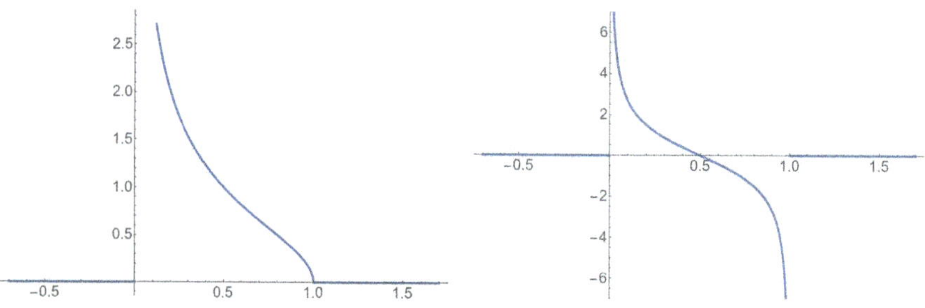

Fig. 9 The functions $w_{1/2}$ and $W_{1/2}^\star$

Appendix F: Proof of (2.21)

Fixed $y \in \mathbb{R}^n \setminus \{0\}$ we let $\mathscr{R}^y$ be a rotation which sends $\frac{y}{|y|}$ into the vector $e_1 = (1, 0, \ldots, 0)$, that is

$$\sum_{k=1}^{n} \mathscr{R}^y_{ik}\, y_k = |y|\delta_{i1}, \tag{F.1}$$

for any $i \in \{1, \ldots, n\}$. We also denote by

$$K(y) := \frac{y}{|y|^2}$$

the so-called Kelvin Transform. We recall that for any $i, j \in \{1, \ldots, n\}$,

$$\partial_{y_i} K_j(y) = \frac{\delta_{ij}}{|y|^2} - \frac{2y_i y_j}{|y|^4}$$

and so, by (F.1),

$$\left(\mathscr{R}^y\ (DK(y))\ (\mathscr{R}^y)^{-1}\right)_{ij} = \sum_{k,h=1}^{n} \mathscr{R}^y_{ik}\, \partial_{y_k} K_h(y)\, \mathscr{R}^y_{jh} = \frac{\delta_{ij}}{|y|^2} - \frac{2\delta_{i1}\delta_{j1}}{|y|^2}.$$

This says that $\mathscr{R}^y\ (DK(y))\ (\mathscr{R}^y)^{-1}$ is a diagonal[6] matrix, with first entry equal to $-\frac{1}{|y|^2}$ and the others equal to $\frac{1}{|y|^2}$.

As a result,

$$\left|\det(DK(y))\right| = \left|\det\left(\mathscr{R}^y\ (DK(y))\ (\mathscr{R}^y)^{-1}\right)\right| = \frac{1}{|y|^{2n}}. \tag{F.2}$$

The Kelvin Transform is also useful to write the Green function of the ball B_1, see e.g. formula (41) on p. 40 and Theorem 13 on p. 35 of [62]. Namely, we take $n \geqslant 3$ for simplicity, and we write

$$\begin{aligned} G(x, y) &:= \text{const} \left(\frac{1}{|y-x|^{n-2}} - \frac{1}{\big| |x|(y - K(x)) \big|^{n-2}} \right) \\ &= \text{const} \left(\frac{1}{|x-y|^{n-2}} - \frac{1}{\big| |y|(x - K(y)) \big|^{n-2}} \right) = G(y, x) \end{aligned}$$

[6]From the geometric point of view, one can also take radial coordinates, compute the derivatives of K along the unit sphere and use scaling.

and, for a suitable choice of the constant, for any $x \in B_1$ we can write the solution of (2.20) in the form

$$u(x) = \int_{B_1} f(y)\, G(x, y)\, dy.$$

see e.g. page 35 in [62].

On this account, we have that, for any $x \in B_1$,

$$\begin{aligned}
|\nabla u(x)| &\leqslant \int_{B_1} |f(y)|\, |\partial_x G(x, y)|\, dy \\
&\leqslant \text{const} \sup_{B_1} |f| \int_{B_1} \left(\frac{1}{|x-y|^{n-1}} + \frac{1}{|y|^{n-2} \big| x - K(y) \big|^{n-1}} \right) dy \\
&\leqslant \text{const} \sup_{B_1} |f| \left(\int_{B_2} \frac{d\zeta}{|\zeta|^{n-1}} + \int_{\mathbb{R}^n \setminus B_1} \frac{d\eta}{|\eta|^{n+2} \big| x - \eta \big|^{n-1}} \right) \\
&\leqslant \text{const} \sup_{B_1} |f| \left(1 + \int_{B_2 \setminus B_1} \frac{d\eta}{|x-\eta|^{n-1}} + \int_{\mathbb{R}^n \setminus B_2} \frac{d\eta}{|\eta|^{n+2}} \right) \\
&\leqslant \text{const} \sup_{B_1} |f|.
\end{aligned}$$

Notice that here we have used the transformations $\zeta := x - y$ and $\eta := K(y)$, exploiting also (F.2). The claim in (2.21) is thus established.

Appendix G: Proof of (2.24) and Probabilistic Insights

We give a proof of (2.24) by taking a derivative of (2.17). To this aim, we claim[7] that

$$\begin{aligned}
&\frac{d}{dx} \int_{\mathbb{R}} \frac{u_{1/2}(x+y) + u_{1/2}(x-y) - 2u_{1/2}(x)}{|y|^2}\, dy \\
&= - \int_{\mathbb{R}} \frac{(x+y)u_{-1/2}(x+y) + (x-y)u_{-1/2}(x-y) - 2xu_{-1/2}(x)}{|y|^2}\, dy.
\end{aligned} \tag{G.1}$$

[7]The difficulty in proving (G.1) is that the function $u_{1/2}$ is not differentiable at ± 1 and the derivative taken inside the integral might produce a singularity (in fact, formula (G.1) exactly says that such derivative can be performed with no harm inside the integral). The reader who is already familiar with the basics of functional analysis can prove (G.1) by using the theory of absolutely continuous functions, see e.g. Theorem 8.21 in [98]. We provide here a direct proof, available to everybody.

To this end, we fix $x \in (-1, 1)$ and $h \in \mathbb{R}$. We define

$$\ell_x := \min\{|x-1|,\ |x+1|\} > 0.$$

In the sequel, we will take $|h|$ as small as we wish in order to compute incremental quotients, hence we can assume that

$$|h| < \frac{\ell_x}{4}. \tag{G.2}$$

We also define

$$I_x(h) := \Big\{ y \in \mathbb{R} \text{ s.t. } \min\{|(x+y)-1|,\ |(x-y)-1|,\ |(x+y)+1|,\ |(x-y)+1|\} \leqslant 2|h| \Big\}. \tag{G.3}$$

Since $I_x(h) \subseteq (x-1-2|h|, x-1+2|h|) \cup (x+1-2|h|, x+1+2|h|) \cup (1-x-2|h|, 1-x+2|h|) \cup (-1-x-2|h|, -1-x+2|h|)$, we have that

$$\text{the measure of } I_x \text{ is less than const}\,|h|. \tag{G.4}$$

Furthermore,

$$I_x(h) \subseteq \left\{ y \in \mathbb{R} \text{ s.t. } |y| \geqslant \frac{\ell_x}{2} \right\}. \tag{G.5}$$

To check this, let $y \in I_x(h)$. Then, by (G.3), there exist $\sigma_{1,x,y}, \sigma_{2,x,y} \in \{-1, 1\}$ such that

$$|x + \sigma_{1,x,y} y + \sigma_{2,x,y}| \leqslant 2|h|$$

and therefore

$$|y| = |\sigma_{1,x,y} y| \geqslant |x + \sigma_{2,x,y}| - |x + \sigma_{1,x,y} y + \sigma_{2,x,y}| \geqslant \ell_x - 2|h| \geqslant \frac{\ell_x}{2},$$

where the last inequality is a consequence of (G.2), and this establishes (G.5).

Now, we introduce the following notation for the incremental quotient

$$Q_h(x,y) := \frac{\Big(\big(u_{1/2}(x+y+h) + u_{1/2}(x-y+h) - 2u_{1/2}(x+h)\big) - \big(u_{1/2}(x+y) + u_{1/2}(x-y) - 2u_{1/2}(x)\big)\Big)}{h}$$

and we observe that, since $u_{1/2}$ is globally Hölder continuous with exponent $1/2$, it holds that

$$\begin{aligned}
|Q_h(x,y)| &\leqslant \frac{\Big(\big|u_{1/2}(x+y+h)-u_{1/2}(x+y)\big|+\big|u_{1/2}(x-y+h)-u_{1/2}(x-y)\big| \\ +2\,\big|u_{1/2}(x+h)-u_{1/2}(x)\big|\Big)}{|h|} \\
&\leqslant \frac{\text{const}\,|h|^{1/2}}{|h|} \\
&= \frac{\text{const}}{|h|^{1/2}},
\end{aligned}$$

for any $x, y \in \mathbb{R}$. Consequently, recalling (G.4) and (G.5), we conclude that

$$\lim_{h\to 0}\left|\int_{I_x(h)}\frac{Q_h(x,y)}{|y|^2}\,dy\right| \leqslant \lim_{h\to 0}\int_{I_x(h)}\frac{\text{const}}{|h|^{1/2}\,\ell_x^2}\,dy \leqslant \lim_{h\to 0}\frac{\text{const}\,|h|}{|h|^{1/2}\,\ell_x^2}=0. \tag{G.6}$$

Now we take derivatives of $u_{1/2}$. For this, we observe that, for any $\xi \in (-1,1)$,

$$u'_{1/2}(\xi) = -\xi(1-\xi^2)^{-1/2} = -\xi u_{-1/2}(\xi).$$

Since the values outside $(-1,1)$ are trivial, this implies that

$$u'_{1/2}(\xi) = -\xi u_{-1/2}(\xi) \qquad \text{for any } \xi \in \mathbb{R}\setminus\{-1,1\}. \tag{G.7}$$

Now, by (G.3), we know that if $y \in \mathbb{R}\setminus I_x(h)$ we have that $x+y+t \in \mathbb{R}\setminus\{-1,1\}$ for all $t \in \mathbb{R}$ with $|t| < |h|$ and therefore we can exploit (G.7) and find that

$$\lim_{h\to 0}\frac{u_{1/2}(x+y+h)-u_{1/2}(x+y)}{h} = -(x+y)u_{-1/2}(x+y).$$

Similar arguments show that, for any $y \in \mathbb{R}\setminus I_x(h)$,

$$\lim_{h\to 0}\frac{u_{1/2}(x-y+h)-u_{1/2}(x-y)}{h} = -(x-y)u_{-1/2}(x-y)$$

$$\text{and}\quad \lim_{h\to 0}\frac{u_{1/2}(x+h)-u_{1/2}(x)}{h} = -xu_{-1/2}(x).$$

Consequently, for any $y \in \mathbb{R}\setminus I_x(h)$,

$$\lim_{h\to 0}\frac{Q_h(x,y)}{|y|^2} = -\frac{(x+y)u_{-1/2}(x+y)+(x-y)u_{-1/2}(x-y)-2xu_{-1/2}(x)}{|y|^2}. \tag{G.8}$$

Now we set

$$\begin{aligned}\Xi_h(x,y) &:= \frac{Q_h(x,y)\,\chi_{\mathbb{R}\setminus I_x(h)}(y)}{|y|^2}\\ &= \frac{1}{h\,|y|^2}\Big(\big(u_{1/2}(x+y+h)+u_{1/2}(x-y+h)-2u_{1/2}(x+h)\big)\\ &\qquad -\big(u_{1/2}(x+y)+u_{1/2}(x-y)-2u_{1/2}(x)\big)\Big)\,\chi_{\mathbb{R}\setminus I_x(h)}(y)\end{aligned}$$

and we claim that

$$|\Xi_h(x,y)| \leqslant C_x\left[\chi_{(-3,3)}(y)\left(\frac{1}{|1-(x+y)^2|^{1/2}}+\frac{1}{|1-(x-y)^2|^{1/2}}\right)+\frac{\chi_{\mathbb{R}\setminus(-3,3)}(y)}{|y|^2}\right], \tag{G.9}$$

for a suitable $C_x > 0$, possibly depending on x. For this, we first observe that if $|y| \geqslant 3$ then $|x \pm y| \geqslant 1$ and also $|x \pm y + h| \geqslant 1$. This implies that if $|y| \geqslant 3$, then $u_{1/2}(x \pm y) = u_{1/2}(x \pm y + h) = 0$ and therefore

$$\Xi_h(x,y) = \frac{1}{h\,|y|^2}\big(2u_{1/2}(x) - 2u_{1/2}(x+h)\big).$$

This and the fact that $u_{1/2}$ is smooth in the vicinity of the fixed $x \in (-1,1)$ imply that (G.9) holds true when $|y| \geqslant 3$. Therefore, from now on, to prove (G.9) we can suppose that

$$|y| < 3. \tag{G.10}$$

We will also distinguish two regimes, the one in which $|y| \leqslant \frac{\ell_x}{4}$ and the one in which $|y| > \frac{\ell_x}{4}$.

If $|y| \leqslant \frac{\ell_x}{4}$ and $|t| \leqslant h$, we have that

$$|(x+y+t)+1| \geqslant |x+1| - |y| - |t| \geqslant \ell_x - |y| - |h| \geqslant \frac{\ell_x}{2},$$

due to (G.2), and similarly $|(x-y+t)-1| \geqslant \frac{\ell_x}{2}$. This implies that

$$|u_{1/2}(x+y+t) + u_{1/2}(x-y+t) - 2u_{1/2}(x+t)| \leqslant C_x\,|y|^2,$$

for some $C_x > 0$ that depends on ℓ_x. Consequently, we find that if $|y| \leqslant \frac{\ell_x}{4}$ then

$$|\Xi_h(x,y)| \leqslant \frac{\text{const}\ C_x\,|y|^2}{|y|^2} = \text{const}\,C_x. \tag{G.11}$$

Conversely, if $y \in \mathbb{R} \setminus I_x(h)$, with $|y| > \frac{\ell_x}{4}$, then we make use of (G.7) and (G.10) to see that

$$\begin{aligned}
|u_{1/2}(x+y+h) - u_{1/2}(x+y)| &\leqslant \int_0^{|h|} |u'_{1/2}(x+y+\tau)|\,d\tau \\
&= \int_0^{|h|} |x+y+\tau|\,|u_{-1/2}(x+y+\tau)|\,d\tau \leqslant 5\int_0^{|h|} |u_{-1/2}(x+y+\tau)|\,d\tau \\
&\leqslant 5\int_0^{|h|} \frac{d\tau}{|1-(x+y+\tau)^2|^{1/2}}.
\end{aligned} \tag{G.12}$$

Also, if $y \in \mathbb{R} \setminus I_x(h)$ we deduce from (G.3) that $|1 \pm (x+y)| \geqslant 2|h|$ and therefore, if $|\tau| \leqslant |h|$, then

$$|1 \pm (x+y+\tau)| \geqslant |1 \pm (x+y)| - |\tau| \geqslant |1 \pm (x+y)| - |h| \geqslant \frac{|1 \pm (x+y)|}{2}.$$

Therefore

$$\begin{aligned}
|1-(x+y+\tau)^2| &= |1+(x+y+\tau)|\,|1-(x+y+\tau)| \\
&\geqslant \frac{1}{4}\,|1+(x+y)|\,|1-(x+y)| = \frac{1}{4}\,|1-(x+y)^2|.
\end{aligned}$$

Hence, we insert this information into (G.12) and we conclude that

$$|u_{1/2}(x+y+h) - u_{1/2}(x+y)| \leqslant \text{const} \int_0^{|h|} \frac{d\tau}{|1-(x+y)^2|^{1/2}} = \frac{\text{const}\,|h|}{|1-(x+y)^2|^{1/2}}. \tag{G.13}$$

Similarly, one sees that

$$|u_{1/2}(x-y+h) - u_{1/2}(x-y)| \leqslant \frac{\text{const}\,|h|}{|1-(x-y)^2|^{1/2}}. \tag{G.14}$$

In view of (G.13) and (G.14), we get that, for any $y \in \mathbb{R} \setminus I_x(h)$ with $|y| > \frac{\ell_x}{4}$,

$$\begin{aligned}
|\Xi_h(x,y)| &\leqslant \frac{1}{h\,|y|^2}\left(\text{const}\,|h| + \frac{\text{const}\,|h|}{|1-(x+y)^2|^{1/2}} + \frac{\text{const}\,|h|}{|1-(x-y)^2|^{1/2}}\right) \\
&\leqslant \frac{\text{const}}{\ell_x^2}\left(1 + \frac{1}{|1-(x+y)^2|^{1/2}} + \frac{1}{|1-(x-y)^2|^{1/2}}\right).
\end{aligned}$$

Combining this with (G.11), we obtain (G.9), up to renaming constants.

Now, we point out that the right hand side of (G.9) belongs to $L^1(\mathbb{R})$. Accordingly, using (G.9) and the Dominated Convergence Theorem, and recalling also (G.7), it follows that

$$\begin{aligned}\lim_{h\to 0}\int_{\mathbb{R}\setminus I_x(h)} &\frac{1}{h|y|^2}\Big(\big(u_{1/2}(x+y+h)+u_{1/2}(x-y+h)-2u_{1/2}(x+h)\big)\\ &-\big(u_{1/2}(x+y)+u_{1/2}(x-y)-2u_{1/2}(x)\big)\Big)\,dy\\ &=\lim_{h\to 0}\int_{\mathbb{R}}\Xi_h(x,y)\,dy\\ &=\int_{\mathbb{R}}\lim_{h\to 0}\Xi_h(x,y)\,dy=\int_{\mathbb{R}}\frac{u'_{1/2}(x+y)+u'_{1/2}(x-y)-2u'_{1/2}(x)}{|y|^2}\,dy\\ &=-\int_{\mathbb{R}}\frac{(x+y)u_{-1/2}(x+y)+(x-y)u_{-1/2}(x-y)-2xu_{-1/2}(x)}{|y|^2}\,dy,\end{aligned}$$

where the last identity is a consequence of (G.8).

From this and (G.6), the claim in (G.1) follows, as desired.

Now, we rewrite (G.1) as

$$\begin{aligned}&\frac{d}{dx}\int_{\mathbb{R}}\frac{u_{1/2}(x+y)+u_{1/2}(x-y)-2u_{1/2}(x)}{|y|^2}\,dy\\ &=-\mathcal{J}(x)-x\int_{\mathbb{R}}\frac{u_{-1/2}(x+y)+u_{-1/2}(x-y)-2u_{-1/2}(x)}{|y|^2}\,dy\\ &\qquad\text{where } \mathcal{J}(x):=\int_{\mathbb{R}}\frac{y\big(u_{-1/2}(x+y)-u_{-1/2}(x-y)\big)}{|y|^2}\,dy\\ &\qquad\qquad=\int_{\mathbb{R}}\frac{u_{-1/2}(x+y)-u_{-1/2}(x-y)}{y}\,dy.\end{aligned} \tag{G.15}$$

We claim that

$$\mathcal{J}(x)=0. \tag{G.16}$$

This follows plainly for $x = 0$, since $u_{-1/2}$ is even. Hence, from here on, to prove (G.16) we assume without loss of generality that $x \in (0, 1)$. Moreover, by changing variable $y \mapsto -y$, we see that

$$-\text{P.V.}\int_{\mathbb{R}}\frac{u_{-1/2}(x-y)}{y}\,dy=\text{P.V.}\int_{\mathbb{R}}\frac{u_{-1/2}(x+y)}{y}\,dy$$

and therefore

$$\begin{aligned}\mathcal{J}(x) &= 2 \text{ P.V.} \int_{\mathbb{R}} \frac{u_{-1/2}(x+y)}{y} \, dy \;=\; 2 \text{ P.V.} \int_{-1-x}^{1-x} \frac{dy}{y\sqrt{1-(x+y)^2}} \\ &= 2 \text{ P.V.} \int_{-1}^{1} \frac{dz}{(z-x)\sqrt{1-z^2}}.\end{aligned} \tag{G.17}$$

Now, we apply the change of variable

$$\xi := \frac{1-\sqrt{1-z^2}}{z}, \qquad \text{hence } z = \frac{2\xi}{1+\xi^2}.$$

We observe that when z ranges in $(-1, 1)$, then ξ ranges therein as well. Moreover,

$$\sqrt{1-z^2} = 1 - \xi z = \frac{1-\xi^2}{1+\xi^2},$$

thus, by (G.17),

$$\begin{aligned}\mathcal{J}(x) &= 2 \text{ P.V.} \int_{-1}^{1} \frac{1}{(\frac{2\xi}{1+\xi^2} - x)\frac{1-\xi^2}{1+\xi^2}} \cdot \frac{2-2\xi^2}{(1+\xi^2)^2} \, d\xi \\ &= 4 \text{ P.V.} \int_{-1}^{1} \frac{d\xi}{2\xi - x(1+\xi^2)} = 4x \text{ P.V.} \int_{-1}^{1} \frac{d\xi}{1-x^2-(1-x\xi)^2}.\end{aligned}$$

We now apply another change of variable

$$\eta := \frac{1-x\xi}{\sqrt{1-x^2}}$$

which gives

$$\mathcal{J}(x) = \frac{4}{\sqrt{1-x^2}} \text{ P.V.} \int_{a_-}^{a^+} \frac{d\eta}{1-\eta^2}, \tag{G.18}$$

where

$$a_+ := \sqrt{\frac{1+x}{1-x}} \text{ and } a_- := \sqrt{\frac{1-x}{1+x}} = \frac{1}{a_+}.$$

Now we notice that

$$\text{P.V.} \int_{a_-}^{a_+} \frac{d\eta}{1-\eta^2} = \frac{1}{2} \ln \left| \frac{(1+a_+)(1-a_-)}{(1-a_+)(1+a_-)} \right| = 0.$$

Inserting this identity into (G.18), we obtain (G.16), as desired.

Then, from (G.15) and (G.16) we get that

$$\frac{d}{dx} \int_{\mathbb{R}} \frac{u_{1/2}(x+y) + u_{1/2}(x-y) - 2u_{1/2}(x)}{|y|^2} \, dy$$
$$= -x \int_{\mathbb{R}} \frac{u_{-1/2}(x+y) + u_{-1/2}(x-y) - 2u_{-1/2}(x)}{|y|^2} \, dy$$

that is

$$\frac{d}{dx} (-\Delta)^{1/2} u_{1/2} = -x \, (-\Delta)^{1/2} u_{-1/2} \qquad \text{in } (-1, 1).$$

From this and (2.17) we infer that $x \, (-\Delta)^{1/2} u_{-1/2} = 0$ and so $(-\Delta)^{1/2} u_{-1/2} = 0$ in $(-1, 1)$.

These consideration establish (2.24), as desired. Now, we give a brief probabilistic insight on it. In probability—and in stochastic calculus—a measurable function $f : \mathbb{R}^n \to \mathbb{R}$ is said to be harmonic in an open set $D \subset \mathbb{R}^n$ if, for any $D_1 \subset D$ and any $x \in D_1$,

$$f(x) = \mathbb{E}_x \left[f(W_{\tau_{D_1}}) \right],$$

where W_t is a Brownian motion and τ_{D_1} is the first exit time from D_1, namely

$$\tau = \inf\{t > 0 : W_t \notin D_1\}. \tag{G.19}$$

Notice that, since W_t has (a.s.) continuous trajectories, then (a.s.) $W_{\tau_{D_1}} \in \partial D_1$. This notion of harmonicity coincides with the analytic one.

If one considers a Lévy-type process X_t in place of the Brownian motion, the definition of harmonicity (with respect to this other process) can be given in the very same way. When X_t is an isotropic $(2s)$-stable process, the definition amounts to having zero fractional Laplacian $(-\Delta)^s$ at every point of D and replace (G.19) by

$$f(x) = \mathbb{E}_x[f(X_{\tau_{D_1}})], \qquad \text{for any } D_1 \subseteq D.$$

In this identity, we can consider a sequence $\{D_j : D_j \subset D, \, j \in \mathbb{N}\}$, with $D_j \nearrow D$, and equality

$$f(x) = \mathbb{E}_x[f(X_{\tau_{D_j}})], \qquad \text{for any } j \in \mathbb{N}. \tag{G.20}$$

When $f = 0$ in $\mathbb{R}^n \setminus D$, the right-hand side of (G.20) can be not 0 (since $X_{\tau_{D_j}}$ may also end up in $D \setminus D_j$), and this leaves the possibility of finding f which satisfies (G.20) without vanish identically (an example of this phenomenon is exactly given by the function $u_{-1/2}$ in (2.24)).

It is interesting to observe that if f vanishes outside D and does not vanish identically, then, the only possibility to satisfy (G.20) is that f diverges along ∂D. Indeed, if $|f| \leqslant \kappa$, since $f(X_{\tau_{D_j}}) \neq 0$ only when $x \in D \setminus D_j$ and $|D \setminus D_j| \searrow 0$ as $j \to \infty$, we would have that

$$\lim_{j\to+\infty} \mathbb{E}_x[f(X_{\tau_{D_j}})] \leqslant \lim_{j\to+\infty} \text{const}\,\kappa\, |D \setminus D_j| = 0,$$

and (G.20) would imply that f must vanish identically.

Of course, the function $u_{-1/2}$ in (2.23) embodies exactly this singular boundary behavior.

Appendix H: Another Proof of (2.24)

Here we give a different proof of (2.24) by using complex analysis and extension methods. We use the principal complex square root introduced in (D.2) and, for any $x \in \mathbb{R}$ and $y > 0$ we define

$$U_{-1/2}(x, y) := \Re\left(\frac{1}{\sqrt{1 - z^2}}\right),$$

where $z := x + iy$.

The function $U_{-1/2}$ is plotted in Fig. 10. We recall that the function $U_{-1/2}$ is well-defined, thanks to (D.6). Also, the denominator never vanishes when $y > 0$ and so $U_{-1/2}$ is harmonic in the halfplane, being the real part of a holomorphic function in such domain.

Furthermore, in light of (D.9), we have that

$$\lim_{y\searrow 0} \frac{1}{\sqrt{1 - z^2}} = \begin{cases} -\dfrac{i}{\sqrt{|1 - x^2|}} & \text{if } x^2 > 1, \\ \dfrac{1}{\sqrt{|1 - x^2|}} & \text{if } x^2 < 1, \\ +\infty & \text{if } x^2 = 1, \end{cases}$$

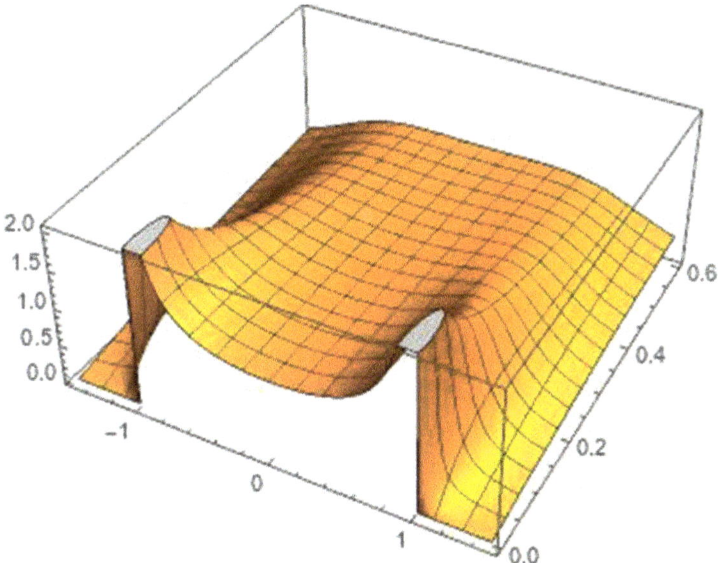

Fig. 10 Harmonic extension in the halfplane of the function $\mathbb{R} \ni x \mapsto (1-x^2)_+^{-1/2}$

and therefore

$$\lim_{y\searrow 0} U_{-1/2}(x,y) = \Re\left(\lim_{y\searrow 0}\frac{1}{\sqrt{1-z^2}}\right) = \begin{cases} 0 & \text{if } x^2>1, \\ \dfrac{1}{\sqrt{|1-x^2|}} & \text{if } x^2<1, \\ +\infty & \text{if } x^2=1, \end{cases}$$

$$= (1-x^2)_+^{-1/2} = u_{-1/2}(x).$$

This gives that $U_{-1/2}$ is the harmonic extension of $u_{-1/2}$ to the halfplane. Therefore, by (2.6), (2.7), and (D.14), for any $x \in (-1,1)$ we have that

$$\begin{aligned} -(-\Delta)^{-1/2}u_{1/2}(x) &= \lim_{y\searrow 0} \partial_y U_{-1/2}(x,y) \\ &= \lim_{y\searrow 0} \partial_y\left(\Re\left(\frac{1}{\sqrt{1-z^2}}\right)\right) \\ &= \lim_{y\searrow 0} \Re\left(\partial_y\left(\frac{1}{\sqrt{1-z^2}}\right)\right) \end{aligned}$$

$$= -\lim_{y\searrow 0} \Re\left(\left(\frac{1}{\sqrt{1-z^2}}\right)^2 \partial_y\left(\sqrt{1-z^2}\right)\right)$$
$$= -\frac{1}{1-x^2}\lim_{y\searrow 0} \Re\left(\partial_y\left(\sqrt{1-z^2}\right)\right)$$
$$= 0,$$

that is (2.24).

Appendix I: Proof of (2.29) (Based on Fourier Methods)

When $n = 1$, we use (2.28) to find that[8]

$$\begin{aligned}\mathcal{G}_{1/2}(x) &= \int_{\mathbb{R}} e^{-|\xi|}\, e^{ix\xi}\, d\xi = \lim_{R\to+\infty}\int_0^R e^{-\xi}\, e^{ix\xi}\, d\xi + \int_{-R}^0 e^{\xi}\, e^{ix\xi}\, d\xi \\ &= \lim_{R\to+\infty} \frac{e^{R(ix-1)}-1}{ix-1} + \frac{1-e^{-R(ix+1)}}{ix+1} = -\frac{1}{ix-1}+\frac{1}{ix+1} = \frac{2}{x^2+1}.\end{aligned} \tag{I.1}$$

This proves (2.28) when $n = 1$.

Let us now deal with the case $n \geqslant 2$. By changing variable $Y := 1/y$, we see that

$$\int_0^{+\infty} e^{-\frac{|\xi|\left(y-\frac{1}{y}\right)^2}{2}}\, dy = \int_0^{+\infty} e^{-\frac{|\xi|\left(Y-\frac{1}{Y}\right)^2}{2}}\, \frac{dY}{Y^2}.$$

Therefore, summing up the left hand side to both sides of this identity and using the transformation $\eta := y - \frac{1}{y}$,

$$\begin{aligned}2\int_0^{+\infty} e^{-\frac{|\xi|\left(y-\frac{1}{y}\right)^2}{2}}\, dy &= \int_0^{+\infty}\left(1+\frac{1}{y^2}\right) e^{-\frac{|\xi|\left(y-\frac{1}{y}\right)^2}{2}}\, dy \\ &= \text{const}\int_0^{+\infty} e^{-\frac{|\xi|\,\eta^2}{2}}\, d\eta \\ &= \frac{\text{const}}{\sqrt{|\xi|}}.\end{aligned}$$

[8]As a historical remark, we recall that $e^{-|\xi|}$ is sometimes called the "Abel Kernel" and its Fourier Transform the "Poisson Kernel", which in dimension 1 reduces to the "Cauchy-Lorentz, or Breit-Wigner, Distribution" (that has also classical geometric interpretations as the "Witch of Agnesi", and so many names attached to a single function clearly demonstrate its importance in numerous applications).

As a result,

$$\begin{aligned}
e^{-|\xi|} = \frac{\text{const}\, e^{-|\xi|}\sqrt{|\xi|}}{\sqrt{|\xi|}} &= \text{const}\, e^{-|\xi|}\sqrt{|\xi|}\int_0^{+\infty} e^{-\frac{|\xi|\left(y-\frac{1}{y}\right)^2}{2}}\,dy \\
&= \text{const}\, e^{-|\xi|}\sqrt{|\xi|}\int_0^{+\infty} e^{-\frac{|\xi|\left(y^2+\frac{1}{y^2}-2\right)}{2}}\,dy \\
&= \text{const}\sqrt{|\xi|}\int_0^{+\infty} e^{-\frac{|\xi|\left(y^2+\frac{1}{y^2}\right)}{2}}\,dy \\
&= \text{const}\int_0^{+\infty} \frac{1}{\sqrt{t}}\, e^{-\frac{t}{2}}\, e^{-\frac{|\xi|^2}{2t}}\,dt,
\end{aligned}$$

where the substitution $t := |\xi|\, y^2$ has been used.

Accordingly, by (2.28), the Gaussian Fourier transform and the change of variable $\tau := t(1+|x|^2)$,

$$\begin{aligned}
\mathcal{G}_{1/2}(x) &= \int_{\mathbb{R}^n} e^{-|\xi|}\, e^{ix\cdot\xi}\,d\xi \\
&= \text{const}\iint_{\mathbb{R}^n\times(0,+\infty)} \frac{1}{\sqrt{t}}\, e^{-\frac{t}{2}}\, e^{-\frac{|\xi|^2}{2t}}\, e^{ix\cdot\xi}\,d\xi\,dt \\
&= \text{const}\int_{(0,+\infty)} t^{\frac{n-1}{2}}\, e^{-\frac{t}{2}}\, e^{-\frac{t|x|^2}{2}}\,dt \\
&= \text{const}\int_{(0,+\infty)} \left(\frac{\tau}{1+|x|^2}\right)^{\frac{n-1}{2}} e^{-\frac{\tau}{2}}\,\frac{d\tau}{1+|x|^2} \\
&= \frac{\text{const}}{\left(1+|x|^2\right)^{\frac{n+1}{2}}}.
\end{aligned}$$

This establishes (2.29).

Appendix J: Another Proof of (2.29) (Based on Extension Methods)

The idea is to consider the fundamental solution in the extended space and take a derivative (the time variable acting as a translation and, to favor the intuition, one may keep in mind that the Poisson kernel is the normal derivative of the Green function). Interestingly, this proof is, in a sense, “conceptually simpler”, and “less technical” than that in Appendix I, thus demonstrating that, at least in some cases, when appropriately used, fractional methods may lead to cultural advantages[9] with respect to more classical approaches.

For this proof, we consider variables $X := (x, y) \in \mathbb{R}^n \times (0, +\infty) \subset \mathbb{R}^{n+1}$ and fix $t > 0$. We let Γ be the fundamental solution in $\mathbb{R}^{n+1}$, namely

$$\Gamma(X) := \begin{cases} -\,\text{const} \log |X| & \text{if } n = 1, \\ \dfrac{\text{const}}{|X|^{n-1}} & \text{if } n \geqslant 2. \end{cases}$$

By construction $\Delta\Gamma$ is the Delta Function at the origin and so, for any $t > 0$, we have that $\tilde{\Gamma}(X; t) = \tilde{\Gamma}(x, y; t) := \Gamma(x, y + t)$ is harmonic for $(x, y) \in \mathbb{R}^n \times (0, +\infty)$. Accordingly, the function $U(x, y; t) := \partial_y \tilde{\Gamma}(x, y; t)$ is also harmonic for $(x, y) \in \mathbb{R}^n \times (0, +\infty)$. We remark that

$$\begin{aligned} U(x, y; t) = \partial_y \Gamma(x, y + t) &= \frac{\text{const}}{|(x, y + t)|^n} \partial_y \sqrt{|x|^2 + (y + t)^2} = \frac{\text{const}\,(y + t)}{|(x, y + t)|^{n+1}} \\ &= \frac{\text{const}\,(y + t)}{\left(|x|^2 + (y + t)^2\right)^{\frac{n+1}{2}}}. \end{aligned}$$

This function is plotted in Fig. 11 (for the model case in the plane). We observe that

$$\lim_{y \searrow 0} U(x, 0; t) = \frac{\text{const } t}{\left(|x|^2 + t^2\right)^{\frac{n+1}{2}}} = \frac{\text{const}}{t^n \left(1 + (|x|/t)^2\right)^{\frac{n+1}{2}}} =: u(x, t).$$

As a consequence, by (2.6) and (2.7) (and noticing that the role played by the variables y and t in the function U is the same),

$$\begin{aligned} -(-\Delta)^{1/2} u(x, t) = \lim_{y \searrow 0} \partial_y U(x, y; t) &= \lim_{y \searrow 0} \partial_y \frac{\text{const}\,(y + t)}{\left(|x|^2 + (y + t)^2\right)^{\frac{n+1}{2}}} \\ &= \lim_{y \searrow 0} \partial_t \frac{\text{const}\,(y + t)}{\left(|x|^2 + (y + t)^2\right)^{\frac{n+1}{2}}} \\ &= \partial_t \frac{\text{const } t}{\left(|x|^2 + t^2\right)^{\frac{n+1}{2}}} = \partial_t u(x, t). \end{aligned}$$

[9] Let us mention another conceptual simplification of nonlocal problems: in this setting, the integral representation often allows the formulation of problems with minimal requirements on the functions involved (such as measurability and possibly minor pointwise or integral bounds). Conversely, in the classical setting, even to just formulate a problem, one often needs assumptions and tools from functional analysis, comprising e.g. Sobolev differentiability, distributions or functions of bounded variations.

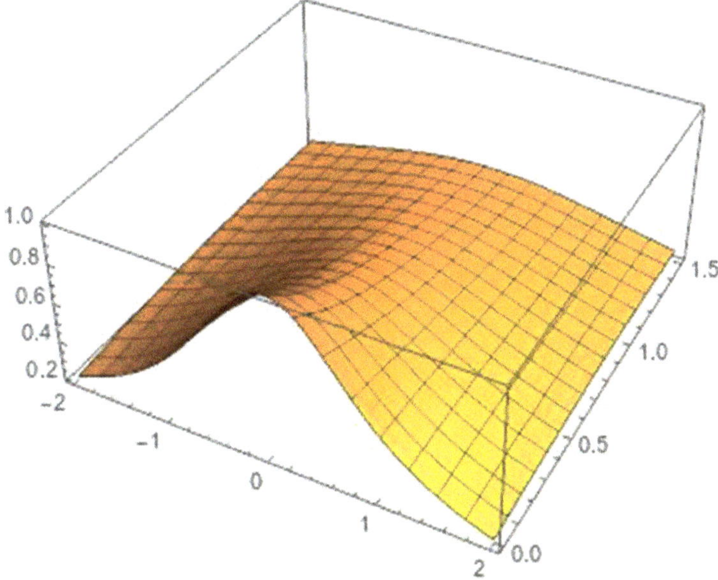

Fig. 11 Harmonic extension in the halfplane of the function $\mathbb{R} \ni x \mapsto \frac{1}{1+|x|^2}$

This shows that u solves the fractional heat equation, with u approaching a Delta function when $t \searrow 0$. Hence

$$\mathcal{G}_{1/2}(x) = u(x, 1) = \frac{\text{const}}{\left(1 + |x|^2\right)^{\frac{n+1}{2}}},$$

that is (2.29).

Appendix K: Proof of (2.36)

First, we construct a useful barrier. Given $A > 1$, we define

$$w(t) := \begin{cases} A & \text{if } |t| \leqslant 1, \\ t^{-1-2s} & \text{if } |t| > 1. \end{cases}$$

We claim that if A is sufficiently large, then

$$(-\Delta)^s w(t) < -3w(t) \quad \text{for all } t \in \mathbb{R} \setminus (-3, 3). \tag{K.1}$$

To prove this, fix $t \geqslant 3$ (the case $t \leqslant -3$ being similar). Then, if $|\xi - t| < 1$, we have that

$$\xi \geqslant t - 1 = \frac{2t}{3} + \frac{t}{3} - 1 \geqslant \frac{2t}{3}.$$

As a consequence, if $|\tau - t| < 1$,

$$\begin{aligned}\left|w(t) - w(\tau) + \dot{w}(t)(\tau - t)\right| &\leqslant \sup_{|\xi - t|<1} |\ddot{w}(\xi)|\, |t - \tau|^2 \\ &\leqslant \text{const} \sup_{\xi \geqslant 2t/3} \xi^{-3-2s}\, |t - \tau|^2 \leqslant \text{const}\, t^{-3-2s}\, |t - \tau|^2.\end{aligned}$$

This implies that

$$\begin{aligned}\int_{\{|\tau - t|<1\}} \frac{w(t) - w(\tau)}{|t - \tau|^{1+2s}}\, d\tau &= \int_{\{|\tau - t|<1\}} \frac{w(t) - w(\tau) + \dot{w}(t)(\tau - t)}{|t - \tau|^{1+2s}}\, d\tau \\ \leqslant \text{const}\, t^{-3-2s} \int_{\{|\tau - t|<1\}} \frac{|t - \tau|^2}{|t - \tau|^{1+2s}}\, d\tau &= \text{const}\, t^{-3-2s} \\ \leqslant \text{const}\, t^{-1-2s} &= \text{const}\, w(t).\end{aligned} \tag{K.2}$$

On the other hand,

$$\int_{\{|\tau - t|\geqslant 1\}\cap\{|\tau|>1\}} \frac{w(t) - w(\tau)}{|t - \tau|^{1+2s}}\, d\tau \leqslant \int_{\{|\tau - t|\geqslant 1\}} \frac{w(t)}{|t - \tau|^{1+2s}}\, d\tau \leqslant \text{const}\, w(t). \tag{K.3}$$

In addition, if $|\tau| \leqslant 1$ then $|\tau - t| \geqslant t - \tau \geqslant 3 - 1 > 1$, hence

$$\{|\tau - t| \geqslant 1\} \cap \{|\tau| \leqslant 1\} = \{|\tau| \leqslant 1\}.$$

Accordingly,

$$\int_{\{|\tau - t|\geqslant 1\}\cap\{|\tau|\leqslant 1\}} \frac{w(t) - w(\tau)}{|t - \tau|^{1+2s}}\, d\tau = \int_{\{|\tau|\leqslant 1\}} \frac{t^{-1-2s} - A}{|t - \tau|^{1+2s}}\, d\tau \leqslant \int_{\{|\tau|\leqslant 1\}} \frac{1 - A}{|t - \tau|^{1+2s}}\, d\tau. \tag{K.4}$$

We also observe that if $|\tau| \leqslant 1$ then $|t - \tau| \leqslant t + 1 \leqslant 2t$ and therefore

$$\int_{\{|\tau|\leqslant 1\}} \frac{d\tau}{|t - \tau|^{1+2s}} \geqslant \frac{\text{const}}{t^{1+2s}} = \text{const}\, w(t).$$

So, we plug this information into (K.4), assuming $A > 1$ and we obtain that

$$\int_{\{|\tau-t|\geqslant 1\}\cap\{|\tau|\leqslant 1\}} \frac{w(t)-w(\tau)}{|t-\tau|^{1+2s}}\,d\tau \leqslant -(A-1)\,\mathrm{const}\,w(t). \tag{K.5}$$

Thus, gathering together the estimates in (K.2), (K.3) and (K.5), we conclude that

$$\int_{\mathbb{R}} \frac{w(t)-w(\tau)}{|t-\tau|^{1+2s}}\,d\tau \leqslant \mathrm{const}\,w(t) - (A-1)\,\mathrm{const}\,w(t) \leqslant -4w(t) < -3w(t),$$

as long as A is sufficiently large. This proves (K.1).

Now, to prove (2.36), we define $v := \dot{u} > 0$. From (2.40), we know that

$$(-\Delta)^s v = (1-3u^2)v \geqslant -3u^2 v \geqslant -3v. \tag{K.6}$$

Given $\varepsilon > 0$, we define

$$w_\varepsilon(t) := \frac{\iota}{A}\,w(t) - \varepsilon, \qquad \text{where } \iota := \min_{t\in[-3,3]} v(t).$$

We claim that

$$w_\varepsilon \leqslant v. \tag{K.7}$$

Indeed, for large ε, it holds that $w_\varepsilon < 0 < v$ and so (K.7) is satisfied. In addition, for any $\varepsilon > 0$,

$$\lim_{t\to+\infty} w_\varepsilon(t) = -\varepsilon < 0 \leqslant \inf_{t\in\mathbb{R}} v(t). \tag{K.8}$$

Suppose now that $\varepsilon_\star > 0$ produces a touching point between $w_{\varepsilon_\star}$ and v, namely $w_{\varepsilon_\star} \leqslant v$ and $w_{\varepsilon_\star}(t_\star) = v(t_\star)$ for some $t_\star \in \mathbb{R}$. Notice that, if $|\tau| \leqslant 3$,

$$w_{\varepsilon_\star}(\tau) \leqslant \frac{\iota}{A}\sup_{t\in\mathbb{R}} w(t) - \varepsilon \leqslant \iota - \varepsilon = \min_{t\in[-3,3]} v(t) - \varepsilon \leqslant v(\tau) - \varepsilon < v(\tau),$$

and therefore $|t_\star| > 3$. Accordingly, if we set $v_\star := v - w_{\varepsilon_\star}$, using (K.1) and (K.6), we see that

$$\begin{aligned} 0 = -3v_\star(t_\star) &= -3v(t_\star) + 3w_{\varepsilon_\star}(t_\star) \leqslant (-\Delta)^s v(t_\star) - (-\Delta)^s w_{\varepsilon_\star}(t_\star) \\ &= (-\Delta)^s v_\star(t_\star) = \int_{\mathbb{R}} \frac{v_\star(t_\star)-v_\star(\tau)}{|t_\star-\tau|^{1+2s}}\,d\tau = -\int_{\mathbb{R}} \frac{v_\star(\tau)}{|t_\star-\tau|^{1+2s}}\,d\tau. \end{aligned}$$

Since the latter integrand is nonnegative, we conclude that $v_\star$ must vanish identically, and thus $w_{\varepsilon_\star}$ must coincide with v. But this is in contradiction with (K.8) and so the proof of (K.7) is complete.

Then, by sending $\varepsilon \searrow 0$ in (K.7) we find that $v \geqslant \frac{L}{A} w$, and therefore, for $t \geqslant 1$ we have that $\dot{u}(t) = v(t) \geqslant \kappa t^{-1-2s}$, for all $t > 1$, for some $\kappa > 0$.

Consequently, for any $t > 1$,

$$1 - u(t) = \lim_{T \to +\infty} u(T) - u(t) = \lim_{T \to +\infty} \int_t^T \dot{u}(\tau)\, d\tau$$

$$= \int_t^{+\infty} \dot{u}(\tau)\, d\tau \geqslant \kappa \int_t^{+\infty} \tau^{-1-2s}\, d\tau = \frac{\kappa}{2s} t^{-2s},$$

and a similar estimates holds for $1 + u(t)$ when $t < -1$.

These considerations establish (2.36), as desired.

Appendix L: Proof of (2.38)

Here we prove that (2.38) is a solution of (2.37). The idea of the proof, as showed in Fig. 12, is to consider the harmonic extension of the function $\mathbb{R} \ni x \mapsto \frac{2}{\pi} \arctan x$ in the halfplane $\mathbb{R} \times (0, +\infty)$ and use the method described in (2.6) and (2.7).

We let

$$U(x, y) := \frac{2}{\pi} \arctan \frac{x}{y + 1}.$$

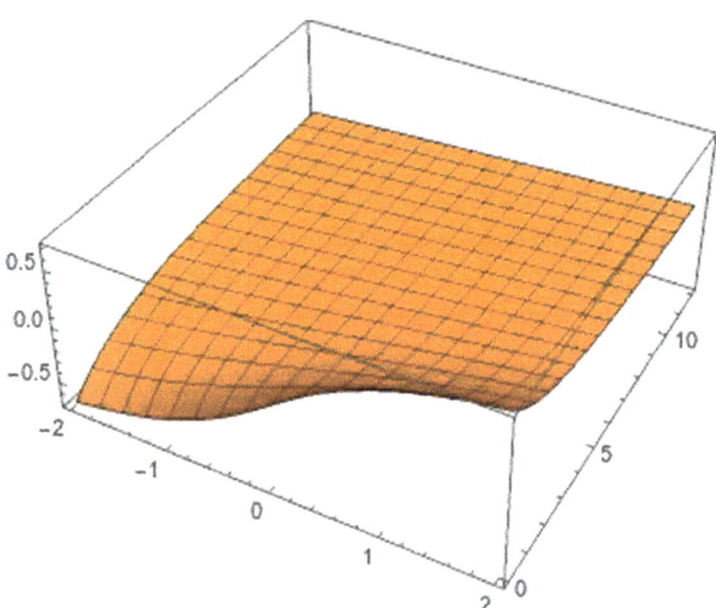

Fig. 12 Harmonic extension in the halfplane of the function $\mathbb{R} \ni x \mapsto \frac{2}{\pi} \arctan x$

The function U is depicted[10] in Fig. 12. Of course, it coincides with u when $y = 0$ and, for any $x \in \mathbb{R}$ and $y > 0$,

$$\frac{\pi}{2}\,\Delta U(x,y) = -\frac{2x(1+y)}{(x^2+(1+y)^2)^2} + \frac{2x(1+y)}{(x^2+(1+y)^2)^2} = 0. \tag{L.1}$$

Hence, the setting in (2.6) is satisfied and so, in light of (2.7). we have

$$(-\Delta)^{1/2}u(x) = -\lim_{y\searrow 0}\partial_y U(x,y) = \frac{2}{\pi}\lim_{y\searrow 0}\frac{x}{x^2+(1+y)^2} = \frac{2x}{\pi\,(x^2+1)} \tag{L.2}$$

Also, by the trigonometric Double-angle Formula, for any $\theta \in \left(-\frac{\pi}{2}, \frac{\pi}{2}\right)$,

$$\sin(2\theta) = 2\sin\theta\cos\theta = \frac{2\tan\theta}{\tan^2\theta+1}.$$

Hence, taking $\theta := \arctan x$,

$$\sin(\pi u(x)) = \sin(2\arctan x) = \frac{2x}{x^2+1}.$$

This and (L.2) show that (2.38) is a solution of (2.37).

Appendix M: Another Proof of (2.38) (Based on (2.29))

This proof of (2.38) is based on the fractional heat kernel in (2.29). This approach has the advantage of being quite general (see e.g. Theorem 3.1 in [27]) and also to relate the two "miraculous" explicit formulas (2.29) and (2.38), which are available only in the special case $s = 1/2$.

---

[10] In complex variables, one can also interpret the function U in terms of the principal argument function

$$\mathrm{Arg}(re^{i\varphi}) = \varphi \in (-\pi, \pi],$$

with branch cut along the nonpositive real axis. Notice indeed that, if $z = x + iy$ and $y > 0$,

$$\mathrm{Arg}(z+i) = \frac{\pi}{2} - \arctan\frac{x}{y+1} = \frac{\pi}{2}\,(1 - U(x,y)).$$

This observation would also lead to (L.1).

For this, we let $P = P(x,t)$ the fundamental solution of the heat flow in (2.25) with $n = 1$ and $s = 1/2$. Notice that, by (2.29), we know that

$$P(x,1) = \mathcal{G}_{1/2}(x) = \frac{c}{1+x^2}, \tag{M.1}$$

with

$$c := \left(\int_{\mathbb{R}} \frac{dx}{1+x^2} \right)^{-1} = \frac{1}{\pi}.$$

Also, by scaling,

$$P(x,t) = t^{-1} P(t^{-1}x, 1) = t^{-1} \mathcal{G}_{1/2}(t^{-1}x). \tag{M.2}$$

For any $x \in \mathbb{R}$ and any $t > 0$, we define

$$U(x,t) := 2 \int_0^x P(\eta, t+1)\, d\eta. \tag{M.3}$$

In light of (M.2), we see that

$$|U(x,t)| \leqslant 2(t+1)^{-1} \int_0^x \mathcal{G}_{1/2}\big((t+1)^{-1}\eta\big)\, d\eta = 2 \int_0^{(t+1)^{-1}x} \mathcal{G}_{1/2}(\zeta)\, d\zeta,$$

which is bounded in $\mathbb{R} \times [0, +\infty)$, and infinitesimal as $t \to +\infty$ for any fixed $x \in \mathbb{R}$.

Notice also that

$$\partial_t^2 P = \partial_t(\partial_t P) = \partial_t (-\Delta)^{1/2} P = (-\Delta)^{1/2} \partial_t P = (-\Delta)^{1/2} (-\Delta)^{1/2} P = -\partial_x^2 P,$$

by (2.5). As a consequence,

$$\begin{aligned} \frac{1}{2}(\partial_x^2 + \partial_t^2) U(x,t) &= \partial_x P(x,t+1) + \int_0^x \partial_t^2 P(\eta, t+1)\, d\eta \\ &= \partial_x P(x,t+1) - \int_0^x \partial_x^2 P(\eta, t+1)\, d\eta \\ &= \partial_x P(0,t+1) \\ &= 0, \end{aligned} \tag{M.4}$$

where the last identity follows from (M.2).

Besides, from (M.2) we have that

$$\partial_t P(x,t) = \partial_t \Big(t^{-1} \mathcal{G}_{1/2}(t^{-1}x) \Big) = -t^{-2} \mathcal{G}_{1/2}(t^{-1}x) - t^{-3} x \mathcal{G}'_{1/2}(t^{-1}x)$$

and so

$$-\partial_t P(x,1) = \mathscr{G}_{1/2}(x) + x\mathscr{G}_{1/2}'(x) = \partial_x\big(x\,\mathscr{G}_{1/2}(x)\big).$$

In view of this, we have that

$$\partial_t U(x,0) = 2\int_0^x \partial_t P(\eta,1)\,d\eta = 2\int_0^x \partial_\eta\big(\eta\,\mathscr{G}_{1/2}(\eta)\big)\,d\eta = 2x\,\mathscr{G}_{1/2}(x). \tag{M.5}$$

Accordingly, from (M.4) and (M.5), using the extension method in (2.6) and (2.7) (with the variable y called t here), we conclude that, if

$$u(x) := U(x,0),$$

then

$$(-\Delta)^{1/2}u(x) = 2x\,\mathscr{G}_{1/2}(x). \tag{M.6}$$

We remark that, by (M.1) and (M.3),

$$u(x) = 2c\int_0^x \frac{d\eta}{1+x^2} = \frac{2}{\pi}\arctan x. \tag{M.7}$$

This, (M.1) and (M.6) give that

$$(-\Delta)^{1/2}u(x) = \frac{1}{\pi}\frac{2x}{1+x^2} = \frac{1}{\pi}\sin(2\arctan x) = \frac{1}{\pi}\sin\big(\pi u(x)\big),$$

that is (2.38), as desired.

Appendix N: Proof of (2.46)

Due to translation invariance, we can reduce ourselves to proving (2.46) at the origin. We consider a measurable $u : \mathbb{R}^n \to \mathbb{R}$ such that

$$\int_{\mathbb{R}^n} \frac{|u(y)|}{1+|y|^{n+2}} < +\infty.$$

Assume first that

$$u(x) = 0 \text{ for any } x \in B_r, \tag{N.1}$$

for some $r > 0$. As a matter of fact, under these assumptions on u, the right-hand side of (2.46) vanishes at 0 regardless the size of r. Indeed,

$$\begin{aligned}
&\int_{\mathbb{R}^n} \frac{u(x+2y)+u(x-2y)-4u(x+y)-4u(x-y)+6u(x)}{|y|^{n+2}}\,dy\Big|_{x=0}\\
&= \int_{\mathbb{R}^n} \frac{u(2y)+u(-2y)-4u(y)-4u(-y)}{|y|^{n+2}}\,dy\\
&= 2\int_{\mathbb{R}^n\setminus B_{r/2}} \frac{u(2y)}{|y|^{n+2}}\,dy - 8\int_{\mathbb{R}^n\setminus B_r} \frac{u(y)}{|y|^{n+2}}\,dy\\
&= 2\int_{\mathbb{R}^n\setminus B_r} \frac{2^{n+2}\,u(Y)}{2^n\,|Y|^{n+2}}\,dY - 8\int_{\mathbb{R}^n\setminus B_r} \frac{u(y)}{|y|^{n+2}}\,dy \;=\; 0.
\end{aligned}$$

This proves (2.46) under the additional assumption in (N.1), that we are now going to remove. To this end, for $r \in (0, 1)$, denote by χ_r the characteristic function of B_r, i.e. $\chi_r(x) = 1$ if $x \in B_r$ and $\chi_r(x) = 0$ otherwise. Consider now $u \in C^{2,\alpha}(B_r)$, for some $\alpha \in (0, 1)$, with

$$u(0) = |\nabla u(0)| = 0 \tag{N.2}$$

for simplicity (note that one can always modify u by considering $\tilde{u}(x) = u(x) - u(0) - \nabla u(0) \cdot x$ and without affecting the operators in (2.46)). Then, the right hand side of (2.46) becomes in this case

$$\begin{aligned}
&\int_{\mathbb{R}^n} \frac{u(2y)+u(-2y)-4u(y)-4u(-y)}{|y|^{n+2}}\,dy \;=\; 2\int_{\mathbb{R}^n} \frac{u(2y)-4u(y)}{|y|^{n+2}}\,dy \;=\\
&= 2\int_{\mathbb{R}^n} \frac{u(2y)\chi_r(2y)-4u(y)\chi_r(y)}{|y|^{n+2}}\,dy\\
&+ 2\int_{\mathbb{R}^n} \frac{u(2y)(1-\chi_r(2y))-4u(y)(1-\chi_r(y))}{|y|^{n+2}}\,dy.
\end{aligned}$$

The second addend is trivial for any $r \in (0, 1)$, in view of the above remark, since $u(1 - \chi_r)$ is constantly equal to 0 in B_r. For the first one, we have

$$\int_{\mathbb{R}^n} \frac{u(2y)\chi_r(2y)-4u(y)\chi_r(y)}{|y|^{n+2}}\,dy \;=\; \int_{B_{r/2}} \frac{u(2y)-4u(y)}{|y|^{n+2}}\,dy - 4\int_{B_r\setminus B_{r/2}} \frac{u(y)}{|y|^{n+2}}\,dy. \tag{N.3}$$

Now, we recall (N.2) and we notice that

$$|u(2y) - 4u(y)| \leqslant \|u\|_{C^{2,\alpha}(B)}|y|^{2+\alpha},$$

which in turn implies that

$$\left| \int_{B_{r/2}} \frac{u(2y) - 4u(y)}{|y|^{n+2}} \, dy \right| \leqslant \text{const} \, \|u\|_{C^{2,\alpha}(B)} r^{\alpha}. \tag{N.4}$$

On the other hand, a Taylor expansion and (N.2) yield

$$\begin{aligned}
\int_{B_r \setminus B_{r/2}} \frac{u(y)}{|y|^{n+2}} \, dy &= \int_{r/2}^{r} \frac{1}{\rho^{n+2}} \int_{\partial B_\rho} u(y) \, dy \, d\rho \\
= \int_{r/2}^{r} \frac{1}{\rho^3} \int_{\partial B_1} u(\rho\theta) \, d\theta \, d\rho &= \int_{r/2}^{r} \frac{1}{2\rho} \int_{\partial B_1} \Big(D^2 u(0)\theta \cdot \theta + \eta(\rho\theta) \Big) \, d\theta \, d\rho \\
&= \text{const} \, \Delta u(0) + \int_{r/2}^{r} \frac{1}{2\rho} \int_{\partial B_1} \eta(\rho\theta) \, d\theta \, d\rho
\end{aligned} \tag{N.5}$$

in view of (1.1), for some $\eta : B_r \to \mathbb{R}$ such that $|\eta(x)| \leqslant c|x|^{\alpha}$. From this, (N.3) and (N.4) we deduce that

$$\begin{aligned}
&\int_{\mathbb{R}^n} \frac{u(2y) + u(-2y) - 4u(y) - 4u(-y)}{|y|^{n+2}} \, dy \\
&= -\,\text{const} \lim_{r \searrow 0} \int_{B_r \setminus B_{r/2}} \frac{u(y)}{|y|^{n+2}} \, dy \;=\; -\,\text{const} \, \Delta u(0)
\end{aligned}$$

which finally justifies (2.46).

It is interesting to remark that the main contribution to prove (2.46) comes in this case from the "intermediate ring" in (N.5).

Appendix O: Proof of (2.48)

Take for instance Ω to be the unit ball and $\bar{u} = 1 - |x|^2$. Suppose that $\|\bar{u} - v_\varepsilon\|_{C^2(\Omega)} \leqslant \varepsilon$. Then, for small ε, if $x \in \mathbb{R}^n \setminus B_{1/2}$ it holds that

$$v_\varepsilon(x) \leqslant \bar{u}(x) + \varepsilon = 1 - |x|^2 + \varepsilon \leqslant \frac{3}{4} + \varepsilon \leqslant \frac{4}{5},$$

while

$$v_\varepsilon(0) \geqslant \bar{u}(0) - \varepsilon = 1 - \varepsilon \geqslant \frac{5}{6}.$$

This implies that there exists $x_\varepsilon \in \overline{B_{1/2}}$ such that

$$v_\varepsilon(x_\varepsilon) = \sup_{B_1} v_\varepsilon \geqslant \frac{5}{6} > \frac{4}{5} \geqslant \sup_{B_1 \setminus B_{1/2}} v_\varepsilon .$$

As a result,

$$\begin{aligned} \text{P.V.} \int_\Omega \frac{v_\varepsilon(x_\varepsilon) - v_\varepsilon(y)}{|x_\varepsilon - y|^{n+2s}}\, dy &\geqslant \int_{B_1 \setminus B_{3/4}} \frac{v_\varepsilon(x_\varepsilon) - v_\varepsilon(y)}{|x_\varepsilon - y|^{n+2s}}\, dy \\ &\geqslant \int_{B_1 \setminus B_{3/4}} \left(\frac{5}{6} - \frac{4}{5}\right) dy \geqslant \text{const}\,. \end{aligned}$$

This says that $(-\Delta)^s v_\varepsilon$ cannot vanish at x_ε and so (2.48) is proved.

Appendix P: Proof of (2.52)

Let us first notice that the identity

$$\lambda^s \;=\; \frac{s}{\Gamma(1-s)} \int_0^\infty \frac{1 - e^{-t\lambda}}{t^{1+s}}\, dt \tag{P.1}$$

holds for any $\lambda > 0$ and $s \in (0, 1)$, because

$$\int_0^\infty \frac{1 - e^{-t}}{t^{1+s}}\, dt = \left.\frac{1 - e^{-t}}{-s\, t^s}\right|_0^\infty + \frac{1}{s} \int_0^\infty \frac{e^{-t}}{t^s}\, dt = \frac{\Gamma(1-s)}{s}.$$

We also observe that when $u \in C_0^\infty(\Omega)$, the coefficients $\hat{u}_j$ decay fast as $j \to \infty$: indeed

$$\hat{u}_j = -\frac{1}{\mu_j} \int_\Omega u\, \Delta\psi_j = -\frac{1}{\mu_j} \int_\Omega \psi_j\, \Delta u = \ldots = (-1)^k \frac{1}{\mu_j^k} \int_\Omega \psi_j\, \Delta^k u.$$

Therefore, applying equality (P.1) to the μ_j's in (2.51) we obtain[11]

$$\begin{aligned} (-\Delta)^s_{N,\Omega} u &= \frac{s}{\Gamma(1-s)} \sum_{j=0}^{+\infty} \int_0^\infty \frac{\hat{u}_j \psi_j - e^{-t\mu_j} \hat{u}_j \psi_j}{t^{1+s}}\, dt \\ &= \frac{s}{\Gamma(1-s)} \int_0^\infty \frac{u - e^{t\Delta_{N,\Omega}} u}{t^{1+s}}\, dt, \quad u \in C_0^\infty(\Omega) \end{aligned} \tag{P.2}$$

[11] The representation in (P.2) makes sense for a larger class of functions with respect to (2.51), so in a sense (P.2) can be interpreted as an extension of definition (2.51).

where $\{e^{t\Delta_{N,\Omega}}\}_{t>0}$ stands for the heat semigroup associated to $\Delta_{N,\Omega}$. i.e. $e^{t\Delta_{N,\Omega}}u$ solves

$$\begin{cases} \partial_t v(x,t) = \Delta v(x,t) \text{ in } \Omega \times (0,\infty) \\ \partial_\nu v(x,t) = 0 \text{ on } \partial\Omega \times (0,\infty) \\ v(x,0) = u(x) \text{ on } \Omega \times \{0\}. \end{cases}$$

To check (P.2), it is sufficient to notice that

$$\partial_t \left(\sum_{j=0}^{+\infty} e^{-t\mu_j} \hat{u}_j \psi_j \right) = -\sum_{j=0}^{+\infty} \mu_j e^{-t\mu_j} \hat{u}_j \psi_j = \sum_{j=0}^{+\infty} e^{-t\mu_j} \hat{u}_j \Delta \psi_j = \Delta \left(\sum_{j=0}^{+\infty} e^{-t\mu_j} \hat{u}_j \psi_j \right)$$

and that

$$\left(\sum_{j=0}^{+\infty} e^{-t\mu_j} \hat{u}_j \psi_j \right) \Bigg|_{t=0} = \sum_{j=0}^{+\infty} \hat{u}_j \psi_j = u.$$

Under suitable regularity assumptions on Ω, write now the heat semigroup in terms of the heat kernel p_N^Ω as

$$e^{t\Delta_{N,\Omega}} u(x) = \int_\Omega p_N^\Omega(t,x,y)\, u(y)\, dy, \qquad x \in \Omega,\ t > 0 \tag{P.3}$$

where the following two-sided estimate on p_N^Ω holds (see, for example, [102, Theorem 3.1])

$$\frac{c_1\, e^{-c_2|x-y|^2/t}}{t^{n/2}} \leqslant p_N^\Omega(t,x,y) \leqslant \frac{c_3\, e^{-c_4|x-y|^2/t}}{t^{n/2}}, \qquad x, y \in \Omega,\ t, c_1, c_2, c_3, c_4 > 0. \tag{P.4}$$

Recall also that $p_N^\Omega(t,x,y) = p_N^\Omega(t,y,x)$ for any $t > 0$ and $x, y \in \Omega$, and that

$$\int_\Omega p_N^\Omega(t,x,y)\, dy = 1, \qquad x \in \Omega,\ t > 0, \tag{P.5}$$

which follows from noticing that, for any $u \in C_0^\infty(\Omega)$,

$$\partial_t \int_\Omega e^{t\Delta_{N,\Omega}} u = \int_\Omega \partial_t e^{t\Delta_{N,\Omega}} u = \int_\Omega \Delta e^{t\Delta_{N,\Omega}} u = -\int_{\partial\Omega} \partial_\nu e^{t\Delta_{N,\Omega}} u = 0$$

and therefore

$$\begin{aligned}\int_{\Omega} u(x)\,dx &= \int_{\Omega} e^{t\Delta_{N,\Omega}} u(x)\,dx = \int_{\Omega}\int_{\Omega} p_N^{\Omega}(t,x,y)\,u(y)\,dy\,dx\\ &= \int_{\Omega} u(y) \int_{\Omega} p_N^{\Omega}(t,x,y)\,dx\,dy.\end{aligned}$$

By (P.5), for any $x \in \Omega$ and $t > 0$,

$$u(x) - e^{t\Delta_{N,\Omega}} u(x) = \int_{\Omega} p_N^{\Omega}(t,x,y)\,(u(x)-u(y))\,dy$$

and, exchanging the order of integration in (P.2) (see below for a justification of this passage), one gets

$$\begin{aligned}(-\Delta)^s_{N,\Omega} u(x) &= \frac{s}{\Gamma(1-s)} \int_0^{\infty} \frac{u(x) - e^{t\Delta_{N,\Omega}} u(x)}{t^{1+s}}\,dy\\ &= \frac{s}{\Gamma(1-s)} \int_0^{\infty} \frac{\int_{\Omega} p_N^{\Omega}(t,x,y)\,(u(x)-u(y))\,dy}{t^{1+s}}\,dt\\ &= \frac{s}{\Gamma(1-s)}\ \text{P.V.} \int_{\Omega} (u(x)-u(y)) \int_0^{\infty} \frac{p_N^{\Omega}(t,x,y)}{t^{1+s}}\,dt\,dy\\ &= \text{P.V.} \int_{\Omega} \frac{(u(x)-u(y))\,k(x,y)}{|x-y|^{n+2s}}\,dy,\end{aligned}$$

where, in view of (P.4), we have

$$\begin{aligned}k(x,y) &:= \frac{s}{\Gamma(1-s)}\,|x-y|^{n+2s} \int_0^{\infty} \frac{p_N^{\Omega}(t,x,y)}{t^{1+s}}\,dt\\ &\simeq |x-y|^{n+2s} \int_0^{\infty} \frac{e^{-|x-y|^2/t}}{t^{n/2+1+s}}\,dt \simeq \int_0^{\infty} \frac{e^{-1/t}}{t^{n/2+1+s}}\,dt \simeq 1.\end{aligned}$$

These considerations establish (2.52). Note however that in the above computations there is a limit exiting the integral in the t variable, namely:

$$\int_0^{\infty} \frac{\int_{\Omega} p_N^{\Omega}(t,x,y)\,(u(x)-u(y))\,dy}{t^{1+s}}\,dt = \lim_{\varepsilon \searrow 0} \int_0^{\infty} \frac{\int_{\Omega \setminus B_{\varepsilon}(x)} p_N^{\Omega}(t,x,y)\,(u(x)-u(y))\,dy}{t^{1+s}}\,dt. \tag{P.6}$$

To properly justify this we are going to build an integrable majorant in t and independent of ε of the integrand

$$\frac{\int_{\Omega\setminus B_\varepsilon(x)} p_N^\Omega(t,x,y)\,(u(x)-u(y))\,dy}{t^{1+s}}. \tag{P.7}$$

To this end, first of all we observe that, by the boundedness of u and (P.5),

$$\left|\frac{\int_{\Omega\setminus B_\varepsilon(x)} p_N^\Omega(t,x,y)\,(u(x)-u(y))\,dy}{t^{1+s}}\right| \leqslant \frac{2\|u\|_{L^\infty(\Omega)}}{t^{1+s}}\int_{\Omega\setminus B_\varepsilon(x)} p_N^\Omega(t,x,y)\,dy \leqslant \frac{2\|u\|_{L^\infty(\Omega)}}{t^{1+s}}$$

which is integrable at infinity. So, to obtain an integrable bound for (P.7), we can now focus on small values of t, say $t \in (0,1)$. For this, we denote by p the heat kernel in $\mathbb{R}^N$ and we write

$$\begin{aligned}
&\int_{\Omega\setminus B_\varepsilon(x)} p_N^\Omega(t,x,y)\,(u(x)-u(y))\,dy \;=\\
&= \int_{\Omega\setminus B_\varepsilon(x)} p(t,x,y)\,(u(x)-u(y))\,dy\\
&- \int_{\Omega\setminus B_\varepsilon(x)} (p_N^\Omega(t,x,y)-p(t,x,y))(u(x)-u(y))\,dy =: A+B.
\end{aligned}$$

We first manipulate A. We reformulate u as

$$u(y) = u(x)+\nabla u(x)\cdot(y-x)+\eta(y)|x-y|^2, \qquad y\in\mathbb{R}^n,\ \|\eta\|_{L^\infty(\mathbb{R}^n)} \leqslant \|u\|_{C^2(\Omega)},$$

so that

$$\begin{aligned}
&\int_{\Omega\setminus B_\varepsilon(x)} p(t,x,y)\,(u(x)-u(y))\,dy = \int_{\mathbb{R}^n\setminus B_\varepsilon(x)} p(t,x,y)\,(u(x)-u(y))\,dy\\
&\qquad - u(x)\int_{\mathbb{R}^n\setminus\Omega} p(t,x,y)\,dy\\
&= \int_{\mathbb{R}^n\setminus B_\varepsilon(x)} p(t,x,y)\nabla u(x)\cdot(x-y)\,dy\\
&\qquad - \int_{\mathbb{R}^n\setminus B_\varepsilon(x)} p(t,x,y)\,\eta(y)|x-y|^2\,dy - u(x)\int_{\mathbb{R}^n\setminus\Omega} p(t,x,y)\,dy.
\end{aligned} \tag{P.8}$$

In the last expression, the first integral on the right-hand side is 0 by odd symmetry, while for the second one

$$\begin{aligned}\left|\int_{\mathbb{R}^n\setminus B_\varepsilon(x)} p(t,x,y)\,\eta(y)|x-y|^2\,dy\right| &\leqslant \|u\|_{C^2(\Omega)}\int_{\mathbb{R}^n\setminus B_\varepsilon(x)} p(t,x,y)|x-y|^2\,dy\\ &\leqslant \text{const}\,\|u\|_{C^2(\Omega)}t^{-/2}\int_{\mathbb{R}^n\setminus B_\varepsilon(x)} e^{-|x-y|^2/(4t)}|x-y|^2\,dy\\ &\leqslant \text{const}\,\|u\|_{C^2(\Omega)}t\int_{\mathbb{R}^n\setminus B_{\varepsilon/\sqrt{4t}}} e^{-|z|^2}|z|^2\,dz\\ &\leqslant \text{const}\,\|u\|_{C^2(\Omega)}t.\end{aligned} \tag{P.9}$$

As for the last integral in (P.8), we have that

$$\begin{aligned}|u(x)|\int_{\mathbb{R}^n\setminus\Omega} p(t,x,y)\,dy &\leqslant \text{const}\,|u(x)|t^{-n/2}\int_{\mathbb{R}^n\setminus\Omega} e^{-|x-y|^2/(4t)}\,dy \leqslant\\ &\leqslant \text{const}\,|u(x)|t^{-n/2}\int_{\mathbb{R}^n\setminus B_{\text{dist}(x,\partial\Omega)}} e^{-|y|^2/(4t)}\,dy \leqslant \text{const}\,|u(x)|\int_{\mathbb{R}^n\setminus B_{\text{dist}(x,\partial\Omega)/\sqrt{4t}}} e^{-|z|^2}\,dz\\ &\leqslant \text{const}\,|u(x)|e^{-\text{dist}(x,\partial\Omega)/\sqrt{4t}}.\end{aligned} \tag{P.10}$$

Equations (P.9) and (P.10) imply that

$$\frac{|A|}{t^{1+s}} \leqslant \text{const}\,t^{-s}, \qquad t\in(0,1),$$

which is integrable for $t\in(0,1)$.

We turn now to the estimation of B which we rewrite as

$$B=\int_\Omega \big(p_N^\Omega(t,x,y)-p(t,x,y)\big)\big(u(x)-u(y)\big)\chi_{\Omega\setminus B_\varepsilon(x)}(y)\,dy$$

where χ_U stands for the characteristic function of a set $U\subset\mathbb{R}^n$. By definition, B solves the heat equation in Ω with zero initial condition. Moreover, since u is supported in a compact subset K of Ω, B is satisfying the (lateral) boundary condition

$$\begin{aligned}\Big|B\big|_{\partial B}\Big| &\leqslant \int_\Omega \big|p_N^\Omega(t,x,y)-p(t,x,y)\big||u(y)|\chi_{\Omega\setminus B_\varepsilon(x)}(y)\,dy\\ &\leqslant \text{const}\,t^{-n/2}\int_K e^{-c_1|x-y|^2/t}|u(y)|\,dy\\ &\leqslant \text{const}\,\|u\|_{L^1(\Omega)}t^{-n/2}e^{-c_2/t}\end{aligned}$$

for some $c_1, c_2 > 0$, in view of (P.4) and that, for $x \in \partial\Omega$ and $y \in K$, $|x - y| \geqslant \operatorname{dist}(K, \partial\Omega) > 0$. Then, by the parabolic maximum principle (see, for example, Section 7.1.4 in [62]),

$$\frac{|B|}{t^{1+s}} \leqslant \text{const}\, t^{-n/2-1-s} e^{-c_2/t},$$

which again is integrable for $t \in (0, 1)$. These observations provide an integrable bound for the integrand in (P.8), thus completing the justification of the claim in (P.6), as desired.

Appendix Q: Proof of (2.53)

If u is periodic, we can write it in Fourier series as

$$u(x) = \sum_{k \in \mathbb{Z}^n} u_k\, e^{2\pi i k \cdot x},$$

and the Fourier basis is also a basis of eigenfunctions. We have that

$$\begin{aligned}
&\int_{\mathbb{R}^n} \frac{u(x+y) + u(x-y) - 2u(x)}{|y|^{n+2s}}\, dy \\
&= \sum_{k \in \mathbb{Z}^n} \int_{\mathbb{R}^n} \frac{u_k e^{2\pi i k\cdot(x+y)} + u_k e^{2\pi i k\cdot(x-y)} - 2u_k e^{2\pi i k\cdot x}}{|y|^{n+2s}}\, dy \\
&= \sum_{k \in \mathbb{Z}^n} u_k e^{2\pi i k\cdot x} \int_{\mathbb{R}^n} \frac{e^{2\pi i k\cdot y} + e^{-2\pi i k\cdot y} - 2}{|y|^{n+2s}}\, dy \\
&= \sum_{k \in \mathbb{Z}^n} u_k e^{2\pi i k\cdot x}\, |k|^{2s} \int_{\mathbb{R}^n} \frac{e^{2\pi i \frac{k}{|k|}\cdot Y} + e^{-2\pi i \frac{k}{|k|}\cdot Y} - 2}{|Y|^{n+2s}}\, dY \\
&= \sum_{k \in \mathbb{Z}^n} u_k e^{2\pi i k\cdot x}\, |k|^{2s} \int_{\mathbb{R}^n} \frac{e^{2\pi i Y_1} + e^{-2\pi i Y_1} - 2}{|Y|^{n+2s}}\, dY \\
&= \text{const} \sum_{k \in \mathbb{Z}^n} u_k e^{2\pi i k\cdot x}\, |k|^{2s}
\end{aligned}$$

and this shows (2.53).

Appendix R: Proof of (2.54)

We fix $\bar{k} \in \mathbb{N}$. We consider the $\bar{k}$th eigenvalue $\lambda_{\bar{k}} > 0$ and the corresponding normalized eigenfunction $\phi_{\bar{k}} =: \bar{u}$. We argue by contradiction and suppose that for any $\varepsilon > 0$ we can find v_ε such that $\|\bar{u} - v_\varepsilon\|_{C^2(B_1)} \leqslant \varepsilon$, with $(-\Delta)^s_{D,\Omega} v_\varepsilon = 0$ in B_1.

Using the notation in (2.49), we have that $\bar{u}_k = \delta_{k\bar{k}}$ and therefore

$$\begin{aligned}\big\|(-\Delta)^s_{D,\Omega}\bar{u} - (-\Delta)^s_{D,\Omega}v_\varepsilon\big\|^2_{L^2(\Omega)} &= \big\|(-\Delta)^s_{D,\Omega}\bar{u}\big\|^2_{L^2(\Omega)} = \big\|(-\Delta)^s_{D,\Omega}\phi_{\bar{k}}\big\|^2_{L^2(\Omega)} \\ &= \left\|\lambda^s_{\bar{k}}\,\phi_{\bar{k}}\right\|^2_{L^2(\Omega)} = \lambda^{2s}_{\bar{k}}.\end{aligned} \tag{R.1}$$

Furthermore

$$\begin{aligned}\big\|(-\Delta)^s_{D,\Omega}\bar{u} - (-\Delta)^s_{D,\Omega}v_\varepsilon\big\|^2_{L^2(\Omega)} &= \big\|(-\Delta)^s_{D,\Omega}(\bar{u} - v_\varepsilon)\big\|^2_{L^2(\Omega)} \\ &= \left\|\sum_{k=0}^{+\infty}\lambda^s_k(\bar{u} - v_\varepsilon)_k\,\phi_k\right\|^2_{L^2(\Omega)} \\ &= \sum_{k=0}^{+\infty}\lambda^{2s}_k(\bar{u} - v_\varepsilon)^2_k \leqslant \text{const}\sum_{k=0}^{+\infty}\lambda^2_k(\bar{u} - v_\varepsilon)^2_k \\ &= \text{const}\,\|\Delta(\bar{u} - v_\varepsilon)\|^2_{L^2(\Omega)} \\ &\leqslant \text{const}\,\|\bar{u} - v_\varepsilon\|^2_{C^2(\Omega)} \leqslant \text{const}\,\varepsilon.\end{aligned}$$

Comparing this with (R.1), we obtain that $\lambda^{2s}_{\bar{k}} \leqslant \text{const}\,\varepsilon$, which is a contradiction for small ε. Hence, the proof of (2.54) is complete.

Appendix S: Proof of (2.60)

Let

$$v(t) := \int_0^t \frac{\dot{u}(\tau)}{(t-\tau)^s}\,d\tau.$$

Then, by (2.59) and writing $\vartheta := \omega\,(t-\tau)$, we see that

$$\begin{aligned}
\mathscr{L}v(\omega) &= \int_0^{+\infty}\left[\int_0^t \frac{\dot u(\tau)}{(t-\tau)^s}\,d\tau\right]e^{-\omega t}\,dt = \int_0^{+\infty}\left[\int_\tau^{+\infty}\frac{\dot u(\tau)e^{-\omega t}}{(t-\tau)^s}\,dt\right]d\tau \\
&= \omega^{s-1}\int_0^{+\infty}\left[\int_0^{+\infty}\frac{\dot u(\tau)e^{-\omega\tau}\,e^{-\vartheta}}{\vartheta^s}\,d\vartheta\right]d\tau \\
&= \Gamma(1-s)\,\omega^{s-1}\int_0^{+\infty}\dot u(\tau)e^{-\omega\tau}\,d\tau \\
&= \Gamma(1-s)\,\omega^{s-1}\int_0^{+\infty}\left(\frac{d}{d\tau}\big(u(\tau)e^{-\omega\tau}\big)+\omega u(\tau)e^{-\omega\tau}\right)d\tau \\
&= \Gamma(1-s)\,\omega^{s-1}\left(-u(0)+\omega\int_0^{+\infty}u(\tau)e^{-\omega\tau}\,d\tau\right) \\
&= \Gamma(1-s)\,\omega^{s-1}\left(-u(0)+\omega\mathscr{L}u(\omega)\right),
\end{aligned}$$

where Γ denotes here the Euler's Gamma Function. This and (2.56) give (2.60), up to neglecting normalizing constants, as desired.

It is also worth pointing out that, as $s \nearrow 1$, formula (2.60) recovers the classical derivative, since, by (2.59),

$$\begin{aligned}
\mathscr{L}\dot u(\omega) &= \int_0^{+\infty}\dot u(t)e^{-\omega t}\,dt \\
&= \int_0^{+\infty}\left(\frac{d}{dt}\big(u(t)e^{-\omega t}\big)+\omega u(t)e^{-\omega t}\right)dt \\
&= -u(0)+\omega\int_0^{+\infty}u(t)e^{-\omega t}\,dt \\
&= -u(0)+\omega\,\mathscr{L}u(\omega),
\end{aligned}$$

which is (2.60) when $s = 1$.

Appendix T: Proof of (2.61)

First, we compute the Laplace Transform of the constant function. Namely, by (2.59), for any $b \in \mathbb{R}$,

$$\mathscr{L}b(\omega) = b\int_0^{+\infty}e^{-\omega t}\,dt = \frac{b}{\omega}. \tag{T.1}$$

We also set

$$\Psi(t) := \int_0^t \frac{f(\tau)}{(t-\tau)^{1-s}}\,d\tau$$

and we use (2.59) and the substitution $\vartheta := \omega\,(t-\tau)$ to calculate that

$$\begin{aligned}
\mathscr{L}\Psi(\omega) &= \int_0^{+\infty} \left[\int_0^t \frac{f(\tau)}{(t-\tau)^{1-s}}\,d\tau\right] e^{-\omega t}\,dt \\
&= \int_0^{+\infty} \left[\int_\tau^{+\infty} \frac{f(\tau)\,e^{-\omega t}}{(t-\tau)^{1-s}}\,dt\right] d\tau \\
&= \omega^{-s} \int_0^{+\infty} \left[\int_0^{+\infty} \frac{f(\tau)\,e^{-\omega\tau}\,e^{-\vartheta}}{\vartheta^{1-s}}\,d\vartheta\right] d\tau \\
&= \Gamma(s)\,\omega^{-s} \int_0^{+\infty} f(\tau)\,e^{-\omega\tau}\,d\tau = \Gamma(s)\,\omega^{-s}\,\mathscr{L}f(\omega),
\end{aligned}$$

where Γ denotes here the Euler's Gamma Function.

Exploiting this and (T.1), and making use also of (2.60), we can write the expression $\partial^s_{C,t} u = f$ in terms of the Laplace Transform as

$$\begin{aligned}
\omega^s\Big(\mathscr{L}u(\omega) - \mathscr{L}b(\omega)\Big) &= \omega^s\,\mathscr{L}u(\omega) - \omega^{s-1}u(0) = \mathscr{L}(\partial^s_{C,t}u)(\omega) \\
&= \mathscr{L}f(\omega) = \frac{\omega^s}{\Gamma(s)}\mathscr{L}\Psi(\omega),
\end{aligned}$$

with $b := u(0)$. Hence, dividing by ω^s and inverting the Laplace Transform, we obtain that

$$u(t) - b = \frac{1}{\Gamma(s)}\Psi(t),$$

which is (2.61).

Appendix U: Proof of (2.62)

We take G to be the fundamental solution of the operator "identity minus Laplacian", namely

$$G - \Delta G = \delta_0 \quad \text{in } \mathbb{R}^n, \tag{U.1}$$

being δ_0 the Dirac's Delta. The study of this fundamental solution can be done by Fourier Transform in the sense of distributions, and this leads to an explicit representation in dimension 1 recalling (I.1); we give here a general argument, valid in any dimension, based on the heat kernel

$$g(x,t) := \frac{1}{(4\pi t)^{n/2}} e^{-\frac{|x|^2}{4t}}.$$

Notice that $\partial_t g = \Delta g$ and $g(\cdot, 0) = \delta_0(\cdot)$. Let also

$$G(x) := \int_0^{+\infty} e^{-t} g(x,t)\, dt. \tag{U.2}$$

Notice that

$$\begin{aligned} \Delta G(x) &= \int_0^{+\infty} e^{-t} \Delta g(x,t)\, dt = \int_0^{+\infty} e^{-t} \partial_t g(x,t)\, dt \\ &= \int_0^{+\infty} \Big(\partial_t (e^{-t} g(x,t)) + e^{-t} g(x,t) \Big)\, dt \\ &= -\delta_0(x) + \int_0^{+\infty} e^{-t} g(x,t)\, dt = -\delta_0(x) + G(x), \end{aligned}$$

hence G, as defined in (U.2) solves (U.1).

Notice also that G is positive and bounded, due to (U.2). We also claim that

$$\text{for any } x \in \mathbb{R}^n \setminus B_1, \text{ it holds that } G(x) \leqslant C e^{-c|x|}, \tag{U.3}$$

for some $c, C > 0$. To this end, let us fix $x \in \mathbb{R}^n \setminus B_1$ and distinguish two regimes. If $t \in [0, |x|]$, we have that $\frac{|x|^2}{t} \geqslant |x|$ and thus

$$g(x,t) \leqslant \frac{1}{(4\pi t)^{n/2}} e^{-\frac{|x|^2}{8t}} e^{-\frac{|x|}{8}}.$$

Consequently, using the substitution $\rho := \frac{|x|^2}{8t}$,

$$\begin{aligned} \int_0^{|x|} e^{-t} g(x,t)\, dt &\leqslant \int_0^{|x|} \frac{1}{(4\pi t)^{n/2}} e^{-\frac{|x|^2}{8t}} e^{-\frac{|x|}{8}}\, dt \\ &= \int_{|x|/8}^{+\infty} \frac{C\rho^{n/2}}{|x|^n} e^{-\rho} e^{-\frac{|x|}{8}} \frac{|x|^2\, d\rho}{\rho^2} \leqslant C|x|\, e^{-\frac{|x|}{8}}, \end{aligned} \tag{U.4}$$

for some $C > 0$ possibly varying from line to line. Furthermore

$$\int_{|x|}^{+\infty} e^{-t} g(x,t)\,dt \leqslant \int_{|x|}^{+\infty} e^{-\frac{|x|}{2}} e^{-\frac{t}{2}} g(x,t)\,dt \leqslant e^{-\frac{|x|}{2}} \int_{1}^{+\infty} \frac{e^{-\frac{t}{2}}}{(4\pi t)^{n/2}}\,dt \leqslant C\,e^{-\frac{|x|}{2}},$$

for some $C > 0$. This and (U.4) give that

$$\int_{0}^{+\infty} e^{-t} g(x,t)\,dt \leqslant C|x|\,e^{-\frac{|x|}{8}},$$

up to renaming C, which implies (U.3) in view of (U.2).

Now we compute the Laplace Transform of t^s: namely, by (2.59),

$$\mathscr{L}(t^s)(\omega) = \int_0^{+\infty} t^s e^{-\omega t}\,dt = \omega^{-1-s} \int_0^{+\infty} \tau^s e^{-\tau}\,d\tau = C\omega^{-1-s}. \tag{U.5}$$

We compare this result with the Laplace Transform of the mean squared displacement related to the diffusion operator in (2.62). For this, we take u to be as in (2.62) and, in the light of (2.42), we consider the function

$$v(\omega) := \mathscr{L}\left(\int_{\mathbb{R}^n} |x|^2 u(x,t)\,dx\right)(\omega) = \int_{\mathbb{R}^n} |x|^2\, \mathscr{L}u(x,\omega)\,dx. \tag{U.6}$$

In addition, by taking the Laplace Transform (in the variable t, for a fixed $x \in \mathbb{R}^n$) of the equation in (2.62), making use of (2.60) we find that

$$\omega^s \mathscr{L}u(x,\omega) - \omega^{s-1}\delta_0(x) = \Delta \mathscr{L}u(x,\omega). \tag{U.7}$$

Now, we let

$$W(x,\omega) := \omega^{1-\frac{sn}{2}}\, \mathscr{L}u(\omega^{-s/2}x,\omega). \tag{U.8}$$

From (U.7), we have that

$$\begin{aligned}
\Delta W(x,\omega) &= \omega^{1-\frac{sn}{2}}\,\omega^{-s}\,\Delta \mathscr{L}u(\omega^{-s/2}x,\omega)\\
&= \omega^{1-\frac{sn}{2}}\,\omega^{-s}\left(\omega^s \mathscr{L}u(\omega^{-s/2}x,\omega) - \omega^{s-1}\delta_0(\omega^{-s/2}x)\right)\\
&= W(x,\omega) - \omega^{-\frac{sn}{2}}\delta_0(\omega^{-s/2}x)\\
&= W(x,\omega) - \delta_0(x),
\end{aligned}$$

and so, comparing with (U.1), we have that $W(x,\omega) = G(x)$.

Accordingly, by (U.8),

$$\mathscr{L}u(x,\omega) = \omega^{\frac{sn}{2}-1}\, W(\omega^{s/2}x,\,\omega) = \omega^{\frac{sn}{2}-1}\, G(\omega^{s/2}x).$$

We insert this information into (U.6) and we conclude that

$$v(\omega) = \omega^{\frac{sn}{2}-1} \int_{\mathbb{R}^n} |x|^2 \, G(\omega^{s/2} x) \, dx = \omega^{-1-s} \int_{\mathbb{R}^n} |y|^2 \, G(y) \, dy.$$

We remark that the latter integral is finite, thanks to (U.3), hence we can write that

$$v(\omega) = C\omega^{-1-s},$$

for some $C > 0$.

Therefore, we can compare this result with (U.5) and use the inverse Laplace Transform to obtain that the mean squared displacement in this case is proportional to t^s, as desired.

Appendix V: Memory Effects of Caputo Type

It is interesting to observe that the Caputo derivative models a simple memory effect that the classical derivative cannot comprise. For instance, integrating a classical derivative of a function u with $u(0) = 0$, one obtains the original function "independently on the past", namely if we set

$$M_u(t) := \int_0^t \dot{u}(\vartheta) \, d\vartheta, \tag{V.1}$$

we just obtain in this case that $M_u(t) = u(t) - u(0) = u(t)$. On the other hand, an expression as in (V.1) which takes into account the Caputo derivative does "remember the past" and takes into account the preceding events in such a way that recent events "weight" more than far away ones. To see this phenomenon, we can modify (V.1) by defining, for every $s \in (0, 1)$,

$$M_u^s(t) := \int_0^t \partial_{C,t}^s u(\vartheta) \, d\vartheta. \tag{V.2}$$

To detect the memory effect, for the sake of concreteness, we take a large time $t := N \in \mathbb{N}$ and we suppose that the function u is constant on unit intervals, that is $u = u_k$ in $[k-1, k)$, for each $k \in \{1, \ldots, N\}$, with $u_k \in \mathbb{R}$, and $u(0) = u_1 = 0$. We see that M_u^s in this case does not produce just the final outcome u_N, but a weighted average of the form

$$M_u^s(N) = \sum_{k=0}^{N-1} c_k \, u_{N-k}, \qquad \text{with } c_j > 0 \text{ decreasing and } c_j \simeq \frac{1}{j^s} \text{ for large } j. \tag{V.3}$$

To check this, we notice that, for all $\tau \in (0, N)$,

$$\dot{u}(\tau) = \sum_{k=2}^{N} (u_k - u_{k-1})\delta_{k-1}(\tau),$$

and hence we exploit (2.56) and (V.2) to see that

$$\begin{aligned}
M_u^s(N) &= \int_0^N \left[\int_0^\vartheta \frac{\dot{u}(\tau)}{(\vartheta - \tau)^s} \, d\tau \right] d\vartheta \\
&= \sum_{k=2}^{N} \int_0^N \left[\int_0^\vartheta (u_k - u_{k-1})\delta_{k-1}(\tau) \frac{d\tau}{(\vartheta - \tau)^s} \right] d\vartheta \\
&= \sum_{k=2}^{N} \int_{k-1}^N \frac{(u_k - u_{k-1})}{(\vartheta - k + 1)^s} \, d\vartheta \\
&= \sum_{k=2}^{N} u_k \int_{k-1}^N \frac{d\vartheta}{(\vartheta - k + 1)^s} - \sum_{k=2}^{N} u_{k-1} \int_{k-1}^N \frac{d\vartheta}{(\vartheta - k + 1)^s} \\
&= \sum_{k=2}^{N} u_k \int_{k-1}^N \frac{d\vartheta}{(\vartheta - k + 1)^s} - \sum_{k=1}^{N-1} u_k \int_{k}^N \frac{d\vartheta}{(\vartheta - k)^s} \\
&= \sum_{k=1}^{N} u_k \left[\int_{k-1}^N \frac{d\vartheta}{(\vartheta - k + 1)^s} - \int_{k}^N \frac{d\vartheta}{(\vartheta - k)^s} \right] \\
&= \sum_{k=1}^{N} u_k \frac{(N - k + 1)^{1-s} - (N - k)^{1-s}}{1 - s} \\
&= \sum_{k=2}^{N} c_{N-k} \, u_k \\
&= \sum_{k=0}^{N-2} c_k \, u_{N-k},
\end{aligned}$$

with

$$c_j := \frac{(j + 1)^{1-s} - j^{1-s}}{1 - s}.$$

This completes the proof of the memory effect claimed in (V.3).

Appendix W: Proof of (3.7)

Since M is bounded and positive and u is bounded, it holds that

$$\int_{\mathbb{R}^n\setminus B_1} \frac{|u(x)-u(x-y)|}{|M(x-y,y)\,y|^{n+2s}}\,dy \leqslant \text{const} \int_{\mathbb{R}^n\setminus B_1} \frac{dy}{|y|^{n+2s}}\,dy \leqslant \frac{\text{const}}{s}. \tag{W.1}$$

Moreover, for $y \in B_1$,

$$u(x-y) = u(x) - \nabla u(x)\cdot y + \frac{1}{2}D^2u(x)\,y\cdot y + O(|y|^3). \tag{W.2}$$

To simplify the notation, we now fix $x \in \mathbb{R}^n$ and we define $\mathcal{M}(y) := M(x-y,y)$. Then, for $y \in B_1$, we have that

$$M(x-y,y)\,y = \mathcal{M}(y)\,y = \mathcal{M}(0)\,y + \sum_{i=1}^{n} \partial_i\mathcal{M}(0)\,y\,y_i + O(|y|^3)$$

and so

$$|M(x-y,y)\,y|^2 = |\mathcal{M}(0)\,y|^2 + 2\sum_{i=1}^{n}(\mathcal{M}(0)\,y)\cdot(\partial_i\mathcal{M}(0)\,y)\,y_i + O(|y|^4).$$

Consequently, since $\mathcal{M}(0) = M(x,0)$ is non-degenerate, we can write

$$\mathcal{E}(y) := 2\sum_{i=1}^{n}(\mathcal{M}(0)\,y)\cdot(\partial_i\mathcal{M}(0)\,y)\,y_i = O(|y|^3)$$

and

$$\begin{aligned}
&|M(x-y,y)\,y|^{-n-2s}\\
&= \Big(|\mathcal{M}(0)\,y|^2 + \mathcal{E}(y) + O(|y|^4)\Big)^{-\frac{n+2s}{2}}\\
&= |\mathcal{M}(0)\,y|^{-n-2s}\Big(1 + |\mathcal{M}(0)\,y|^{-2}\,\mathcal{E}(y) + O(|y|^2)\Big)^{-\frac{n+2s}{2}}\\
&= |\mathcal{M}(0)\,y|^{-n-2s}\Big(1 - \frac{n+2s}{2}\,|\mathcal{M}(0)\,y|^{-2}\,\mathcal{E}(y) + O(|y|^2)\Big)\\
&= |\mathcal{M}(0)\,y|^{-n-2s} - \frac{n+2s}{2}\,|\mathcal{M}(0)\,y|^{-n-2s-2}\,\mathcal{E}(y) + O(|y|^{2-n-2s}).
\end{aligned} \tag{W.3}$$

Hence (for smooth and bounded functions u, and $y \in B_1$) we obtain that

$$\begin{aligned}&\frac{u(x)-u(x-y)}{|M(x-y,y)\,y|^{n+2s}}\\&=\frac{u(x)-u(x-y)}{|\mathcal{M}(0)\,y|^{n+2s}}-\frac{n+2s}{2}\,\frac{\big(u(x)-u(x-y)\big)\,\mathcal{E}(y)}{|\mathcal{M}(0)\,y|^{n+2s+2}}+O(|y|^{3-n-2s}).\end{aligned}$$

Thus, since the map $y\mapsto \frac{\nabla u(x)\cdot y}{|\mathcal{M}(0)\,y|^{n+2s}}$ is odd, recalling (W.2) we conclude that

$$\begin{aligned}&\int_{B_1}\frac{u(x)-u(x-y)}{|M(x-y,y)\,y|^{n+2s}}\,dy\\&=\int_{B_1}\left(\frac{u(x)-u(x-y)}{|\mathcal{M}(0)\,y|^{n+2s}}-\frac{n+2s}{2}\,\frac{\big(u(x)-u(x-y)\big)\,\mathcal{E}(y)}{|\mathcal{M}(0)\,y|^{n+2s+2}}+O(|y|^{3-n-2s})\right)dy\\&=\int_{B_1}\left(\frac{u(x)-u(x-y)-\nabla u(x)\cdot y}{|\mathcal{M}(0)\,y|^{n+2s}}-\frac{n+2s}{2}\,\frac{\big(u(x)-u(x-y)\big)\,\mathcal{E}(y)}{|\mathcal{M}(0)\,y|^{n+2s+2}}\right.\\&\qquad\left.+\,O(|y|^{3-n-2s})\right)dy\\&=\int_{B_1}\left(-\frac{D^2u(x)\,y\cdot y}{2\,|\mathcal{M}(0)\,y|^{n+2s}}-(n+2s)\sum_{i=1}^{n}\frac{(\nabla u(x)\cdot y)\big((\mathcal{M}(0)\,y)\cdot(\partial_i\mathcal{M}(0)\,y)\big)\,y_i}{|\mathcal{M}(0)\,y|^{n+2s+2}}\right.\\&\qquad\left.+\,O(|y|^{3-n-2s})\right)dy \qquad\qquad \text{(W.4)}\end{aligned}$$

Now we observe that, for any $\alpha \geqslant 0$,

$$\begin{aligned}&\text{if } \varphi \text{ is positively homogeneous of degree } 2+\alpha \text{ and } T\in \mathrm{Mat}(n\times n), \text{ then}\\&(1-s)\int_{B_1}\frac{\varphi(y)}{|Ty|^{n+2s+\alpha}}\,dy=\frac{1}{2}\int_{S^{n-1}}\frac{\varphi(\omega)}{|T\omega|^{n+2s+\alpha}}\,d\mathcal{H}^{n-1}_{\omega}. \qquad\qquad \text{(W.5)}\end{aligned}$$

Indeed, using polar coordinates and the fact that $\varphi(\rho\omega)=\rho^{2+\alpha}\varphi(\omega)$, for any $\rho\geqslant 0$ and $\omega\in S^{n-1}$, thanks to the homogeneity, we see that

$$\begin{aligned}\int_{B_1}\frac{\varphi(y)}{|Ty|^{n+2s+\alpha}}\,dy&=\iint_{(0,1)\times S^{n-1}}\frac{\rho^{n-1}\,\varphi(\rho\omega)}{\rho^{n+2s+\alpha}\,|T\omega|^{n+2s+\alpha}}\,d\rho\,d\mathcal{H}^{n-1}_{\omega}\\&=\iint_{(0,1)\times S^{n-1}}\frac{\rho^{1-2s}\varphi(\omega)}{|T\omega|^{n+2s+\alpha}}\,d\rho\,d\mathcal{H}^{n-1}_{\omega}=\frac{1}{2(1-s)}\int_{S^{n-1}}\frac{\varphi(\omega)}{|T\omega|^{n+2s+\alpha}}\,d\mathcal{H}^{n-1}_{\omega},\end{aligned}$$

which implies (W.5).

Using (W.5) (with $\alpha := 0$ and $\alpha := 2$), we obtain that

$$\lim_{s\nearrow 1}(1-s)\int_{B_1}\frac{D^2u(x)\,y\cdot y}{|\mathcal{M}(0)\,y|^{n+2s}}\,dy=\frac{1}{2}\int_{S^{n-1}}\frac{D^2u(x)\,\omega\cdot\omega}{|\mathcal{M}(0)\,\omega|^{n+2}}\,d\mathcal{H}^{n-1}_\omega$$

and

$$\begin{aligned}&\lim_{s\nearrow 1}(1-s)\int_{B_1}\frac{(\nabla u(x)\cdot y)\big((\mathcal{M}(0)\,y)\cdot(\partial_i\mathcal{M}(0)\,y)\big)\,y_i}{|\mathcal{M}(0)\,y|^{n+2s+2}}\,dy\\ &\qquad=\frac{1}{2}\int_{S^{n-1}}\frac{(\nabla u(x)\cdot\omega)\big((\mathcal{M}(0)\,\omega)\cdot(\partial_i\mathcal{M}(0)\,\omega)\big)\,\omega_i}{|\mathcal{M}(0)\,\omega|^{n+4}}\,d\mathcal{H}^{n-1}_\omega\end{aligned}$$

Thanks to this, (W.1) and (W.4), we find that

$$\begin{aligned}&\lim_{s\nearrow 1}\int_{\mathbb{R}^n}\frac{u(x)-u(x-y)}{|M(x-y,y)\,y|^{n+2s}}\,dy\\ &=\lim_{s\nearrow 1}\int_{B_1}\frac{u(x)-u(x-y)}{|M(x-y,y)\,y|^{n+2s}}\,dy\\ &=-\frac{1}{4}\int_{S^{n-1}}\frac{D^2u(x)\,\omega\cdot\omega}{|\mathcal{M}(0)\,\omega|^{n+2}}\,d\mathcal{H}^{n-1}_\omega\\ &\quad-\frac{n+2}{2}\sum_{i-1}^{n}\int_{S^{n-1}}\frac{(\nabla u(x)\cdot\omega)\big((\mathcal{M}(0)\,\omega)\cdot(\partial_i\mathcal{M}(0)\,\omega)\big)\,\omega_i}{|\mathcal{M}(0)\,\omega|^{n+4}}\,d\mathcal{H}^{n-1}_\omega\\ &=-\sum_{i,j=1}^{n}a_{ij}(x)\partial^2_{ij}u(x)-\sum_{j=1}^{n}b_j\partial_ju(x),\end{aligned}\tag{W.6}$$

with

$$a_{ij}(x):=\frac{1}{4}\int_{S^{n-1}}\frac{\omega_i\,\omega_j}{|\mathcal{M}(0)\,\omega|^{n+2}}\,d\mathcal{H}^{n-1}_\omega=\frac{1}{4}\int_{S^{n-1}}\frac{\omega_i\,\omega_j}{|M(x,0)\,\omega|^{n+2}}\,d\mathcal{H}^{n-1}_\omega$$

$$\text{and}\quad b_j(x):=\frac{n+2}{2}\sum_{i=1}^{n}\int_{S^{n-1}}\frac{\omega_i\,\omega_j\big((\mathcal{M}(0)\,\omega)\cdot(\partial_i\mathcal{M}(0)\,\omega)\big)}{|\mathcal{M}(0)\,\omega|^{n+4}}\,d\mathcal{H}^{n-1}_\omega.$$

We observe that

$$b_j=\sum_{i=1}^{n}\partial_i a_{ij}(x).\tag{W.7}$$

To check this, we first compute that

$$\begin{aligned}\sum_{i=1}^{n}\partial_i a_{ij}(x) &= \frac{1}{4}\sum_{i=1}^{n}\partial_{x_i}\left(\int_{S^{n-1}}\frac{\omega_i\,\omega_j}{|M(x,0)\,\omega|^{n+2}}\,d\mathscr{H}^{n-1}_\omega\right)\\ &= -\frac{n+2}{4}\sum_{i=1}^{n}\int_{S^{n-1}}\frac{\omega_i\,\omega_j\,\big((M(x,0)\,\omega)\cdot(\partial_{x_i}M(x,0)\,\omega)\big)}{|M(x,0)\,\omega|^{n+4}}\,d\mathscr{H}^{n-1}_\omega.\end{aligned} \tag{W.8}$$

Now, we write a Taylor expansion of $M(x,y)$ in the variable y of the form

$$M_{\ell m}(x,y) = A_{\ell m}(x) + B_{\ell m}(x)\cdot y + O(y^2),$$

for some $A_{\ell m}:\mathbb{R}^n\to\mathbb{R}$ and $B_{\ell m}:\mathbb{R}^n\to\mathbb{R}^n$. We notice that

$$\partial_{x_i}M_{\ell m}(x,0) = \partial_{x_i}A_{\ell m}(x). \tag{W.9}$$

Also,

$$\begin{aligned}\partial_i\mathscr{M}_{\ell m}(0) &= \lim_{y\to 0}\partial_{y_i}\big(M_{\ell m}(x-y,y)\big)\\ &= \lim_{y\to 0}\partial_{y_i}\big(A_{\ell m}(x-y) + B_{\ell m}(x-y)\cdot y + O(y^2)\big)\\ &= -\partial_{x_i}A_{\ell m}(x) + B_{\ell m}(x)\cdot e_i.\end{aligned} \tag{W.10}$$

Furthermore, we use the structural assumption (3.6), and we see that

$$\begin{aligned}A_{\ell m}(x) - B_{\ell m}(x)\cdot y + O(y^2) &= M(x,-y)\\ &= M(x-y,y) = A_{\ell m}(x-y) + B_{\ell m}(x-y)\cdot y + O(y^2)\\ &= A_{\ell m}(x) - \nabla A_{\ell m}(x)\cdot y + B_{\ell m}(x)\cdot y + O(y^2).\end{aligned}$$

Comparing the linear terms, this gives that

$$2B_{\ell m}(x) = \nabla A_{\ell m}(x).$$

This and (W.10) imply that

$$\partial_i\mathscr{M}_{\ell m}(0) = -\partial_{x_i}A_{\ell m}(x) + \frac{1}{2}\nabla A_{\ell m}(x)\cdot e_i = -\frac{1}{2}\partial_{x_i}A_{\ell m}(x).$$

Comparing this with (W.9), we see that

$$\partial_{x_i}M_{\ell m}(x,0) = -2\partial_i\mathscr{M}_{\ell m}(0).$$

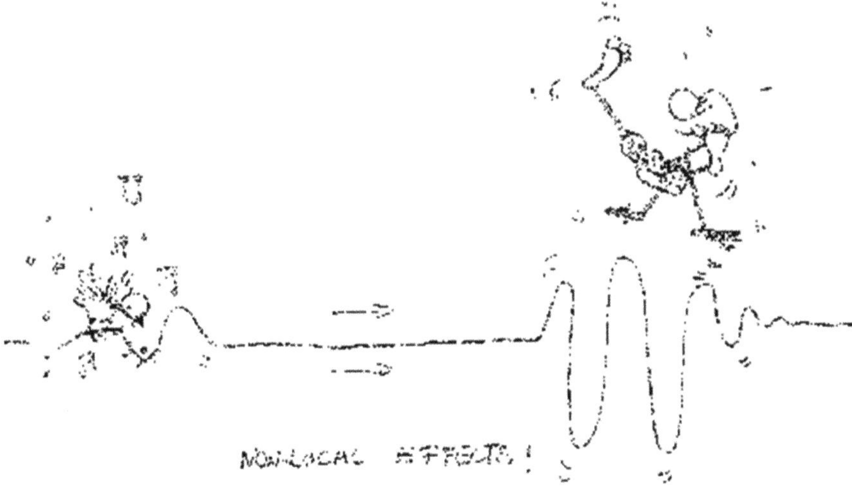

Fig. 13 A nice representation of nonlocal effects

So, we insert this information into (W.8) and we conclude that

$$\sum_{i=1}^{n} \partial_i a_{ij}(x) = \frac{n+2}{2} \sum_{i=1}^{n} \int_{S^{n-1}} \frac{\omega_i\, \omega_j\, \big((M(x,0)\,\omega)\cdot(\partial_i \mathcal{M}(0)\,\omega)\big)}{|M(x,0)\,\omega|^{n+4}}\, d\mathcal{H}^{n-1}_{\omega}.$$

This establishes (W.7), as desired.

Then, plugging (W.7) into (W.6), we obtain the equation in divergence form[12] which was claimed in (3.7).

[12]A slightly different approach as that in (3.7) is to consider the energy functional in (3.9) and prove, e.g. by Taylor expansion, that it converges to the energy functional

$$\text{const} \int_{\mathbb{R}^n} a_{ij}(x)\, \partial_i u(x)\, \partial_j u(x)\, dx.$$

On the other hand, a different proof of (3.7), that was nicely pointed out to us by Jonas Hirsch (who has also acted as a skilled cartoonist for Fig. 13) after a lecture, can be performed by taking into account the weak form of the operator in (3.5), i.e. integrating such expression against a test function $\varphi \in C_0^\infty(\mathbb{R}^n)$, thus finding

$$\begin{aligned} &(1-s) \iint_{\mathbb{R}^n\times\mathbb{R}^n} \frac{\big(u(x)-u(x-y)\big)\,\varphi(x)}{|M(x-y,y)\,y|^{n+2s}}\,dx\,dy \\ &= (1-s) \iint_{\mathbb{R}^n\times\mathbb{R}^n} \frac{\big(u(x)-u(z)\big)\,\varphi(x)}{|M(z,x-z)\,(x-z)|^{n+2s}}\,dx\,dz \end{aligned}$$

Appendix X: Proof of (3.12)

First we observe that

$$\int_{\mathbb{R}^n\setminus B_1} \frac{|u(x)-u(x-y)|}{|M(x,y)\,y|^{n+2s}}\,dy \leqslant \text{const}\int_{\mathbb{R}^n\setminus B_1} \frac{dy}{|y|^{n+2s}} \leqslant \frac{\text{const}}{s}. \tag{X.1}$$

Furthermore, for $y \in B_1$,

$$M(x,y)\,y = M(x,0)\,y + O(|y|^2).$$

Consequently,

$$|M(x,y)\,y|^2 = |M(x,0)\,y|^2 + O(|y|^3)$$

and so, from the non-degeneracy of $M(\cdot,\cdot)$,

$$\begin{aligned} |M(x,y)\,y|^{-n-2s} &= \big(|M(x,0)\,y|^2 + O(|y|^3)\big)^{-\frac{n+2s}{2}} \\ &= |M(x,0)\,y|^{-n-2s}\big(1+O(|y|)\big)^{-\frac{n+2s}{2}} = |M(x,0)\,y|^{-n-2s}\big(1-O(|y|)\big). \end{aligned}$$

Using this and the expansion in (W.2), we see that, for $y \in B_1$,

$$\begin{aligned} &\frac{u(x)-u(x-y)-\nabla u(x)\cdot y}{|M(x,y)\,y|^{n+2s}} \\ &= |M(x,0)\,y|^{-n-2s}\big(1-O(|y|)\big)\left(-\frac{1}{2}D^2u(x)\,y\cdot y + O(|y|^3)\right) \\ &= |M(x,0)\,y|^{-n-2s}\left(-\frac{1}{2}D^2u(x)\,y\cdot y + O(|y|^3)\right). \end{aligned}$$

$$\begin{aligned} &= (1-s)\iint_{\mathbb{R}^n\times\mathbb{R}^n} \frac{\big(u(z)-u(x)\big)\,\varphi(z)}{|M(x,z-x)\,(x-z)|^{n+2s}}\,dx\,dz \\ &= -(1-s)\iint_{\mathbb{R}^n\times\mathbb{R}^n} \frac{\big(u(x)-u(z)\big)\,\varphi(z)}{|M(z,x-z)\,(x-z)|^{n+2s}}\,dx\,dz, \end{aligned}$$

where the structural condition (3.6) has been used in the last line. This means that the weak formulation of the operator in (3.5) can be written as

$$\frac{1-s}{2}\iint_{\mathbb{R}^n\times\mathbb{R}^n} \frac{\big(u(x)-u(z)\big)\big(\varphi(x)-\varphi(z)\big)}{|M(z,x-z)\,(x-z)|^{n+2s}}\,dx\,dz.$$

So one can expand this expression and take the limit as $s \nearrow 1$, to obtain

$$\text{const}\int_{\mathbb{R}^n} a_{ij}(x)\,\partial_i u(x)\,\partial_j \varphi(x)\,dx,$$

which is indeed the weak formulation of the classical divergence form operator.

Thus, since, in the light of (3.11), we know that the map $y \mapsto \frac{\nabla u(x)\cdot y}{|M(x,y)\,y|^{n+2s}}$ is odd, we can write that

$$\begin{aligned}\int_{B_1}\frac{u(x)-u(x-y)}{|M(x,y)\,y|^{n+2s}}\,dy &= \int_{B_1}\frac{u(x)-u(x-y)-\nabla u(x)\cdot y}{|M(x,y)\,y|^{n+2s}}\,dy\\ &= -\frac{1}{2}\int_{B_1}\frac{D^2u(x)\,y\cdot y}{|M(x,0)\,y|^{n+2s}}\,dy+\frac{O(1)}{3-2s}\\ &= -\frac{\mathrm{const}}{1-s}\int_{S^{n-1}}\frac{D^2u(x)\,\omega\cdot\omega}{|M(x,0)\,\omega|^{n+2s}}\,d\mathscr{H}^{n-1}_{\omega}+\frac{O(1)}{3-2s},\end{aligned}$$

where the last identity follows by using (W.5) (with $\alpha := 0$). From this and (X.1) we obtain that

$$\begin{aligned}\lim_{s\nearrow 1}(1-s)\int_{\mathbb{R}^n}\frac{u(x)-u(x-y)}{|M(x,y)\,y|^{n+2s}}\,dy &= \lim_{s\nearrow 1}(1-s)\int_{B_1}\frac{u(x)-u(x-y)}{|M(x,y)\,y|^{n+2s}}\,dy\\ &= -\,\mathrm{const}\int_{S^{n-1}}\frac{D^2u(x)\,\omega\cdot\omega}{|M(x,0)\,\omega|^{n+2}}\,d\mathscr{H}^{n-1}_{\omega}\\ &= -\,\mathrm{const}\sum_{i,j=1}^{n}\int_{S^{n-1}}\frac{\omega_i\,\omega_j}{|M(x,0)\,\omega|^{n+2}}\,d\mathscr{H}^{n-1}_{\omega}\,\partial^2_{ij}u(x),\end{aligned}$$

which gives (3.12).

References

1. N. Abatangelo, Large s-harmonic functions and boundary blow-up solutions for the fractional Laplacian. Discrete Contin. Dyn. Syst. **35**(12), 5555–5607 (2015). https://doi.org/10.3934/dcds.2015.35.5555. MR3393247
2. N. Abatangelo, L. Dupaigne, Nonhomogeneous boundary conditions for the spectral fractional Laplacian. Ann. Inst. H. Poincaré Anal. Non Linéaire **34**(2), 439–467 (2017). https://doi.org/10.1016/j.anihpc.2016.02.001. MR3610940
3. N. Abatangelo, S. Jarohs, A. Saldaña, Positive powers of the Laplacian: from hypersingular integrals to boundary value problems. Commun. Pure Appl. Anal. **17**(3), 899–922 (2018). https://doi.org/10.3934/cpaa.2018045. MR3809107
4. N. Abatangelo, S. Jarohs, A. Saldaña, Green function and Martin kernel for higher-order fractional Laplacians in balls. Nonlinear Anal. **175**, 173–190 (2018). https://doi.org/10.1016/j.na.2018.05.019. MR3830727
5. N. Abatangelo, S. Jarohs, A. Saldaña, On the loss of maximum principles for higher-order fractional Laplacians. Proc. Am. Math. Soc. **146**(11), 4823–4835 (2018). https://doi.org/10.1090/proc/14165. MR3856149
6. E. Affili, S. Dipierro, E. Valdinoci, Decay estimates in time for classical and anomalous diffusion. arXiv e-prints (2018), available at 1812.09451
7. M. Allen, L. Caffarelli, A. Vasseur, A parabolic problem with a fractional time derivative. Arch. Ration. Mech. Anal. **221**(2), 603–630 (2016). https://doi.org/10.1007/s00205-016-0969-z. MR3488533

8. F. Andreu-Vaillo, J.M. Mazón, J.D. Rossi, J.J. Toledo-Melero, *Nonlocal Diffusion Problems*, Mathematical Surveys and Monographs, vol. 165 (American Mathematical Society, Providence, 2010); Real Sociedad Matemática Española, Madrid, 2010. MR2722295
9. D. Applebaum, *Lévy Processes and Stochastic Calculus*. Cambridge Studies in Advanced Mathematics, 2nd edn., vol. 116 (Cambridge University Press, Cambridge, 2009). MR2512800
10. V.E. Arkhincheev, É.M. Baskin, Anomalous diffusion and drift in a comb model of percolation clusters. J. Exp. Theor. Phys. **73**, 161–165 (1991)
11. A.V. Balakrishnan, Fractional powers of closed operators and the semigroups generated by them. Pac. J. Math. **10**, 419–437 (1960). MR0115096
12. R. Bañuelos, K. Bogdan, Lévy processes and Fourier multipliers. J. Funct. Anal. **250**(1), 197–213 (2007). https://doi.org/10.1016/j.jfa.2007.05.013. MR2345912
13. B. Barrios, I. Peral, F. Soria, E. Valdinoci, A Widder's type theorem for the heat equation with nonlocal diffusion. Arch. Ration. Mech. Anal. **213**(2), 629–650 (2014). https://doi.org/10.1007/s00205-014-0733-1. MR3211862
14. R.F. Bass, D.A. Levin, Harnack inequalities for jump processes. Potential Anal. **17**(4), 375–388 (2002). https://doi.org/10.1023/A:1016378210944. MR1918242
15. A. Bendikov, Asymptotic formulas for symmetric stable semigroups. Expo. Math. **12**(4), 381–384 (1994). MR1297844
16. J. Bertoin, *Lévy Processes*. Cambridge Tracts in Mathematics, vol. 121 (Cambridge University Press, Cambridge, 1996). MR1406564
17. R.M. Blumenthal, R.K. Getoor, Some theorems on stable processes. Trans. Am. Math. Soc. **95**, 263–273 (1960). https://doi.org/10.2307/1993291. MR0119247
18. K. Bogdan, T. Byczkowski, Potential theory for the α-stable Schrödinger operator on bounded Lipschitz domains. Stud. Math. **133**(1), 53–92 (1999). MR1671973
19. K. Bogdan, T. Żak, On Kelvin transformation. J. Theor. Probab. **19**(1), 89–120 (2006). MR2256481
20. M. Bonforte, A. Figalli, J.L. Vázquez, Sharp global estimates for local and nonlocal porous medium-type equations in bounded domains. Anal. PDE **11**(4), 945–982 (2018). https://doi.org/10.2140/apde.2018.11.945. MR3749373
21. L. Brasco, S. Mosconi, M. Squassina, Optimal decay of extremals for the fractional Sobolev inequality. Calc. Var. Partial Differ. Equ. **55**(2), 23, 32 (2016). https://doi.org/10.1007/s00526-016-0958-y. MR3461371
22. C. Bucur, Some observations on the Green function for the ball in the fractional Laplace framework. Commun. Pure Appl. Anal. **15**(2), 657–699 (2016). https://doi.org/10.3934/cpaa.2016.15.657. MR3461641
23. C. Bucur, Local density of Caputo-stationary functions in the space of smooth functions. ESAIM Control Optim. Calc. Var. **23**(4), 1361–1380 (2017). https://doi.org/10.1051/cocv/2016056. MR3716924
24. C. Bucur, E. Valdinoci, *Nonlocal Diffusion and Applications*. Lecture Notes of the Unione Matematica Italiana, vol. 20 (Springer, Cham, 2016); Unione Matematica Italiana, Bologna. MR3469920
25. C. Bucur, L. Lombardini, E. Valdinoci, Complete stickiness of nonlocal minimal surfaces for small values of the fractional parameter. Ann. Inst. H. Poincaré Anal. Non Linéaire **36**(3), 655–703 (2019)
26. X. Cabré, M. Cozzi, A gradient estimate for nonlocal minimal graphs. Duke Math. J. **168**(5), 775–848 (2019)
27. X. Cabré, Y. Sire, Nonlinear equations for fractional Laplacians II: existence, uniqueness, and qualitative properties of solutions. Trans. Am. Math. Soc. **367**(2), 911–941 (2015). https://doi.org/10.1090/S0002-9947-2014-05906-0. MR3280032
28. L.A. Caffarelli, Further regularity for the Signorini problem. Commun. Partial Differ. Equ. **4**(9), 1067–1075 (1979). https://doi.org/10.1080/03605307908820119. MR542512
29. L. Caffarelli, F. Charro, On a fractional Monge-Ampère operator. Ann. PDE **1**(1), 4, 47 (2015). MR3479063

30. L. Caffarelli, L. Silvestre, An extension problem related to the fractional Laplacian. Commun. Partial Differ. Equ. **32**(7–9), 1245–1260 (2007). https://doi.org/10.1080/03605300600987306. MR2354493
31. L. Caffarelli, L. Silvestre, Regularity theory for fully nonlinear integro-differential equations. Commun. Pure Appl. Math. **62**(5), 597–638 (2009). MR2494809
32. L. Caffarelli, L. Silvestre, Hölder regularity for generalized master equations with rough kernels, in *Advances in Analysis: The Legacy of Elias M. Stein*. Princeton Mathematical Series, vol. 50 (Princeton University Press, Princeton, 2014), pp. 63–83. MR3329847
33. L.A. Caffarelli, J.L. Vázquez, Asymptotic behaviour of a porous medium equation with fractional diffusion. Discrete Contin. Dyn. Syst. **29**(4), 1393–1404 (2011). MR2773189
34. L. Caffarelli, F. Soria, J.L. Vázquez, Regularity of solutions of the fractional porous medium flow. J. Eur. Math. Soc. (JEMS) **15**(5), 1701–1746 (2013). https://doi.org/10.4171/JEMS/401. MR3082241
35. M. Caputo, Linear models of dissipation whose Q is almost frequency independent. II. Fract. Calc. Appl. Anal. **11**(1), 4–14 (2008). Reprinted from Geophys. J. R. Astr. Soc. **13**(1967), no. 5, 529–539. MR2379269
36. A. Carbotti, S. Dipierro, E. Valdinoci, *Local Density of Solutions to Fractional Equations*. Graduate Studies in Mathematics (De Gruyter, Berlin, 2019)
37. A. Carbotti, S. Dipierro, E. Valdinoci, Local density of Caputo-stationary functions of any order. Complex Var. Elliptic Equ. (to appear). https://doi.org/10.1080/17476933.2018.1544631
38. R. Carmona, W.C. Masters, B. Simon, Relativistic Schrödinger operators: asymptotic behavior of the eigenfunctions. J. Funct. Anal. **91**(1), 117–142 (1990). https://doi.org/10.1016/0022-1236(90)90049-Q. MR1054115
39. A. Cesaroni, M. Novaga, *Symmetric self-shrinkers for the fractional mean curvature flow*. ArXiv e-prints (2018), available at 1812.01847
40. A. Cesaroni, S. Dipierro, M. Novaga, E. Valdinoci, Fattening and nonfattening phenomena for planar nonlocal curvature flows. Math. Ann. (to appear). https://doi.org/10.1007/s00208-018-1793-6
41. E. Cinti, C. Sinestrari, E. Valdinoci, Neckpinch singularities in fractional mean curvature flows. Proc. Am. Math. Soc. **146**(6), 2637–2646 (2018). https://doi.org/10.1090/proc/14002. MR3778164
42. E. Cinti, J. Serra, E. Valdinoci, Quantitative flatness results and BV-estimates for stable nonlocal minimal surfaces. J. Differ. Geom. (to appear)
43. J. Coville, Harnack type inequality for positive solution of some integral equation. Ann. Mat. Pura Appl. **191**(3), 503–528 (2012). https://doi.org/10.1007/s10231-011-0193-2. MR2958346
44. J.C. Cox, The valuation of options for alternative stochastic processes. J. Finan. Econ. **3**(1–2), 145–166 (1976). https://doi.org/10.1016/0304-405X(76)90023-4
45. M. Cozzi, E. Valdinoci, On the growth of nonlocal catenoids. Atti Accad. Naz. Lincei Rend. Lincei Mat. Appl. (to appear)
46. J. Dávila, M. del Pino, J. Wei, Nonlocal s-minimal surfaces and Lawson cones. J. Differ. Geom. **109**(1), 111–175 (2018). https://doi.org/10.4310/jdg/1525399218. MR3798717
47. C.-S. Deng, R.L. Schilling, Exact Asymptotic Formulas for the Heat Kernels of Space and Time-Fractional Equations, ArXiv e-prints (2018), available at 1803.11435
48. A. de Pablo, F. Quirós, A. Rodríguez, J.L. Vázquez, A fractional porous medium equation. Adv. Math. **226**(2), 1378–1409 (2011). https://doi.org/10.1016/j.aim.2010.07.017. MR2737788
49. E. Di Nezza, G. Palatucci, E. Valdinoci, Hitchhiker's guide to the fractional Sobolev spaces. Bull. Sci. Math. **136**(5), 521–573 (2012). https://doi.org/10.1016/j.bulsci.2011.12.004. MR2944369
50. S. Dipierro, H.-C. Grunau, Boggio's formula for fractional polyharmonic Dirichlet problems. Ann. Mat. Pura Appl. **196**(4), 1327–1344 (2017). https://doi.org/10.1007/s10231-016-0618-z. MR3673669

51. S. Dipierro, E. Valdinoci, A simple mathematical model inspired by the Purkinje cells: from delayed travelling waves to fractional diffusion. Bull. Math. Biol. **80**(7), 1849–1870 (2018). https://doi.org/10.1007/s11538-018-0437-z. MR3814763
52. S. Dipierro, G. Palatucci, E. Valdinoci, Dislocation dynamics in crystals: a macroscopic theory in a fractional Laplace setting. Commun. Math. Phys. **333**(2), 1061–1105 (2015). https://doi.org/10.1007/s00220-014-2118-6. MR3296170
53. S. Dipierro, O. Savin, E. Valdinoci, Graph properties for nonlocal minimal surfaces. Calc. Var. Partial Differ. Equ. **55**(4), 86, 25 (2016). https://doi.org/10.1007/s00526-016-1020-9. MR3516886
54. S. Dipierro, X. Ros-Oton, E. Valdinoci, Nonlocal problems with Neumann boundary conditions. Rev. Mat. Iberoam. **33**(2), 377–416 (2017). https://doi.org/10.4171/RMI/942. MR3651008
55. S. Dipierro, O. Savin, E. Valdinoci, All functions are locally s-harmonic up to a small error. J. Eur. Math. Soc. (JEMS) **19**(4), 957–966 (2017). https://doi.org/10.4171/JEMS/684. MR3626547
56. S. Dipierro, O. Savin, E. Valdinoci, Boundary behavior of nonlocal minimal surfaces. J. Funct. Anal. **272**(5), 1791–1851 (2017). https://doi.org/10.1016/j.jfa.2016.11.016. MR3596708
57. S. Dipierro, N. Soave, E. Valdinoci, On stable solutions of boundary reaction-diffusion equations and applications to nonlocal problems with Neumann data. Indiana Univ. Math. J. **67**(1), 429–469 (2018). https://doi.org/10.1512/iumj.2018.67.6282. MR3776028
58. S. Dipierro, O. Savin, E. Valdinoci, Local approximation of arbitrary functions by solutions of nonlocal equations. J. Geom. Anal. **29**(2), 1428–1455 (2019)
59. S. Dipierro, O. Savin, E. Valdinoci, Definition of fractional Laplacian for functions with polynomial growth. Rev. Mat. Iberoam (to appear)
60. S. Dipierro, J. Serra, E. Valdinoci, Improvement of flatness for nonlocal phase transitions. Amer. J. Math. (to appear)
61. B. Dyda, Fractional calculus for power functions and eigenvalues of the fractional Laplacian. Fract. Calc. Appl. Anal. **15**(4), 536–555 (2012). https://doi.org/10.2478/s13540-012-0038-8. MR2974318
62. L.C. Evans, *Partial Differential Equations*. Graduate Studies in Mathematics, vol. 19 (American Mathematical Society, Providence, 1998). MR1625845
63. M.M. Fall, T. Weth, Nonexistence results for a class of fractional elliptic boundary value problems. J. Funct. Anal. **263**(8), 2205–2227 (2012). https://doi.org/10.1016/j.jfa.2012.06.018. MR2964681
64. M.M. Fall, T. Weth, Liouville theorems for a general class of nonlocal operators. Potential Anal. **45**(1), 187–200 (2016). https://doi.org/10.1007/s11118-016-9546-1. MR3511811
65. A. Farina, E. Valdinoci, The state of the art for a conjecture of De Giorgi and related problems, in *Recent Progress on Reaction-Diffusion Systems and Viscosity Solutions* (World Scientific Publishing, Hackensack, 2009), pp. 74–96. https://doi.org/10.1142/9789812834744_0004. MR2528756
66. P. Felmer, A. Quaas, Boundary blow up solutions for fractional elliptic equations. Asymptot. Anal. **78**(3), 123–144 (2012). MR2985500
67. A. Figalli, J. Serra, On stable solutions for boundary reactions: a De Giorgi-type result in dimension 4 + 1, preprint at arXiv:1705.02781 (2017, submitted)
68. R.L. Frank, E. Lenzmann, L. Silvestre, Uniqueness of radial solutions for the fractional Laplacian. Commun. Pure Appl. Math. **69**(9), 1671–1726 (2016). https://doi.org/10.1002/cpa.21591. MR3530361
69. R.K. Getoor, First passage times for symmetric stable processes in space. Trans. Am. Math. Soc. **101**, 75–90 (1961). https://doi.org/10.2307/1993412. MR0137148
70. T. Ghosh, M. Salo, G. Uhlmann, The Calderón problem for the fractional Schrödinger equation. ArXiv e-prints (2016), available at 1609.09248
71. D. Gilbarg, N.S. Trudinger, *Elliptic Partial Differential Equations of Second Order*. Classics in Mathematics (Springer, Berlin, 2001). Reprint of the 1998 edition. MR1814364

72. E. Giusti, *Direct Methods in the Calculus of Variations* (World Scientific Publishing, River Edge, 2003). MR1962933
73. Q. Han, F. Lin, *Elliptic Partial Differential Equations*. Courant Lecture Notes in Mathematics, 2nd edn., vol. 1 (Courant Institute of Mathematical Sciences/American Mathematical Society, New York/Providence, 2011). MR2777537
74. N. Jacob, *Pseudo-Differential Operators and Markov Processes*. Mathematical Research, vol. 94 (Akademie Verlag, Berlin, 1996). MR1409607
75. M. Kaßmann, *Harnack-Ungleichungen Für nichtlokale Differentialoperatoren und Dirichlet-Formen* (in German). Bonner Mathematische Schriften [Bonn Mathematical Publications], vol. 336 (Universität Bonn, Mathematisches Institut, Bonn, 2001). Dissertation, Rheinische Friedrich-Wilhelms-Universität Bonn, Bonn, 2000. MR1941020
76. M. Kaßmann, A new formulation of Harnack's inequality for nonlocal operators. C. R. Math. Acad. Sci. Paris **349**(11–12), 637–640 (2011). https://doi.org/10.1016/j.crma.2011.04.014 (English, with English and French summaries). MR2817382
77. M. Kaßmann, Jump processes and nonlocal operators, in *Recent Developments in Nonlocal Theory* (De Gruyter, Berlin, 2018), pp. 274–302. MR3824215
78. V. Kolokoltsov, Symmetric stable laws and stable-like jump-diffusions. Proc. Lond. Math. Soc. **80**(3), 725–768 (2000). https://doi.org/10.1112/S0024611500012314. MR1744782
79. N.V. Krylov, On the paper "All functions are locally s-harmonic up to a small error" by Dipierro, Savin, and Valdinoci. ArXiv e-prints (2018), available at 1810.07648
80. T. Kuusi, G. Mingione, Y. Sire, Nonlocal equations with measure data. Commun. Math. Phys. **337**(3), 1317–1368 (2015). https://doi.org/10.1007/s00220-015-2356-2. MR3339179
81. M. Kwaśnicki, Ten equivalent definitions of the fractional Laplace operator. Fract. Calc. Appl. Anal. **20**(1), 7–51 (2017). https://doi.org/10.1515/fca-2017-0002. MR3613319
82. N.S. Landkof, *Foundations of Modern Potential Theory* (Springer, New York, 1972). Translated from the Russian by A. P. Doohovskoy; Die Grundlehren der mathematischen Wissenschaften, Band 180. MR0350027
83. H.C. Lara, G. Dávila, Regularity for solutions of non local parabolic equations. Calc. Var. Partial Differ. Equ. **49**(1–2), 139–172 (2014). https://doi.org/10.1007/s00526-012-0576-2. MR3148110
84. F. Mainardi, *Fractional Calculus and Waves in Linear Viscoelasticity* (Imperial College Press, London, 2010). An introduction to mathematical models. MR2676137
85. F. Mainardi, Y. Luchko, G. Pagnini, The fundamental solution of the space-time fractional diffusion equation. Fract. Calc. Appl. Anal. **4**(2), 153–192 (2001). MR1829592
86. F. Mainardi, P. Paradisi, R. Gorenflo, Probability distributions generated by fractional diffusion equations. preprint at arXiv:0704.0320v1 (2007, submitted)
87. B. Mandelbrot, The variation of certain speculative prices. J. Bus. **36**, 394 (1963)
88. J.M. Mazón, J.D. Rossi, J. Toledo, The heat content for nonlocal diffusion with non-singular kernels. Adv. Nonlinear Stud. **17**(2), 255–268 (2017). https://doi.org/10.1515/ans-2017-0005. MR3641640
89. R. Metzler, J. Klafter, The restaurant at the end of the random walk: recent developments in the description of anomalous transport by fractional dynamics. J. Phys. A **37**(31), R161–R208 (2004). https://doi.org/10.1088/0305-4470/37/31/R01. MR2090004
90. E. Montefusco, B. Pellacci, G. Verzini, Fractional diffusion with Neumann boundary conditions: the logistic equation. Discrete Contin. Dyn. Syst. Ser. B **18**(8), 2175–2202 (2013). https://doi.org/10.3934/dcdsb.2013.18.2175. MR3082317
91. R. Musina, A.I. Nazarov, On fractional Laplacians. Commun. Partial Differ. Equ. **39**(9), 1780–1790 (2014). https://doi.org/10.1080/03605302.2013.864304. MR3246044
92. G. Palatucci, O. Savin, E. Valdinoci, Local and global minimizers for a variational energy involving a fractional norm. Ann. Mat. Pura Appl. **192**(4), 673–718 (2013). https://doi.org/10.1007/s10231-011-0243-9. MR3081641
93. V. Pareto, *Cours D'économie Politique*, vol. I/II (F. Rouge, Lausanne, 1896)

94. I. Podlubny, *Fractional Differential Equations*. Mathematics in Science and Engineering, vol. 198 (Academic Press, San Diego, CA, 1999). An introduction to fractional derivatives, fractional differential equations, to methods of their solution and some of their applications. MR1658022
95. G. Pólya, On the zeros of an integral function represented by Fourier's integral. Messenger **52**, 185–188 (1923)
96. M. Riesz, L'intégrale de Riemann-Liouville et le problème de Cauchy (French). Acta Math. **81**, 1–223 (1949). MR0030102
97. X. Ros-Oton, J. Serra, The Dirichlet problem for the fractional Laplacian: regularity up to the boundary. J. Math. Pures Appl. **101**(3), 275–302 (2014). https://doi.org/10.1016/j.matpur.2013.06.003 (English, with English and French summaries). MR3168912
98. W. Rudin, *Real and Complex Analysis* (McGraw-Hill, New York, 1966). MR0210528
99. A. Rüland, M. Salo, The fractional Calderón problem: low regularity and stability. ArXiv e-prints (2017), available at 1708.06294
100. A. Rüland, M. Salo, Quantitative approximation properties for the fractional heat equation. ArXiv e-prints (2017), available at 1708.06300
101. L.A. Sakhnovich, *Levy Processes, Integral Equations, Statistical Physics: Connections and Interactions*. Operator Theory: Advances and Applications, vol. 225 (Birkhäuser/Springer, Basel, 2012). MR2963050
102. L. Saloff-Coste, The heat kernel and its estimates, in *Probabilistic Approach to Geometry*. Advanced Studies in Pure Mathematics, vol. 57 (Mathematical Society of Japan, Tokyo, 2010), pp. 405–436. MR2648271
103. S. Salsa, *Equazioni a Derivate Parziali. Metodi, Modelli e Applicazioni* (Italian), 2nd edn. (Springer, Milano, 2010)
104. S.G. Samko, A.A. Kilbas, O.I. Marichev, *Fractional Integrals and Derivatives* (Gordon and Breach Science Publishers, Yverdon, 1993). Theory and applications; Edited and with a foreword by S. M. Nikolʹskiĭ; Translated from the 1987 Russian original; Revised by the authors. MR1347689
105. T. Sandev, A. Schulz, H. Kantz, A. Iomin, Heterogeneous diffusion in comb and fractal grid structures. Chaos Solitons Fractals **114**, 551–555 (2018). https://doi.org/10.1016/j.chaos.2017.04.041. MR3856678
106. R. Servadei, E. Valdinoci, Mountain pass solutions for non-local elliptic operators. J. Math. Anal. Appl. **389**(2), 887–898 (2012). https://doi.org/10.1016/j.jmaa.2011.12.032. MR2879266
107. R. Servadei, E. Valdinoci, On the spectrum of two different fractional operators. Proc. R. Soc. Edinburgh Sect. A **144**(4), 831–855 (2014). https://doi.org/10.1017/S0308210512001783. MR3233760
108. E.M. Stein, *Singular Integrals and Differentiability Properties of Functions*. Princeton Mathematical Series, vol. 30 (Princeton University Press, Princeton, 1970). MR0290095
109. P.R. Stinga, J.L. Torrea, Extension problem and Harnack's inequality for some fractional operators. Commun. Partial Differ. Equ. **35**(11), 2092–2122 (2010). https://doi.org/10.1080/03605301003735680. MR2754080
110. J.F. Toland, The Peierls-Nabarro and Benjamin-Ono equations. J. Funct. Anal. **145**(1), 136–150 (1997). https://doi.org/10.1006/jfan.1996.3016. MR1442163
111. E. Valdinoci, From the long jump random walk to the fractional Laplacian. Bol. Soc. Esp. Mat. Apl. SeMA **49**, 33–44 (2009). MR2584076
112. E. Valdinoci, All functions are (locally) s-harmonic (up to a small error)—and applications, in *Partial Differential Equations and Geometric Measure Theory*. Lecture Notes in Mathematics, vol. 2211 (Springer, Cham, 2018), pp. 197–214. MR3790948
113. J.L. Vázquez, Recent progress in the theory of nonlinear diffusion with fractional Laplacian operators. Discrete Contin. Dyn. Syst. Ser. S **7**(4), 857–885 (2014). https://doi.org/10.3934/dcdss.2014.7.857. MR3177769

114. J.L. Vázquez, The Dirichlet problem for the fractional p-Laplacian evolution equation. J. Differ. Equ. **260**(7), 6038–6056 (2016). https://doi.org/10.1016/j.jde.2015.12.033. MR3456825
115. J.L. Vázquez, The mathematical theories of diffusion: nonlinear and fractional diffusion, in *Nonlocal and Nonlinear Diffusions and Interactions: New Methods and Directions*. Lecture Notes in Mathematics, vol. 2186 (Springer, Cham, 2017), pp. 205–278. MR3588125

Dirichlet Problems for Fully Nonlinear Equations with "Subquadratic" Hamiltonians

Isabeau Birindelli, Françoise Demengel, and Fabiana Leoni

Abstract For a class of fully nonlinear equations having second order operators which may be singular or degenerate when the gradient of the solutions vanishes, and having first order terms with power growth, we prove the existence and uniqueness of suitably defined viscosity solutions of Dirichlet problems and we further show that it is a Lipschitz continuous function.

Keywords Inhomogenous equations · Degenerate and singular fully nonlinear elliptic PDE · Dirichlet problem

2010 Mathematical Subject Classification 35J70, 35J75

1 Content of the Paper

In this paper we prove some existence, uniqueness and Lipschitz regularity results for solutions of the following class of Dirichlet problems

$$\begin{cases} -F(\nabla u, D^2u) + b(x)|\nabla u|^\beta + \lambda |u|^\alpha u = f & \text{in } \Omega \\ u = \varphi \quad \text{on } \partial\Omega \end{cases} \tag{1.1}$$

where $\Omega \subset \mathbb{R}^N$ is a bounded, $\mathcal{C}^2$ open set and the operator F satisfies the structural assumptions

I. Birindelli (✉) · F. Leoni
Dipartimento di Matematica, Sapienza Università di Roma, Rome, Italy
e-mail: isabeau@mat.uniroma1.it

F. Demengel
Département de Mathématiques, Université de Cergy-Pontoise, Cergy-Pontoise, France

S. Dipierro (ed.), *Contemporary Research in Elliptic PDEs and Related Topics*, Springer INdAM Series 33, https://doi.org/10.1007/978-3-030-18921-1_2

(H1) $F : \mathbb{R}^N \setminus \{0\} \times \mathcal{S}_N \to \mathbb{R}$ is a continuous function, $\mathcal{S}_N$ being the set of $N \times N$ symmetric matrices;

(H2) $F(p, M)$ is homogeneous of degree $\alpha > -1$ with respect to p, positively homogeneous of degree 1 with respect to M, and it satisfies, for some constants $A \geq a > 0$,

$$a|p|^\alpha \operatorname{tr}(N) \leq F(p, M+N) - F(p, M) \leq A|p|^\alpha \operatorname{tr}(N) \tag{1.2}$$

for any $M, N \in \mathcal{S}_N$, with $N \geq 0$, and, for some $c > 0$,

$$|F(p, M) - F(q, M)| \leq c|M| \, ||p|^\alpha - |q|^\alpha| \tag{1.3}$$

for any $p, q \in \mathbb{R}^N \setminus \{0\}$, and $M \in \mathcal{S}_N$.

We further assume that $\lambda > 0$, the first order coefficient b is Lipschitz continuous, the forcing term f is bounded and continuous, and the boundary datum φ is Lipschitz continuous.

The novelty lies in the choice of $\beta \in (0, \alpha + 2]$. In the case $\alpha = 0$, the condition on β reduces to $0 < \beta \leq 2$, hence the terminology "subquadratic".

The definition of viscosity solution we adopt must be clarified, at least in the case $\alpha < 0$, since the usual one given in [8] cannot be applied in the singular case. We use the definition firstly introduced in [4], which is equivalent to the usual one in the case $\alpha \geq 0$, and which allows not to test points where the gradient of the test function is zero, except in the locally constant case, when $\alpha < 0$.

Definition 1.1 Let Ω be an open set in $\mathbb{R}^N$, let $f : \Omega \times \mathbb{R} \to \mathbb{R}$ be a continuous function. An upper (lower) semicontinuous function u a is a subsolution (supersolution) of

$$-F(\nabla u, D^2 u) + b(x)|\nabla u|^\beta = f(x, u) \quad \text{in } \Omega$$

if for any $x_0 \in \Omega$, either u is locally constant around x_0 and $f(x_o, u(x_0)) \geq 0(\leq)$, or for any test function ϕ of class C^2 around x_0 such that $u - \phi$ has a local maximum (minimum) point in x_0 and $\nabla\phi(x_0) \neq 0$, then

$$-F(\nabla\phi(x_0), D^2\phi(x_0)) + b(x_0)|\nabla\phi(x_0)|^\beta \leq (\geq) f(x_0, u(x_o)).$$

A solution is a continuous function which is both a supersolution and a subsolution.

For a nice discussion on the different kind of definition of viscosity solutions for singular operators see the work of Attouchi and Ruosteenoja [1].

The results of the present paper have been announced in [6], where we consider solutions $u = u_\lambda$ of problem (1.1) with either $\varphi = 0$ or $\varphi = +\infty$ and the behaviour of u_λ as $\lambda \to 0$ is studied. Clearly, the first step to perform this analysis is a detailed description of the existence, uniqueness and regularity properties of the solutions of problems (1.1) in the case $\lambda > 0$, which is precisely the object of this paper. Our results can be summarized in the following main theorem.

Theorem 1.2 *Under the above assumptions, problem* (1.1) *has a unique viscosity solution, which is Lipschitz continuous up to the boundary.*

We recall that the restrictions on the exponent β in order to have existence of solutions already appear in the non singular nor degenerate case $\alpha = 0$, when generalized solutions satisfying the boundary conditions in the viscosity sense can be constructed, but loss of boundary conditions may occur, see [2, 7].

Theorem 1.2 is obtained as a consequence of the classical Perron's method, which in turn relies on the comparison principle given by Theorem 3.3. Observe that when $\alpha < 0$ and the gradient of the involved test functions is 0, the information on the solution is recovered through the result in Lemma 3.2, which is analogous to one used in [4] in the "sublinear" case i.e. $\beta \leq \alpha + 1$. Moreover when b is not constant it is necessary to check that the gradient of the test functions be uniformly bounded. This comes from a priori interior Lipschitz estimates, proved in Theorem 2.1, which are of independent interest.

We prove Lipschitz estimates up to the boundary for solutions of the Dirichlet problem (1.1). Our proof follows the Ishii-Lions technique, see [11], which has to be adapted to the present singular/degenerate case. We borrow ideas from [3, 5], and we first prove Hölder estimates and then push the argument up to the Lipschitz result.

Regularity results for degenerate elliptic equations have also been obtained in [9], where equations having only a principal term of the form $F(\nabla u, D^2 u)$ have been considered. However, we observe that any scaling argument relying on the homogeneity of the operator is not applicable in our case when $\beta \neq \alpha + 1$, due to the different homogeneities of the terms in the equation.

We also emphasize that we make no assumptions on the sign of the first order coefficient b, meaning that the estimates we obtain rely only on the ellipticity properties of the second order term, despite its singularity or degeneracy. These estimates are radically different from the local Lipschitz regularity result proved in [6], where only positive hamiltonians with "superlinear" exponent $\beta > \alpha + 1$ are considered. In that case, the obtained Lipschitz estimates, which do not depend on the L^∞ norm of the solution, are consequence of the coercivity of the first order term and require that the forcing term f be Lipschitz continuous.

The Lipschitz estimates are proved in Sect. 2. In Sect. 3, after the construction of sub and supersolutions vanishing on the boundary, we prove the comparison principle and obtain the proof of Theorem 1.2 in the case $\varphi = 0$. A Strong Maximum Principle and Hopf Principle are also included. Finally, the changes to be done in order to prove Theorem 1.2 for any boundary datum φ are detailed in Sect. 4.

2 Lipschitz Estimates

The main result in this section is the following Lipschitz type estimate, where we denote by B_1 the unit ball in $\mathbb{R}^N$.

Theorem 2.1 *Suppose that F satisfies (H1) and (H2) with $\alpha > -1$ and suppose that $\beta \in (0, \alpha + 2]$. Let u be a bounded viscosity subsolution of*

$$-F(\nabla u, D^2 u) + b(x)|\nabla u|^\beta \le g(x) \quad \text{in } B_1$$

and v be a bounded viscosity supersolution of

$$-F(\nabla v, D^2 v) + b(x)|\nabla v|^\beta \ge f(x) \quad \text{in } B_1\,,$$

with f and g bounded and b Lipschitz continuous. Then, for all $r < 1$, there exists c_r such that for all $(x, y) \in B_r^2$

$$u(x) - v(y) \le \sup_{B_1}(u - v) + c_r|x - y|.$$

In order to prove Theorem 2.1 we first obtain the following Hölder estimate:

Lemma 2.2 *Under the hypothesis of Theorem 2.1, for any $\gamma \in (0, 1)$, there exists $c_{r,\gamma} > 0$ such that for all $(x, y) \in B_r^2$*

$$u(x) - v(y) \le \sup_{B_1}(u - v) + c_{r,\gamma}|x - y|^\gamma. \tag{2.1}$$

Proof of Lemma 2.2 We borrow ideas from [3, 5, 11]. Fix $x_o \in B_r$, and define

$$\phi(x, y) = u(x) - v(y) - \sup_{B_1}(u - v) - M|x - y|^\gamma - L(|x - x_o|^2 + |y - x_o|^2)$$

with $L = \frac{4(|u|_\infty + |v|_\infty)}{(1-r)^2}$ and $M = \frac{(|u|_\infty + |v|_\infty)+1}{\delta^\gamma}$, δ will be chosen later small enough depending only on the data and on universal constants. We want to prove that $\phi(x, y) \le 0$ in B_1 which will imply the result, taking first $x = x_o$ and making x_o vary.

We argue by contradiction and suppose that $\sup_{B_1} \phi(x, y) > 0$. By the previous assumptions on M and L the supremum is achieved on $(\bar{x}, \bar{y})$ which belongs to $B^2_{\frac{1+r}{2}}$ and it is such that $0 < |\bar{x} - \bar{y}| \le \delta$.

By Ishii's Lemma [10], for all $\epsilon > 0$ there exist X_ϵ and Y_ϵ in S such that $(q^x, X_\epsilon) \in J^{2,+}u(\bar{x})$, $(q^y, -Y_\epsilon) \in J^{2,-}v(\bar{y})$ with

$$q^x = \gamma M|\bar{x} - \bar{y}|^{\gamma-2}(\bar{x} - \bar{y}) + 2L(\bar{x} - x_o)$$

$$q^y = \gamma M|\bar{x} - \bar{y}|^{\gamma-2}(\bar{x} - \bar{y}) - 2L(\bar{y} - x_o).$$

Hence

$$-F(q^x, X_\epsilon) + b(\bar{x})|q^x|^\beta \le g(\bar{x}), \quad -F(q^y, -Y_\epsilon) + b(\bar{y})|q^y|^\beta \ge f(\bar{y}) \tag{2.2}$$

Furthermore,

$$-\left(\frac{1}{\epsilon}+|B|\right)\begin{pmatrix} I & 0 \\ 0 & I \end{pmatrix} \leq \begin{pmatrix} X_\epsilon & 0 \\ 0 & Y_\epsilon \end{pmatrix} - 2L\begin{pmatrix} I & 0 \\ 0 & I \end{pmatrix} \tag{2.3}$$
$$\leq \begin{pmatrix} B+2\epsilon B^2 & -B-2\epsilon B^2 \\ -B-2\epsilon B^2 & B+2\epsilon B^2 \end{pmatrix}$$

with $B = M\gamma|\bar{x}-\bar{y}|^{\gamma-2}\left(I-(2-\gamma)\frac{(\bar{x}-\bar{y})\otimes(\bar{x}-\bar{y})}{|\bar{x}-\bar{y}|^2}\right)$. It is immediate to see that, as soon as δ is small enough, with our choice of M and L, there exists c_1 and c_2, such that

$$c_1\delta^{-\gamma}|\bar{x}-\bar{y}|^{\gamma-1} \leq |q^x|, |q^y| \leq c_2\delta^{-\gamma}|\bar{x}-\bar{y}|^{\gamma-1}.$$

We take $\epsilon = \frac{1}{8|B|+1} = \frac{1}{8M\gamma|\bar{x}-\bar{y}|^{\gamma-2}(N-\gamma)+1}$ and drop the index ϵ for X_ϵ and Y_ϵ. By standard considerations on the eigenvalues of B, in particular note that $B+\epsilon B^2$ has a negative eigenvalue less than $-\frac{3\gamma(1-\gamma)}{4}M|\bar{x}-\bar{y}|^{\gamma-2}$, and, using (2.3), the following holds

$$X+Y \leq (2L+\epsilon)I \text{ and } \inf\lambda_i(X+Y) \leq 4\inf\lambda_i(B+\epsilon B^2) \leq -3\gamma(1-\gamma)M|\bar{x}-\bar{y}|^{\gamma-2}.$$

Hence, as soon as δ is small enough, for some constant c depending only on a, A, γ and N,

$$\begin{aligned} F(q^x,X)-F(q^x,-Y) &\leq |q^x|^\alpha\left(A\textstyle\sum_{\lambda_i>0}\lambda_i(X+Y)+a\sum_{\lambda_i<0}\lambda_i(X+Y)\right) \\ &\leq |q^x|^\alpha\left(2AN(2L+1)-3aM\gamma(1-\gamma)\right)|\bar{x}-\bar{y}|^{\gamma-2} \\ &\leq -c\delta^{-\gamma(\alpha+1)}|\bar{x}-\bar{y}|^{\gamma(\alpha+1)-(\alpha+2)}. \end{aligned} \tag{2.4}$$

And the following standard inequalities hold:

$$\text{If } 0\leq\alpha<1,\ ||q^y|^\alpha-|q^x|^\alpha| \leq |q^x-q^y|^\alpha \leq cL^\alpha$$

$$\text{If } \alpha\geq 1,\ ||q^y|^\alpha-|q^x|^\alpha| \leq |\alpha||q^x-q^y|(|q^x|+|q^y|)^{\alpha-1} \leq c(\delta^{-\gamma}|\bar{x}-\bar{y}|^{(\gamma-1)})^{(\alpha-1)}$$

$$\begin{aligned} \text{If } -1<\alpha<0,\ ||q^y|^\alpha-|q^x|^\alpha| &\leq |\alpha||q^x-q^y|\inf(|q^x|,|q^y|)^{\alpha-1} \\ &\leq c(\delta^{-\gamma}|\bar{x}-\bar{y}|^{(\gamma-1)})^{(\alpha-1)} \end{aligned}$$

This implies that, for any $\alpha>-1$,

$$|F(q^x,X)-F(q^y,X)| \leq c\max(\delta^{-\gamma}|\bar{x}-\bar{y}|^{\gamma-2}, \delta^{-\alpha\gamma}|\bar{x}-\bar{y}|^{\alpha\gamma-\alpha-1}). \tag{2.5}$$

Next, we need to evaluate $|b(\bar{x})|q^x|^\beta - b(\bar{y})|q^y|^\beta|$. One easily has, for some constant which depends only on β and universal constants,

$$|b(\bar{x})|q^x|^\beta - b(\bar{y})|q^y|^\beta| \leq \text{Lip } b\, c\delta^{-\gamma\beta}|\bar{x} - \bar{y}|^{\gamma\beta-\beta+1} \qquad (2.6)$$
$$+ c|b|_\infty \max(1, \delta^{-\gamma(\beta-1)}|\bar{x} - \bar{y}|^{\gamma(\beta-1)-\beta+1}).$$

Observe that choosing δ small, the terms in (2.5) and in (2.6) are of lower order with respect to (2.4) i.e.

$$\max(\delta^{-\gamma}|\bar{x} - \bar{y}|^{\gamma-2}, \delta^{-\alpha\gamma}|\bar{x} - \bar{y}|^{\alpha\gamma-\alpha-1}) + \text{Lip } b\, \delta^{-\gamma\beta}|\bar{x} - \bar{y}|^{\gamma\beta-\beta+1}$$
$$+|b|_\infty \max(1, \delta^{-\gamma(\beta-1)}|\bar{x} - \bar{y}|^{\gamma(\beta-1)-\beta+1}) << \delta^{-\gamma(\alpha+1)}|\bar{x} - \bar{y}|^{\gamma(\alpha+1)-(\alpha+2)}$$

We then have, for some constant c,

$$\begin{aligned}
-g(\bar{x}) &\leq F(q^x, X) - b(\bar{x})|q^x|^\beta \\
&\leq F(q^y, -Y) - b(\bar{y})|q^y|^\beta - c\delta^{-\gamma(1+\alpha)}|\bar{x} - \bar{y}|^{\gamma(\alpha+1)-(2+\alpha)} \\
&\leq -f(\bar{y}) - c\delta^{-\gamma(1+\alpha)}|\bar{x} - \bar{y}|^{\gamma(\alpha+1)-(2+\alpha)}
\end{aligned}$$

This is a contradiction with the fact that f and g are bounded, as soon as δ is small enough. □

We are now in a position to prove the Lipschitz estimate. The proof proceeds analogously to the Hölder estimate above, but with a modification in the term depending on $|x - y|$ in the function $\phi(x, y)$.

Proof of Theorem 2.1 For $\alpha \leq 0$ we choose $\tau \in (0, \frac{1}{2})$ while for $\alpha > 0$ we fixe $\tau \in (0, \frac{\inf(1,\alpha)}{2}))$. For $s_o = (1 + \tau)^{\frac{1}{\tau}}$, we define

$$\omega(s) = s - \frac{s^{1+\tau}}{2(1+\tau)} \text{ for } s \in (0, s_o),\ \omega(s) = \omega(s_o) \text{for } s \geq s_o. \qquad (2.7)$$

Note that $\omega(s)$ is $\mathcal{C}^2$ on $s > 0$, $s < s_o$ and satisfies $\omega' > 0$, $\omega'' < 0$ on $]0, 1[$, and $s > \omega(s) \geq \frac{s}{2}$.

As before in the Hölder case, with $L = \frac{4(|u|_\infty+|v|_\infty)+1}{(1-r)^2}$ and $M = \frac{(|u|_\infty+|v|_\infty)+1}{\delta}$ we define

$$\phi(x, y) = u(x) - v(y) - \sup_{B_1}(u - v) - M\omega(|x - y|) - L(|x - x_o|^2 + |y - x_o|^2).$$

Classically, as before, we suppose that there exists a maximum point $(\bar{x}, \bar{y})$ such that $\phi(\bar{x}, \bar{y}) > 0$, then by the assumptions on M, and L, $\bar{x}$, $\bar{y}$ belong to $B(x_o, \frac{1-r}{2})$, hence they are interior points. This implies, using (2.1) in Lemma 2.2 with $\gamma < 1$ such that $\frac{\gamma}{2} > \frac{\tau}{\inf(1,\alpha)}$ for $\alpha > 0$ and $\frac{\gamma}{2} > \tau$ when $\alpha \le 0$ that, for some constant c_r,

$$L|\bar{x} - x_o|^2 \le c_r|\bar{x} - \bar{y}|^\gamma. \tag{2.8}$$

and then one has $|\bar{x} - x_o| \le \left(\frac{c_r}{L}\right)^{\frac{1}{2}} |\bar{x} - \bar{y}|^{\frac{\gamma}{2}}$.

Furthermore, for all $\epsilon > 0$, there exist X_ϵ and Y_ϵ in S such that $(q^x, X_\epsilon) \in J^{2,+}u(\bar{x})$, $(q^y, -Y_\epsilon) \in J^{2,-}v(\bar{y})$ with

$$q^x = M\omega'(|\bar{x} - \bar{y}|)\frac{\bar{x} - \bar{y}}{|\bar{x} - \bar{y}|} + L(\bar{x} - x_o),\ q^y = M\omega'(|\bar{x} - \bar{y}|)\frac{\bar{x} - \bar{y}}{|\bar{x} - \bar{y}|} - L(\bar{y} - x_o).$$

While X_ϵ and Y_ϵ satisfy (2.3) with

$$B = M\left(\frac{\omega'(|\bar{x} - \bar{y}|)}{|\bar{x} - \bar{y}|}(I - \frac{\bar{x} - \bar{y} \otimes \bar{x} - \bar{y}}{|\bar{x} - \bar{y}|^2}) + \omega''(|\bar{x} - \bar{y}|)\frac{\bar{x} - \bar{y} \otimes \bar{x} - \bar{y}}{|\bar{x} - \bar{y}|^2}\right).$$

Note that as soon as δ is small enough $\frac{M}{2} \le |q^x|, |q^y| \le \frac{3M}{2}$. Also, $M\omega''(|\bar{x} - \bar{y}|) = -M\frac{\tau}{2}|\bar{x} - \bar{y}|^{\tau-1}$ is an eigenvalue of B which is large negative as soon as δ is small enough.

Taking $\epsilon = \frac{1}{8|B|+1}$ and dropping the ϵ in the notations arguing as in the above proof, we get that there exists some constant c such that

$$F(q^x, X) - F(q^x, -Y) \le -c\delta^{-(1+\alpha)}|\bar{x} - \bar{y}|^{\tau-1}. \tag{2.9}$$

Note that by (2.3) using the explicit value of B one has

$$|X| + |Y| \le c(M|\bar{x} - \bar{y}|^{-1} + 4(L + 1)N) \le c\delta^{-1}|\bar{x} - \bar{y}|^{-1} \tag{2.10}$$

as soon δ is small enough.

On the other hand, using the mean value theorem, that for some universal constant

$$\text{if } \alpha \ge 1, \text{ or } \alpha < 0, \quad ||q^x|^\alpha - |q^y|^\alpha| \le c\delta^{-\alpha+1}|\bar{x} - \bar{y}|^{\frac{\gamma}{2}},$$

while

$$\text{if } 0 < \alpha \le 1,\ ||q^x|^\alpha - |q^y|^\alpha| \le c|\bar{x} - \bar{y}|^{\frac{\gamma\alpha}{2}}.$$

In each of these cases one easily obtains, using (2.10)

$$||q^x|^\alpha - |q^y|^\alpha|\,|X| \le c\delta^{-1-\alpha}|\bar{x} - \bar{y}|^{-1+\frac{\gamma}{2}} \text{ if } \alpha \le 0, \text{ or } \alpha > 1,$$

$$||q^x|^\alpha - |q^y|^\alpha|\,|X| \le c\delta^{-1}|\bar{x} - \bar{y}|^{-1+\frac{\gamma\alpha}{2}} \text{ if } \alpha \in]0, 1[$$

We have obtained, with the above choice of γ that, as soon as δ is small enough, $|F(q^x, X) - F(q^y, X)|$ is small with respect to $c\delta^{-1-\alpha}|\bar{x} - \bar{y}|^{\tau-1}$. We now treat the terms involving b with similar considerations. If $\beta \geq 1$,

$$|b(\bar{x})||q^x|^\beta - |q^y|^\beta| \leq |b|_\infty |q^x - q^y| M^{\beta-1} \leq c\delta^{-\beta+1}|\bar{x} - \bar{y}|^{\frac{\gamma}{2}},$$

while if $\beta \leq 1$,

$$|b(\bar{x})||q^x|^\beta - |q^y|^\beta| \leq |b|_\infty c|\bar{x} - \bar{y}|^{\frac{\gamma\beta}{2}}.$$

Observe also that

$$|b(\bar{x}) - b(\bar{y})||q^x|^\beta \leq c \text{ lip } b|\bar{x} - \bar{y}|\delta^{-\beta}.$$

Finally,

$$\begin{aligned} |b(\bar{x}) - b(\bar{y})||q^x|^\beta &\leq c|\bar{x} - \bar{y}|^{\tau-1}\delta^{-\beta+2-\tau} \\ &= c|\bar{x} - \bar{y}|^{\tau-1}\delta^{-1-\alpha}\delta^{-\beta+3-\tau+\alpha} \end{aligned}$$

Since $3 + \alpha - \beta - \tau > 0$, this term is also small with respect to $\delta^{-1-\alpha}|\bar{x} - \bar{y}|^{\tau-1}$.

Putting all the estimates together we get

$$\begin{aligned} -g(\bar{x}) &\leq F(q^x, X) - b(\bar{x})|q^x|^\beta \\ &\leq F(q^y, -Y) - b(\bar{y})|q^y|^\beta - c\delta^{-1-\alpha}|\bar{x} - \bar{y}|^{-1+\tau} \\ &\leq -f(\bar{y}) - c\delta^{-1-\alpha}|\bar{x} - \bar{y}|^{-1+\tau}. \end{aligned}$$

This is clearly a contradiction as soon as δ is small enough since f and g are bounded. □

3 Existence and Uniqueness Results for Homogenous Dirichlet Conditions

The existence of solutions for (1.1) with the boundary condition $\varphi = 0$ will be classically obtained as a consequence of the existence of sub- and supersolutions, of some comparison result, and the Perron's method.

We start with a result on the existence of sub and supersolutions of equations involving the Pucci's operators

$$\mathcal{M}^+(M) = A \sum_{\lambda_i > 0} \lambda_i + a \sum_{\lambda_i < 0} \lambda_i$$

$$\mathcal{M}^-(M) = a \sum_{\lambda_i > 0} \lambda_i + A \sum_{\lambda_i < 0} \lambda_i$$

defined for any $M \in \mathcal{S}_N$, $M \sim \text{diag}(\lambda_1, \ldots, \lambda_N)$.

Proposition 3.1 *Let $\lambda > 0$ and $b > 0$ be given and $\beta \in (0, \alpha+2]$. For all $M > 0$, there exists a continuous function $\varphi \geq 0$ supersolution of*

$$\begin{cases} -|\nabla\varphi|^\alpha \mathcal{M}^+(D^2\varphi) - b|\nabla\varphi|^\beta + \lambda\varphi^{1+\alpha} \geq M & \text{in } \Omega \\ \varphi = 0 & \text{on } \partial\Omega \end{cases} \tag{3.1}$$

and symmetrically, for all $M > 0$, there exists a continuous function $\varphi \leq 0$ subsolution

$$\begin{cases} -|\nabla\varphi|^\alpha \mathcal{M}^-(D^2\varphi) + b|\nabla\varphi|^\beta + \lambda|\varphi|^\alpha\varphi \leq -M & \text{in } \Omega \\ \varphi = 0 & \text{on } \partial\Omega. \end{cases}$$

Proof of Proposition 3.1 We shall construct explicitly a positive supersolution, the subsolution is just the negative of it.

Since Ω is a smooth domain, $d(x)$, the function distance from the boundary, is $\mathcal{C}^2$ in the neighbourhood $\{d(x) < \delta_0\}$ for some $\delta_0 > 0$, hence we will suppose to extend it to a $\mathcal{C}^2$ function in Ω, which is greater than δ_0 outside of a δ_o neighbourhood of the boundary.

Let us fix a positive constant κ satisfying $\lambda \log(1+\kappa)^{1+\alpha} > M$ and, for $C > \frac{2\kappa}{\delta_0}$, consider the function $\varphi(x) = \log(1 + Cd(x))$. Observe that $|\nabla d| = 1$ in $\{d(x) < \delta_0\}$.

Suppose that φ is showed to be a supersolution in the set $\{Cd(x) < 2\kappa\}$. Then, the function

$$\phi(x) := \begin{cases} \log(1+Cd(x)) \text{ in } \{Cd(x) < \kappa\} \\ \quad \log(1+\kappa) \quad \text{ in } \{Cd(x) \geq \kappa\} \end{cases}$$

is the required supersolution, being the minimum of two supersolutions.

We now prove that in $\{Cd(x) < 2\kappa\}$, φ is a supersolution. Easily, one gets

$$\nabla\varphi = \frac{C\nabla d}{(1+Cd)}, \quad D^2\varphi = \frac{CD^2 d}{1+Cd} - \frac{C^2\nabla d \otimes \nabla d}{(1+Cd)^2}.$$

We suppose first that $\beta < \alpha + 2$. Let C_1 be such that $D^2 d \leq C_1 \mathrm{Id}$. For C satisfying furthermore

$$\frac{aC}{2(1+2\kappa)} \geq ANC_1, \quad \frac{a}{2}\left(\frac{C}{1+2\kappa}\right)^{2+\alpha-\beta} > 2b, \quad \frac{aC^{2+\alpha}}{4(1+2\kappa)^{2+\alpha}} > M,$$

one gets that

$$|\nabla\varphi|^\alpha \mathcal{M}^+(D^2\varphi) + b|\nabla\varphi|^\beta \leq -M,$$

from which the conclusion follows.

We now suppose that $\beta = \alpha + 2$. We begin to observe that the calculations above can be extended to $\beta = \alpha + 2$ as soon as $b < \frac{a}{4}$. So suppose that $\epsilon = \frac{a}{4b}$, that φ is some barrier for the equation

$$-|\nabla\varphi|^\alpha \mathcal{M}^+(D^2\varphi) + \frac{a}{4}|\nabla\varphi|^\beta \geq M\epsilon^{1+\alpha}$$

Then $\psi = \frac{\varphi}{\epsilon}$ satisfies

$$-|\nabla\psi|^\alpha \mathcal{M}^+(D^2\psi) + b|\nabla\psi|^\beta \geq M.$$

□

Recall that, for $\alpha < 0$, the equation we are considering are singular. Hence, we need to treat differently the solutions when the test function has a vanishing gradient. That is the object of the following lemma, which is proved on the model of [4] where we treat the case $\beta = \alpha + 1$. We give the detail of the proof for the convenience of the reader.

Lemma 3.2 *Let γ and b be continuous functions. Assume that v is a supersolution of*

$$-|\nabla v|^\alpha \mathcal{M}^-(D^2 v) + b(x)|\nabla v|^\beta + \gamma(v) \geq f \quad in \quad \Omega$$

such that, for some $C > 0$ and $q \geq \frac{\alpha+2}{\alpha+1}$, $\bar{x} \in \Omega$ is a strict local minimum of $v(x) + C|x - \bar{x}|^q$. Then

$$f(\bar{x}) \leq \gamma(v(\bar{x})).$$

Proof Without loss of generality we can suppose that $\bar{x} = 0$.

If v is locally constant around 0 the conclusion is the definition of viscosity supersolutions. If v is not locally constant, since $q > 1$, for any $\delta > 0$ sufficiently small, there exist $(z_\delta, t_\delta) \in B_\delta^2$ such that

$$v(t_\delta) > v(z_\delta) + C|z_\delta - t_\delta|^q. \tag{3.2}$$

The idea of the proof is to construct a test function near 0 whose gradient is not zero.

Since 0 is a strict minimum point of $v(x) + C|x|^q$, there exist $R > 0$ and, for any $0 < \eta < R$, $\epsilon(\eta) > 0$ satisfying

$$\min_{\eta \leq |x| \leq R} (v(x) + C|x|^q) \geq v(0) + \epsilon(\eta).$$

Let $\eta > 0$ be fixed. We choose $\delta = \delta(\eta) > 0$ such that $C\delta^q \leq \frac{\epsilon(\eta)}{4}$. Note that $\delta \to 0$ as $\eta \to 0$. With this choice of δ, we have

$$\min_{|x|\leq R} (v(x) + C|x - t_\delta|^q) \leq v(0) + C|t_\delta|^q \leq v(0) + \frac{\epsilon(\eta)}{4}.$$

On the other hand, restricting further δ such that $q\delta C(R+1)^{q-1} < \frac{\epsilon(\eta)}{4}$, we get

$$\min_{\eta\leq|x|\leq R} (v(x)+C|x-t_\delta|^q) \geq \min_{\eta\leq|x|\leq R} (v(x)+C|x|^q)-qC|t_\delta|(R+|t_\delta|)^{q-1} \geq v(0)+\frac{3\epsilon(\eta)}{4}.$$

This implies that $\min_{|x|\leq R}(v(x) + C|x - t_\delta|^q)$ is achieved in B_η. Furthermore, it cannot be achieved in t_δ by (3.2). Hence, there exists $y_\delta \in B_\eta$, $y_\delta \neq t_\delta$, such that

$$v(y_\delta) + C|y_\delta - t_\delta|^q = \min_{|x|\leq R} (v(x) + C|x - t_\delta|^q).$$

Let us now consider the test function

$$\varphi(z) = v(y_\delta) + C|y_\delta - t_\delta|^q - C|z - t_\delta|^q\,,$$

that touches v from below at y_δ. Since v is a supersolution, we obtain

$$NA(q-1)q^{\alpha+2}C^{\alpha+1}|y_\delta-t_\delta|^{q(\alpha+1)-(\alpha+2)}+C^\beta|b(y_\delta)||y_\delta-t_\delta|^{(q-1)\beta}+\gamma(v(y_\delta)) \geq f(y_\delta). \tag{3.3}$$

On the other hand, we observe that

$$v(y_\delta) \leq v(y_\delta) + C|y_\delta - t_\delta|^q \leq v(0) + C|t_\delta|^q \leq v(0) + C\delta^q.$$

By the lower semicontinuity of v, this implies that

$$v(y_\delta) \to v(0) \quad \text{as } \eta \to 0\,.$$

Thus, letting $\eta \to 0$ in (3.3), by the continuity of γ and f it follows that

$$\gamma(v(0)) \geq f(0)\,.$$

□

We are now in a position to prove the following comparison principle that will be essential to the proof of the existence of the solution.

Theorem 3.3 *Suppose that F and β are as in Theorem 2.1, that Ω is a bounded domain and that γ is a non decreasing function. Assume that, in Ω, u is an upper semicontinuous bounded from above viscosity subsolution of*

$$-F(\nabla u, D^2 u) + b(x)|\nabla u|^\beta + \gamma(u) \le g$$

and that v is a lower semicontinuous bounded from below viscosity supersolution of

$$-F(\nabla v, D^2 v) + b(x)|\nabla v|^\beta + \gamma(v) \ge f\,,$$

with f and g bounded and b Lipschitz continuous.

Suppose furthermore that

- *either $g \le f$ and γ is increasing,*
- *or $g < f$.*

Then

$$u \le v \ on\ \partial\Omega \implies u \le v \ in\ \Omega.$$

Proof **The case** $\alpha \ge 0$. We use classically the doubling of variables. So we define for all $j \in N$, $\psi_j(x,y) = u(x) - v(y) - \frac{j}{2}|x-y|^2$. Suppose by contradiction that $u > v$ somewhere, then the supremum of $u - v$ is strictly positive and achieved inside Ω.

Then one also has $\sup \psi_j > 0$ for j large enough and it is achieved on $(x_j, y_j) \in \Omega^2$. Using Ishii's lemma, [10], there exist X_j and Y_j in S such that $(j(x_j - y_j), X_j) \in \overline{J}^{2,+} u(x_j)$, $(j(x_j - y_j), -Y_j) \in \overline{J}^{2,-} v(y_j)$. It is clear that Theorem 2.1 in section 2 can be extended to the case where Ω replaces $B(0,1)$ and $\Omega' \subset\subset \Omega$ replaces $B(0,r)$. Since (x_j, y_j) converges to $(\bar{x}, \bar{x})$, both of them belong, for j large enough, to some Ω'. We use Theorem 2.1 to obtain that $j|x_j - y_j|$ is bounded.

Indeed $u(x_j) - v(y_j) - \frac{j}{2}|x_j - y_j|^2 \ge \sup(u - v)$, hence

$$\frac{j}{2}|x_j - y_j|^2 \le u(x_j) - v(y_j) - \sup(u-v) \le \sup(u-v) + c|x_j - y_j| - \sup(u-v).$$

We obtain

$$\begin{aligned} g(x_j) - \gamma(u(x_j)) &\ge -F(j(x_j - y_j), X_j) + b(x_j)|j(x_j - y_j)|^\beta \\ &\ge -F(j(x_j - y_j), -Y_j) + b(y_j)|j(x_j - y_j)|^\beta - o(x_j - y_j)(j|x_j - y_j|)^\beta \\ &\ge f(y_j) + o(j|x_j - y_j|^2)|(j|x_j - y_j|)^\alpha - \gamma(v(y_j)). \end{aligned} \tag{3.4}$$

By passing to the limit on j one gets that up to a subsequence $(x_j, y_j) \to (\bar{x}, \bar{x})$ and (3.4) becomes

$$g(\bar{x}) - \gamma(u(\bar{x})) \geq f(\bar{x}) - \gamma(v(\bar{x}))$$

and we have obtained a contradiction in each of the cases “$f > g$ and γ is non decreasing”, or “$f \geq g$ and γ is increasing”.

The case $\alpha < 0$. We recall that in this case one must use a convenient definition of viscosity solutions, see [4].

We suppose by contradiction that $\sup(u - v) > 0$ then it is achieved inside Ω, and taking $q > \frac{\alpha+2}{\alpha+1} \geq 2$ the function

$$\psi_j(x, y) = u(x) - v(y) - \frac{j}{q}|x - y|^q$$

has also a local maximum on (x_j, y_j). Then, there are $X_j, Y_j \in \mathcal{S}^N$ such that

$$(j|x_j - y_j|^{q-2}(x_j - y_j), X_j) \in J^{2,+}u(x_j)$$

$$(j|x_j - y_j|^{q-2}(x_j - y_j), -Y_j) \in J^{2,-}v(y_j)$$

and

$$-4jk_j \begin{pmatrix} I & 0 \\ 0 & I \end{pmatrix} \leq \begin{pmatrix} X_j & 0 \\ 0 & Y_j \end{pmatrix} \leq 3jk_j \begin{pmatrix} I & -I \\ -I & I \end{pmatrix}$$

where

$$k_j = 2^{q-3}q(q-1)|x_j - y_j|^{q-2}.$$

In order to use the fact that u and v are respectively sub and super solutions we need to prove the following

Claim For j large enough, we can choose $x_j \neq y_j$. Note that :

(i) from the boundedness of u and v one deduces that $|x_j - y_j| \to 0$ as $j \to \infty$. Thus up to subsequence $(x_j, y_j) \to (\bar{x}, \bar{x})$.
(ii) One has $\liminf \psi_j(x_j, y_j) \geq \sup(u - v)$;
(iii) $\limsup \psi_j(x_j, y_j) \leq \limsup u(x_j) - v(y_j) = u(\bar{x}) - v(\bar{x})$
(iv) Thus $j|x_j - y_j|^q \to 0$ as $j \to +\infty$ and $\bar{x}$ is a maximum point for $u - v$.

Furthermore by the Lipschitz estimates in Theorem 2.1 $j|x_j - y_j|^q \leq u(x_j) - v(y_j) - \sup(u - v) \leq c|x_j - y_j|$ since (x_j, y_j) belong to a compact set inside Ω. This implies that $j|x_j - y_j|^{q-1}$ is bounded.

Suppose by contradiction that $x_j = y_j$. Then one would have

$$\psi_j(x_j, x_j) = u(x_j) - v(x_j) \geq u(x_j) - v(y) - \frac{j}{q}|x_j - y|^q;$$

$$\psi_j(x_j, x_j) = u(x_j) - v(x_j) \geq u(x) - v(x_j) - \frac{j}{q}|x - x_j|^q;$$

and then x_j would be a local minimum for $\Phi := v(y) + \frac{j}{q}|x_j - y|^q$, and similarly a local maximum for $\Psi := u(x) - \frac{j}{q}|x_j - x|^q$.

We first exclude that these extrema are both strict. Indeed in that case, by Lemma 3.2

$$\gamma(v(x_j)) \geq f(x_j), \text{ and } \gamma(u(x_j)) \leq g(x_j).$$

This is a contradiction with the assumptions on γ and f and g, once we pass to the limit when j goes to ∞ since we get

$$\gamma(v(\bar{x})) \geq f(\bar{x}) \geq g(\bar{x}) \geq \gamma(u(\bar{x})).$$

Hence x_j cannot be both a strict minimum for Φ and a strict maximum for Ψ. Suppose without loss of generality that x_j is not a strict minimum for Φ, then there exist $\delta > 0$ and $R > \delta$ such that $B(x_j, R) \subset \Omega$ and

$$v(x_j) = \min_{\delta \leq |x - x_j| \leq R} \{v(x) + \frac{j}{q}|x - x_j|^q\}.$$

Then if y_j such that $\delta \leq |y - x_j| \leq R$, is a point on which the minimum above is achieved, one has

$$v(x_j) = v(y_j) + \frac{j}{q}|x_j - y_j|^q,$$

and (x_j, y_j) is still a maximum point for ψ_j.

Recalling that $j|x_j - y_j|^{q-1}$ is bounded, we can now conclude, using the fact that u and v are respectively sub and super solution to obtain:

$$\begin{aligned}
g(x_j) &\geq F(j|x_j - y_j|^{q-2}(x_j - y_j), X_j) + b(x_j)|j(x_j - y_j|^{q-1}|^\beta + \gamma(u(x_j))\\
&\geq F(j|x_j - y_j|^{q-2}(x_j - y_j), -Y_j) + b(y_j)|j(x_j - y_j|^{q-1}|^\beta\\
&- \text{lip}b|x_j - y_j||j|x_j - y_j|^{q-1}|^\beta + \gamma(v(y_j)) + \big(-\gamma(v(y_j)) + \gamma(u(x_j))\big)\\
&\geq f(y_j) - \gamma(v(y_j)) + \gamma(u(x_j)) + O(x_j - y_j)
\end{aligned}$$

Passing to the limit the following inequality holds

$$g(\bar{x}) \geq f(\bar{x}) - \gamma(v(\bar{x})) + \gamma(u(\bar{x})).$$

This is a contradiction in each of the cases "$f > g$ and γ is non decreasing" or "$f \geq g$ and γ is increasing". □

We derive from the construction of barriers and the comparison theorem the following existence result for Dirichlet homogeneous boundary conditions

Theorem 3.4 *Let Ω be a bounded C^2 domain, $\lambda > 0$, $f \in C(\overline{\Omega})$ and $b \in W^{1,\infty}(\Omega)$. Under the assumptions (H1) and (H2), there exists a unique u which satisfies*

$$\begin{cases} -F(\nabla u, D^2 u) + b(x)|\nabla u|^\beta + \lambda |u|^\alpha u = f & \text{in } \Omega \\ u = 0 & \text{on } \partial\Omega. \end{cases} \tag{3.5}$$

Furthermore, u is Lipschitz continuous up to the boundary.

Proof The existence result is an easy consequence of the comparison principle and Perron's method adapted to our framework, as it is done in [4]. In order to prove that u is Lipschitz continuous up to the boundary, observe that, by construction, there exist $C > c > 0$ such that $cd \leq u \leq Cd$, where d is the distance function from $\partial\Omega$.

Then, consider the function

$$\phi(x, y) = u(x) - u(y) - M\omega(|x - y|),$$

where ω has been defined in (2.7), and suppose that $M > 2C$. Arguing as in the proof of Lemma 2.2 and Theorem 2.1, we assume by contradiction that ϕ is positive somewhere, say on $(\bar{x}, \bar{y})$. Then, neither $\bar{x}$ nor $\bar{y}$ belong to the boundary, due to the inequality, for $y \in \partial\Omega$,

$$u(x) \leq Cd(x) \leq \frac{M}{2} d(x) \leq M\omega(|x - y|).$$

A similar reasoning proves that $\bar{x}$ cannot belong to the boundary. The rest of the proof runs as in Theorem (2.1), with even simpler computations, since we do not need the additional term $|x - x_o|^2 + |y - y_o|^2$ in the auxiliary function ϕ. □

We end this section with some Strong maximum principle and Hopf principle.

Theorem 3.5 *Suppose that u is a non negative solution in Ω of*

$$-|\nabla u|^\alpha \mathcal{M}^-(D^2 u) + b(x)|\nabla u|^\beta \geq 0.$$

Then either $u > 0$ inside Ω, or $u \equiv 0$. Moreover if $\bar{x}$ is in $\partial\Omega$ so that an interior sphere condition holds and $u(\bar{x}) = 0$, then "$\partial_n u(\bar{x})$" < 0.

Proof Suppose that there exists an interior point x_1 such that $u(x_1) = 0$. We can choose x_1 such that there exists x_o satisfying $|x_o - x_1| = R$, the ball $B(x_o, 2R) \subset \Omega$ and $u > 0$ in $B(x_o, R)$. Since u is lower semi-continuous, let $\delta < \inf(1, \inf_{B(x_o, \frac{R}{2})} u)$, and define in the crown $B(x_o, 2R) \setminus B(x_o, \frac{R}{2})$.

$v(r) = \delta(e^{-cr} - e^{-cR})$ where $c > \frac{2(N-1)A}{Ra}$, and $\frac{a}{2}c^{2+\alpha-\beta} > |b|_\infty$, if $\beta < \alpha+2$, (which implies that $\delta^{1+\alpha}\frac{a}{2}c^{2+\alpha-\beta} > \delta^\beta |b|_\infty$), if $\beta = \alpha + 2$, note that we can take δ so that $\delta^{1+\alpha-\beta}\frac{a}{2} > |b|_\infty$. Then one has $v \le u$ on the boundary of the crown, and

$$-|v'|^\alpha(av'' + A\frac{N-1}{r}v') + |b|_\infty |v'|^\beta < 0.$$

Using the comparison principle one gets that $u \ge v$. Then v is a C^2 function which achieves u on below on x_1 and this is a contradiction with the fact that u is a super-solution. The last statement is proved. Suppose that $\bar{x}$ is on the boundary and consider an interior sphere $B(x_o, R) \subset \Omega$ with $|x_o - x_1| = R$, and the function v as above. We still have by construction that $v \ge u$. Then taking for $h > 0$ small, $x_h = hx_o + (1-h)x_1$ one has $|x_h - xo| = (1-h)R$ and

$$\frac{u(x_h) - u(x_1)}{x_h - x_1} \ge \frac{v(x_h) - v(x_1)}{x_h - x_1} \ge cRe^{-cR} > 0$$

which implies the desired Hopf's principle. □

4 Non Homogeneous Boundary Conditions

In order to obtain solutions for the non homogeneous boundary condition, we need the construction of barriers, and a Lipschitz estimate near the boundary, as follows

Lemma 4.1 *Let B' be the unit ball in $\mathbb{R}^{N-1}$, and let $\varphi \in W^{1,\infty}(B')$. Let $\eta \in C^2(B')$ such that $\eta(0) = 0$ and $\nabla\eta(0) = 0$. Let d be the distance to the hypersurface $\{x_N = \eta(x')\}$.*

Then, for all $r < 1$ and for all $\gamma < 1$, there exists δ_o depending on $\|f\|_\infty$, a, A, $\|b\|_\infty$, Ω, r and Lipφ, such that for all $\delta < \delta_o$, if u be a USC bounded by above sub-solution of

$$\begin{cases} -F(\nabla u, D^2u) + b|\nabla u|^\beta \le f & \text{in } B \cap \{x_N > \eta(x')\} \\ u \le \varphi & \text{on } B \cap \{x_N = \eta(x')\} \end{cases} \tag{4.1}$$

then it satisfies

$$u(x', x_N) \le \varphi(x') + (\sup u)\log(1 + \frac{2}{\delta}d) \ \text{ in } B_r(0) \cap \{x_N = \eta(x')\}.$$

We have a symmetric result for supersolutions bounded by below.

Proof **First case.** Without loss of generality we will suppose that $u \leq 1$. We suppose that $\beta < \alpha + 2$. We also write the details of the proof for $\varphi = 0$. The changes to bring in the case $\varphi \neq 0$ will be given at the end of the proof, the detailed calculation being left to the reader.

It is sufficient to consider the set where $d(x) < \delta$ since the assumption $u \leq 1$ implies the result elsewhere.

We begin by choosing $\delta < \delta_1$, such that on $d(x) < \delta_1$ the distance is $\mathcal{C}^2$ and satisfies $|D^2 d| \leq C_1$. We shall also later choose δ smaller depending of $(a, A, \|f\|_\infty, \|b\|_\infty, N)$.

In order to use the comparison principle we want to construct w a super solution of

$$-|\nabla w|^\alpha \mathcal{M}^+(D^2 w) - b|\nabla w|^\beta \geq \|f\|_\infty, \text{ in } B \cap \{x_N > \eta(x'),\ d(x) < \delta\} \tag{4.2}$$

such that $w \geq u$ on $\partial(B \cap \{x_N > \eta(x'),\ d(x) < \delta\})$.

We then suppose that $\delta < \frac{1-r}{9}$, define $C = \frac{2}{\delta}$ and

$$w(x) = \begin{cases} \log(1 + Cd) & \text{for } |y| < r \\ \log(1 + Cd) + \frac{1}{(1-r)^3}(|x| - r)^3 & \text{for } 1 \geq |x| \geq r. \end{cases}$$

In order to prove the boundary condition, let us observe that,

on $\{d(x) = \delta\}$, $w \geq \log 3 \geq 1 \geq u$,

on $\{|x| = 1\} \cap \{d(x) < \delta\}$, $w \geq \frac{1}{(1-r)^3}(1 - r)^3 \geq u$ and finally

on $B \cap \{x_N = \eta(x')\}$, $w \geq 0 = u$.

We need to check that w is a super solution. For that aim, we compute

$$\nabla w = \begin{cases} \frac{C}{1+Cd}\nabla d & \text{when } |x| < r \\ \frac{C}{1+Cd}\nabla d + \frac{x}{|x|}\frac{3}{(1-r)^3}(|x| - r)^2 & \text{if } |x| > r. \end{cases}$$

Note that $|\nabla w| \geq \frac{C}{2(1+Cd)} \geq \frac{1}{3\delta}$ since $\delta \leq \frac{1-r}{9}$ and that $|\nabla w| \leq \frac{3C}{2(1+Cd)}$. By construction w is $\mathcal{C}^2$ and

$$D^2 w = \frac{CD^2 d}{1 + Cd} - \frac{C^2 \nabla d \otimes \nabla d}{(1 + Cd)^2} + H(x)$$

where $|H(x)| \leq \frac{6}{(1-r)^2} + \frac{3N}{r(1-r)}$. In particular

$$-\mathcal{M}^+(D^2w) \geq a\frac{C^2}{(1+Cd)^2} - A\frac{C|D^2d|_\infty}{1+Cd} - A\left(\frac{6}{(1-r)^2} + \frac{3N}{r(1-r)}\right)$$
$$\geq a\frac{C^2}{(1+Cd)^2} - AC|D^2d|_\infty - A\left(\frac{6}{(1-r)^2} + \frac{3N}{r(1-r)}\right)$$
$$\geq a\frac{C^2}{4(1+Cd)^2}$$

as soon as δ is small enough, depending only on r, A, a.

Hence (we do the computations for $\alpha > 0$ and leave to the reader the case $\alpha < 0$, which can easily be deduced since $\frac{C}{2(1+Cd)} \leq |\nabla w| \leq \frac{3C}{2(1+Cd)}$).

$$-|\nabla w|^\alpha \mathcal{M}^+(D^2w) - b|\nabla w|^\beta \geq a\frac{C^{2+\alpha}}{2^{2\alpha+2}(1+Cd)^{2+\alpha}} - b\left(\frac{C}{2(1+Cd)}\right)^\beta$$
$$\geq a\frac{C^{2+\alpha}}{2^{4+\alpha}(1+Cd)^{2+\alpha}} \geq |f|_\infty$$

as soon as δ is small enough in order that

$$b < a2^{\beta-2\alpha-4}\left(\frac{C}{3}\right)^{\alpha+2-\beta} \tag{4.3}$$

and so that $a\frac{C^{2+\alpha}}{2^{4+\alpha}(1+Cd)^{2+\alpha}} > |f|_\infty$.

By the comparison principle, Theorem 3.3, $u \leq w$ in $B \cap \{x_N > \eta(x')\} \cap \{d(x) < \delta\}$.

Furthermore the desired lower bound on u is easily deduced by considering $-w$ in place of w in the previous computations and restricting to $B_r \cap \{x_N > \eta(x')\}$. This ends the case $\varphi \equiv 0$.

To treat the case where φ is non zero, let ψ be a solution of

$$\mathcal{M}^+(\psi) = 0, \ \psi = \varphi \text{ on } \partial\Omega.$$

It is known that $\psi \in \mathcal{C}^2(\Omega)$, ψ is Lipschitz, $|\nabla\psi| \leq K|\nabla\varphi|_\infty$. Then in the previous calculation it is sufficient to define as soon as δ is small enough in order that $\frac{1}{3\delta} > 2K|\nabla\varphi|_\infty$,

$$w(y) = \begin{cases} \log(1+Cd) + \psi(x) & \text{for } |x| < r \\ \log(1+Cd) + \frac{1}{(1-r)^3}(|x|-r)^3 + \psi(x) & \text{for } |x| \geq r. \end{cases}$$

And we choose $C = \frac{2}{\delta}$ large enough in order that $\frac{C}{2} > 2(|\nabla\psi|_\infty + \frac{1}{1-r})$ and also large enough in order that $-|\nabla w|^\alpha \mathcal{M}^+(D^2 w) - b|\nabla w|^\beta > |f|_\infty$ which can be easily done by the same argument as before, using

$$-|\nabla w|^\alpha \mathcal{M}^+(D^2 w) - b|\nabla w|^\beta \geq -2^{-|\alpha|}|\nabla(\log(1+Cd)|^\alpha \mathcal{M}^+(\log(1+Cd)) - b2^{-\beta}|\nabla(\log(1+Cd)|^\beta.$$

Second case Suppose now that $\beta = \alpha + 2$. It is sufficient to construct a convenient super-solution w.

Recall that the previous proof used (4.3), which reduces when $\alpha + 2 = \beta$, to the condition $b < a2^{\beta-2\alpha-4}$. Let then $\epsilon = \frac{a2^{\beta-2\alpha-4}}{b}$, and let w which equals $\frac{\varphi}{\epsilon}$ on the boundary, and satisfies

$$-|\nabla w|^\alpha \mathcal{M}^+(D^2 w) - \epsilon b|\nabla w|^\beta \geq \epsilon^{1+\alpha}|f|_\infty$$

Then we obtain by the comparison principle that $\frac{u}{\epsilon} \leq w$ and then $u \leq \epsilon w$. □

This enables us to prove the following Lipschitz estimate up to the boundary.

Proposition 4.2 *Let φ be a Lipschitz continuous function on the part $T = B(0,1) \cap \{x_N = \eta(x')\}$, that $f \in \mathcal{C}(\overline{\Omega})$ and b is Lipschitz continuous. Suppose that u and v are respectively sub and supersolution which satisfy (4.1) in $B(0,1) \cap \{x_N > \eta(x')\}$, with $u = \psi = v$ on T.*

Then, for all $r < 1$, there exists $c_r > 0$ depending on r, a, A, Lip(b) and N such that, for all x and y in $B_r \cap \{x_N \geq \eta(x')\}$),

$$u(x) - v(y) \leq \sup_{B_1 \cap \{y_N \geq \eta(y')\})} (u - v) + c_r \left(|f|_\infty^{\frac{1}{1+\alpha}} + |u|_\infty + |\psi|_{W^{1,\infty}(T)} \right) |x - y|$$

In particular, when u is a solution we have a Lipschitz local estimate up to the boundary.

Proof The proof is similar to the proof of Theorem 2.1. Hence we must first prove that, for $\gamma < 1$,

$$u(x) - v(y) \leq \sup_{B_1 \cap \{y_N \geq \eta(y')\})} (u - v) + c_r \left(|f|_\infty^{\frac{1}{1+\alpha}} + |u|_\infty + |\psi|_{W^{1,\infty}(T)} \right) |x - y|^\gamma.$$

We do not give the details, but this is done introducing the function

$$\phi(x,y) = u(x) - v(y) - \sup_{B_1 \cap \{x_N > \eta(x')\}} (u - v) - M\omega(|x-y|) - L|x - x_o|^2 - L|y - x_o|^2,$$

where ω is as in the proof of Lemma 2.2 and x and y are in $B(0,1) \cap \{x_N > \eta(x')\}$.

We want to prove that $\phi \leq 0$, which will classically imply the result. We argue by contradiction and need to prove first that if $(\bar{x}, \bar{y})$ is a maximum point for ϕ, then neither $\bar{x}$ nor $\bar{y}$ belongs to $x_N = \eta(x')$. We notice that, if $\bar{y} \in \{x_N = \eta(x')\}$, then

$$\begin{aligned} u(\bar{x}) - \psi(\bar{y}) &\geq \sup_{B_1 \cap \{x_N > \eta(x')\}} (u - v) + M\omega(|x - y|) \\ &\geq u(\bar{y}) - v(\bar{y}) + M\omega(|\bar{x} - \bar{y}|) \end{aligned}$$

which contradicts Lemma 4.1 for M large enough. The case when $\bar{x} \in \{x_N = \eta(x')\}$ is excluded in the same manner. The rest of the proof follows the lines of Lemma 2.2 and Theorem 2.1. □

Remark 4.3 When the boundary condition is prescribed on the whole boundary, we have a simpler proof, as we noticed in the homogeneous case, taking

$$\phi(x, y) = u(x) - v(y) - \sup_{B_1 \cap \{x_N > \eta(x')\}} (u - v) - M\omega(|x - y|)\,.$$

In this case the localizing terms are not needed, since it is immediate to exclude that the maximum point $(\bar{x}, \bar{y})$ is on the boundary.

Acknowledgements Part of this work has been done while the first and third authors were visiting the University of Cergy-Pontoise and the second one was visiting Sapienza University of Rome, supported by INDAM-GNAMPA and Laboratoire AGM Research Center in Mathematics of the University of Cergy-Pontoise.

References

1. A. Attouchi, E. Ruosteenoja, Remarks on regularity for p-Laplacian type equations in non-divergence form. J. Differ. Equ. **265**, 1922–1961 (2018)
2. G. Barles, F. Da Lio, On the generalized Dirichlet problem for viscous Hamilton–Jacobi equations. J. Math. Pures Appl. (9) **83**(1), 53–75 (2004)
3. G. Barles, E. Chasseigne, C. Imbert, Hölder continuity of solutions of second-order non-linear elliptic integro-differential equations. J. Eur. Math. Soc. **13**, 1–26 (2011)
4. I. Birindelli, F. Demengel, First eigenvalue and maximum principle for fully nonlinear singular operators. Adv. Differ. Equ. **11**(1), 91–119 (2006)
5. I. Birindelli, F. Demengel, $\mathcal{C}^{1,\beta}$ regularity for Dirichlet problems associated to fully nonlinear degenerate elliptic equations. ESAIM Control Optim. Calc. Var. **20**(40), 1009–1024 (2014)
6. I. Birindelli, F. Demengel, F. Leoni, Ergodic pairs for singular or degenerates fully nonlinear operators. ESAIM Control Optim. Calc. Var. (forthcoming). https://doi.org/10.1051/cocv/2018070
7. I. Capuzzo Dolcetta, F. Leoni, A. Porretta, Hölder's estimates for degenerate elliptic equations with coercive Hamiltonian. Trans. Am. Math. Soc. **362**(9), 4511–4536 (2010)
8. M.G. Crandall, H. Ishii, P.L. Lions, User's guide to viscosity solutions of second order partial differential equations. Bull. Amer. Math. Soc. (N.S.) **27**(1), 1–67 (1992)
9. C. Imbert, L. Silvestre, $C^{1,\alpha}$ regularity of solutions of some degenerate fully nonlinear elliptic equations. Adv. Math. **233**, 196–206 (2013)

10. H. Ishii, Viscosity solutions of nonlinear partial differential equations. Sugaku Expositions, 144–151 (1996)
11. H. Ishii, P.L. Lions, Viscosity solutions of fully-nonlinear second order elliptic partial differential equations. J. Differ. Equ. **83**, 26–78 (1990)

Monotonicity Formulas for Static Metrics with Non-zero Cosmological Constant

Stefano Borghini and Lorenzo Mazzieri

Abstract In this paper we adopt the approach presented in Agostiniani and Mazzieri (J Math Pures Appl 104:561–586, 2015; Commun Math Phys 355:261–301, 2017) to study non-singular vacuum static space-times with non-zero cosmological constant. We introduce new integral quantities, and under suitable assumptions we prove their monotonicity along the level set flow of the static potential. We then show how to use these properties to derive a number of sharp geometric and analytic inequalities, whose equality case can be used to characterize the rotational symmetry of the underlying static solutions. As a consequence, we are able to prove some new uniqueness statements for the de Sitter and the anti-de Sitter metrics. In particular, we show that the de Sitter solution has the least possible surface gravity among three-dimensional static metrics with connected boundary and positive cosmological constant.

Keywords Static metrics · Splitting theorem · (Anti)-de Sitter solution · Overdetermined boundary value problems

MSC (2010) 35B06, 53C21, 83C57, 35N25

1 Introduction

Throughout this paper we let (M, g_0) be an n-dimensional Riemannian manifold, $n \geq 3$, with (possibly empty) smooth compact boundary ∂M.

S. Borghini · L. Mazzieri (✉)
Università degli Studi di Trento, Povo, TN, Italy
e-mail: stefano.borghini@unitn.it; lorenzo.mazzieri@unitn.it

S. Dipierro (ed.), *Contemporary Research in Elliptic PDEs and Related Topics*, Springer INdAM Series 33, https://doi.org/10.1007/978-3-030-18921-1_3

1.1 Static Einstein System

Consider positive functions $u \in \mathscr{C}^\infty(M)$ such that the triple (M, g_0, u) satisfies the *static Einstein system*

$$\begin{cases} u\,\mathrm{Ric} = \mathrm{D}^2 u + \dfrac{2\Lambda}{n-1}\, u\, g_0, & \text{in } M \\[2ex] \Delta u = -\dfrac{2\Lambda}{n-1}\, u, & \text{in } M \end{cases} \tag{1.1}$$

where Ric, D, and Δ represent the Ricci tensor, the Levi-Civita connection, and the Laplace–Beltrami operator of the metric g_0, respectively, and $\Lambda \in \mathbb{R}$ is a constant called *cosmological constant*. Note that a consequence of the above equations is that the scalar curvature is

$$\mathrm{R} = 2\Lambda\,.$$

We notice that the equations in (1.1) are assumed to be satisfied in the whole M in the sense that they hold in $M \setminus \partial M$ in the classical sense and if we take the limits of both the left hand side and the right hand side, they coincide at the boundary. In the rest of the paper the metric g_0 and the function u will be referred to as *static metric* and *static potential*, respectively, whereas the triple (M, g_0, u) will be called a *static solution*. A classical computation shows that if (M, g_0, u) satisfies (1.1), then the Lorentzian metric $\gamma = -u^2\, dt \otimes dt + g_0$ satisfies the *vacuum Einstein equations*

$$\mathrm{Ric}_\gamma \;=\; \frac{2\Lambda}{n-1}\,\gamma \qquad \text{in} \quad \mathbb{R} \times (M \setminus \partial M)\,.$$

Throughout this work we will be interested to the case $\Lambda \neq 0$ (see [4] for the case $\Lambda = 0$). If $\Lambda > 0$ (respectively $\Lambda < 0$) we can rescale the metric to obtain $\Lambda = \frac{1}{2}n(n-1)$ (respectively $\Lambda = -\frac{1}{2}n(n-1)$). We recall that the simplest solutions of the rescaled problem (1.1) are given by the *de Sitter solution* [18]

$$(M, g_0, u) = \left(\mathbb{D}^n\,,\; g_D = \frac{d|x| \otimes d|x|}{1-|x|^2} + |x|^2 g_{\mathbb{S}^{n-1}}\,,\; u_D = \sqrt{1-|x|^2}\right), \tag{1.2}$$

where $\mathbb{D}^n := \{x \in \mathbb{R}^n \,:\, |x| < 1\}$ is the n-disc, when the cosmological constant is *positive*, and by the *anti-de Sitter solution*

$$(M, g_0, u) = \left(\mathbb{R}^n\,,\; g_A = \frac{d|x| \otimes d|x|}{1+|x|^2} + |x|^2 g_{\mathbb{S}^{n-1}}\,,\; u_A = \sqrt{1+|x|^2}\right), \tag{1.3}$$

when the cosmological constant is *negative*.

1.2 Setting of the Problem and Statement of the Main Results (Case $\Lambda > 0$)

In the case $\Lambda > 0$ it seems physically reasonable (see for instance [8, 25]) to suppose that M is compact with non-empty boundary, and that $u \in \mathscr{C}^\infty(M)$ is a nonnegative function (strictly positive in $\mathrm{int}(M)$) which solves the problem

$$\begin{cases} u\,\mathrm{Ric} = \mathrm{D}^2 u + n\, u\, g_0, & \text{in } M \\ \Delta u = -n\, u, & \text{in } M \\ u = 0, & \text{on } \partial M \end{cases} \tag{1.4}$$

We notice that the first two equations coincide with the equations of the rescaled problem (1.1) in the case of a positive cosmological constant.

Normalization 1 *Since the problem is invariant under a multiplication of u by a positive constant, without loss of generality we will suppose from now on $\max_M(u) = 1$. We also let*

$$MAX(u) = \{p \in M \,:\, u(p) = 1\}$$

be the set of the points that realize the maximum.

Recall that, since $u = 0$ on ∂M, the first equation of problem (1.4) implies that $\mathrm{D}^2 u = 0$ on ∂M. Therefore, $|\mathrm{D}u|$ is constant (and different from zero, see [8, Lemma 3]) on each connected component of ∂M. The positive constant value of $|\mathrm{D}u|$ on a connected component of ∂M is known in the literature as the *surface gravity* of the connected component. It is easily seen that the surface gravity of the boundary of the *de Sitter solution* (1.2) is equal to 1. Thus, it makes sense to consider the following hypothesis, that will play a fundamental role in what follows.

Assumption 1 *The surface gravity on each connected component of the boundary is less than or equal to 1, namely, $|\mathrm{D}u| \leq 1$ on ∂M.*

We notice that the de Sitter triple $(\mathbb{D}^n, g_D, u_D)$ defined by (1.2) is still a static solution of the rescaled problem (1.4) and satisfies Normalization 1 and Assumption 1. On the other hand, Assumption 1 rules out other known solutions of (1.4), such as the de Sitter–Schwarzschild triple [26]

$$\left(M = [r_1(m), r_2(m)] \times \mathbb{S}^{n-1}\,,\; g_0 = \frac{dr \otimes dr}{1 - r^2 - 2mr^{2-n}} + r^2 g_{\mathbb{S}^{n-1}}\,,\; u = \sqrt{1 - r^2 - 2mr^{2-n}}\right), \tag{1.5}$$

where $m \in \left(0, \sqrt{\frac{(n-2)^{n-2}}{n^n}}\right)$ and $r_1(m), r_2(m)$ are the two positive solutions of $1 - r^2 - 2mr^{2-n} = 0$ (once u is rescaled according to Normalization 1, it can be seen that the surface gravity of the event horizon $r = r_1(m)$ is greater than 1 for all m), and the Nariai solution [31]

$$\Big(M = [0, \pi] \times \mathbb{S}^{n-1}\,,\ g_0 = \frac{1}{n}\left[dr \otimes dr + (n-2)\, g_{\mathbb{S}^{n-1}}\right],\ u = \sin(r)\Big) \tag{1.6}$$

which has $|\mathrm{D}u| = \sqrt{n}$ at both its boundaries.

Proceeding in analogy with [4], we are now ready to introduce, for all $p \geq 0$, the functions $U_p : [0, 1) \to \mathbb{R}$ given by

$$t \longmapsto U_p(t) = \left(\frac{1}{1-t^2}\right)^{\frac{n+p-1}{2}} \int\limits_{\{u=t\}} |\mathrm{D}u|^p \, \mathrm{d}\sigma. \tag{1.7}$$

It is worth noticing that the functions $t \mapsto U_p(t)$ are well defined, since the integrands are globally bounded and the level sets of u have finite hypersurface area. In fact, since u is analytic (see [13, 41]), the level sets of u have locally finite $\mathscr{H}^{n-1}$-measure by the results in [21, 30] (see also [27, Theorem 6.3.3]). Moreover, they are compact and thus their hypersurface area is finite. To give further insights about the definition of the functions $t \mapsto U_p(t)$, we note that, using the explicit formulæ (1.2), one easily realizes that the quantities

$$M \ni x \longmapsto \frac{|\mathrm{D}u|}{\sqrt{1-u^2}}(x) \quad \text{and} \quad [0,1) \ni t \longmapsto U_0(t) = \int\limits_{\{u=t\}} \left(\frac{1}{1-u^2}\right)^{\frac{n-1}{2}} \mathrm{d}\sigma \tag{1.8}$$

are constant on the de Sitter solution. In the following, via a conformal reformulation of problem (1.4), we will be able to give a more geometric interpretation of this fact. On the other hand, we notice that the function $t \mapsto U_p(t)$ can be rewritten in terms of the above quantities as

$$U_p(t) = \int\limits_{\{u=t\}} \left(\frac{|\mathrm{D}u|}{\sqrt{1-u^2}}\right)^p \left(\frac{1}{1-u^2}\right)^{\frac{n-1}{2}} \mathrm{d}\sigma. \tag{1.9}$$

Hence, thanks to (1.8), we have that for every $p \geq 0$ the function $t \mapsto U_p(t)$ is constant on the de Sitter solution. Our main result illustrates how the functions $t \mapsto U_p(t)$ can be also used to detect the rotational symmetry of the *static solution* (M, g_0, u). In fact, for $p \geq 3$, they are nonincreasing and the monotonicity is strict unless (M, g_0, u) is isometric to the de Sitter solution.

Theorem 1.1 (Monotonicity-Rigidity Theorem, Case $\Lambda > 0$) *Let (M, g_0, u) be a static solution to problem* (1.4) *satisfying Normalization 1 and Assumption 1. Then*

$$|\mathrm{D}u|^2 \leq 1 - u^2, \quad \textit{in } M. \tag{1.10}$$

Moreover, the functions $U_p : [0,1) \to \mathbb{R}$ defined in (1.7) satisfy the following properties.

(i) *For every $p \geq 1$, the function U_p is continuous.*
(ii) *The function U_1 is monotonically nonincreasing. Moreover, if $U_1(t_1) = U_1(t_2)$ for some $t_1 \neq t_2$, then (M, g_0, u) is isometric to the de Sitter solution.*
(iii) *For every $p \geq 3$, the function U_p is differentiable and the derivative satisfies, for every $t \in [0, 1)$,*

$$\begin{aligned} U_p'(t) &= -(p-1)t\left(\frac{1}{1-t^2}\right)^{\frac{n+p-1}{2}} \\ &\quad \times \int_{\{u=t\}} |\mathrm{D}u|^{p-2}\left[\left|\frac{\mathrm{D}u}{u}\right|\mathrm{H} + \left(\frac{np}{p-1}\right) - \left(\frac{n+p-1}{p-1}\right)\left(\frac{|\mathrm{D}u|^2}{1-u^2}\right)\right]\mathrm{d}\sigma \\ &= -(p-1)t\left(\frac{1}{1-t^2}\right)^{\frac{n+p-1}{2}} \\ &\quad \times \int_{\{u=t\}} |\mathrm{D}u|^{p-2}\left[(n-1) - \mathrm{Ric}(\nu,\nu) + \left(\frac{n+p-1}{p-1}\right)\left(1 - \frac{|\mathrm{D}u|^2}{1-u^2}\right)\right]\mathrm{d}\sigma \\ &\leq -(p-1)t\left(\frac{1}{1-t^2}\right)^{\frac{n+p-1}{2}} \int_{\{u=t\}} |\mathrm{D}u|^{p-2}\left(\frac{n}{p-1}\right)\left(1 - \frac{|\mathrm{D}u|^2}{1-u^2}\right)\mathrm{d}\sigma \leq 0, \end{aligned} \tag{1.11}$$

where H is the mean curvature of the level set $\{u = t\}$ and $\nu = \mathrm{D}u/|\mathrm{D}u|$ is the unit normal to the set $\{u = t\}$. Moreover, if there exists $t \in (0, 1)$ such that $U_p'(t) = 0$ for some $p \geq 3$, then the static solution (M, g_0, u) is isometric to the de Sitter solution.
(iv) *For every $p \geq 3$, we have $U_p'(0) := \lim_{t\to 0^+} U_p'(t) = 0$ and, setting $U_p''(0) := \lim_{t\to 0^+} U_p'(t)/t$, it holds*

$$\begin{aligned} U_p''(0) &= -(p-1) \\ &\quad \times \int_{\partial M} |\mathrm{D}u|^{p-2}\left[\frac{\mathrm{R}^{\partial M} - (n-1)(n-2)}{2} + \left(\frac{n+p-1}{p-1}\right)\left(1 - |\mathrm{D}u|^2\right)\right]\mathrm{d}\sigma \\ &\leq -(p-1)\int_{\partial M} |\mathrm{D}u|^{p-2}\left(\frac{n}{p-1}\right)\left(1 - |\mathrm{D}u|^2\right)\mathrm{d}\sigma \leq 0, \end{aligned} \tag{1.12}$$

where $\mathrm{R}^{\partial M}$ is the scalar curvature of the metric $g_{\partial M}$ induced by g_0 on ∂M. Moreover, if $U_p''(0) = 0$ for some $p \geq 3$, then the static solution (M, g_0, u) is isometric to the de Sitter solution.

Remark 1 Notice that formula (1.11) is well-posed also in the case where $\{u = t\}$ is not a regular level set of u. In fact, one has from [13, 41] that u is analytic, hence we can use the results from [21, 30] to conclude that the $(n-1)$-dimensional Hausdorff measure of the level sets of u is finite. More than that, it follows from [30] (see also [27, Theorem 6.3.3]) that the set $\mathrm{Crit}(\varphi) = \{x \in M : \nabla\varphi(x) = 0\}$ contains an open $(n-1)$-submanifold N such that $\mathscr{H}^{n-1}(\mathrm{Crit}(\varphi) \setminus N) = 0$. In particular, the unit normal vector field to the level set is well defined $\mathscr{H}^{n-1}$-almost everywhere and so does the mean curvature H. In turn, the integrand in (1.11) is well defined $\mathscr{H}^{n-1}$-almost everywhere. Finally, we observe that where $|\mathrm{D}u| \neq 0$ it holds

$$|\mathrm{D}u|^{p-1}\mathrm{H} \;=\; |\mathrm{D}u|^{p-2}\,\Delta u - \;|\mathrm{D}u|^{p-4}\,\mathrm{D}^2u(\mathrm{D}u, \mathrm{D}u) \;=\; -\,u\,|\mathrm{D}u|^{p-2}\,\mathrm{Ric}(\nu, \nu)\,,$$

where $\nu = \mathrm{D}u/|\mathrm{D}u|$ as usual. It is also clear that $|\mathrm{D}u|^{p-1}\mathrm{H} = -\,u\,|\mathrm{D}u|^{p-2}$ $\mathrm{Ric}(\nu, \nu) = 0$ on the whole N for every $p > 2$. Since $|\mathrm{Ric}|$ is uniformly bounded on M, this shows that the integrand in (1.11) is essentially bounded and thus summable on every level set of u, provided $p > 2$.

The analytic and geometric implications of Theorem 1.1 will be discussed in full details in Sect. 2. However, we have decided to collect the more significant among them in Theorem 1.3 below. Before giving the statement, it is worth noticing that, combining Theorem 1.1 with some approximations near the extremal points of u, we are able to characterize the set $\mathrm{MAX}(u)$ and to estimate the behavior of the $U_p(t)$'s as t approaches 1.

Theorem 1.2 *Let (M, g_0, u) be a solution of* (1.4) *satisfying Assumption 1. The set* $\mathrm{MAX}(u)$ *is discrete (and finite) and, for every $p \leq n-1$, it holds*

$$\liminf_{t \to 1^-} U_p(t) \;\geq\; |\mathrm{MAX}(u)|\,|\mathbb{S}^{n-1}|\,, \tag{1.13}$$

where $|\mathrm{MAX}(u)|$ *is the cardinality of the set* $\mathrm{MAX}(u)$.

For the detailed proof of this result, we refer the reader to Theorem A.1 in the Appendix A.

Remark 2 The above result is false without Assumption 1. In fact, we can easily find solutions (that does not satisfy our assumption) such that the set $\mathrm{MAX}(u)$ is very large. For instance, the set of the maximum points of the de Sitter–Schwarzschild solution (1.5) has non-zero $\mathscr{H}^{n-1}$-measure, and the same holds for the maximum points of the Nariai solution (1.6).

Now we are ready to state the main consequences of Theorem 1.1 on the geometry of the boundary of M.

Theorem 1.3 (Geometric Inequalities, Case $\Lambda > 0$) *Let (M, g_0, u) be a static solution to problem* (1.4) *satisfying Normalization 1 and Assumption 1. Then the following properties are satisfied.*

(i) (Area bound) *The inequality*

$$|\mathrm{MAX}(u)|\,|\mathbb{S}^{n-1}| \;\le\; |\partial M|\,, \tag{1.14}$$

holds true. Moreover, the equality is fulfilled if and only if the static solution (M, g_0, u) *is isometric to the de Sitter solution.*

(ii) (Willmore-type inequality) *The inequality*

$$|\mathrm{MAX}(u)|\,|\mathbb{S}^{n-1}| \;\le\; \int\limits_{\partial M} \left| \frac{\mathrm{R}^{\partial M} - n\,(n-3)}{2} \right|^{n-1} \mathrm{d}\sigma \tag{1.15}$$

holds true. Moreover, the equality is fulfilled if and only if the static solution (M, g_0, u) *is isometric to the de Sitter solution.*

(iii) *The inequality*

$$|\mathrm{MAX}(u)|\,|\mathbb{S}^{n-1}| \;\le\; \int\limits_{\partial M} \frac{\mathrm{R}^{\partial M}}{(n-1)(n-2)}\,\mathrm{d}\sigma \tag{1.16}$$

holds true. Moreover, the equality is fulfilled if and only if the static solution (M, g_0, u) *is isometric to the de Sitter solution.*

(iv) (Uniqueness Theorem) *Let* $n = 3$. *If* ∂M *is connected, then* (M, g_0, u) *is isometric to the de Sitter solution. If* ∂M *is not connected, then* $3\,|\mathrm{MAX}(u)| < \pi_0(\partial M)$, *in particular,* ∂M *must have at least four connected components.*

We conclude this subsection observing that point (iv) in the above statement can be rephrased by saying that (after normalization) the de Sitter solution has the least possible surface gravity among three-dimensional solutions to problem (1.4) with connected boundary.

1.3 *Setting of the Problem and Statement of the Main Results (Case* $\Lambda < 0$*)*

Suppose that M has empty boundary and at least one end, and consider positive functions $u \in \mathscr{C}^\infty(M)$ such that the triple (M, g_0, u) satisfies the system

$$\begin{cases} u\,\mathrm{Ric} = \mathrm{D}^2 u - n\,u\,g_0, & \text{in } M \\ \Delta u = n\,u, & \text{in } M \\ u(x) \to +\infty & \text{as } x \to \infty \end{cases} \tag{1.17}$$

We notice that the first two equations coincide with the equations of the rescaled problem (1.1) in the case of a negative cosmological constant.

Normalization 2 *Since the problem is invariant under a multiplication of* u *by a positive constant, without loss of generality we will suppose from now on* $\min_M(u) = 1$. *We also let*

$$\mathrm{MIN}(u) = \{p \in M \,:\, u(p) = 1\}$$

be the set of the points that realize the minimum.

For future convenience, we introduce the following classical definition, originally introduced by Penrose in [33] (see also [24] and the references therein).

Definition 1 (Conformally Compact Static Solution) A static solution (M, g_0, u) of problem (1.17) is said to be *conformally compact* if the following conditions are satisfied:

(i) The manifold M is diffeomorphic to the interior of a compact manifold with boundary $\overline{M}$.
(ii) There exists a compact $K \subset M$ and a function $r \in \mathscr{C}^\infty(\overline{M} \setminus K)$ such that $r \neq 0$ on M, $r = 0$ on $\partial\overline{M}$, $dr \neq 0$ on $\partial\overline{M}$ and the metric $\bar{g} = r^2 g_0$ extends smoothly to a metric on $\overline{M} \setminus K$.

In the following, we will call $\partial\overline{M}$ the *conformal boundary* of M and, in order to simplify the notation, we will set

$$\partial M \,:=\, \partial\overline{M}\,.$$

We will refer to a function with the same properties of r in (ii) as to a *defining function* for ∂M.

We are now ready to introduce the analogous of Assumption 1 in the case of a negative cosmological constant.

Assumption 2 *The triple* (M, g_0, u) *is conformally compact, the function* $1/\sqrt{u^2 - 1}$ *is a defining function for* ∂M *and* $\lim_{x \to \bar{x}} \left(u^2 - 1 - |\mathrm{D}u|^2\right) \geq 0$ *for every* $\bar{x} \in \partial M$.

Some comments are in order to justify these requirements. First we notice that the requirement of $1/\sqrt{u^2 - 1}$ being a defining function is not unusual, in the sense that it has already appeared in various articles like [34, 38, 39]. Moreover, we notice that, if $1/\sqrt{u^2 - 1}$ is a defining function, then the limit in Assumption 2 exists and is finite (see Lemma A.8-(i) in the Appendix A).

Finally, we observe that the anti-de Sitter triple $(\mathbb{R}^n, g_A, u_A)$ defined by (1.3) indeed verifies all our hypothesis, namely it is a conformally compact static solution of problem (1.17) satisfying Normalization 2 and Assumption 2.

Proceeding in analogy with [4], for all $p \geq 0$ we introduce the functions $U_p : (1, +\infty) \to \mathbb{R}$ defined as

$$t \longmapsto U_p(t) = \left(\frac{1}{t^2 - 1}\right)^{\frac{n+p-1}{2}} \int\limits_{\{u=t\}} |\mathrm{D}u|^p \,\mathrm{d}\sigma. \tag{1.18}$$

It is worth noticing that the functions $t \mapsto U_p(t)$ are well defined, since the integrands are globally bounded and the level sets of u have finite hypersurface area. In fact, since u is analytic (see [13, 41]), the level sets of u have locally finite $\mathscr{H}^{n-1}$-measure by the results in[21, 30] (see also [27, Theorem 6.3.3]). Moreover, they are compact and thus their hypersurface area is finite. Another important observation comes from the fact that, using the explicit formulæ (1.3), one easily realizes that the quantities

$$M \ni x \longmapsto \frac{|\mathrm{D}u|}{\sqrt{u^2-1}}(x) \quad \text{and} \quad [0,1) \ni t \longmapsto U_0(t) = \int\limits_{\{u=t\}} \left(\frac{1}{u^2-1}\right)^{\frac{n-1}{2}} d\sigma \tag{1.19}$$

are constant on the anti-de Sitter solution. In the following, via a conformal reformulation of problem (1.17), we will be able to give a more geometric interpretation of this fact. On the other hand, we notice that the function $t \mapsto U_p(t)$ can be rewritten in terms of the above quantities as

$$U_p(t) = \int\limits_{\{u=t\}} \left(\frac{|\mathrm{D}u|}{\sqrt{u^2-1}}\right)^p \left(\frac{1}{u^2-1}\right)^{\frac{n-1}{2}} d\sigma. \tag{1.20}$$

Hence, thanks to (1.19), we have that for every $p \geq 0$ the function $t \mapsto U_p(t)$ is constant on the anti-de Sitter solution. Our main result illustrates how the functions $t \mapsto U_p(t)$ can be used to detect the rotational symmetry of the *static solution* (M, g_0, u). In fact, for $p \geq 3$, they are nondecreasing and the monotonicity is strict unless (M, g_0, u) is isometric to the anti-de Sitter solution.

Theorem 1.4 (Monotonicity-Rigidity Theorem, Case $\Lambda < 0$)

Let (M, g_0, u) be a conformally compact static solution to problem (1.17) *in the sense of Definition 1. Suppose moreover that (M, g_0, u) satisfies Normalization 2 and Assumption 2. Then*

$$|\mathrm{D}u|^2 \leq u^2 - 1\,, \tag{1.21}$$

on the whole manifold M. Moreover, for every $p \geq 1$ let $U_p : (1, +\infty) \to \mathbb{R}$ be the function defined in (1.18)*. Then, the following properties hold true.*

(i) *For every $p \geq 1$, the function U_p is continuous.*
(ii) *The function U_1 is monotonically nondecreasing. Moreover, if $U_1(t_1) = U_1(t_2)$ for some $t_1 \neq t_2$, then (M, g_0, u) is isometric to the anti-de Sitter solution.*

(iii) *For every $p \geq 3$, the function U_p is differentiable and the derivative satisfies, for every $t \in (1, +\infty)$,*

$$\begin{aligned} U_p'(t) &= (p-1)t\left(\frac{1}{t^2-1}\right)^{\frac{n+p-1}{2}} \\ &\quad\times \int\limits_{\{u=t\}} |\mathrm{D}u|^{p-2}\left[-\left|\frac{\mathrm{D}u}{u}\right|\mathrm{H} + \left(\frac{np}{p-1}\right) - \left(\frac{n+p-1}{p-1}\right)\left(\frac{|\mathrm{D}u|^2}{u^2-1}\right)\right]\mathrm{d}\sigma \\ &= (p-1)t\left(\frac{1}{t^2-1}\right)^{\frac{n+p-1}{2}} \\ &\quad\times \int\limits_{\{u=t\}} |\mathrm{D}u|^{p-2}\left[(n-1) + \mathrm{Ric}(\nu,\nu) + \left(\frac{n+p-1}{p-1}\right)\left(1 - \frac{|\mathrm{D}u|^2}{u^2-1}\right)\right]\mathrm{d}\sigma \\ &\geq (p-1)t\left(\frac{1}{t^2-1}\right)^{\frac{n+p-1}{2}} \int\limits_{\{u=t\}} |\mathrm{D}u|^{p-2}\left(\frac{n}{p-1}\right)\left(1 - \frac{|\mathrm{D}u|^2}{u^2-1}\right)\mathrm{d}\sigma \geq 0, \end{aligned} \tag{1.22}$$

where H *is the mean curvature of the level set $\{u = t\}$ and $\nu = \mathrm{D}u/|\mathrm{D}u|$ is the unit normal to the level set $\{u = t\}$. Moreover, if there exists $t \in (1, +\infty)$ such that $U_p'(t) = 0$ for some $p \geq 3$, then the static solution (M, g_0, u) is isometric to the anti-de Sitter solution.*

(iv) *For our next result it is convenient to see $U_p(t)$ as a function of the defining function $r = 1/\sqrt{u^2-1}$, that is, we consider the function $V_p(r) = U_p(\sqrt{1+1/r^2})$. We have that, for every $p \geq 3$, it holds*

$$\begin{aligned} \lim_{r\to 0^+} V_p''(r) &= -(p-1) \\ &\quad\times \int\limits_{\partial M} \left[\frac{(n-1)(n-2) - \mathrm{R}_g^{\partial M}}{2(n-1)} + \frac{n(p+1)}{2(p-1)}\left(u^2 - 1 - |\mathrm{D}u|^2\right)\right]\mathrm{d}\sigma_g \\ &\leq -(p-1)\int\limits_{\partial M}\left(\frac{n}{p-1}\right)\left(u^2 - 1 - |\mathrm{D}u|^2\right)\mathrm{d}\sigma_g \leq 0, \end{aligned} \tag{1.23}$$

where $g = g_0/(u^2-1)$ and $\mathrm{R}_g^{\partial M}$ is the scalar curvature of the metric $g_{\partial M}$ induced by g on ∂M. The integrands in (1.23) have to be thought as the limits of the corresponding functions as $x \to \bar{x}$, with $\bar{x} \in \partial M$.

Remark 3 Using the same arguments of Remark 1, one observes that formula (1.22) is well posed even when the set $\{u = t\}$ is not a regular level set of u. Notice that the integrands in (1.23) are finite functions, as it has been stated in the discussion below Assumption 2 (see also Lemma A.8-(i)).

Remark 4 Note that, unlike the case $\Lambda > 0$, the rigidity statement does not hold for point (iii) of Theorem 1.4. The reason for this will be clear later (see the discussion at the end of Sect. 6.4).

In general, as it will become apparent in Sect. 2, the analysis of the static solutions is more delicate in the case $\Lambda < 0$. In particular, we will see that, in the case $\Lambda < 0$, in order to obtain results that are comparable with the ones for $\Lambda > 0$, it will be useful to require some extra hypotheses on the behavior of the static solution near the conformal boundary (namely Assumption 2-bis in Sect. 2.4). Still, some of the consequences for $\Lambda > 0$ will have no analogue in the case $\Lambda < 0$ (compare Theorem 1.3 with Theorem 1.6 below).

The analytic and geometric implications of Theorem 1.4 will be discussed in full details in Sect. 2. However, we have decided to collect the more significant among them in Theorem 1.6 below. Before giving the statement, it is worth noticing that, combining Theorem 1.4 with some approximations near the extremal points of the static potential u, we are able to characterize the set $\mathrm{MIN}(u)$ and to estimate the behavior of the $U_p(t)$'s as t approaches 1.

Theorem 1.5 *Let* (M, g_0, u) *be a conformally compact solution of* (1.17) *satisfying Normalization 2 and Assumption 2. Then the set* $\mathrm{MIN}(u)$ *is discrete (and finite) and, for every* $p \leq n-1$, *it holds*

$$\liminf_{t \to 1^+} U_p(t) \;\geq\; |\mathrm{MIN}(u)|\, |\mathbb{S}^{n-1}|\,, \tag{1.24}$$

where $|\mathrm{MIN}(u)|$ *is the cardinality of the set* $\mathrm{MIN}(u)$.

For the detailed proof of this result, we refer the reader to Theorem A.7 in the Appendix A.

Theorem 1.6 (Geometric Inequalities, Case $\Lambda < 0$) *Let* (M, g_0, u) *be a conformally compact static solution to problem* (1.17) *in the sense of Definition 1. Suppose moreover that* (M, g_0, u) *satisfies Normalization 2 and Assumption 2. Then the metric* $g = g_0/(u^2 - 1)$ *extends to the conformal boundary and the following properties are satisfied.*

(i) (Area bound) *The inequality*

$$|\mathrm{MIN}(u)|\, |\mathbb{S}^{n-1}| \;\leq\; |\partial M|_g\,, \tag{1.25}$$

holds true. Moreover, the equality is fulfilled if and only if the static solution (M, g_0, u) *is isometric to the anti-de Sitter solution.*

(ii) (Willmore-type inequality) *Suppose that* $\lim_{t\to+\infty}(u^2 - 1 - |\mathrm{D}u|^2) = 0$. *Then the inequality*

$$|\mathrm{MIN}(u)|\, |\mathbb{S}^{n-1}| \;\leq\; \int\limits_{\partial M} \left| \frac{\mathrm{R}_g^{\partial M} - (n+1)(n-2)}{2(n-2)} \right|^{n-1} \mathrm{d}\sigma_g \tag{1.26}$$

holds true. Moreover, the equality is fulfilled if and only if the static solution (M, g_0, u) *is isometric to the anti-de Sitter solution.*

We underline the similarity between this result and statements (i), (ii) of Theorem 1.3. Unfortunately, we are not able to provide analogues of points (iii), (iv).

1.4 Strategy of the Proof

To describe the strategy of the proof, we focus our attention on the rigidity statements in Theorems 1.1-(iii), 1.4-(iii) and for simplicity, we let $p = 3$. At the same time, we provide an heuristic for the monotonicity statement. In this introductory section, we treat the two cases $\Lambda > 0$ and $\Lambda < 0$ at the same time, in order to emphasize the similarities between them. For a more specific and precise analysis we address the reader to Sect. 3 and following.

The method employed is based on the conformal splitting technique introduced in [1], which consists of two main steps. The first step is the construction of the so called *cylindrical ansatz* and amounts to find an appropriate conformal deformation g of the *static metric* g_0 in terms of the *static potential* u. In the case under consideration, the natural deformation is given by

$$g = \frac{g_0}{1-u^2} \qquad \text{(case } \Lambda > 0),$$

$$g = \frac{g_0}{u^2-1} \qquad \text{(case } \Lambda < 0),$$

defined on $M^* := M \setminus \mathrm{MAX}(u)$ (respectively $M^* := M \setminus \mathrm{MIN}(u)$) if $\Lambda > 0$ (respectively $\Lambda < 0$). The manifold M^* has the same boundary as M and each point of $\mathrm{MAX}(u)$ (respectively $\mathrm{MIN}(u)$) corresponds to an end of M^*. When (M, g_0, u) is the *de Sitter solution* (respectively the *anti-de Sitter solution*), the metric g obtained through the above formula is immediately seen to be the cylindrical one. In general, the *cylindrical ansatz* leads to a conformal reformulation of problems (1.4), (1.17) in which the conformally related metric g obeys the quasi-Einstein type equation

$$\begin{aligned}\mathrm{Ric}_g - \left[\frac{1-(n-1)\tanh^2(\varphi)}{\tanh(\varphi)}\right]\nabla^2\varphi + (n-2)d\varphi\otimes d\varphi \\ = \Big(n-2+2(1-|\nabla\varphi|_g^2)\Big)\, g\,, \qquad \text{(case } \Lambda > 0)\end{aligned}$$

$$\begin{aligned}\mathrm{Ric}_g - \left[\frac{1-(n-1)\coth^2(\varphi)}{\coth(\varphi)}\right]\nabla^2\varphi + (n-2)d\varphi\otimes d\varphi \\ = \Big(n-2+2(1-|\nabla\varphi|_g^2)\Big)\, g\,, \qquad \text{(case } \Lambda < 0)\end{aligned}$$

where ∇ is the Levi-Civita connection of g and the function φ is defined by

$$\varphi = \frac{1}{2}\log\left(\frac{1+u}{1-u}\right), \qquad (\text{case } \Lambda > 0)$$

$$\varphi = \frac{1}{2}\log\left(\frac{u+1}{u-1}\right), \qquad (\text{case } \Lambda < 0)$$

and satisfies

$$\Delta_g \varphi = -n\tanh(\varphi)\left(1 - |\nabla\varphi|_g^2\right), \quad (\text{case } \Lambda > 0)$$

$$\Delta_g \varphi = -n\coth(\varphi)\left(1 - |\nabla\varphi|_g^2\right), \quad (\text{case } \Lambda > 0)$$

where Δ_g is the Laplace–Beltrami operator of the metric g. Before proceeding, it is worth pointing out that taking the trace of the quasi-Einstein type equation gives

$$\frac{\mathrm{R}_g}{n-1} = (n-2) + \left(n\tanh^2(\varphi) + 2\right)\left(1 - |\nabla\varphi|_g^2\right), \qquad (\text{case } \Lambda > 0)$$

$$\frac{\mathrm{R}_g}{n-1} = (n-2) + \left(n\coth^2(\varphi) + 2\right)\left(1 - |\nabla\varphi|_g^2\right), \qquad (\text{case } \Lambda < 0)$$

where R_g is the scalar curvature of the conformal metric g. On the other hand, it is easy to see that $|\nabla\varphi|_g^2$ is proportional to the first term in (1.8) (respectively (1.19)). In fact, if (M, g_0) is the *de Sitter solution* (respectively *anti-de Sitter solution*), then (M^*, g) is a round cylinder with constant scalar curvature. Furthermore, the second term appearing in (1.8) (respectively (1.19)) is (proportional to) the hypersurface area of the level sets of φ computed with respect to the metric induced on them by g. Again, in the cylindrical situation such a function is expected to be constant.

The second step of our strategy consists in proving via a splitting principle that the metric g has indeed a product structure, provided the hypotheses of the Rigidity statement are satisfied. More precisely, we use the above conformal reformulation of the original system combined with the Bochner identity to deduce the equation

$$\Delta_g|\nabla\varphi|_g^2 - \left(\frac{1+(n+1)\tanh^2(\varphi)}{\tanh(\varphi)}\right)\langle\nabla|\nabla\varphi|_g^2 \,|\, \nabla\varphi\rangle_g =$$
$$= 2\,|\nabla^2\varphi|_g^2 + 2n\tanh^2(\varphi)\,|\nabla\varphi|_g^2\left(1 - |\nabla\varphi|_g^2\right), \qquad (\text{case } \Lambda > 0)$$

$$\Delta_g|\nabla\varphi|_g^2 - \left(\frac{1+(n+1)\coth^2(\varphi)}{\coth(\varphi)}\right)\langle\nabla|\nabla\varphi|_g^2 \,|\, \nabla\varphi\rangle_g =$$
$$= 2\,|\nabla^2\varphi|_g^2 + 2n\coth^2(\varphi)\,|\nabla\varphi|_g^2\left(1 - |\nabla\varphi|_g^2\right). \qquad (\text{case } \Lambda < 0)$$

Observing that the drifted Laplacian appearing on the left hand side is formally self-adjoint with respect to the weighted measure

$$\frac{d\mu_g}{\sinh(\varphi)\cosh^{n+1}(\varphi)}, \qquad (\text{case } \Lambda > 0)$$

$$\frac{d\mu_g}{\sinh^{n+1}(\varphi)\cosh(\varphi)}, \qquad (\text{case } \Lambda < 0)$$

we integrate by parts and we obtain, for every $s \geq 0$, the integral identity

$$\int_{\{\varphi=s\}} \left[\frac{|\nabla\varphi|_g^2 \mathrm{H}_g - |\nabla\varphi|_g \Delta_g \varphi}{\sinh(\varphi)\cosh^{n+1}(\varphi)}\right] d\sigma_g$$

$$= \int_{\{\varphi>s\}} \left[\frac{|\nabla^2\varphi|_g^2 + n\tanh^2(\varphi)|\nabla\varphi|_g^2\left(1-|\nabla\varphi|_g^2\right)}{\sinh(\varphi)\cosh^{n+1}(\varphi)}\right] d\mu_g, \ (\Lambda > 0)$$

$$\int_{\{\varphi=s\}} \left[\frac{|\nabla\varphi|_g^2 \mathrm{H}_g - |\nabla\varphi|_g \Delta_g \varphi}{\sinh^{n+1}(\varphi)\cosh(\varphi)}\right] d\sigma_g$$

$$= \int_{\{\varphi>s\}} \left[\frac{|\nabla^2\varphi|_g^2 + n\coth^2(\varphi)|\nabla\varphi|_g^2\left(1-|\nabla\varphi|_g^2\right)}{\sinh^{n+1}(\varphi)\cosh(\varphi)}\right] d\mu_g, \ (\Lambda > 0)$$

where H_g is the mean curvature of the level set $\{\varphi = s\}$ inside the ambient (M^*, g) (notice that the same considerations as in Remark 1 apply here). We then observe that, up to a negative function of s, the left hand side is closely related to U_3' (see formulæ (3.32) and (3.34)). On the other hand, we will prove that, under suitable assumptions, the right hand side is always nonnegative. This will easily imply the Monotonicity statement. Also, under the hypotheses of the Rigidity statement, the left hand side of the above identity vanishes and thus the Hessian of φ must be zero in an open region of M. In turn, by analyticity, it vanishes everywhere. Translating this information back in terms of the hessian of u, we are able to conclude using Obata's theorem.

1.5 Summary

The paper is organized as follows. In Sect. 2 we describe the geometric consequences of Theorems 1.1, 1.4, obtaining several sharp inequalities for which the equality is satisfied if and only if the solution to system (1.4) or (1.17) is rotationally symmetric. We distinguish the consequences of Theorems 1.1-(iii), 1.4-(iii) on the geometry of a generic level set of u (see Sects. 2.1 and 2.3), from the consequences of Theorems 1.1-(iv), 1.4-(iv) on the geometry of the boundary of M (see Sects. 2.2 and 2.4).

In Sect. 2.2, we deduce some sharp inequalities for static solutions of problem (1.4) and we use them to obtain some corollaries on the uniqueness of the de Sitter metric (see Theorem 2.9 and the discussion below). In particular, we show that, if Assumption 1 holds, then the only 3-dimensional static solution of problem (1.4) with a connected boundary is the de Sitter solution. For $n \geq 4$, we are not able to prove such a general result. Nevertheless, we discuss some geometric conditions under which the uniqueness statement holds in every dimension. The analogous consequences in Sect. 2.4 are less strong. In any case, we are still able to state a result (Theorem 2.19) that extends the classical Uniqueness Theorems of the anti-de Sitter metric proved in [12, 34, 39].

In Sect. 3, we reformulate problem (1.4) and (1.17) in terms of a quasi-Einstein type metric g and a function φ satisfying system (3.23) (*cylindrical ansatz*), according to the strategy described in Sect. 1.4. In this new framework, both Theorem 1.1 and Theorem 1.4 results to be equivalent to Theorem 3.2 in Sect. 3.5 below, as we will prove in detail in Sect. 4. Theorem 3.2 will be proven in Sect. 6 with the help of the integral identities proved in Sect. 5.

Finally, in Appendix B we discuss a different approach to the study of problems (1.4) and (1.17), that does not rely on the machinery of Sects. 3, 4, 5, 6 and provides some consequences that are comparable with the ones discussed in Sect. 2. In the case $\Lambda > 0$, the results that we show in this section are known (see [12, 14]), but in the case $\Lambda < 0$ they appear to be new.

1.6 Added Note

The results presented in this paper have been improved by the authors in the works [9, 10]. In light of these new results, the present work may be interpreted as a preliminary study of problem (1.1), which have had the relevance of providing the basic framework where to build our analysis. In fact, this work provides the heuristic behind the definition of the monotonic function U used in [10] as one of the key tools of the proofs.

For these reasons, we stress that the main focus of the present paper is on the method employed in the proofs, which is based, as already explained in Sect. 1.4, on the application of an appropriate *cylindrical ansatz*. As already mentioned, this technique has been already used in [4] in order to study static solutions with zero cosmological constant. The reader may notice that the results shown in Sect. 2 share some analogies with the ones presented in [4]. It is also worth mentioning that a similar analysis has been employed in [1, 2] in order to study the geometric aspects of potential theory in the Euclidean space (see also [22] and [6] for the natural extensions to the non linear and non flat setting, respectively). Although this may appear to be a completely different problem, it actually shares some strong analogies with the study of static spacetimes with zero cosmological constant, as discussed in [3]. A different but related approach to the study of the electrostatic potential, based on a *spherical ansatz*, has been developed in [11]. Finally, we mention that static metrics with nonzero cosmological constant also admit a Euclidean analogue, that is the well known torsion problem (for some standard references, see [36, 40]). In [5] we will study this Euclidean problem, and in particular we will discuss some overdetermining conditions that force the rotational symmetry of the solution.

2 Consequences

In this section we discuss some consequences of Theorems 1.1 and 1.4, distinguishing the two cases $\Lambda > 0$ and $\Lambda < 0$.

2.1 Consequences on a Generic Level Set of u (Case $\Lambda > 0$)

Since, as already observed, the functions $t \mapsto U_p(t)$ defined in (1.7) are constant on the de Sitter solution, we obtain, as an immediate consequence of Theorem 1.1 and formula (1.11), the following characterizations of the rotationally symmetric solutions to system (1.4).

Theorem 2.1 *Let* (M, g_0, u) *be a solution to problem* (1.4) *satisfying Normalization 1 and Assumption 1. Then, for every* $p \geq 3$ *and every* $t \in [0, 1)$*, it holds*

$$\int_{\{u=t\}} |\mathrm{D}u|^{p-2}\left[(n-1) - \mathrm{Ric}(\nu,\nu) + \left(1 - \frac{|\mathrm{D}u|^2}{1-u^2}\right)\right] \mathrm{d}\sigma \geq 0 .$$

Moreover, the equality is fulfilled for some $p \geq 3$ *and some* $t \in [0, 1)$ *if and only if the static solution* (M, g_0, u) *is isometric to the de Sitter solution.*

Setting $t = 0$ in the above formula, and using the Gauss–Codazzi equation, one gets

$$\int_{\partial M} |\mathrm{D}u|^{p-2}\left[\frac{\mathrm{R}^{\partial M} - (n-1)(n-2)}{2} + \left(1 - \frac{|\mathrm{D}u|^2}{1-u^2}\right)\right] \mathrm{d}\sigma \geq 0 ,$$

This inequality is just a rewriting of formula (1.23), whose consequences will be discussed in Sect. 2.2 (see Theorem 2.6 and below). Another way to rewrite formula (1.11) is the following.

Theorem 2.2 *Let (M, g_0, u) be a solution to problem* (1.4) *satisfying Normalization 1 and Assumption 1. Then, for every $p \geq 3$ and every $t \in [0, 1)$, the inequality*

$$\int_{\{u=t\}} \left(\frac{|\mathrm{D}u|}{\sqrt{1-u^2}}\right)^p \mathrm{d}\sigma \leq \int_{\{u=t\}} \left(\frac{|\mathrm{D}u|}{\sqrt{1-u^2}}\right)^{p-2} \Big[\mathrm{H}\,|\mathrm{D}\log u| + n\Big]\,\mathrm{d}\sigma \tag{2.1}$$

holds true, where H *is the mean curvature of the level set $\{u = t\}$. Moreover, the equality is fulfilled for some $p \geq 3$ and some $t \in (0, 1)$ if and only if the static solution (M, g_0, u) is isometric to the de Sitter solution.*

To illustrate other implications of Theorem 2.2, let us observe that, applying Hölder inequality to the right hand side of (2.1) with conjugate exponents $p/(p-2)$ and $p/2$, one gets

$$\begin{aligned} &\int_{\{u=t\}} \left(\frac{|\mathrm{D}u|}{\sqrt{1-u^2}}\right)^{p-2} \Big[\mathrm{H}|\mathrm{D}\log u| + n\Big]\mathrm{d}\sigma \\ &\leq \left(\int_{\{u=t\}} \left(\frac{|\mathrm{D}u|}{\sqrt{1-u^2}}\right)^p \mathrm{d}\sigma\right)^{\frac{p-2}{p}} \left(\int_{\{u=t\}} \Big|\mathrm{H}|\mathrm{D}\log u| + n\Big|^{\frac{p}{2}} \mathrm{d}\sigma\right)^{\frac{2}{p}}. \end{aligned}$$

This implies on every level set of u the following sharp L^p-bound for the gradient of the *static potential*.

Corollary 2.3 *Let (M, g_0, u) be a solution to problem* (1.4) *satisfying Normalization 1 and Assumption 1. Then, for every $p \geq 3$ and every $t \in [0, 1)$ the inequality*

$$\left\|\frac{\mathrm{D}u}{\sqrt{1-u^2}}\right\|_{L^p(\{u=t\})} \leq \sqrt{\left\|\mathrm{H}\,|\mathrm{D}\log u| + n\right\|_{L^{p/2}(\{u=t\})}}, \tag{2.2}$$

holds true, where H *is the mean curvature of the level set $\{u = t\}$. Moreover, the equality is fulfilled for some $p \geq 3$ and some $t \in (0, 1)$ if and only if the static solution (M, g_0, u) is isometric to the de Sitter solution.*

It is worth pointing out that the right hand side in (2.2) may possibly be unbounded. However, for regular level sets of the *static potential* the L^p-norm of the mean curvature is well defined and finite (see Remark 1). We also observe that letting $p \to +\infty$, we deduce, under the same hypothesis of Corollary 2.3, the following L^∞-bound

$$\left\|\frac{\mathrm{D}u}{\sqrt{1-u^2}}\right\|_{L^\infty(\{u=t\})} \leq \sqrt{\left\|\mathrm{H}\,|\mathrm{D}\log u| + n\right\|_{L^\infty(\{u=t\})}}, \tag{2.3}$$

for every $t \in [0, 1)$. Unfortunately, in this case we do not know whether the rigidity statement holds true or not. However, the equality is satisfied on the de Sitter solution and this makes the inequality sharp.

Now we will combine inequality (2.2) in Corollary 2.3 with the observation that for every $t \in [0, 1)$ and every $3 \leq p \leq n-1$ (we need to take $n \geq 4$) it holds

$$U_p(t) \geq |\mathrm{MAX}(u)|\, |\mathbb{S}^{n-1}|\,, \tag{2.4}$$

where the latter estimate follows from estimate (1.13) in Theorem 1.2 and from the monotonicity of the U_p's stated in Theorem 1.1-(iii). Recalling the explicit expression (1.7) of the U_p's, we obtain the following.

Theorem 2.4 *Let $n \geq 4$. Let (M, g_0, u) be a solution to problem* (1.4) *satisfying Normalization 1 and Assumption 1. Then, for every $3 \leq p \leq n-1$ and every $t \in [0, 1)$, the inequalities*

$$(1-t^2)^{\frac{n-1}{2}}\, |\mathrm{MAX}(u)|\, |\mathbb{S}^{n-1}| \leq \Big\| \mathrm{H}\, |\mathrm{D} \log u| + n \Big\|^{\frac{p}{2}}_{L^{p/2}(\{u=t\})} \tag{2.5}$$

hold true. Moreover, the equality is fulfilled for some $t \in (0, 1)$ and some $3 \leq p \leq n-1$, if and only if the static solution (M, g_0, u) is isometric to the de Sitter solution.

To give a geometric interpretation of the above theorem, we recall the identity

$$\mathrm{H}\, |\mathrm{D} \log u| + n = n - \mathrm{Ric}(\nu, \nu)\,,$$

and we observe that the quantity $(1-t^2)^{\frac{n-1}{2}}\, |\mathbb{S}^{n-1}|$ corresponds to the hypersurface area of the level set $\{u_D = t\}$ in the de Sitter solution (1.2). Combining together these two facts, we arrive at the following corollary.

Corollary 2.5 *Let $n \geq 4$. Let (M, g_0, u) be a solution to problem* (1.4) *satisfying Normalization 1 and Assumption 1. Then, for every $3 \leq p \leq n-1$ and every $t \in [0, 1)$, the inequality*

$$\frac{|\mathrm{MAX}(u)|\, |\{u_D = t\}|}{|\{u = t\}|} \leq \fint_{\{u=t\}} \big| n - \mathrm{Ric}(\nu, \nu) \big|^{\frac{p}{2}}\, \mathrm{d}\sigma \tag{2.6}$$

holds true. Moreover, the equality is fulfilled for some $t \in (0, 1)$ and some $3 \leq p \leq n-1$, if and only if the static solution (M, g_0, u) is isometric to the de Sitter solution. In particular, for every $t \in (0, 1)$, it holds

$$1 \leq \Big\| n - \mathrm{Ric}(\nu, \nu) \Big\|_{L^\infty(\{u=t\})}\,. \tag{2.7}$$

2.2 The Geometry of ∂M (Case $\Lambda > 0$)

We pass now to describe some consequences of the behaviour of the *static solution* (M, g_0, u) at the boundary ∂M, as prescribed by Theorem 1.1-(iv). We remark that $|Du|$ is constant on every connected component of ∂M, and that ∂M is a totally geodesic hypersurface inside (M, g_0). In particular, also the mean curvature H vanishes at ∂M. Hence, formula (1.11) implies that $U_p'(0) = 0$.

The following theorem is a rephrasing of formula (1.12) and is the analog of Theorem 2.2.

Theorem 2.6 *Let (M, g_0, u) be a solution to problem* (1.4) *satisfying Normalization 1 and Assumption 1. Then, for every $p \geq 3$, it holds*

$$\int_{\partial M} |\mathrm{D}u|^{p-2}\Big[(n-1)(n-2) - \mathrm{R}^{\partial M} - 2\big(1 - |\mathrm{D}u|^2\big) \Big] \mathrm{d}\sigma \leq 0 , \tag{2.8}$$

where $\mathrm{R}^{\partial M}$ denotes the scalar curvature of the metric induced by g_0 on ∂M. Moreover, the equality holds for some $p \geq 3$ if and only if (M, g_0, u) is isometric to the de Sitter solution. In particular, the boundary of M has only one connected component and it is isometric to a $(n-1)$-dimensional sphere.

Since the quantity $|\mathrm{D}u|$ is constant on each connected component of ∂M (because $\mathrm{D}^2 u = 0$ on ∂M, as it follows from the first equation in problem (1.4)), if we assume that the boundary is connected, formula (2.8) can be replaced by

$$\int_{\partial M} \Big[(n-1)(n-2) - \mathrm{R}^{\partial M} - 2\big(1 - |\mathrm{D}u|^2\big) \Big] \mathrm{d}\sigma \leq 0 . \tag{2.9}$$

Remark 5 In the case $p = 3$, Theorem 2.6 is a weaker version of Corollary B.2 in the Appendix B. This corollary is not new, but it has been proved in [14] generalizing some early computations in [12] and [29]. In particular, in the case of a connected boundary, from Corollary B.2 it follows the inequality

$$\int_{\partial M} \Big[(n-1)(n-2) - \mathrm{R}^{\partial M} \Big] \mathrm{d}\sigma \leq 0 , \tag{2.10}$$

that is strictly better than our formula (2.9), and is proved without the need of Assumption 1. Note that from inequality (2.10) it follows the remarkable result that the only static solution of (1.4) whose boundary is isometric to a sphere with its standard metric, is the de Sitter solution. This is not a direct consequence of our Theorem 2.6.

To illustrate some other consequences of Theorem 2.6, we rewrite formula (2.8) as

$$\int_{\partial M} |\mathrm{D}u|^p \, \mathrm{d}\sigma \ \leq \ \int_{\partial M} |\mathrm{D}u|^{p-2} \left[\frac{\mathrm{R}^{\partial M} - n\,(n-3)}{2} \right] \mathrm{d}\sigma \,. \tag{2.11}$$

Then we apply Hölder inequality to the right hand side with conjugate exponents $p/(p-2)$ and $p/2$, obtaining

$$\int_{\partial M} |\mathrm{D}u|^{p-2} \left[\frac{\mathrm{R}^{\partial M} - n(n-3)}{2} \right] \mathrm{d}\sigma \leq \left(\int_{\partial M} |\mathrm{D}u|^p \mathrm{d}\sigma \right)^{(p-2)/p} \left(\int_{\partial M} \left| \frac{\mathrm{R}^{\partial M} - n\,(n-3)}{2} \right|^{p/2} \mathrm{d}\sigma \right)^{2/p} .$$

This immediately implies the following corollary, that should be compared with Corollary 2.3.

Corollary 2.7 *Let* (M, g_0, u) *be a solution to problem* (1.4) *satisfying Normalization 1 and Assumption 1. Then, for every* $p \geq 3$*, the inequality*

$$\| \, \mathrm{D}u \, \|_{L^p(\partial M)} \ \leq \ \sqrt{ \left\| \frac{\mathrm{R}^{\partial M} \ - n\,(n-3)}{2} \right\|_{L^{p/2}(\partial M)} } \tag{2.12}$$

holds true, where $\mathrm{R}^{\partial M}$ *denotes the scalar curvature of the metric induced by* g_0 *on* ∂M*. Moreover, the equality holds for some* $p \geq 3$ *if and only if* (M, g_0, u) *is isometric to the de Sitter solution. In particular, the boundary of* M *has only one connected component and it is isometric to a* $(n-1)$*-dimensional sphere.*

Letting $p \to +\infty$ in formula (2.12), we obtain, under the hypotheses of the above corollary, the L^∞-bound

$$\| \, \mathrm{D}u \, \|_{L^\infty(\partial M)} \leq \ \sqrt{ \left\| \frac{\mathrm{R}^{\partial M} \ - n\,(n-3)}{2} \right\|_{L^{\infty}(\partial M)} } \,. \tag{2.13}$$

For our next result, we are going to combine the monotonicity of the U_p's, as stated by Theorem 1.1, together with the estimate (1.13) given in Theorem 1.2.

Theorem 2.8 *Let* (M, g_0, u) *be a solution to problem* (1.4) *satisfying Normalization 1 and Assumption 1. Then, it holds*

$$|\mathrm{MAX}(u)| \, |\mathbb{S}^{n-1}| \ \leq \ \int_{\partial M} |\mathrm{D}u|^p \mathrm{d}\sigma \ \leq \ |\partial M| \,. \tag{2.14}$$

for $0 \leq p \leq 1$ *if* $n = 3$ *and for* $0 \leq p \leq n-1$ *if* $n \geq 4$. *Moreover, the equality* $|\mathrm{MAX}(u)|\,|\mathbb{S}^{n-1}| = |\partial M|$ *holds if and only if* (M, g_0, u) *is isometric to the de Sitter solution.*

Proof First, recalling Assumption 1, it is clear that

$$U_p(0) = \int_{\partial M} |\mathrm{D}u|^p \, \mathrm{d}\sigma \leq |\partial M|,$$

for every $p \geq 0$.

Now consider the case $n \geq 4$ and let $3 \leq p \leq n-1$. From formula (1.13) and Theorem 1.1-(iii), we obtain

$$|\mathrm{MAX}(u)|\,|\mathbb{S}^{n-1}| \leq U_p(0),$$

and the equality holds if and only if $U_p(t)$ is constant, that is, if and only if (M, g_0, u) is isometric to the de Sitter solution. Combining this with the inequality above, we obtain the thesis for $3 \leq p \leq n-1$. If $0 \leq p \leq 3$ instead, to conclude it is enough to observe that $\int_{\partial M} |\mathrm{D}u|^p \mathrm{d}\sigma \geq \int_{\partial M} |\mathrm{D}u|^3 \mathrm{d}\sigma$, thanks to Assumption 1.

In the case $n = 3$, we can repeat the argument above using $U_1(t)$, that we know to be monotonic thanks to Theorem 1.1-(ii). □

Remark 6 The result above is particularly effective in dimension $n = 3$. In that case, it is known from [12] that any solution (M, g_0, u) of problem (1.4) with a connected boundary satisfies $|\partial M| \leq 4\pi$. Since formula (2.14) gives the opposite inequality, we conclude that the only 3-dimensional static solution to problem (1.4) with ∂M connected and satisfying Normalization 1 and Assumption 1 is the de Sitter solution. A direct proof of this fact will be given later (see Theorem 2.11). Note that the same thesis does not hold without Assumption 1. An explicit example of a non-trivial 3-dimensional static solution with a connected boundary diffeomorphic to $\mathbb{S}^2$ (which does not satisfy Assumption 1) can be constructed via a quotient of the Nariai solution (1.6) (see [8, Section 7]).

In the case $n \geq 4$ we are not able to provide such a general result, and the situation seems much wilder. For instance, for any $4 \leq n \leq 8$, one can prove the existence of a countable family of non-trivial static solutions of (1.4) with ∂M connected and diffeomorphic to a sphere or to a product of spheres (see [23]). However, looking at the numerical approximations of some of these solutions, it appears that they do not satisfy our hypotheses, thus the question of the uniqueness of the de Sitter solution under our assumptions seems still open.

Using Corollary 2.7 in place of Corollary 2.3 we obtain the following analog of Corollary 2.5.

Theorem 2.9 *Let* (M, g_0, u) *be a solution to problem* (1.4) *satisfying Normalization 1 and Assumption 1. Then, the following statements hold true.*

(i) *For every $p \geq 2$, the inequality*

$$\frac{|\mathrm{MAX}(u)|\,|\mathbb{S}^{n-1}|}{|\partial M|} \leq \fint_{\partial M} \left| \frac{\mathrm{R}^{\partial M} - n\,(n-3)}{2} \right|^{\frac{p}{2}} \mathrm{d}\sigma \tag{2.15}$$

holds true. Moreover, the equality is fulfilled for some $p \geq 2$, if and only if the static solution (M, g_0, u) is isometric to the de Sitter solution. In particular, it holds

$$2 \leq \left\| \mathrm{R}^{\partial M} - n\,(n-3) \right\|_{L^\infty(\{u=t\})}. \tag{2.16}$$

(ii) *The inequality*

$$\frac{|\mathrm{MAX}(u)|\,|\mathbb{S}^{n-1}|}{|\partial M|} \leq \fint_{\partial M} \frac{\mathrm{R}^{\partial M}}{(n-1)(n-2)}\,\mathrm{d}\sigma \tag{2.17}$$

holds true. Moreover, the equality is fulfilled if and only if the static solution (M, g_0, u) is isometric to the de Sitter solution.

Proof For $n \geq 4$ and $3 \leq p \leq n-1$, statements (i) and (ii) can be derived from inequality (2.12) in Corollary 2.7 and formula (2.14) in Theorem 2.8. In general, we need to use inequality (B.6) in Corollary B.2, proved in the Appendix B.

To prove statement (i), we rewrite formula (B.6) as

$$\int_{\partial M} |\mathrm{D}u|\,\mathrm{d}\sigma \leq \int_{\partial M} |\mathrm{D}u| \left[\frac{\mathrm{R}^{\partial M} - n(n-3)}{2} \right] \mathrm{d}\sigma\,. \tag{2.18}$$

Compare this inequality with (2.11), which holds for every $p \geq 3$ but is weaker than (2.18) in the case $p = 3$.

Using Hölder inequality, we have

$$\int_{\partial M} |\mathrm{D}u| \left[\frac{\mathrm{R}^{\partial M} - n(n-3)}{2} \right] \mathrm{d}\sigma \leq \left[\int_{\partial M} \left| \frac{\mathrm{R}^{\partial M} - n(n-3)}{2} \right|^{\frac{p}{2}} \mathrm{d}\sigma \right]^{\frac{2}{p}} \left(\int_{\partial M} |\mathrm{D}u|^{\frac{p}{p-2}}\,\mathrm{d}\sigma \right)^{\frac{p-2}{p}}.$$

Moreover, since $|\mathrm{D}u| \leq 1$ on ∂M thanks to Assumption 1, we have $|\mathrm{D}u|^{\frac{p}{p-2}} \leq |\mathrm{D}u|$ for every $p \geq 2$. Substituting in (2.18), with some easy computations we find

$$\frac{\|\mathrm{D}u\|_{L^1(\partial M)}}{|\partial M|} \leq \fint_{\partial M} \left| \frac{\mathrm{R}^{\partial M} - n\,(n-3)}{2} \right|^{\frac{p}{2}} \mathrm{d}\sigma$$

Now using inequality (2.14) we obtain (2.15). Moreover, if the equality holds in (2.15), then also (2.18) is an equality, thus by Corollary B.2 we have that (M, g_0, u) is the de Sitter solution.

Statement (ii), is an immediate consequence of formula (2.18) and inequality (2.14) in Theorem 2.8. □

If we set $p = 2(n-1)$ in Theorem 2.9-(i), we obtain the following nicer statement.

Corollary 2.10 (Willmore-Type Inequality) *Let (M, g_0, u) be a static solution to problem (1.4), satisfying Normalization 1 and Assumption 1. Then, it holds*

$$\left(|\mathrm{MAX}(u)|\,|\mathbb{S}^{n-1}|\right)^{\frac{1}{n-1}} \leq \left\| \frac{\mathrm{R}^{\partial M} - n\,(n-3)}{2} \right\|_{L^{n-1}(\partial M)},$$

where $\mathrm{R}^{\partial M}$ denotes the scalar curvature of the metric induced by g_0 on ∂M. Moreover, the equality holds if and only if (M, g_0, u) is isometric to the de Sitter solution. In particular, the boundary of M has only one connected component and it is isometric to a $(n-1)$-dimensional sphere.

The result above should be compared with [4, Theorem 2.11-(ii)] where a similar inequality is provided for the Schwarzschild metric [35].

For our next result we restrict to dimension $n = 3$, and we use the Gauss-Bonnet Formula to prove that, in the hypothesis of a connected boundary, the equality is achieved in formula (2.17).

Theorem 2.11 (Uniqueness Theorem) *Let (M, g_0, u) be a 3-dimensional static solution to problem (1.4), satisfying Normalization 1 and Assumption 1. If ∂M is connected, then (M, g_0, u) is isometric to the de Sitter solution. More generally, let $\partial M = \sqcup_{i=1}^{r} \Sigma_i$, where $\Sigma_1, \dots, \Sigma_r$ are connected surfaces. Then*

$$2\,|\mathrm{MAX}(u)| \leq \sum_{i=1}^{r} k_i\, \chi(\Sigma_i)\,, \tag{2.19}$$

where k_i is the surface gravity of Σ_i, that is, the constant value of $|\mathrm{D}u|$ on Σ_i. Moreover, the equality holds if and only if ∂M is connected and (M, g_0, u) is isometric to the de Sitter solution.

Proof Again, it is useful to use Corollary B.2, proved in the Appendix B. Setting $n = 3$ in formula (B.6), we obtain

$$2\,\|\mathrm{D}u\|_{L^1(\partial M)} \leq \int_{\partial M} |\mathrm{D}u|\,\mathrm{R}^{\partial M}\, \mathrm{d}\sigma\,,$$

where the equality holds if and only if (M, g_0, u) is isometric to the de Sitter solution. Recalling formula (2.14), we obtain

$$8\pi\,|\mathrm{MAX}(u)| \;\leq\; \sum_{i=1}^{r} k_i \int_{\Sigma_i} \mathrm{R}^{\Sigma_i}\, d\sigma\,.$$

The thesis is now a consequence of the equalities

$$\int_{\Sigma_i} \mathrm{R}^{\Sigma_i}\, d\sigma \;=\; 4\pi\,\chi(\Sigma_i)\,, \qquad \text{for all } i = 1, \ldots, r\,,$$

which follow from the Gauss-Bonnet theorem. □

Combining the theorem above with the results in [8], we obtain the following strengthening of formula (2.19).

Corollary 2.12 *Let (M, g_0, u) be a 3-dimensional static solution to problem* (1.4), *satisfying Normalization 1 and Assumption 1. If (M, g_0, u) is not isometric to the de Sitter solution, then*

$$3\,|\mathrm{MAX}(u)| \;<\; \sum_{i=1}^{r} k_i \;\leq\; \pi_0(\partial M)\,,$$

where $k_1, \ldots, k_r$ are the surface gravities of the connected components $\Sigma_1, \ldots, \Sigma_r$ of ∂M. In particular, a non-trivial 3-dimensional static solution satisfying Normalization 1 and Assumption 1, must have a boundary with at least four connected components.

Proof Let $(\tilde{M}, \tilde{g}_0) \xrightarrow{\pi} (M, g_0)$ be the universal covering. Clearly the triple $(\tilde{M}, \tilde{g}_0, \tilde{u} = u \circ \pi)$ is still a solution of problem (1.4) and satisfies Assumption 1 and Normalization 1. From [8, Theorem B], we know that $(\tilde{M}, \tilde{g}_0)$ is compact. In particular, the degree d of the covering π is a finite number and $|\mathrm{MAX}(\tilde{u})| = d\,|\mathrm{MAX}(u)|$.

Let $\partial\tilde{M} = \sqcup_{i=1}^{s} \tilde{\Sigma}_i$, where $\tilde{\Sigma}_1, \ldots, \tilde{\Sigma}_s$ are connected. From [8, Theorem C], we have that $(\tilde{M}, \tilde{g}_0, \tilde{u})$ is isometric to the de Sitter triple or

$$\sum_{i=1}^{s} \tilde{k}_i\,|\tilde{\Sigma}_i| \;<\; \frac{4\pi}{3} \sum_{i=1}^{s} \tilde{k}_i\,, \tag{2.20}$$

where $\tilde{k}_i$ is the surface gravity of $\tilde{\Sigma}_i$ for all $i = 1, \ldots, s$.

If $(\tilde{M}, \tilde{g}_0, \tilde{u})$ is isometric to the de Sitter triple, then $\partial\tilde{M}$ is connected, hence also ∂M is connected. Recalling Theorem 2.11, we deduce that (M, g_0, u) is isometric to the de Sitter solution, against our hypotheses.

Therefore, formula (2.20) must hold. Recalling Theorem 2.14, we obtain the following chain of inequalities

$$4\pi\, d\,|\mathrm{MAX}(u)| \,=\, 4\pi\,|\mathrm{MAX}(\tilde{u})| \,\leq\, \int\limits_{\partial\tilde{M}} |\tilde{\mathrm{D}}\tilde{u}|\,\mathrm{d}\tilde{\sigma} \,=\, \sum_{i=1}^{s} \tilde{k}_i\,|\tilde{\Sigma}_i| \,<\, \frac{4\pi}{3}\sum_{i=1}^{s}\tilde{k}_i\,.$$

Since each connected component of ∂M lifts to at most d connected components of $\partial\tilde{M}$, we have

$$\sum_{i=1}^{s}\tilde{k}_i \,\leq\, d\sum_{i=1}^{r}k_i\,.$$

This proves the first part of the statement. The inequality $\sum_{i=1}^{r}k_i \leq \pi_0(\partial M)$ is a consequence of Assumption 1. □

2.3 *Consequences on a Generic Level Set of u (Case* $\Lambda < 0$*)*

Now we start to discuss the consequences in the case of a negative cosmological constant. Since, as already observed, the functions $t \mapsto U_p(t)$ defined in (1.18) are constant on the anti-de Sitter solution, we obtain from Theorem 1.4 and formula (1.22), the following characterizations of the rotationally symmetric solutions to system (1.17).

Theorem 2.13 *Let* (M, g_0, u) *be a solution to problem* (1.17) *satisfying Normalization 2 and Assumption 2. Then, for every* $p \geq 3$ *and every* $t \in (1, +\infty)$*, it holds*

$$\int\limits_{\{u=t\}} |\mathrm{D}u|^{p-2}\Big[\,(n-1) + \mathrm{Ric}(\nu,\nu) + \Big(1 - \frac{|\mathrm{D}u|^2}{1-u^2}\Big)\Big]\,\mathrm{d}\sigma \,\geq\, 0\,,$$

where $\nu = \mathrm{D}u/|\mathrm{D}u|$*. Moreover, the equality is fulfilled for some* $p \geq 3$ *and some* $t \in (1, +\infty)$ *if and only if the static solution* (M, g_0, u) *is isometric to the anti-de Sitter solution.*

Theorem 2.14 *Let* (M, g_0, u) *be a solution to problem* (1.17) *satisfying Normalization 2 and Assumption 2. Then, for every* $p \geq 3$ *and every* $t \in (1, +\infty)$*, the inequality*

$$\int\limits_{\{u=t\}} \Big(\frac{|\mathrm{D}u|}{\sqrt{u^2-1}}\Big)^{p}\mathrm{d}\sigma \,\leq\, \int\limits_{\{u=t\}} \Big(\frac{|\mathrm{D}u|}{\sqrt{u^2-1}}\Big)^{p-2}\Big[-\mathrm{H}\,|\mathrm{D}\log u| + n\Big]\,\mathrm{d}\sigma \tag{2.21}$$

holds true, where H *is the mean curvature of the level set* $\{u = t\}$. *Moreover, the equality is fulfilled for some* $p \geq 3$ *and some* $t \in (1, +\infty)$ *if and only if the static solution* (M, g_0, u) *is isometric to the anti-de Sitter solution.*

To illustrate other implications of Theorem 2.14, let us observe that, applying Hölder inequality to the right hand side of (2.21) with conjugate exponents $p/(p-2)$ and $p/2$, one gets

$$\int_{\{u=t\}} \left(\frac{|\mathrm{D}u|}{\sqrt{u^2-1}}\right)^{p-2} \Big[-\mathrm{H}|\mathrm{D}\log u| + n \Big] \mathrm{d}\sigma$$

$$\leq \left(\int_{\{u=t\}} \left(\frac{|\mathrm{D}u|}{\sqrt{u^2-1}}\right)^{p} \mathrm{d}\sigma \right)^{\frac{p-2}{p}} \left(\int_{\{u=t\}} \Big| -\mathrm{H}|\mathrm{D}\log u| + n \Big|^{\frac{p}{2}} \mathrm{d}\sigma \right)^{\frac{2}{p}}.$$

This implies on every level set of u the following sharp L^p-bound for the gradient of the *static potential* in terms of the L^p-norm of the mean curvature of the level set.

Corollary 2.15 *Let* (M, g_0, u) *be a solution to problem* (1.17) *satisfying Normalization 2 and Assumption 2. Then, for every* $p \geq 3$ *and every* $t \in (1, +\infty)$ *the inequality*

$$\left\| \frac{\mathrm{D}u}{\sqrt{u^2-1}} \right\|_{L^p(\{u=t\})} \leq \sqrt{\Big\| -\mathrm{H}\,|\mathrm{D}\log u| + n \Big\|_{L^{p/2}(\{u=t\})}}, \tag{2.22}$$

holds true, where H *is the mean curvature of the level set* $\{u = t\}$. *Moreover, the equality is fulfilled for some* $p \geq 3$ *and some* $t \in (1, +\infty)$ *if and only if the static solution* (M, g_0, u) *is isometric to the anti-de Sitter solution.*

It is worth pointing out that the right hand side in (2.22) may possibly be unbounded. However, for regular level sets of the *static potential* the L^p-norm of the mean curvature is well defined and finite (see Remark 3). We also observe that letting $p \to +\infty$, we deduce, under the same hypothesis of Corollary 2.15, the following L^∞-bound

$$\left\| \frac{\mathrm{D}u}{\sqrt{u^2-1}} \right\|_{L^\infty(\{u=t\})} \leq \sqrt{\Big\| -\mathrm{H}\,|\mathrm{D}\log u| + n \Big\|_{L^{\infty}(\{u=t\})}}, \tag{2.23}$$

for every $t \in (1, +\infty)$. Unfortunately, in this case we do not know whether the rigidity statement holds true or not. However, the equality is satisfied on the anti-de Sitter solution and this makes the inequality sharp.

Now we will combine inequality (2.22) in Corollary 2.15 with the observation that for every $t \in (1, +\infty)$ and every $3 \le p \le n-1$ (we need to take $n \ge 4$) it holds

$$U_p(t) \ge |\mathrm{MIN}(u)|\, |\mathbb{S}^{n-1}|\,, \tag{2.24}$$

where the latter estimate follows from estimate (1.24) in Theorem 1.5 and the monotonicity of the U_p's stated in Theorem 1.4-(iii). Recalling the explicit expression (1.18) of the U_p's, we obtain the following analogue of Theorem 2.4.

Theorem 2.16 *Let $n \ge 4$. Let (M, g_0, u) be a solution to problem (1.17) satisfying Normalization 2 and Assumption 2. Then, for every $3 \le p \le n-1$ and every $t \in (1, +\infty)$, the inequalities*

$$(t^2-1)^{\frac{n-1}{2}}\, |\mathrm{MIN}(u)|\, |\mathbb{S}^{n-1}| \le \Big\| -\mathrm{H}\,|\mathrm{D}\log u| + n \Big\|^{\frac{p}{2}}_{L^{p/2}(\{u=t\})} \tag{2.25}$$

hold true. Moreover, the equality is fulfilled for some $t \in (1, +\infty)$ and some $3 \le p \le n-1$, if and only if the static solution (M, g_0, u) is isometric to the anti-de Sitter solution.

To give a geometric interpretation of the above theorem, we recall the identity

$$-\mathrm{H}\,|\mathrm{D}\log u| + n = n + \mathrm{Ric}(\nu, \nu)\,,$$

and we observe that the quantity $(t^2-1)^{\frac{n-1}{2}}\, |\mathbb{S}^{n-1}|$ corresponds to the hypersurface area of the level set $\{u_A = t\}$ in the anti-de Sitter solution (1.3). Combining together these two facts, we arrive at the following corollary, that should be compared with Corollary 2.5.

Corollary 2.17 *Let $n \ge 4$. Let (M, g_0, u) be a solution to problem (1.17) satisfying Normalization 2 and Assumption 2. Then, for every $p \ge 3$ and every $t \in (1, +\infty)$, the inequality*

$$\frac{|\mathrm{MIN}(u)|\, |\{u_A = t\}|}{|\{u=t\}|} \le \fint_{\{u=t\}} \big| n + \mathrm{Ric}(\nu, \nu) \big|^{\frac{p}{2}}\, \mathrm{d}\sigma \tag{2.26}$$

holds true. Moreover, the equality is fulfilled for some $t \in (1, +\infty)$ and some $p \ge 3$, if and only if the static solution (M, g_0, u) is isometric to the anti-de Sitter solution. In particular, for every $t \in (1, +\infty)$, it holds

$$1 \le \big\| n + \mathrm{Ric}(\nu, \nu) \big\|_{L^\infty(\{u=t\})}\,. \tag{2.27}$$

2.4 The Geometry of ∂M (Case $\Lambda < 0$)

We pass now to describe some consequences of the behaviour of the *static solution* (M, g_0, u) at the conformal boundary ∂M, as prescribed by Theorem 1.4-(iv). We remark that the conformal boundary of M is a totally geodesic hypersurface inside $(\overline{M}, g)$ (see Lemma A.8-(ii) in the Appendix A).

The following theorem is a rephrasing of formula (1.23) and is the analogue of Theorem 2.6.

Theorem 2.18 *Let* (M, g_0, u) *be a solution to problem* (1.17) *satisfying Normalization 2 and Assumption 2, and let* $g = g_0/(u^2 - 1)$. *Then it holds*

$$\int_{\partial M} \left[(n-1)(n-2) - \mathrm{R}_g^{\partial M} + n(n-1)\left(u^2 - 1 - |\mathrm{D}u|^2\right)\right] \mathrm{d}\sigma_g \geq 0\,, \qquad (2.28)$$

where $\mathrm{R}_g^{\partial M}$ *denotes the scalar curvature of the metric induced by* g *on* ∂M.

Note that inequality (2.28) is sharp, but the rigidity statement does not hold for Theorem 2.18. Moreover, unlike Theorem 2.6, formula (2.28) does not depend on p and we are not able to find an analogue of Corollary 2.7 for the case $\Lambda < 0$.

We can still provide the following result, that should be compared with Theorem 2.8.

Theorem 2.19 *Let* (M, g_0, u) *be a static solution to problem* (1.17), *satisfying Normalization 2 and Assumption 2. Let* $|\mathrm{MIN}(u)|$ *be the cardinality of the set* $\mathrm{MIN}(u)$ *of the points where* u *attains its minimum and let* $g = g_0/(u^2 - 1)$. *Then*

$$|\mathrm{MIN}(u)|\, |\mathbb{S}^{n-1}| \leq |\partial M|_g\,, \qquad (2.29)$$

and the equality holds if and only if (M, g_0, u) *is isometric to the anti-de Sitter solution.*

Proof From formula (1.24) and Theorem 1.4-(ii) we obtain

$$|\mathrm{MIN}(u)|\, |\mathbb{S}^{n-1}| \leq \lim_{t \to +\infty} U_1(t)\,,$$

and the equality holds if and only if $U_1(t)$ is constant, that is, if and only if (M, g_0, u) is isometric to the anti-de Sitter solution. On the other hand

$$\lim_{t \to +\infty} U_p(t) = \lim_{t \to +\infty} \int_{\{u=t\}} \frac{|\mathrm{D}u|}{(u^2-1)^{\frac{n}{2}}}\, \mathrm{d}\sigma = \lim_{t \to +\infty} \int_{\{u=t\}} \sqrt{\frac{|\mathrm{D}u|^2}{u^2-1}}\, \mathrm{d}\sigma_g \leq |\partial M|_g\,,$$

where the last inequality follows from Assumption 2. □

An immediate corollary of Theorem 2.19 above is the following uniqueness result.

Corollary 2.20 *Let (M, g_0, u) be a conformally compact static solution to problem* (1.17) *satisfying Normalization 2 and Assumption 2. If the conformal boundary is isometric to the sphere $(\mathbb{S}^{n-1}, g_{\mathbb{S}^{n-1}})$, then (M, g_0, u) is isometric to the anti-de Sitter solution.*

The result above should be compared with the uniqueness theorems in [15, 34, 39], where the same thesis is obtained for $n \leq 7$ or M spin, without the need of Assumption 2.

In order to have a clearer exposition, and to highlight the analogies between the results in this section and the ones in Sect. 2.2, for the rest of this section we will assume the following stronger version of Assumption 2.

Assumption 2-bis *The triple (M, g_0, u) is conformally compact, the function $1/\sqrt{u^2-1}$ is a defining function for ∂M and $\lim_{x\to\bar{x}}\left(u^2-1-|\mathrm{D}u|^2\right) = 0$ for every $\bar{x} \in \partial M$.*

First, we observe that with this additional hypothesis, formula (2.29) in Theorem 2.18 becomes

$$\int_{\partial M} \left[(n-1)(n-2) - \mathrm{R}_g^{\partial M}\right] d\sigma_g \geq 0\,, \tag{2.30}$$

Now we use formula (2.30) to prove the analogue of Theorem 2.9-(i).

Theorem 2.21 *Let (M, g_0, u) be a solution to problem* (1.17) *satisfying Normalization 2 and Assumption 2-bis, and let $g = g_0/(u^2-1)$. Then for every $p \geq 2$ it holds*

$$\frac{|\mathrm{MIN}(u)|\,|\mathbb{S}^{n-1}|}{|\partial M|_g} \leq \fint_{\partial M} \left|\frac{\mathrm{R}_g^{\partial M} - (n+1)(n-2)}{2\,(n-2)}\right|^{\frac{p}{2}} d\sigma_g\,, \tag{2.31}$$

where $\mathrm{R}_g^{\partial M}$ denotes the scalar curvature of the metric induced by g on ∂M.

Proof First, we rearrange formula (2.30) in the following way

$$|\partial M|_g \leq \int_{\partial M} \left[\frac{-\mathrm{R}_g^{\partial M} + (n+1)(n-2)}{2(n-2)}\right] d\sigma_g\,.$$

Now we rewrite the right hand side of the above formula, using Jensen Inequality. We obtain

$$|\partial M|_g \leq |\partial M|_g^{\frac{p-2}{p}} \left[\int_{\partial M} \left|\frac{\mathrm{R}_g^{\partial M} - (n+1)(n-2)}{2(n-2)}\right|^{\frac{p}{2}} d\sigma_g\right]^{\frac{2}{p}},$$

that may be rewritten as

$$|\partial M|_g \leq \int_{\partial M} \left| \frac{\mathrm{R}_g^{\partial M} - (n+1)(n-2)}{2(n-2)} \right|^{\frac{p}{2}} \mathrm{d}\sigma_g .$$

Now the thesis is an immediate consequence of Theorem 2.19. □

Finally, setting $p = 2(n-1)$ in Theorem 2.21 above, we obtain the analogue of Corollary 2.10.

Corollary 2.22 (Willmore-Type Inequality) *Let (M, g_0, u) be a solution to problem* (1.17) *satisfying Normalization 2 and Assumption 2-bis, and let $g = g_0/(u^2 - 1)$. Then it holds*

$$\left(|\mathrm{MIN}(u)|\, |\mathbb{S}^{n-1}| \right)^{\frac{1}{n-1}} \leq \left\| \frac{\mathrm{R}_g^{\partial M} - (n+1)(n-2)}{2\,(n-2)} \right\|_{L^{n-1}(\partial M)}, \qquad (2.32)$$

where $\mathrm{R}_g^{\partial M}$ denotes the scalar curvature of the metric induced by g on ∂M.

3 A Conformally Equivalent Formulation of the Problem

The aim of this section is to reformulate system (1.4) and system (1.17) in a conformally equivalent setting.

3.1 A Conformal Change of Metric (Case $\Lambda > 0$)

First of all, we notice that if (M, g_0, u) is a solution of problem (1.4) and satisfies Normalization 1, then one has that $1 - u^2 > 0$ everywhere in $M^* = M \setminus \mathrm{MAX}(u)$.

Motivated by the explicit formulæ (1.2) of the de Sitter solution, we are led to consider the following conformal change of metric

$$g = \frac{g_0}{1 - u^2} . \qquad (3.1)$$

on the manifold M^*. It is immediately seen that when u and g_0 are as in (1.2) then g is a cylindrical metric. Hence, we will refer to the conformal change (3.1) as to a *cylindrical ansatz*.

Our next task is to reformulate problem (1.4) in terms of g. To this aim we fix local coordinates $\{y^\alpha\}_{\alpha=1}^n$ in M^* and using standard formulæ for conformal changes of metrics, we deduce that the Christoffel symbols $\Gamma^{\gamma}_{\alpha\beta}$ and $G^{\gamma}_{\alpha\beta}$, of the metric g and g_0 respectively, are related to each other via the identity

$$\Gamma^{\gamma}_{\alpha\beta} = G^{\gamma}_{\alpha\beta} + \frac{u}{1-u^2}\left(\delta^{\gamma}_{\alpha}\,\partial_{\beta}u + \delta^{\gamma}_{\beta}\,\partial_{\alpha}u - (g_0)_{\alpha\beta}(g_0)^{\gamma\eta}\partial_{\eta}u\right) \tag{3.2}$$

Comparing the local expressions for the Hessians of a given function $w \in \mathscr{C}^2(M^*)$ with respect to the metrics g and g_0, namely $\nabla^2_{\alpha\beta}w = \partial^2_{\alpha\beta}w - \Gamma^{\gamma}_{\alpha\beta}\partial_{\gamma}w$ and $D^2_{\alpha\beta}w = \partial^2_{\alpha\beta}w - G^{\gamma}_{\alpha\beta}\partial_{\gamma}w$, one gets

$$\nabla^2_{\alpha\beta}w = D^2_{\alpha\beta}w - \frac{u}{1-u^2}\left(\partial_{\alpha}u\,\partial_{\beta}w + \partial_{\alpha}w\,\partial_{\beta}u - \langle Du \,|\, Dw\rangle\, g^{(0)}_{\alpha\beta}\right)$$

$$\Delta_g w = (1-u^2)\Delta w + (n-2)u\,\langle Du \,|\, Dw\rangle_{g_0}$$

We note that in the above expressions as well as in the following ones, the notations ∇ and Δ_g represent the Levi-Cita connection and the Laplace–Beltrami operator of the metric g. In particular, letting $w = u$ and using $\Delta u = -n\,u$, one has

$$\nabla^2_{\alpha\beta}u \;=\; D^2_{\alpha\beta}u \;-\; \frac{u}{1-u^2}\left(2\,\partial_{\alpha}u\,\partial_{\beta}u \;-\; |Du|^2\,g^{(0)}_{\alpha\beta}\right), \tag{3.3}$$

$$\Delta_g u \;=\; -n\,u\,(1-u^2) + \,(n-2)\;|Du|^2\,. \tag{3.4}$$

To continue, we observe that the Ricci tensor $\mathrm{Ric}_g = R^{(g)}_{\alpha\beta}\,dy^{\alpha}\otimes dy^{\beta}$ of the metric g can be expressedin terms of the Ricci tensor $\mathrm{Ric} = R^{(0)}_{\alpha\beta}\,dy^{\alpha}\otimes dy^{\beta}$ of the metric g_0 as

$$R^{(g)}_{\alpha\beta} = R^{(0)}_{\alpha\beta} - \frac{(n-2)u}{1-u^2}\,D^2_{\alpha\beta}u - \frac{n-2}{(1-u^2)^2}\,\partial_{\alpha}u\,\partial_{\beta}u - \left(\frac{u\,\Delta u}{1-u^2} + \frac{(n-1)u^2+1}{(1-u^2)^2}\,|Du|^2\right)g^{(0)}_{\alpha\beta}. \tag{3.5}$$

If we plug equations $\Delta u = -n\,u$ and $u\,\mathrm{Ric} = D^2u + nug_0$ in the above formula we obtain:

$$R^{(g)}_{\alpha\beta} = \frac{1-(n-1)u^2}{u(1-u^2)}D^2_{\alpha\beta}u - \frac{n-2}{(1-u^2)^2}\,\partial_{\alpha}u\,\partial_{\beta}u + \left(\frac{n}{1-u^2} - \frac{(n-1)u^2+1}{(1-u^2)^2}\,|Du|^2\right)g^{(0)}_{\alpha\beta} \tag{3.6}$$

In order to obtain nicer formulæ, it is convenient to introduce the new variable

$$\varphi = \frac{1}{2}\log\left(\frac{1+u}{1-u}\right) \qquad \Longleftrightarrow \qquad u = \tanh(\varphi). \tag{3.7}$$

As a consequence, we have that

$$\partial_\alpha \varphi = \frac{1}{1-u^2} \partial_\alpha u \tag{3.8}$$

$$\nabla^2_{\alpha\beta} \varphi = \frac{1}{1-u^2} \, \mathrm{D}^2_{\alpha\beta} u + \frac{u}{(1-u^2)^2} \, |\mathrm{D}u|^2 \, g^{(0)}_{\alpha\beta} \tag{3.9}$$

For future convenience, we report the relation between $|\nabla\varphi|^2_g$ and $|\mathrm{D}u|^2$ as well as the one between $|\nabla^2\varphi|^2_g$ and $|\mathrm{D}^2 u|^2$, namely

$$\begin{aligned} |\nabla\varphi|^2_g &= \frac{|\mathrm{D}u|^2}{1-u^2}, \\ |\nabla^2\varphi|^2_g &= |\mathrm{D}^2 u|^2 + n\, u^2 \, \frac{|\mathrm{D}u|^2}{1-u^2} \left(\frac{|\mathrm{D}u|^2}{1-u^2} - 2 \right). \end{aligned} \tag{3.10}$$

Combining expressions (3.3), (3.4), (3.6) together with (3.8), (3.9), we are now in the position to reformulate problem (1.4) as

$$\begin{cases} \mathrm{Ric}_g = (\coth(\varphi) - (n-1)\tanh(\varphi))\, \nabla^2\varphi - (n-2) d\varphi \otimes d\varphi + \left(n - 2|\nabla\varphi|^2_g\right) g, & \text{in } M^* \\ \Delta_g \varphi = -n \tanh(\varphi) \left(1 - |\nabla\varphi|^2_g\right), & \text{in } M^* \\ \varphi = 0, & \text{on } \partial M^* \\ \varphi \to +\infty & \text{as } x \to * \end{cases} \tag{3.11}$$

Here we recall that M^* is the manifold $M \setminus \mathrm{MAX}(u)$. The notation $x \to *$ means that $x \to p$, where p is a point of $\mathrm{MAX}(u)$, with respect to the topology induced by M on M^*.

3.2 A Conformal Change of Metric (Case $\Lambda < 0$)

First of all, we notice that if (M, g_0, u) is a solution of problem (1.17) and satisfies Normalization 2, then one has that $u^2 - 1 > 0$ everywhere in $M^* = M \setminus \mathrm{MIN}(u)$. Motivated by the explicit formulæ (1.3) of the anti-de Sitter solution, we are led to consider the following conformal change of metric

$$g = \frac{g_0}{u^2 - 1}. \tag{3.12}$$

on the manifold M^*. Notice that, if Assumption 2 holds, the function $1/\sqrt{u^2-1}$ is a defining function, hence the metric g extends to the conformal boundary. In particular the volume of ∂M with respect to g is finite, that is

$$|\partial M|_g = \lim_{t\to+\infty} \int_{\{u=t\}} \mathrm{d}\sigma_g < +\infty .$$

It is immediately seen that when u and g_0 are as in (1.3) then g is a cylindrical metric. Hence, we will refer to the conformal change (3.12) as to a *cylindrical ansatz*.

Our next task is to reformulate problem (1.17) in terms of g. To this aim we fix local coordinates $\{y^\alpha\}_{\alpha=1}^n$ in M^* and using standard formulæ for conformal changes of metrics, we deduce that the Christoffel symbols $\Gamma_{\alpha\beta}^\gamma$ and $\mathrm{G}_{\alpha\beta}^\gamma$, of the metric g and g_0 respectively, are related to each other via the identity

$$\Gamma_{\alpha\beta}^\gamma = \mathrm{G}_{\alpha\beta}^\gamma - \frac{u}{u^2-1}\Big(\delta_\alpha^\gamma\,\partial_\beta u + \delta_\beta^\gamma\,\partial_\alpha u - (g_0)_{\alpha\beta}(g_0)^{\gamma\eta}\partial_\eta u\Big) \tag{3.13}$$

Comparing the local expressions for the Hessians of a given function $w \in \mathscr{C}^2(M^*)$ with respect to the metrics g and g_0, namely $\nabla^2_{\alpha\beta} w = \partial^2_{\alpha\beta} w - \Gamma^\gamma_{\alpha\beta}\partial_\gamma w$ and $\mathrm{D}^2_{\alpha\beta} w = \partial^2_{\alpha\beta} w - \mathrm{G}^\gamma_{\alpha\beta}\partial_\gamma w$, one gets

$$\nabla^2_{\alpha\beta} w = \mathrm{D}^2_{\alpha\beta} w + \frac{u}{u^2-1}\Big(\partial_\alpha u\,\partial_\beta w + \partial_\alpha w\,\partial_\beta u - \langle \mathrm{D}u \,|\, \mathrm{D}w\rangle\, g^{(0)}_{\alpha\beta}\Big)$$

$$\Delta_g w = (u^2-1)\Delta w - (n-2)u\,\langle \mathrm{D}u \,|\, \mathrm{D}w\rangle_{g_0}$$

We note that in the above expressions as well as in the following ones, the notations ∇ and Δ_g represent the Levi-Cita connection and the Laplace–Beltrami operator of the metric g. In particular, letting $w = u$ and using $\Delta u = n\,u$, one has

$$\nabla^2_{\alpha\beta} u \;=\; \mathrm{D}^2_{\alpha\beta} u \;+\; \frac{u}{u^2-1}\Big(2\,\partial_\alpha u\,\partial_\beta u \;-\; |\mathrm{D}u|^2\, g^{(0)}_{\alpha\beta}\Big), \tag{3.14}$$

$$\Delta_g u \;=\; n\,u\,(u^2-1) - \;(n-2)\;|\mathrm{D}u|^2 . \tag{3.15}$$

To continue, we observe that the Ricci tensor $\mathrm{Ric}_g = \mathrm{R}^{(g)}_{\alpha\beta}\, dy^\alpha \otimes dy^\beta$ of the metric g can be expressedin terms of the Ricci tensor $\mathrm{Ric} = \mathbb{R}^{(0)}_{\alpha\beta}\, dy^\alpha \otimes dy^\beta$ of the metric g_0 as

$$\mathrm{R}^{(g)}_{\alpha\beta} = \mathrm{R}^{(0)}_{\alpha\beta} + \frac{(n-2)u}{u^2-1}\,\mathrm{D}^2_{\alpha\beta} u - \frac{n-2}{(u^2-1)^2}\,\partial_\alpha u\,\partial_\beta u + \left(\frac{u\,\Delta u}{u^2-1} - \frac{(n-1)u^2+1}{(u^2-1)^2}\,|\mathrm{D}u|^2\right) g^{(0)}_{\alpha\beta}. \tag{3.16}$$

If we plug equations $\Delta u = n\,u$ and $u\,\mathrm{Ric} = \mathrm{D}^2 u - n u g_0$ in the above formula we obtain

$$\mathrm{R}^{(g)}_{\alpha\beta} = \frac{(n-1)u^2 - 1}{u(u^2-1)} \mathrm{D}^2_{\alpha\beta} u - \frac{n-2}{(u^2-1)^2}\, \partial_\alpha u\, \partial_\beta u + \left(\frac{n}{u^2-1} - \frac{(n-1)u^2+1}{(u^2-1)^2}\, |\mathrm{D}u|^2 \right) g^{(0)}_{\alpha\beta} \tag{3.17}$$

In order to obtain nicer formulæ, it is convenient to introduce the new variable

$$\varphi = \frac{1}{2}\log\left(\frac{u+1}{u-1}\right) \qquad \Longleftrightarrow \qquad u = \coth(\varphi). \tag{3.18}$$

As a consequence, we have that

$$\partial_\alpha \varphi = -\frac{1}{u^2-1}\, \partial_\alpha u \tag{3.19}$$

$$\nabla^2_{\alpha\beta}\varphi = -\frac{1}{u^2-1}\, \mathrm{D}^2_{\alpha\beta} u + \frac{u}{(u^2-1)^2}\, |\mathrm{D}u|^2\, g^{(0)}_{\alpha\beta} \tag{3.20}$$

For future convenience, we report the relation between $|\nabla\varphi|^2_g$ and $|\mathrm{D}u|^2$ as well as the one between $|\nabla^2\varphi|^2_g$ and $|\mathrm{D}^2 u|^2$, namely

$$\begin{aligned} |\nabla\varphi|^2_g &= \frac{|\mathrm{D}u|^2}{u^2-1}, \\ |\nabla^2\varphi|^2_g &= |\mathrm{D}^2 u|^2 + n\,u^2\, \frac{|\mathrm{D}u|^2}{u^2-1}\left(\frac{|\mathrm{D}u|^2}{u^2-1} - 2\right). \end{aligned} \tag{3.21}$$

Combining expressions (3.14), (3.15), (3.17) together with (3.19), (3.20), we are now in the position to reformulate problem (1.17) as

$$\begin{cases} \mathrm{Ric}_g = (\tanh(\varphi) - (n-1)\coth(\varphi))\, \nabla^2\varphi - (n-2) d\varphi \otimes d\varphi + \left(n - 2|\nabla\varphi|^2_g\right) g\,, & \text{in } M^* \\ \Delta_g \varphi = -n\, \coth(\varphi)\, \left(1 - |\nabla\varphi|^2_g\right), & \text{in } M^* \\ \varphi = 0\,, & \text{on } \partial M^* \\ \varphi \to +\infty & \text{as } x \to *\,. \end{cases} \tag{3.22}$$

Here we recall that M^* is the manifold $M \setminus \mathrm{MIN}(u)$ and that ∂M^* is the conformal boundary of M^*. The notation $x \to *$, means that $x \to p$, where p is a point of $\mathrm{MIN}(u)$, with respect to the topology induced by M on M^*.

3.3 A Unifying Formalism

We recall that the relation between u and φ is given by (3.7) if $\Lambda > 0$ and by (3.18) if $\Lambda < 0$. In both cases, $u = u(\varphi)$ obeys the equation

$$\frac{du}{d\varphi} = 1 - u^2 .$$

Since this is the only formal property of u that will be needed in the following, we proceed by noticing that both systems (3.11) and (3.22) can be rewritten in the form

$$\begin{cases} \mathrm{Ric}_g = \left(\frac{1}{u} - (n-1)u\right)\nabla^2\varphi - (n-2)d\varphi \otimes d\varphi + \left(n - 2 + 2(1 - |\nabla\varphi|_g^2)\right) g\,, & \text{in } M^* \\ \Delta_g\varphi = -n\,u\,\left(1 - |\nabla\varphi|_g^2\right), & \text{in } M^* \\ \varphi = 0\,, & \text{on } \partial M^* \\ \varphi \to +\infty\,, & \text{as } x \to *\,, \end{cases} \tag{3.23}$$

where eventually $u = \tanh(\varphi)$ or $\coth(\varphi)$.

To describe the idea that will lead us throughout the analysis of system (3.23), we note that taking the trace of the first equation one gets

$$\frac{\mathrm{R}_g}{n-1} = (n-2) + (nu^2 + 2)\left(1 - |\nabla\varphi|_g^2\right), \tag{3.24}$$

where R_g is the scalar curvature of the metric g. It is important to observe that in the cylindrical situation, which is the conformal counterpart of the (anti-)de Sitter solution, R_g has to be constant. In this case, the above formula implies that also $|\nabla\varphi|_g$ has to be constant and equal to 1. For these reasons, also in the situation, where we do not know a priori if g is cylindrical, it is natural to think of $\nabla\varphi$ as to a candidate splitting direction and to investigate under which conditions this is actually the case.

Now we rephrase Assumptions 1 and Assumption 2 in terms of φ.

Assumption 3 *We require the following*

(i) *In the case* $\Lambda > 0$, *we assume* $1 - |\nabla\varphi|_g^2 \geq 0$ *on* ∂M.
(ii) *In the case* $\Lambda < 0$, *we suppose that* $\lim_{x\to\bar{x}} u^2(1 - |\nabla\varphi|_g^2) \geq 0$ *for every point* $\bar{x} \in \partial M$.

This assumption allows to estimate the behavior of $|\nabla\varphi|_g$ on the whole manifold M^*.

Lemma 3.1 *Let* (M^*, g, φ) *be a solution of problem* (3.23) *satisfying Assumption 3. Then the following condition holds on the whole manifold* M^*

$$1 - |\nabla\varphi|_g^2 \geq 0\,.$$

Proof From the Bochner formula and the equations in (3.23), we get

$$\begin{aligned}\Delta_g|\nabla\varphi|_g^2 &= 2\left|\nabla^2\varphi\right|_g^2 + 2\,\mathrm{Ric}_g(\nabla\varphi,\nabla\varphi) + 2\left\langle\nabla\Delta_g\varphi\,\middle|\,\nabla\varphi\right\rangle_g \\ &= 2\left|\nabla^2\varphi\right|_g^2 + \left(\frac{1}{u} + (n+1)u\right)\langle\nabla|\nabla\varphi|_g^2\,|\,\nabla\varphi\rangle_g \,+\, 2\,n\,u^2\,|\nabla\varphi|_g^2\,\left(1 - |\nabla\varphi|_g^2\right).\end{aligned} \tag{3.25}$$

Now we turn to the computation of the gradient and Laplacian of the function

$$w \,=\, \beta\,\left(1 - |\nabla\varphi|_g^2\right)\,,$$

where $\beta = \beta(\varphi)$ is an arbitrary $\mathscr{C}^1$ function. Using (3.25) and (3.23) again, we get

$$\nabla w \,=\, \frac{\dot\beta}{\beta}\,w\,\nabla\varphi \,-\, \beta\,\nabla|\nabla\varphi|_g^2\,.$$

$$\begin{aligned}\Delta_g w &= -\beta\Delta_g|\nabla\varphi|_g^2 - \frac{\dot\beta}{\beta}\langle\beta\nabla|\nabla\varphi|_g^2|\nabla\varphi\rangle_g + \frac{\dot\beta}{\beta}\langle\nabla w|\nabla\varphi\rangle_g \\ &\quad + \left(\frac{\ddot\beta}{\beta} - \left(\frac{\dot\beta}{\beta}\right)^2\right)w|\nabla\varphi|_g^2 + \frac{\dot\beta}{\beta}w\Delta_g\varphi \\ &= -2\beta|\nabla^2\varphi|_g^2 + \left(2\frac{\dot\beta}{\beta} + \frac{1}{u} + (n+1)u\right)\langle\nabla w|\nabla\varphi\rangle_g \\ &\quad + \left(\frac{\ddot\beta}{\beta} - 2\left(\frac{\dot\beta}{\beta}\right)^2 - \frac{\dot\beta}{\beta}\left(\frac{1}{u} + (n+1)u\right) - 2nu^2\right)w|\nabla\varphi|_g^2 - \dot\beta\frac{(\Delta_g\varphi)^2}{nu}\,.\end{aligned}$$

We find that the right choice in order to simplify the expression above is to define the function β as the solution of the differential equation

$$\frac{\dot\beta}{\beta} + 2u \,=\, 0\,.$$

More explicitly:

$$\beta(\varphi) = \begin{cases}\dfrac{1}{\cosh^2(\varphi)} & (\text{case } \Lambda > 0)\,,\\[2ex] \dfrac{1}{\sinh^2(\varphi)} & (\text{case } \Lambda < 0)\,.\end{cases}$$

With this choice of β, the equation above may be rewritten in the simplified form:

$$\Delta_g w - \left(\frac{1}{u} + (n-3)u\right) \langle \nabla w \,|\, \nabla \varphi \rangle_g = -2\beta \left(|\nabla^2 \varphi|_g^2 - \frac{(\Delta_g \varphi)^2}{n} \right). \tag{3.26}$$

We notice that the term on the right of Eq. (3.26) is always nonpositive, thus the elliptic operator

$$\mathrm{L}(\,\cdot\,) = \Delta_g \,\cdot\, - \left(\frac{1}{u} + (n-3)u\right) \langle \nabla \,\cdot\, | \nabla \varphi \rangle.$$

satisfies

$$\mathrm{L}(w) \leq 0 \quad \text{on } M^*.$$

Thanks to Assumption 3, it holds $w \geq 0$ on ∂M. Let us suppose for the moment that $w \to 0$ as $\varphi \to +\infty$. Then, recalling that, since φ is analytic, its singular values are discrete (see [37]), we can choose $s > 0$ small enough and $S > 0$ big enough in such a way that the level sets $\{\varphi = s\}$ and $\{\varphi = S\}$ are regular. Thus the set $\{s \leq \varphi \leq S\}$ is a (compact) manifold and we can use the weak maximum principle to obtain

$$\inf_{\{s \leq \varphi \leq S\}} w = \inf_{\partial\{s \leq \varphi \leq S\}} w = \min_{\{\varphi = s\} \cup \{\varphi = S\}} w\,.$$

Hence, since $\min_{\{\varphi=S\}} w \to 0$ as $S \to +\infty$, and $\min_{\{\varphi=s\}} w \to \min_{\partial M} w \geq 0$ as $s \to 0^+$, we easily find that $w \geq 0$ on $\{0 \leq \varphi < +\infty\} = M^*$. This immediately gives the thesis.

It remains to prove that $\lim_{\varphi \to +\infty} w = 0$. It is convenient to rewrite the limit in terms of u, g_0. In the case $\Lambda > 0$, the limit above is equivalent to $\lim_{u \to 1^-} (1 - u^2 - |\mathrm{D}u|^2) = 0$, while in the case $\Lambda < 0$ it is equivalent to $\lim_{u \to 1^+} (u^2 - 1 - |\mathrm{D}u|^2) = 0$. In both cases, since the points at which $u = 1$ are extremals, we have $|\mathrm{D}u| \to 0$ as $u \to 1$ and so the limits above are verified. □

3.4 *The Geometry of the Level Sets of φ*

In the forthcoming analysis a crucial role is played by the study of the geometry of the level sets of φ, which coincide with the level sets of u, by definition. Hence, we pass now to describe the second fundamental form and the mean curvature of the regular level sets of φ (or equivalently of u) in both the original Riemannian context (M, g_0) and the conformally related one (M^*, g). To this aim, we fix a regular level set $\{\varphi = s_0\}$ and we construct a suitable set of coordinates in a neighborhood of it. Note that $\{\varphi = s_0\}$ must be compact, by the properness of φ. In particular, there exists a real number $\delta > 0$ such that in the tubular neighborhood $\mathcal{U}_\delta = \{s_0 - \delta < \varphi < s_0 + \delta\}$ we have $|\nabla\varphi|_g > 0$ so that $\mathcal{U}_\delta$ is foliated by regular level sets of φ. As a consequence, $\mathcal{U}_\delta$ is diffeomorphic to $(s_0 - \delta, s_0 + \delta) \times \{\varphi = s_0\}$ and the function φ can be regarded as a coordinate in $\mathcal{U}_\delta$. Thus, one can choose a local system of coordinates $\{\varphi, \vartheta^1, \ldots, \vartheta^{n-1}\}$, where $\{\vartheta^1, \ldots, \vartheta^{n-1}\}$ are local coordinates on $\{\varphi = s_0\}$. In such a system, the metric g can be written as

$$g = \frac{d\varphi \otimes d\varphi}{|\nabla\varphi|_g^2} + g_{ij}(\varphi, \vartheta^1, \ldots, \vartheta^{n-1})\, d\vartheta^i \otimes d\vartheta^j ,$$

where the Latin indices vary between 1 and $n - 1$. We now fix in $\mathcal{U}_\delta$ the g_0-unit vector field $\nu = \mathrm{D}u/|\mathrm{D}u| = \mathrm{D}\varphi/|\mathrm{D}\varphi|$ and the g-unit vector field $\nu_g = \nabla u/|\nabla u|_g = \nabla\varphi/|\nabla\varphi|_g$. Accordingly, the second fundamental forms of the regular level sets of u or φ with respect to ambient metric g_0 and the conformally-related ambient metric g are respectively given by

$$\mathrm{h}_{ij}^{(0)} = \frac{\mathrm{D}^2_{ij} u}{|\mathrm{D}u|} = \frac{\mathrm{D}^2_{ij}\varphi}{|\mathrm{D}\varphi|} \qquad \text{and} \qquad \mathrm{h}_{ij}^{(g)} = \frac{\nabla^2_{ij} u}{|\nabla u|_g} = \frac{\nabla^2_{ij}\varphi}{|\nabla\varphi|_g}, \qquad \text{for} \quad i, j = 1, \ldots, n-1.$$

Taking the traces of the above expressions with respect to the induced metrics we obtain the following expressions for the mean curvatures in the two ambients

$$\mathrm{H} = \frac{\Delta u}{|\mathrm{D}u|} - \frac{\mathrm{D}^2 u(\mathrm{D}u, \mathrm{D}u)}{|\mathrm{D}u|^3}, \qquad \mathrm{H}_g = \frac{\Delta_g\varphi}{|\nabla\varphi|_g} - \frac{\nabla^2\varphi(\nabla\varphi, \nabla\varphi)}{|\nabla\varphi|_g^3}. \tag{3.27}$$

Taking into account expressions (3.8), (3.19) and (3.9), (3.20), one can show that the second fundamental forms are related by

$$\mathrm{h}_{ij}^{(g)} = \begin{cases} \dfrac{1}{\sqrt{1-u^2}} \left[\mathrm{h}_{ij}^{(0)} + \dfrac{u\,|\mathrm{D}u|}{1-u^2}\, g_{ij}^{(0)} \right] & (\text{case } \Lambda > 0), \\[2ex] \dfrac{1}{\sqrt{u^2-1}} \left[-\mathrm{h}_{ij}^{(0)} + \dfrac{u\,|\mathrm{D}u|}{u^2-1}\, g_{ij}^{(0)} \right] & (\text{case } \Lambda < 0). \end{cases} \tag{3.28}$$

The analogous formula for the mean curvatures reads

$$\mathrm{H}_g = \begin{cases} \sqrt{1-u^2} \left[\mathrm{H} + (n-1)\, \dfrac{u\,|\mathrm{D}u|}{1-u^2} \right] & (\text{case } \Lambda > 0), \\[2ex] \sqrt{u^2-1} \left[-\mathrm{H} + (n-1)\, \dfrac{u\,|\mathrm{D}u|}{u^2-1} \right] & (\text{case } \Lambda < 0). \end{cases} \tag{3.29}$$

Concerning the nonregular level sets of φ, we first observe that φ is analytic (see [13, 41]), thus by the results in [21, 30] (see also [27, Theorem 6.3.3]), one has that the $(n-1)$-dimensional Hausdorff measure of the level sets of φ is locally finite. Hence, the properness of φ forces the level sets to have finite $(n-1)$-dimensional

Hausdorff measure. Using the results in [30] (see also [27, Theorem 6.3.3]), we know that there exists a submanifold $N \subseteq \mathrm{Crit}(\varphi)$ such that $\mathscr{H}^{n-1}(\mathrm{Crit}(\varphi)\setminus N)=0$. In particular, the unit normal to a level set is well-defined $\mathscr{H}^{n-1}$-almost everywhere, and so are the second fundamental form h_g and the mean curvature H_g. We will prove now that formulæ (3.28) and (3.29) hold also at any point $y_0 \in N$, and therefore they hold $\mathscr{H}^{n-1}$-almost everywhere on any level set. We do it in the case $\Lambda > 0$ (the case $\Lambda < 0$ is analogous). Let ν, ν_g be the unit normal vector fields to N at y_0 with respect to g_0, g respectively. Since $|\nu_g|_g^2 = 1 = |\nu|^2 = (1-u^2)\,|\nu|_g^2$, we deduce that $\nu_g = \sqrt{1-u^2}\,\nu$. Let $(\partial/\partial x^1,\ldots,\partial/\partial x^{n-1})$ be a basis of $T_{y_0}N$, so that in particular $(\partial/\partial x^1,\ldots,\partial/\partial x^{n-1},\nu_g)$ is a basis of $T_{y_0}M$. Recalling (3.2) and observing that the derivatives of u in y_0 are all zero since $y_0 \in \mathrm{Crit}(\varphi)$, we have

$$\mathrm{h}^{(g)}_{ij} = \Big\langle \nabla_i \frac{\partial}{\partial x^j} \,\Big|\, \nu_g \Big\rangle_g = \Gamma^n_{ij} = \mathrm{G}^n_{ij} = \Big\langle \mathrm{D}_i \frac{\partial}{\partial x^j} \,\Big|\, \nu_g \Big\rangle_g = \frac{1}{\sqrt{1-u^2}} \Big\langle \mathrm{D}_i \frac{\partial}{\partial x^j} \,\Big|\, \nu \Big\rangle = \frac{1}{\sqrt{1-u^2}}\,\mathrm{h}^{(0)}_{ij}\,.$$

This proves that formula (3.28) holds also on N, and taking its trace we deduce that also (3.29) is verified.

3.5 A Conformal Version of the Monotonicity-Rigidity Theorem

We conclude this section by introducing the conformal analog of the functions $U_p(t)$ introduced in (1.7) ($\Lambda > 0$) and (1.18) ($\Lambda < 0$). To this aim, we let (M^*, g, φ) be a solution to problem (3.23) and we define, for $p \geq 0$, the functions $\Phi_p : [0,+\infty) \longrightarrow \mathbb{R}$ as

$$s \longmapsto \Phi_p(s) = \int\limits_{\{\varphi=s\}} |\nabla\varphi|_g^p \, \mathrm{d}\sigma_g\,. \tag{3.30}$$

As for the U_p's, we observe that the Φ_p's are well defined. This is because the hypersurface area of the level sets is finite, due to the analyticity and properness of φ. Before proceeding, it is worth noticing that, when $p=0$, the function

$$\Phi_0(s) = \int\limits_{\{\varphi=s\}} \mathrm{d}\sigma_g = |\{\varphi=s\}|_g,$$

coincides with the hypersurface area functional $|\{\varphi = s\}|_g$ for the level sets of φ inside the ambient manifold (M^*, g). For future convenience, we observe that the functions U_p and Φ_p and their derivatives (when defined) are related as follows

$$U_p(t) = \Phi_p\Big(\frac{1}{2}\log\Big|\frac{1+t}{1-t}\Big|\Big), \tag{3.31}$$

$$U'_p(t) = \frac{1}{1-t^2}\,\Phi'_p\Big(\frac{1}{2}\log\Big|\frac{1+t}{1-t}\Big|\Big), \tag{3.32}$$

$$U''_p(t) = \frac{1}{(1-t^2)^2}\left[2t\,\Phi'_p\Big(\frac{1}{2}\log\Big|\frac{1+t}{1-t}\Big|\Big) + \Phi''_p\Big(\frac{1}{2}\log\Big|\frac{1+t}{1-t}\Big|\Big)\right], \tag{3.33}$$

Using the above relationships, both the Monotonicity-Rigidity Theorems 1.1 and 1.4 can be rephrased in terms of the functions $s \mapsto \Phi_p(s)$ as follows.

Theorem 3.2 (Monotonicity-Rigidity Theorem—Conformal Version) *Let (M^*, g, φ) be a solution to problem* (3.23)*, satisfying Assumption 3. For every $p \geq 0$ we let $\Phi_p : [0, +\infty) \longrightarrow \mathbb{R}$ be the function defined in* (3.30)*. Then, the following properties hold true.*

(i) *For every $p \geq 1$, the function Φ_p is continuous.*
(ii) *The function $\Phi_1(s)$ is monotonically nonincreasing. Moreover, if $\Phi_1(s_1) = \Phi_1(s_2)$ for some $s_1 \neq s_2$, then (M^*, g, φ) is isometric to one half round cylinder with totally geodesic boundary.*
(iii) *For every $p \geq 3$, the function Φ_p is differentiable and the derivative satisfies, for every $s \in (0, +\infty)$,*

$$\Phi'_p(s) = \int_{\{\varphi=s\}} \Big[-(p-1)|\nabla\varphi|_g^{p-1}\mathrm{H}_g + p|\nabla\varphi|_g^{p-2}\Delta_g\varphi\Big]\,d\sigma_g \leq \int_{\{\varphi=s\}} |\nabla\varphi|_g^{p-2}\Delta_g\varphi\,d\sigma_g \leq 0. \tag{3.34}$$

where H_g is the mean curvature of the level set $\{\varphi = s\}$. Moreover, if the first equality in (3.34) *holds, for some $s \in (0, +\infty)$ and some $p \geq 3$, then (M^*, g, φ) is isometric to an half round cylinder with totally geodesic boundary.*
(iv) *It holds $\Phi'_p(0) = \lim_{s\to 0}\Phi'_p(s) = 0$, for every $p \geq 3$. In particular, setting $\Phi''_p(0) = \lim_{s\to 0^+}\Phi'_p(s)/s$, we have that for every $p \geq 3$, the following formulæ hold*

$$\begin{aligned}
-n\int_{\partial M} |\nabla\varphi|_g^{p-2}\Big(1-|\nabla\varphi|_g^2\Big)\,d\sigma_g &\geq \Phi''_p(0)\\
&= \int_{\partial M} |\nabla\varphi|_g^{p-2}\Big[(p-1)\mathrm{Ric}_g(\nu_g,\nu_g) - np\Big(1-|\nabla\varphi|_g^2\Big)\Big]\,d\sigma_g, \quad (\Lambda > 0)\\
-n\int_{\partial M} u^2\Big(1-|\nabla\varphi|_g^2\Big)\,d\sigma_g &\geq \Phi''_p(0)\\
&= \int_{\partial M} -\Big[\Big(\frac{p-1}{n-1}\Big)\mathrm{Ric}_g(\nu_g,\nu_g) + nu^2(1-|\nabla\varphi|_g^2)\Big]\,d\sigma_g\,, \quad (\Lambda < 0)
\end{aligned} \tag{3.35}$$

where $\nu_g = \nabla\varphi/|\nabla\varphi|_g$ *is the inward pointing unit normal of the boundary* ∂M. *Moreover, in the case* $\Lambda > 0$, *if the equality is fulfilled for some* $p \geq 3$, *then* (M^*, g, φ) *is isometric to an half round cylinder with totally geodesic boundary.*

4 Proof of Theorems 1.1 and 1.4 After Theorem 3.2

In this section we show how to recover Theorems 1.1, 1.4 from Theorem 3.2. The proof of Theorem 3.2 will be the argument of Sect. 6.

4.1 Case $\Lambda > 0$: Theorem 3.2 Implies Theorem 1.1

In the hypotheses of Theorem 1.1, Lemma 3.1 is in charge, hence $|\nabla\varphi|_g^2 \leq 1$ on the whole manifold M. Thus, formula (1.10) is an immediate consequence of the identity

$$|\nabla\varphi|_g^2 = \frac{|\mathrm{D}u|^2}{1-u^2}\,.$$

The equivalence between Theorem 1.1-(i),(ii) and Theorem 3.2-(i),(ii) is also straightforward.

We pass now to prove the equivalence between (1.11) and (3.34). To do this, it is enough to translate (3.34) in terms of the conformally related quantities u, g_0. Recalling the second equation in (3.23) and formulæ (3.29), (3.32), we have the following chain of equalities.

$$\begin{aligned} U_p'(t) &= \frac{1}{1-t^2}\,\Phi_p'\left(\frac{1}{2}\log\left(\frac{1+t}{1-t}\right)\right) \\ &= \frac{1}{1-t^2}\int\limits_{\{u=t\}} |\nabla\varphi|_g^{p-2}\Big[-(p-1)|\nabla\varphi|_g\,\mathrm{H}_g + p\,\Delta_g\varphi\Big]\mathrm{d}\sigma_g \\ &= \int\limits_{\{u=t\}} \frac{|\mathrm{D}u|^{p-2}}{(1-u^2)^{p/2}}\left[-(p-1)|\mathrm{D}u|\left(\mathrm{H} + (n-1)\,\frac{u\,|\mathrm{D}u|}{1-u^2}\right) - n\,p\,u\left(1-\frac{|\mathrm{D}u|^2}{1-u^2}\right)\right]\mathrm{d}\sigma_g \\ &= \left(\frac{1}{1-t^2}\right)^{\frac{p}{2}}\int\limits_{\{u=t\}} |\mathrm{D}u|^{p-2}\left[-(p-1)|\mathrm{D}u|\,\mathrm{H} + (n+p-1)\,\frac{u\,|\mathrm{D}u|^2}{1-u^2} - n\,p\,u\right]\mathrm{d}\sigma_g\,. \end{aligned} \tag{4.1}$$

Now we use formula (3.1) to deduce that the volume elements $\mathrm{d}\sigma$, $\mathrm{d}\sigma_g$ are related by

$$\mathrm{d}\sigma_g = \Big(\frac{1}{1-u^2}\Big)^{\frac{n-1}{2}} \mathrm{d}\sigma\,.$$

Hence equality (4.1) can be rewritten as

$$\begin{aligned} U_p'(t) &= -(p-1)t\Big(\frac{1}{1-t^2}\Big)^{\frac{n+p-1}{2}} \\ &\quad\times \int\limits_{\{u=t\}} |\mathrm{D}u|^{p-2}\Big[\Big|\frac{\mathrm{D}u}{u}\Big|\mathrm{H} + \Big(\frac{np}{p-1}\Big) - \Big(\frac{n+p-1}{p-1}\Big)\Big(\frac{|\mathrm{D}u|^2}{1-u^2}\Big)\Big]\,\mathrm{d}\sigma \\ &= -(p-1)t\Big(\frac{1}{1-t^2}\Big)^{\frac{n+p-1}{2}} \\ &\quad\times \int\limits_{\{u=t\}} |\mathrm{D}u|^{p-2}\Big[(n-1) - \mathrm{Ric}(\nu,\nu) + \Big(\frac{n+p-1}{p-1}\Big)\Big(1-\frac{|\mathrm{D}u|^2}{1-u^2}\Big)\Big]\,\mathrm{d}\sigma, \end{aligned}$$

where in the second equality $\nu = \mathrm{D}u/|\mathrm{D}u|$ is the unit normal to $\{u = t\}$ and we have used the identity

$$\mathrm{Ric}(\nu,\nu) = -\Big|\frac{\mathrm{D}u}{u}\Big|\,\mathrm{H}\,,$$

which is a consequence of the first equation in problem (1.4). On the other hand, since from (3.34) it holds

$$\Phi_p'(s) \le \int\limits_{\{\varphi=s\}} |\nabla\varphi|_g^{p-2}\Delta_g\varphi\; \mathrm{d}\sigma_g\,,$$

we have

$$\begin{aligned} U_p'(t) &\le -\frac{n\,t}{1-t^2}\int\limits_{\{u=t\}}\Big(\frac{|\mathrm{D}u|}{\sqrt{1-u^2}}\Big)^{p-2}\Big(1-\frac{|\mathrm{D}u|^2}{1-u^2}\Big)\,\mathrm{d}\sigma_g \\ &= -n\,t\Big(\frac{1}{1-t^2}\Big)^{\frac{n+p-1}{2}}\int\limits_{\{u=t\}}|\mathrm{D}u|^{p-2}\Big(1-\frac{|\mathrm{D}u|^2}{1-u^2}\Big)\,\mathrm{d}\sigma\,. \end{aligned} \tag{4.2}$$

This proves formula (1.11). Moreover, we recall from Theorem 3.2 that the equality holds in (4.2) if and only if (M^*, g, φ) is isometric to one half round cylinder, that is, if and only if (M, g_0, u) is isometric to the de Sitter solution. This proves Theorem 1.1-(iii).

It remains to prove the equivalence of Theorem 1.1-(iv) and Theorem 3.2-(iv). We first observe, from formula (3.5), that the identity

$$\mathrm{Ric}_g(\nu_g, \nu_g) \;=\; \mathrm{Ric}(\nu, \nu) \;-\; (n-1)\,|\mathrm{D}u|^2\,,$$

holds on ∂M. Then we apply the Gauss–Codazzi equation to obtain

$$\mathrm{Ric}_g(\nu_g, \nu_g) \;=\; \frac{\mathrm{R} - \mathrm{R}^{\partial M}}{2} - (n-1)|\mathrm{D}u|^2 \;=\; \frac{(n-1)(n-2) - \mathrm{R}^{\partial M}}{2} + (n-1)\,(1-|\mathrm{D}u|^2)\,.$$

Now, we recall formulæ (3.33) and (3.35) and we compute

$$\begin{aligned}
U_p''(0) = \Phi_p''(0) &= \int_{\partial M} |\nabla\varphi|_g^{p-2}\left[(p-1)\mathrm{Ric}_g(\nu_g,\nu_g) - n\,p\left(1-|\nabla\varphi|_g^2\right)\right]\mathrm{d}\sigma_g \\
&= \int_{\partial M} |\mathrm{D}u|^{p-2}\left[(p-1)\Big(\frac{(n-1)(n-2)-\mathrm{R}^{\partial M}}{2}\Big) - (n+p-1)\left(1-|\mathrm{D}u|^2\right)\right]\mathrm{d}\sigma \\
&= -(p-1)\int_{\partial M} |\mathrm{D}u|^{p-2}\left[\frac{\mathrm{R}^{\partial M}-(n-1)(n-2)}{2} + \Big(\frac{n+p-1}{p-1}\Big)\left(1-|\mathrm{D}u|^2\right)\right]\mathrm{d}\sigma.
\end{aligned}$$

Moreover, again from formula (3.35), we have

$$U_p''(0) \;=\; \Phi_p''(0) \;\leq\; -n\int_{\partial M} |\nabla\varphi|_g^{p-2}\left(1-|\nabla\varphi|_g^2\right)\mathrm{d}\sigma_g = -n\int_{\partial M} |\mathrm{D}u|^{p-2}\left(1-|\mathrm{D}u|^2\right)\mathrm{d}\sigma\,,$$

and the equality holds if and only if (M, g_0, u) is isometric to the de Sitter solution. This concludes the proof of Theorem 1.1-(iv).

4.2 *Case* $\Lambda < 0$*: Theorem 3.2 Implies Theorem 1.4*

In the hypotheses of Theorem 1.4, Lemma 3.1 is in charge, hence $|\nabla\varphi|_g^2 \leq 1$ on the whole manifold M. Thus, formula (1.21) is an immediate consequence of the identity

$$|\nabla\varphi|_g^2 \;=\; \frac{|\mathrm{D}u|^2}{u^2-1}\,.$$

The equivalence between Theorem 1.4-(i),(ii) and Theorem 3.2-(i),(ii) is also straightforward.

We pass now to prove the equivalence between (1.22) and (3.34). To do this, it is enough to translate (3.34) in terms of the conformally related quantities u, g_0.

Recalling the second equation in (3.23) and formulæ (3.29), (3.32), we have the following chain of equalities.

$$
\begin{aligned}
U_p'(t) &= -\frac{1}{t^2-1}\,\Phi_p'\left(\frac{1}{2}\log\left(\frac{1+t}{t-1}\right)\right)\\
&= -\frac{1}{t^2-1}\int_{\{u=t\}} |\nabla\varphi|_g^{p-2}\Big[-(p-1)|\nabla\varphi|_g\,\mathrm{H}_g + p\,\Delta_g\varphi\Big]\mathrm{d}\sigma_g\\
&= -\int_{\{u=t\}} \frac{|\mathrm{D}u|^{p-2}}{(u^2-1)^{p/2}}\left[-(p-1)|\mathrm{D}u|\left(-\mathrm{H}+(n-1)\frac{u\,|\mathrm{D}u|}{u^2-1}\right)-n\,p\,u\left(1-\frac{|\mathrm{D}u|^2}{u^2-1}\right)\right]\mathrm{d}\sigma_g\\
&= \left(\frac{1}{t^2-1}\right)^{\frac{p}{2}}\int_{\{u=t\}} |\mathrm{D}u|^{p-2}\left[-(p-1)|\mathrm{D}u|\,\mathrm{H}-(n+p-1)\frac{u\,|\mathrm{D}u|^2}{u^2-1}+n\,p\,u\right]\mathrm{d}\sigma_g\,.
\end{aligned}
\tag{4.3}
$$

Now we use formula (3.12) to deduce that the volume elements $\mathrm{d}\sigma$, $\mathrm{d}\sigma_g$ are related by

$$
\mathrm{d}\sigma_g = \left(\frac{1}{u^2-1}\right)^{\frac{n-1}{2}}\mathrm{d}\sigma\,.
$$

Hence equality (4.3) can be rewritten as

$$
\begin{aligned}
U_p'(t) &= (p-1)t\left(\frac{1}{t^2-1}\right)^{\frac{n+p-1}{2}}\\
&\quad\times\int_{\{u=t\}} |\mathrm{D}u|^{p-2}\left[-\left|\frac{\mathrm{D}u}{u}\right|\mathrm{H}+\left(\frac{np}{p-1}\right)-\left(\frac{n+p-1}{p-1}\right)\left(\frac{|\mathrm{D}u|^2}{1-u^2}\right)\right]\mathrm{d}\sigma\\
&= (p-1)t\left(\frac{1}{t^2-1}\right)^{\frac{n+p-1}{2}}\\
&\quad\times\int_{\{u=t\}} |\mathrm{D}u|^{p-2}\left[(n-1)+\mathrm{Ric}(\nu,\nu)+\left(\frac{n+p-1}{p-1}\right)\left(1-\frac{|\mathrm{D}u|^2}{1-u^2}\right)\right]\mathrm{d}\sigma,
\end{aligned}
$$

where in the second equality $\nu = \mathrm{D}u/|\mathrm{D}u|$ is the unit normal to $\{u=t\}$ and we have used the identity

$$
\mathrm{Ric}(\nu,\nu) = -\left|\frac{\mathrm{D}u}{u}\right|\mathrm{H}\,,
$$

which is a consequence of the first equation in problem (1.17). On the other hand, since from (3.34) it holds

$$\Phi'_p(s) \leq \int_{\{\varphi=s\}} |\nabla\varphi|_g^{p-2} \Delta_g\varphi \, d\sigma_g \,,$$

we have

$$\begin{aligned} U'_p(t) &\geq \frac{n\,t}{t^2-1} \int_{\{u=t\}} \Big(\frac{|Du|}{\sqrt{u^2-1}}\Big)^{p-2}\Big(1-\frac{|Du|^2}{u^2-1}\Big)\, d\sigma_g \\ &= n\,t \Big(\frac{1}{t^2-1}\Big)^{\frac{n+p-1}{2}} \int_{\{u=t\}} |Du|^{p-2}\Big(1-\frac{|Du|^2}{u^2-1}\Big)\, d\sigma \,. \end{aligned} \tag{4.4}$$

This proves formula (1.22). Moreover, we recall from Theorem 3.2 that the equality holds in (4.4) if and only if (M^*, g, φ) is isometric to one half round cylinder, that is, if and only if (M, g_0, u) is isometric to the anti-de Sitter solution. This proves Theorem 1.4-(iii).

It remains to prove the equivalence of Theorem 1.4-(iv) and Theorem 3.2-(iv). Let $r = \sqrt{u^2-1}$ and $V_p(r) = U_p(\sqrt{1+1/r^2})$. An easy computation shows that

$$V'_p(r) = -\frac{1}{r^2\sqrt{1+r^2}}\, U'_p(\sqrt{1+1/r^2})\,.$$

Since $\lim_{s\to 0^+}\Phi'_p(s) = 0$, we deduce from (3.32) that $\lim_{t\to+\infty} U'_p(t) = 0$, which implies $\lim_{r\to 0^+} V'_p(r) = 0$. Hence we set $V''_p(0) = \lim_{r\to 0^+} V'_p(r)/r$ and we compute

$$\begin{aligned} V''_p(0) &= \lim_{r\to 0^+} -\frac{1}{r^3}\, U'_p(\sqrt{1+1/r^2}) \\ &= \lim_{t\to+\infty} -t^3\, U'_p(t) \\ &= \lim_{t\to+\infty} t\,\Phi'_p\Big(\frac{1}{2}\log\Big(\frac{t+1}{t-1}\Big)\Big), \end{aligned}$$

where in the last equality we have used formula (3.32). Recalling that $\Phi''_p(0) = \lim_{s\to 0^+}\Phi'_p(s)/s$, we conclude that $V''_p(0) = \Phi''_p(0)$. Hence, from formula (3.35) we obtain

$$V''_p(0) = -\int_{\partial M} \Big[\Big(\frac{p-1}{n-1}\Big)\,\mathrm{Ric}_g(\nu_g,\nu_g) + n\,u^2\Big(1-|\nabla\varphi|_g^2\Big)\Big]\, d\sigma_g\,. \tag{4.5}$$

Recalling that $|\nabla\varphi|_g^2 = |Du|^2/(u^2-1)$, it is easily seen that

$$\lim_{s\to 0^+} u^2(1-|\nabla\varphi|_g^2) \;=\; \lim_{t\to+\infty}\left(u^2-1-|Du|^2\right).$$

Moreover, from the Gauss–Codazzi equation on ∂M and formula (3.24) we find that the identity

$$\mathrm{Ric}_g(\nu_g,\nu_g) \;=\; \frac{\mathrm{R}_g-\mathrm{R}_g^{\partial M}}{2} \;=\; \frac{(n-1)(n-2)-\mathrm{R}_g^{\partial M}}{2}+\frac{(n-1)\,n\,u^2(1-|\nabla\varphi|_g^2)}{2}$$

holds on the conformal boundary ∂M. Therefore, formula (4.5) rewrites as

$$V_p''(0) = -(p-1)\int_{\partial M}\left[\frac{(n-1)(n-2)-\mathrm{R}_g^{\partial M}}{2(n-1)}+\frac{n(p+1)}{2(p-1)}\left(u^2-1-|Du|^2\right)\right]\mathrm{d}\sigma_g\,.$$

Finally, again from formula (3.35), we have

$$V_p''(0) \;=\; \Phi_p''(0) \;\le\; n\int_{\partial M} u^2\left(1-|\nabla\varphi|_g^2\right)\mathrm{d}\sigma_g \;=\; n\int_{\partial M}\left(u^2-1-|Du|^2\right)\mathrm{d}\sigma_g\,.$$

This concludes the proof of Theorem 1.4-(iv).

5 Integral Identities

In this section, we derive some integral identities that will be used to analyze the properties of the functions $s\mapsto\Phi_p(s)$ introduced in (3.30).

5.1 First Integral Identity

To obtain our first identity, we are going to exploit the equation $\Delta_g\varphi = -nu(1-|\nabla\varphi|_g^2)$ in problem (3.23).

Proposition 5.1 *Let (M^*, g, φ) be a solution to problem* (3.23)*. Then, for every $p\ge 1$ and for every $s\in(0,+\infty)$, we have*

$$\int_{\{\varphi=s\}}\frac{|\nabla\varphi|_g^p}{\sinh^n(s)}\mathrm{d}\sigma_g = \int_{\{\varphi>s\}}\frac{|\nabla\varphi|_g^{p-3}\Big(n\coth(\varphi)|\nabla\varphi|_g^4-|\nabla\varphi|_g^2\Delta_g\varphi-(p-1)\nabla^2\varphi(\nabla\varphi,\nabla\varphi)\Big)}{\sinh^n(\varphi)}\mathrm{d}\mu_g. \tag{5.1}$$

Remark 7 Arguing as in Remark 1, it is easy to realize that the integral on the left hand side of (5.1) is well defined also when s is a singular value of φ.

Proof To prove identity (5.1) when $s > 0$ is a regular value of φ, we start from the formula

$$\operatorname{div}_g\left(\frac{|\nabla\varphi|_g^{p-1}\nabla\varphi}{\sinh^n(\varphi)}\right) = \frac{|\nabla\varphi|_g^{p-3}\Big((p-1)\,\nabla^2\varphi(\nabla\varphi,\nabla\varphi) + |\nabla\varphi|_g^2\Delta_g\varphi - n\coth(\varphi)|\nabla\varphi|_g^4\Big)}{\sinh^n(\varphi)}, \tag{5.2}$$

which follows from a direct computation. Since φ is analytic, its singular values are discrete (see [37]). In particular all the big enough values are regular. Hence, we integrate the above formula by parts using the Divergence Theorem in $\{s < \varphi < S\}$, where S is large enough so that we are sure that the level set $\{\varphi = S\}$ is regular. This gives

$$\int\limits_{\{s<\varphi<S\}} \frac{|\nabla\varphi|_g^{p-3}\Big((p-1)\,\nabla^2\varphi(\nabla\varphi,\nabla\varphi) + |\nabla\varphi|_g^2\,\Delta_g\varphi - n\,\coth(\varphi)\,|\nabla\varphi|_g^4\Big)}{\sinh^n(\varphi)}\,d\mu_g =$$

$$= \int\limits_{\{\varphi=S\}} \frac{|\nabla\varphi|_g^{p-1}\big\langle\nabla\varphi\,|\,\mathrm{n}_g\big\rangle_g}{\sinh^n(\varphi)}\,d\sigma_g + \int\limits_{\{\varphi=s\}} \frac{|\nabla\varphi|^{p-1}\big\langle\nabla\varphi\,|\,\mathrm{n}_g\big\rangle_g}{\sinh^n(\varphi)}\,d\sigma_g\,, \tag{5.3}$$

where n_g is the outer unit normal. In particular, one has that $\mathrm{n}_g = -\nabla\varphi/|\nabla\varphi|_g$ on $\{\varphi = s\}$ and $\mathrm{n}_g = \nabla\varphi/|\nabla\varphi|_g$ on $\{\varphi = S\}$. Therefore, if we prove that

$$\lim_{S\to+\infty} \int\limits_{\{\varphi=S\}} \frac{|\nabla\varphi|_g^p}{\sinh^n(\varphi)}\,d\sigma_g = 0\,, \tag{5.4}$$

then the statement of the proposition will follow at once. Form Lemma 3.1, we know that $|\nabla\varphi|_g \leq 1$, hence it is enough to prove that

$$\lim_{S\to+\infty} \int\limits_{\{\varphi=S\}} \frac{d\sigma_g}{\sinh^n(\varphi)} = 0\,,$$

Rewriting this last limit in terms of u, g_0, we find the equalities

$$\lim_{t\to1} \int\limits_{\{u=t\}} \frac{\sqrt{1-u^2}}{u^n}\,d\sigma = 0 \quad (\text{case } \Lambda > 0)\,, \qquad \lim_{t\to1} \int\limits_{\{u=t\}} \sqrt{u^2-1}\,d\sigma = 0 \quad (\text{case } \Lambda < 0)$$

which are easily verified, since the level sets $\{u = t\}$ have finite $\mathscr{H}^{n-1}$-measure (because u is analytic). This proves the limit (5.4) and the thesis in the case in which s is a regular value.

In the case where $s > 0$ is a singular value of φ, we need to apply a refined version of the Divergence Theorem in order to perform the integration by parts which leads to identity (5.3), namely Theorem A.9 in the Appendix A. The rest of the proof is then identical to what we have done for the regular case.

According to the notations of Theorem A.9, we set

$$X = \frac{|\nabla\varphi|_g^{p-1}\nabla\varphi}{\sinh^n(\varphi)} \qquad \text{and} \qquad E = \{s < \varphi < S\}.$$

so that $\partial E = \{\varphi = s\} \sqcup \{\varphi = S\}$. It is clear that the vector field X is Lipschitz for $p \geq 1$ and, by the results of [21, 30] (see also [27, Theorem 6.3.3]), we know that $\mathscr{H}^{n-1}(\partial E)$ is finite. Moreover, from [30] (see also [27, Theorem 6.3.3]), we know that there exists an open $(n-1)$-submanifold $N \subseteq \mathrm{Crit}(\varphi)$ such that $\mathscr{H}^{n-1}(\partial E \setminus N) = 0$. Set $\Sigma = \partial E \cap (\mathrm{Crit}(\varphi) \setminus N)$ and $\Gamma = \partial E \setminus \Sigma$. We have $\mathscr{H}^{n-1}(\Sigma) = 0$ by definition, while Γ is the union of the regular part of ∂E and of N, which are open $(n-1)$-submanifolds. Therefore, the hypotheses of Theorem A.9 are satisfied, hence we can apply it to conclude that (5.3) holds also on the non regular level sets.

□

5.2 *Second Integral Identity*

Now we want to exploit Lemma 3.1 in order to obtain an integral inequality analogous to [4, Prop. 4.2]. We rewrite Eq. (3.25) as

$$\Delta_g|\nabla\varphi|_g^2 - \left(\frac{1}{u} + (n+1)u\right)\langle\nabla|\nabla\varphi|_g^2 \,|\, \nabla\varphi\rangle_g = 2\,|\nabla^2\varphi|_g^2 + 2\,n\,u^2\,|\nabla\varphi|_g^2\,\left(1 - |\nabla\varphi|_g^2\right). \tag{5.5}$$

For every $p \geq 3$, we compute

$$\nabla|\nabla\varphi|_g^{p-1} = \left(\frac{p-1}{2}\right)|\nabla\varphi|_g^{p-3}\,\nabla|\nabla\varphi|_g^2\,,$$

$$\Delta_g|\nabla\varphi|_g^{p-1} = \left(\frac{p-1}{2}\right)|\nabla\varphi|_g^{p-3}\,\Delta_g|\nabla\varphi|^2 + (p-1)(p-3)\,|\nabla\varphi|_g^{p-3}\,\left|\nabla|\nabla\varphi|_g\right|_g^2.$$

We notice en passant that whenever $|\nabla\varphi|_g > 0$ the above formulæ make sense for every $p \geq 0$. These identities, combined with (5.5), lead to

$$\Delta_g|\nabla\varphi|_g^{p-1} - \left(\frac{1}{u} + (n+1)u\right)\langle\nabla|\nabla\varphi|_g^{p-1}\,|\,\nabla\varphi\rangle_g =$$
$$= (p-1)|\nabla\varphi|_g^{p-3}\left(|\nabla^2\varphi|_g^2 + (p-3)\,\big|\nabla|\nabla\varphi|_g\big|_g^2 + n\,u^2\,|\nabla\varphi|_g^2\left(1-|\nabla\varphi|_g^2\right)\right). \tag{5.6}$$

Obviously, for $p = 3$, the above formula coincides with (5.5). If we define the function

$$\gamma = \gamma(\varphi) = \begin{cases} \dfrac{1}{\sinh(\varphi)\cosh^{n+1}(\varphi)} & (\text{case } \Lambda > 0)\,, \\[2ex] \dfrac{1}{\cosh(\varphi)\sinh^{n+1}(\varphi)} & (\text{case } \Lambda < 0)\,, \end{cases}$$

then the equation above can be written as

$$\mathrm{div}_g\left(\gamma(\varphi)\,\nabla|\nabla\varphi|_g^{p-1}\right) =$$
$$= (p-1)|\nabla\varphi|_g^{p-3}\,\gamma(\varphi)\left(|\nabla^2\varphi|_g^2 + (p-3)\,\big|\nabla|\nabla\varphi|_g\big|_g^2 + n\,u^2\,|\nabla\varphi|_g^2\left(1-|\nabla\varphi|_g^2\right)\right). \tag{5.7}$$

Note that the term on the right is always positive, thanks to Lemma 3.1.

Integrating by parts identity (5.7), we obtain the following proposition, which is the main result of this section.

Proposition 5.2 *Let (M^*, g, φ) be a solution to problem* (3.23) *satisfying Assumption 3. Then, for every $s \in [0, +\infty)$ and $p \geq 3$*

$$\gamma(s)\int\limits_{\{\varphi=s\}}\left(|\nabla\varphi|_g^{p-1}\,H_g - |\nabla\varphi|_g^{p-2}\,\Delta_g\varphi\right)d\sigma_g =$$
$$= \int\limits_{\{\varphi>s\}}\gamma(\varphi)\,|\nabla\varphi|_g^{p-3}\left(|\nabla^2\varphi|_g^2 + (p-3)\,\big|\nabla|\nabla\varphi|_g\big|_g^2 + n\,u^2\,|\nabla\varphi|_g^2\left(1-|\nabla\varphi|_g^2\right)\right)d\mu_g\,. \tag{5.8}$$

Moreover, if there exists $s_0 \in (0, +\infty)$ such that

$$\int\limits_{\{\varphi=s_0\}}\left(|\nabla\varphi|_g^{p-1}\,\mathrm{H}_g - |\nabla\varphi|_g^{p-2}\Delta_g\varphi\right)\mathrm{d}\sigma_g \leq 0\,, \tag{5.9}$$

then the manifold (M^*, g) is isometric to one half round cylinder with totally geodesic boundary.

Remark 8 Translating Remark 1 in terms of the conformally related quantities, it is easy to realize that the integral on the left hand side of (5.8) is well defined also when s is a singular value of φ.

For the seek of clearness, we rewrite more explicitly Proposition 5.2, distinguishing the two cases $\Lambda > 0$, $\Lambda < 0$.

Corollary 5.3 (Case $\Lambda > 0$) *Let (M, g_0, u) be a static solution to problem* (1.4) *satisfying Normalization 1 and Assumption 1. Let g be the metric defined in* (3.1) *and φ be the smooth function defined in* (3.7)*. Then, for every $s \in [0, +\infty)$*

$$\int_{\{\varphi=s\}} \frac{|\nabla\varphi|_g^{p-1} H_g - |\nabla\varphi|_g^{p-2} \Delta_g\varphi}{\sinh(s)\cosh^{n+1}(s)} d\sigma_g =$$

$$= \int_{\{\varphi>s\}} |\nabla\varphi|_g^{p-3} \frac{|\nabla^2\varphi|_g^2 + (p-3)\left|\nabla|\nabla\varphi|_g\right|_g^2 + n\tanh^2(\varphi)\,|\nabla\varphi|_g^2\left(1-|\nabla\varphi|_g^2\right)}{\sinh(\varphi)\cosh^{n+1}(\varphi)} d\mu_g\,.$$

Moreover, if there exists $s_0 \in (0, +\infty)$ such that

$$\int_{\{\varphi=s_0\}} \left(|\nabla\varphi|_g^{p-1}\,\mathrm{H}_g - |\nabla\varphi|_g^{p-2}\Delta_g\varphi\right) \mathrm{d}\sigma_g \leq 0\,,$$

then (M, g_0, u) is isometric to the de Sitter solution.

Corollary 5.4 (Case $\Lambda < 0$) *Let (M, g_0, u) be a conformally compact static solution to problem* (1.17) *satisfying Normalization 2 and Assumption 2. Let g be the metric defined in* (3.12)*, and φ be the smooth function defined in* (3.18)*. Then, for every $s \in [0, +\infty)$*

$$\int_{\{\varphi=s\}} \frac{|\nabla\varphi|_g^{p-1} H_g - |\nabla\varphi|_g^{p-2} \Delta_g\varphi}{\sinh^{n+1}(s)\cosh(s)} d\sigma_g =$$

$$= \int_{\{\varphi>s\}} |\nabla\varphi|_g^{p-3} \frac{|\nabla^2\varphi|_g^2 + (p-3)\left|\nabla|\nabla\varphi|_g\right|_g^2 + n\coth^2(\varphi)\,|\nabla\varphi|_g^2\left(1-|\nabla\varphi|_g^2\right)}{\sinh^{n+1}(\varphi)\cosh(\varphi)} d\mu_g\,.$$

Moreover, if there exists $s_0 \in (0, +\infty)$ such that

$$\int_{\{\varphi=s_0\}} \left(|\nabla\varphi|_g^{p-1}\,\mathrm{H}_g - |\nabla\varphi|_g^{p-2}\Delta_g\varphi\right) \mathrm{d}\sigma_g \leq 0\,,$$

then (M, g_0, u) is isometric to the anti-de Sitter solution.

Proof of Proposition 5.2 We start by considering the case where the level set $\{\varphi = s\}$ is regular. Arguing as in the proof of Proposition 5.1, we find that we can choose S large enough to be sure that $\{\varphi = S\}$ is regular. Integrating by parts identity (5.7) in $\{s < \varphi < S\}$, we obtain

$$(p-1)\int_{\{s<\varphi<S\}} \gamma(\varphi)\,|\nabla\varphi|_g^{p-3}\Big(|\nabla^2\varphi|_g^2 + (p-3)\,\big|\nabla|\nabla\varphi|_g\big|_g^2 + n\,u^2\,|\nabla\varphi|_g^2\,\big(1-|\nabla\varphi|_g^2\big)\Big)\mathrm{d}\mu_g =$$

$$= \int_{\{\varphi=S\}} \gamma(\varphi)\,\langle\nabla|\nabla\varphi|_g^{p-1}\,|\,\mathrm{n}_g\rangle_g\,\mathrm{d}\sigma_g + \int_{\{\varphi=s\}} \gamma(\varphi)\,\langle\nabla|\nabla\varphi|_g^{p-1}\,|\,\mathrm{n}_g\rangle_g\,\mathrm{d}\sigma_g\,.$$

where n is the outer g-unit normal of the set $\{s \leq \varphi \leq S\}$ at its boundary. In particular, one has that $\mathrm{n}_g = -\nabla\varphi/|\nabla\varphi|_g$ on $\{\varphi = s\}$ and $\mathrm{n}_g = \nabla\varphi/|\nabla\varphi|_g$ on $\{\varphi = S\}$. On the other hand, from the second formula in (3.27) it is easy to deduce that

$$\langle\nabla|\nabla\varphi|_g^{p-1}|\nabla\varphi\rangle_g = \frac{p-1}{2}|\nabla\varphi|_g^{p-3}\langle\nabla|\nabla\varphi|_g^2|\nabla\varphi\rangle_g =$$

$$= (p-1)\,|\nabla\varphi|_g^{p-3}\nabla^2\varphi(\nabla\varphi,\nabla\varphi) = (p-1)\,|\nabla\varphi|_g^{p-3}\Big(-|\nabla\varphi|_g^3\,\mathrm{H}_g + |\nabla\varphi|_g^2\,\Delta_g\varphi\Big)\,.$$

Therefore, we have obtained

$$\int_{\{s<\varphi<S\}} \gamma(\varphi)\,|\nabla\varphi|_g^{p-3}\Big(|\nabla^2\varphi|_g^2 + (p-3)\,\big|\nabla|\nabla\varphi|_g\big|_g^2 + n\,u^2\,|\nabla\varphi|_g^2\,\big(1-|\nabla\varphi|_g^2\big)\Big)\mathrm{d}\mu_g =$$

$$= \gamma(s)\int_{\{\varphi=s\}}\Big(|\nabla\varphi|_g^{p-1}\,\mathrm{H}_g - |\nabla\varphi|_g^{p-2}\Delta_g\varphi\Big)\mathrm{d}\sigma_g - \gamma(S)\int_{\{\varphi=S\}}\Big(|\nabla\varphi|_g^{p-1}\,\mathrm{H}_g - |\nabla\varphi|_g^{p-2}\Delta_g\varphi\Big)\mathrm{d}\sigma_g\,. \tag{5.10}$$

In order to obtain identity (5.8) it is sufficient to show that the last term on the right hand side tends to zero as $S \to +\infty$. To this end, we first compute

$$\lim_{S\to+\infty}\gamma(S)\int_{\{\varphi=S\}}\Big(|\nabla\varphi|_g^{p-1}\,\mathrm{H}_g - |\nabla\varphi|_g^{p-2}\Delta_g\varphi\Big)\mathrm{d}\sigma_g = \lim_{S\to+\infty} -\gamma(S)\int_{\{\varphi=S\}}|\nabla\varphi|_g^{p-2}\nabla^2\varphi(\mathrm{n}_g,\mathrm{n}_g)\,\mathrm{d}\sigma_g\,.$$

Now we recall that $|\nabla\varphi|_g \leq 1$ thanks to Lemma 3.1, and we use formulæ (3.9), (3.20) to rewrite the limit above in terms of u, g_0. In both the cases $\Lambda > 0$ and $\Lambda < 0$, we find that it is enough to prove

$$\lim_{t\to1}\int_{\{u=t\}}\sqrt{|1-u^2|}\bigg[\Big(\frac{1-u^2}{u}\Big)\mathrm{D}^2u(\mathrm{n},\mathrm{n}) + |\mathrm{D}u|^2\bigg]\,\mathrm{d}\sigma = 0\,, \tag{5.11}$$

where $\mathrm{n} = \mathrm{D}u/|\mathrm{D}u|$. Note that, for t near enough to 1, the vector n is well defined. In fact, since the singular values of an analytic function are discrete (see [37]), it is clear that the values near enough to 1 are regular.

Since u is analytic, the level set $\{u = t\}$ has finite $\mathscr{H}^{n-1}$-measure (see [21, 27, 30])), thus the equality (5.11) is straightforward. This completes the proof of the first part of the statement in the case where $\{\varphi = s\}$ is regular.

In the case where $s > 0$ is a singular value of φ, we need to apply a slightly refined version of the Divergence Theorem, namely Theorem A.9 in the Appendix A, in order to perform the integration by parts which leads to identity (5.10). The rest of the proof is identical to what we have done for the regular case. We set

$$X = \gamma(\varphi)\,\nabla|\nabla\varphi|_g^{p-1} = \Big(\frac{p-1}{2}\Big)\,\gamma(\varphi)\,|\nabla\varphi|_g^{p-3}\,\nabla|\nabla\varphi|_g^2 \qquad \text{and} \qquad E = \{s < \varphi < S\}.$$

so that $\partial E = \{\varphi = s\} \sqcup \{\varphi = S\}$. As we have already observed, φ is proper and analytic, hence the $(n-1)$-dimensional Hausdorff measure of ∂E is finite. Moreover, it is clear that X is Lipschitz for $p \geq 3$. From the results in [30] (see also [27, Theorem 6.3.3]), we know that there exists an open $(n-1)$-submanifold $N \subseteq \mathrm{Crit}(\varphi)$ such that $\mathscr{H}^{n-1}(\partial E \setminus N) = 0$. Set $\Sigma = \partial E \cap (\mathrm{Crit}(\varphi) \setminus N)$ and $\Gamma = \partial E \setminus \Sigma$. We have $\mathscr{H}^{n-1}(\Sigma) = 0$ by definition, while Γ is the union of the regular part of ∂E and of N, which are open $(n-1)$-submanifolds. Therefore the hypotheses of Theorem A.9 are satisfied, hence, taking into account Remark 8 and expression (5.7), we have that identity (5.10) holds true also in the case where s is a singular value of φ.

To prove the second part of the statement, we observe that from (5.8) and (5.9) one immediately gets $\nabla^2\varphi \equiv 0$ in $\{\varphi \geq s_0\}$. Since φ is analytic, then $\nabla^2\varphi \equiv 0$ on the whole M^*. In particular, $\Delta_g\varphi = 0$ and, from the second equation in (3.23), we find $|\nabla\varphi|_g^2 \equiv 1$ on M^*.

Consider now the case $\Lambda > 0$. Substituting $\nabla^2\varphi = 0$ and $|\nabla\varphi|_g = 1$ in equality (3.9), we find $\mathrm{D}^2 u = -u\, g_0$ on M^*. Since u is analytic, the set $\mathrm{MAX}(u)$ is negligible, hence the equality $\mathrm{D}^2 u = -u\, g_0$ holds on the whole $M = M^* \cup \mathrm{MAX}(u)$. Therefore, using the same arguments as in [32], we deduce that (M, g_0) is an half-sphere, and translating this back in terms of the conformally related quantities, we easily find that (M^*, g) is isometric to an half round cylinder.

In the case $\Lambda < 0$ we proceed in a similar way. Substituting in equality (3.20), we find $\mathrm{D}^2 u = u g_0$ on M^* and, with the same argument as above, we deduce that the same equation holds on the whole $M = M^* \cup \mathrm{MIN}(u)$. Then we can use [34, Lemma 3.3] to conclude that (M, g_0) is isometric to the hyperbolic space, from which we deduce that (M^*, g) is an half round cylinder. □

6 Proof of Theorem 3.2

Building on the analysis of the previous section, we are now in the position to prove Theorem 3.2, which in turn implies Theorems 1.1 and 1.4.

6.1 Continuity

We claim that under the hypotheses of Theorem 3.2 the function Φ_p is continuous, for $p \geq 1$.

We first observe that since we are assuming that the boundary ∂M is a regular level set of φ, the function $s \mapsto \Phi_p(s)$ can be described in term of an integral depending on the parameter s, provided $s \in [0, 2\varepsilon)$ with $\varepsilon > 0$ sufficiently small. In this case, the continuous dependence on the parameter s can be easily checked using standard results from classical differential calculus. Thus, we leave the details to the interested reader and we pass to consider the case where $s \in (\varepsilon, +\infty)$. Thanks to Proposition 5.1 one can rewrite expression (3.30) as

$$\Phi_p(s) = -\sinh^n(s) \int_{\{\varphi > s\}} \frac{|\nabla\varphi|_g^{p-3}\Big((p-1)\,\nabla^2\varphi(\nabla\varphi,\nabla\varphi) + |\nabla\varphi|_g^2\,\Delta_g\varphi - n\coth(\varphi)\,|\nabla\varphi|_g^4\Big)}{\sinh^n(\varphi)}\, d\mu_g. \tag{6.1}$$

It is now convenient to set

$$\mu_g^{(p)}(E) = \int_E \frac{|\nabla\varphi|_g^{p-3}\Big((p-1)\,\nabla^2\varphi(\nabla\varphi,\nabla\varphi) + |\nabla\varphi|_g^2\,\Delta_g\varphi - n\coth(\varphi)\,|\nabla\varphi|_g^4\Big)}{\sinh^n(\varphi)}\, d\mu_g\,, \tag{6.2}$$

for every μ_g-measurable set $E \subseteq \{\varphi > \varepsilon\}$. It is then clear that for $p \geq 1$ the measure $\mu_g^{(p)}$ is absolutely continuous with respect to μ_g, since $|\nabla\varphi|_g^{p-3}\,\nabla^2\varphi(\nabla\varphi,\nabla\varphi) \leq |\nabla\varphi|_g^{p-1}|\nabla^2\varphi|_g$ and $|\nabla^2\varphi|_g$ is bounded (this is an easy consequence of equalities (3.10), (3.21)).

In view of (6.1), the function $s \mapsto \Phi_p(s)$ can be interpreted as the repartition function of the measure defined in (6.2), up to the smooth factor $-\sinh^n(s)$. Thus, $s \mapsto \Phi_p(s)$ is continuous if and only if the assignment

$$s \longmapsto \mu_g^{(p)}(\{\varphi > s\})$$

is continuous. Thanks to [7, Proposition 2.6] and thanks to the fact that $\mu_g^{(p)}$ is absolutely continuous with respect to μ_g, proving the continuity of the above assignment is equivalent to checking that $\mu_g(\{\varphi = s\}) = 0$ for every $s > \varepsilon$. On the other hand, the Hausdorff dimension of the level sets of φ is at most $n-1$, as it follows from the results in [21, 27, 30]). Hence, they are negligible with respect to the full n-dimensional measure. This proves the continuity of Φ_p for $p \geq 1$ under the hypotheses of Theorem 3.2.

6.2 *Monotonicity of* $\Phi_1(s)$

From the second equation in problem (3.23) and from Lemma 3.1, we get

$$\Delta_g \varphi = -nu\,(1 - |\nabla \varphi|_g^2) \leq 0\,.$$

Integrating this inequality in $\{s \leq \varphi \leq S\}$, we get

$$\int_{\{s \leq \varphi \leq S\}} \Delta_g \varphi \, \mathrm{d}\sigma_g \leq 0\,. \tag{6.3}$$

Suppose that $\{\varphi = s\}$ and $\{\varphi = S\}$ are regular levels (the case in which they are singular can be handled in the same way as in the proofs of Propositions 5.1 and 5.2). Then, applying the divergence theorem to inequality (6.3), we easily obtain $\Phi_1(S) \leq \Phi_1(s)$, for every $s < S$.

Moreover, if the equality holds for some values of s, S, then $|\nabla \varphi|_g \equiv 1$ on $\{s \leq \varphi \leq S\}$ and, since φ is analytic, we have $|\nabla \varphi|_g \equiv 1$ on the whole M^*. Plugging this information inside formula (3.25), we find that $\nabla^2 \varphi \equiv 0$ on M^*. With the same argument used in the proof of the rigidity statement in Proposition 5.2, we deduce that (M^*, g, φ) is an half round cylinder. This proves Theorem 3.2-(ii).

6.3 *Differentiability*

We now turn our attention to the issue of the differentiability of the functions $s \mapsto \Phi_p(s)$. As already observed in the previous subsection, we are assuming that the boundary ∂M is a regular level set of φ so that the function $s \mapsto \Phi_p(s)$ can be described in term of an integral depending on the parameter s, provided $s \in [0, 2\varepsilon)$ with $\varepsilon > 0$ sufficiently small. Again, the differentiability in the parameter s can be easily checked in this case, using standard results from classical differential calculus. Leaving the details to the interested reader, we pass to consider the case where $s \in (\varepsilon, +\infty)$. We start by noticing that for every $p \geq 2$ the function

$$\frac{|\nabla \varphi|_g^{p-4} \Big((p-1)\, \nabla^2 \varphi(\nabla \varphi, \nabla \varphi) + |\nabla \varphi|_g^2\, \Delta_g \varphi \; - \; n \coth(\varphi)\, |\nabla \varphi|_g^4 \Big)}{\sinh^n(\varphi)}$$

has finite integral in $\{\varphi > s\}$, for every $s > \varepsilon$. Hence, we can apply the coarea formula to expression (6.1), obtaining

$$\begin{aligned}
\Phi_p(s) &= -\sinh^n(s) \\
&\quad\times \int\limits_{\{\tau>s\}}\int\limits_{\{\varphi=\tau\}} \frac{(p-1)|\nabla\varphi|_g^{p-4}\nabla^2\varphi(\nabla\varphi,\nabla\varphi)+|\nabla\varphi|_g^{p-2}\Delta_g\varphi - n\coth(\varphi)|\nabla\varphi|_g^p}{\sinh^n(\varphi)}\,\mathrm{d}\sigma_g\,\mathrm{d}\tau \\
&= \sinh^n(s)\int\limits_{\{\tau>s\}}\int\limits_{\{\varphi=\tau\}} \frac{(p-1)|\nabla\varphi|_g^{p-1}\mathrm{H}_g - p|\nabla\varphi|_g^{p-2}\Delta_g\varphi + n\coth(\varphi)|\nabla\varphi|_g^p}{\sinh^n(\varphi)}\,\mathrm{d}\sigma_g\,\mathrm{d}\tau \\
&= \sinh^n(s)\int\limits_{\{\tau>s\}}\Bigg(\int\limits_{\{\varphi=\tau\}} \frac{(p-1)|\nabla\varphi|_g^{p-1}\mathrm{H}_g - p|\nabla\varphi|_g^{p-2}\Delta_g\varphi}{\sinh^n(\tau)}\,\mathrm{d}\sigma_g + n\frac{\coth(\tau)}{\sinh^n(\tau)}\Phi_p(\tau)\Bigg)\mathrm{d}\tau,
\end{aligned} \tag{6.4}$$

where in the second equality we have used (3.27) and in the third equality we have used the definition of Φ_p given by formula (3.30). By the Fundamental Theorem of Calculus, we have that if the function

$$\tau \longmapsto \int\limits_{\{\varphi=\tau\}} \frac{(p-1)|\nabla\varphi|_g^{p-1}\,\mathrm{H}_g \;-\; p\,|\nabla\varphi|_g^{p-2}\,\Delta_g\varphi}{\sinh^n(\tau)}\,\mathrm{d}\sigma_g \;+\; n\,\frac{\coth(\tau)}{\sinh^n(\tau)}\,\Phi_p(\tau)$$

is continuous, then Φ_p is differentiable. Since we have already discussed in Sect. 6.1 the continuity of $s \mapsto \Phi_p(s)$, we only need to discuss the continuity of the assignment

$$\begin{aligned}
\tau \longmapsto &\int\limits_{\{\varphi=\tau\}} \frac{(p-1)|\nabla\varphi|_g^{p-1}\,\mathrm{H}_g \;-\; p\,|\nabla\varphi|_g^{p-2}\,\Delta_g\varphi}{\sinh^n(\tau)}\,\mathrm{d}\sigma_g \\
&= (p-1)\frac{1}{\gamma(\tau)\sinh^n(\tau)} \\
&\quad\times \int\limits_{\{\varphi>\tau\}} \gamma(\varphi)\,|\nabla\varphi|_g^{p-3}\Big(\big|\nabla^2\varphi\big|_g^2 + (p-3)\,\big|\nabla|\nabla\varphi|_g\big|_g^2 - u\,|\nabla\varphi|_g^2\,\Delta_g\varphi\Big)\,\mathrm{d}\mu_g + \\
&\quad + n\,\frac{u(\tau)}{\sinh^n(\tau)}\big(\Phi_{p-2}(\tau) - \Phi_p(\tau)\big)\,.
\end{aligned} \tag{6.5}$$

We note that the above equality follows from formula $\Delta_g\varphi = -nu(1-|\nabla\varphi|_g^2)$ in problem (3.23) and from the integral identity (5.8) in Proposition 5.2, which is in force under the hypotheses of Theorem 3.2-(iii). In analogy with (6.2) it is natural to set

$$\bar{\mu}_g^{(p)}(E) \;=\; \int\limits_E \gamma(\varphi)\,|\nabla\varphi|_g^{p-3}\Big(\big|\nabla^2\varphi\big|_g^2 + (p-3)\,\big|\nabla|\nabla\varphi|_g\big|_g^2 - u\,|\nabla\varphi|_g^2\,\Delta_g\varphi\Big)\,\mathrm{d}\mu_g\,,$$

for every μ_g-measurable set $E \subseteq \{\varphi > \varepsilon\}$. It is now clear that for $p \geq 3$ the measure $\overline{\mu}_g^{(p)}$ is absolutely continuous with respect to μ_g. Hence, using the same reasoning as in Sect. 6.1, we deduce that the assignment (6.5) is continuous. In turn, we obtain the differentiability of Φ_p for $p \geq 3$, under the hypotheses of Theorem 3.2. Finally, using (6.4) and (5.8), a direct computation shows that

$$\begin{aligned}\Phi'_p(s) &= \int\limits_{\{\varphi=s\}} \Big(-(p-1)|\nabla\varphi|_g^{p-1}\mathrm{H}_g + p|\nabla\varphi|_g^{p-2}\Delta_g\varphi\Big)\,\mathrm{d}\sigma_g \\ &= -(p-1)\int\limits_{\{\varphi>s\}} \frac{\gamma(\varphi)}{\gamma(s)}|\nabla\varphi|_g^{p-3} \\ &\quad\times\Big(\big|\nabla^2\varphi\big|_g^2 + (p-3)\big|\nabla|\nabla\varphi|_g\big|_g^2 - u|\nabla\varphi|_g^2\,\Delta_g\varphi\Big)\mathrm{d}\mu_g + \\ &\quad+\int\limits_{\{\varphi=s\}} |\nabla\varphi|_g^{p-2}\Delta_g\varphi\mathrm{d}\sigma_g. \end{aligned} \tag{6.6}$$

The monotonicity and the rigidity statements in Theorem 3.2-(iii) are now consequences of Proposition 5.2.

6.4 *The Second Derivative*

To complete our analysis, we need to prove statement (iii) in Theorem 3.2. To this aim, we observe from (3.27) and the first equation of problem (3.23) that

$$\begin{aligned}\frac{\Phi'_p(s)}{s} &= \frac{1}{s}\int\limits_{\{\varphi=s\}}\Big(-(p-1)|\nabla\varphi|_g^{p-1}\,\mathrm{H}_g + p\,|\nabla\varphi|_g^{p-2}\,\Delta_g\varphi\Big)\,\mathrm{d}\sigma_g \\ &= \frac{u(s)}{s\left[1-(n-1)\,u(s)^2\right]} \\ &\quad\times\int\limits_{\{\varphi=s\}}|\nabla\varphi|_g^{p-2}\Big[(p-1)\,\mathrm{Ric}_g(\nu_g,\nu_g) - n\big(p-(n-1)u^2\big)\Big(1-|\nabla\varphi|_g^2\Big)\Big]\,\mathrm{d}\sigma_g\,. \end{aligned}$$

Taking the limit as $s \to 0^+$ and using Assumption 3, we obtain (3.35). In the case $\Lambda < 0$, one also needs to recall that $\lim_{s\to 0^+}|\nabla\varphi|_g = 1$. This is an easy consequence of formula (3.24) (see also the proof of Lemma A.8-(i)). To prove the rigidity statement in the case $\Lambda > 0$, we observe that

$$
\begin{aligned}
\Phi_p''(0) &= \lim_{s\to 0^+} \frac{1}{s} \int_{\{\varphi=s\}} \Big(-(p-1)|\nabla\varphi|_g^{p-1}\, \mathrm{H}_g + p\, |\nabla\varphi|_g^{p-2}\, \Delta_g\varphi \Big)\, d\sigma_g \\
&= \lim_{s\to 0^+} \Bigg[-\left(\frac{p-1}{s}\right) \int_{\{\varphi=s\}} \Big(|\nabla\varphi|_g^{p-1}\, \mathrm{H}_g - |\nabla\varphi|_g^{p-2}\, \Delta_g\varphi \Big)\, d\sigma_g \\
&\quad + \frac{1}{s} \int_{\{\varphi=s\}} |\nabla\varphi|_g^{p-2}\, \Delta_g\varphi \, d\sigma_g \Bigg] \\
&= \lim_{s\to 0^+} \Bigg[-(p-1) \int_{\{\varphi=s\}} \left(\frac{|\nabla\varphi|_g^{p-1}\, \mathrm{H}_g - |\nabla\varphi|_g^{p-2}\, \Delta_g\varphi}{\sinh(s)\cosh^{n+1}(s)} \right) d\sigma_g \\
&\quad - n\, \frac{u(s)}{s} \int_{\{\varphi=s\}} |\nabla\varphi|_g^{p-2} \left(1 - |\nabla\varphi|_g^2\right)\, d\sigma_g \Bigg],
\end{aligned}
$$

and we conclude using Proposition 5.2.

To understand why a similar rigidity statement does not hold in the case of a negative cosmological constant, we observe that a computation analogous to the one above gives

$$
\begin{aligned}
\Phi_p''(0) &= \lim_{s\to 0^+} \Bigg[-(p-1)\, s^n \int_{\{\varphi=s\}} \left(\frac{|\nabla\varphi|_g^{p-1}\, \mathrm{H}_g - |\nabla\varphi|_g^{p-2}\, \Delta_g\varphi}{\sinh(s)^{n+1}\cosh(s)} \right) d\sigma_g \\
&\quad - n\, \frac{u(s)}{s} \int_{\{\varphi=s\}} |\nabla\varphi|_g^{p-2} \left(1 - |\nabla\varphi|_g^2\right)\, d\sigma_g \Bigg].
\end{aligned}
$$

Therefore, if the equality holds in (3.35), one inferes

$$
\lim_{s\to 0^+} s^n \int_{\{\varphi=s\}} \left(\frac{|\nabla\varphi|_g^{p-1}\, \mathrm{H}_g - |\nabla\varphi|_g^{p-2}\, \Delta_g\varphi}{\sinh(s)^{n+1}\cosh(s)} \right) d\sigma_g = 0\,,
$$

and this is not sufficient to use Proposition 5.2 to deduce the rotational symmetry of the solution.

Appendix A: Technical Results

This appendix will be dedicated to the proof of the technical results that we have used in our work. Specifically, we will give a complete proof of Theorems 1.2, 1.5 (for the ease of reference, we have restated them here as Theorem A.1 and Theorem A.7), we will prove an estimate on the static solution near the conformal boundary in the case $\Lambda < 0$, and we will state the version of the divergence theorem that we have used in the proofs of Propositions 5.1, 5.2.

Theorem A.1 *Let* (M, g_0, u) *be a solution of* (1.4) *satisfying Assumption 1. The set* $\mathrm{MAX}(u)$ *is discrete (and finite) and*

$$\liminf_{t\to 1^-} U_p(t) \ \geq \ |\mathrm{MAX}(u)|\,|\mathbb{S}^{n-1}|\,, \tag{A.1}$$

for every $0 \leq p \leq n-1$.

In the proof of this theorem, we will need the following result, that will be proven later.

Proposition A.2 *Let* (M, g_0, u) *be a solution of* (1.4) *and let* $y_0 \in \mathrm{MAX}(u)$. *Then for every* $d > 0$ *it holds*

$$\liminf_{t\to 1^-} \left(\frac{1}{1-t^2}\right)^{n-1} \int_{\{u=t\}\cap B_d(y_0)} |\mathrm{D}u|^{n-1}\, \mathrm{d}\sigma \ \geq \ |\mathbb{S}^{n-1}| \tag{A.2}$$

We first show how to use this result to prove Theorem A.1.

Proof of Theorem A.1 First we notice that the functions $U_p(t)$ can be written as follows

$$U_p(t) \ = \ \left(\frac{1}{1-t^2}\right)^{\frac{n-1}{2}} \int_{\{u=t\}} \left[\frac{|\mathrm{D}u|^2}{1-u^2}\right]^{\frac{p}{2}} \mathrm{d}\sigma\,.$$

From formula (1.10) in Theorem 1.1, we have that the term in square bracket is less or equal to 1. Thus, for every $p \leq n-1$, we have

$$\left[\frac{|\mathrm{D}u|^2}{1-u^2}\right]^{\frac{p}{2}} \ \geq \ \left[\frac{|\mathrm{D}u|^2}{1-u^2}\right]^{\frac{n-1}{2}},$$

hence $U_p(t) \geq U_{n-1}(t)$ and, in particular

$$\liminf_{t\to 1^-} U_p(t) \geq \liminf_{t\to 1^-} U_{n-1}(t)\,, \tag{A.3}$$

so it is enough to prove the inequality (A.1) for $p = n - 1$.

Now we pass to analyze the set $\mathrm{MAX}(u)$. Suppose that it contains an infinite number of points. Then for each $k \in \mathbb{N}$ we can consider k points in $\mathrm{MAX}(u)$. Let $2d$ be the minimum of the distances between our points. Applying Proposition A.2 in a neighborhood of radius d of each of these points, we obtain

$$\liminf_{t \to 1^-} U_{n-1}(t) \geq k\,|\mathbb{S}^{n-1}|\,.$$

Since this is true for every $k \in \mathbb{N}$, we conclude that $\lim_{t\to 1^-} U_{n-1}(t) = +\infty$ and, using (A.3), we find that $\lim_{t\to 1^-} U_1(t) = +\infty$. But this is impossible, since from the monotonicity of $U_1(t)$ (stated in Theorem 1.1-(ii)) we know that

$$\lim_{t \to 1^-} U_1(t) \leq U_1(0) = |\partial M|\,.$$

Therefore $\mathrm{MAX}(u)$ contains only a finite number of points. Repeating the argument above with $k = |\mathrm{MAX}(u)|$ we obtain the inequality in the thesis. □

Now we turn to the proof of Proposition A.2, that will be done in various steps. Our strategy consists in choosing a suitable neighborhood of the point y_0 where we are able to control the quantities in our limit, and then proceed to estimate them.

Notation 1 *Here and throughout the paper, we agree that for $f \in \mathscr{C}^\infty(M)$, $\tau \in \mathbb{R}$ and $k \in \mathbb{N}$ it holds*

$$f = o_k(|x|^{-\tau}) \quad \Longleftrightarrow \quad \sum_{|J|\leq k} |x|^{\tau+|J|}\left|\partial^J f\right| = o(1)\,, \qquad \text{as } |x| \to +\infty\,,$$

where the J's are multi-indexes.

Consider a normal set of coordinates $(x^1, \dots, x^n)$ in $B_d(y_0)$, that diagonalize the hessian in y_0. Note that, since y_0 is a maximum of u, the derivatives $\partial^2_\alpha u_{|y_0}$ are non positive numbers for all $\alpha = 1, \dots, n$, hence it makes sense to introduce the quantities $\lambda_\alpha^2 = -\partial^2_\alpha u_{|y_0}$ for $\alpha = 1, \dots, n$. Since $\Delta u = -nu$, we have $\sum_{\alpha=1}^n \lambda_\alpha^2 = n$. In particular, at least one of the λ_α's is different from zero. We have the following Taylor expansion of u in a neighborhood of y_0

$$u = 1 - \frac{1}{2}\sum_{\alpha=1}^{n}\left[\lambda_\alpha^2 \cdot |x^\alpha|^2\right] + o_2\Big(|x|^2\Big)\,, \tag{A.4}$$

From (A.4) we easily compute

$$|\mathrm{D}u|^2 = \sum_{i=1}^{n} \left[\lambda_\alpha^4 \cdot |x^\alpha|^2\right] + o_1\left(|x|^2\right). \tag{A.5}$$

Now we consider polar coordinates $(r, \theta^1, \dots, \theta^{n-1})$, where $\theta = (\theta^1, \dots, \theta^{n-1}) \in \mathbb{R}^{n-1}$ are stereographic coordinates on $\mathbb{S}^{n-1} \setminus \{north\ pole\}$.

Lemma A.3 *With respect to the coordinates $(r, \theta) = (r, \theta^1, \dots, \theta^{n-1})$, the metric g_0 writes as*

$$g_0 = dr \otimes dr + r^2 g_{\mathbb{S}^{n-1}} + \sigma\, dr \otimes dr + \sigma_i \left(dr \otimes d\theta^i + d\theta^i \otimes dr\right) + \sigma_{ij}\, d\theta^i \otimes d\theta^j \tag{A.6}$$

where $\sigma = o_2(r)$, $\sigma_i = o_2(r^2)$, $\sigma_{ij} = o_2(r^3)$, as $r \to 0^+$.

Proof To ease the notation, in this proof we use the Einstein summation convention. It is known that, with respect to the normal coordinates $(x^1, \dots, x^n)$ the metric g_0 writes as

$$g_0 = (\delta_{\alpha\beta} + \eta_{\alpha\beta}) dx^\alpha \otimes dx^\beta$$

where $\delta_{\alpha\beta}$ is the Kronecker delta and $\eta_{\alpha\beta} = o_2(r)$ (actually, the term $\eta_{\alpha\beta}$ can be estimated better, but this is enough for our purposes).

Moreover, it is easy to check that the quantities $\phi^\alpha = x^\alpha / r$ are smooth functions of the coordinates $(\theta^1, \dots, \theta^{n-1})$ only, and that

$$\delta_{\alpha\beta} \frac{\partial \phi^\alpha}{\partial \theta^i} \frac{\partial \phi^\beta}{\partial \theta^j} d\theta^i \otimes d\theta^j = g_{\mathbb{S}^{n-1}}.$$

From $r^2 = \delta_{\alpha\beta}\, x^\alpha x^\beta$ one also finds the equality $\delta_{\alpha\beta}\, \phi^\alpha \phi^\beta = 1$. Deriving it with respect to θ^i we get

$$\delta_{\alpha\beta} \frac{\partial \phi^\alpha}{\partial \theta^i} \phi^\beta = 0, \qquad \text{for all } i = 1, \dots, n-1$$

We are now ready to compute

$$\begin{aligned}
g_0 &= (\delta_{\alpha\beta} + \eta_{\alpha\beta})\, dx^\alpha \otimes dx^\beta \\
&= (1 + \eta_{\alpha\beta} \phi^\alpha \phi^\beta) dr \otimes dr + (\delta_{\alpha\beta} + \eta_{\alpha\beta}) r^2 \frac{\partial \phi^\alpha}{\partial \theta^i} \frac{\partial \phi^\beta}{\partial \theta^j} d\theta^i \otimes d\theta^j \\
&\quad + \eta_{\alpha\beta} \phi^\alpha \frac{\partial \phi^\beta}{\partial \theta^i}\, r \left(dr \otimes d\theta^i + d\theta^i \otimes dr\right) \\
&= dr \otimes dr + r^2 g_{\mathbb{S}^{n-1}} + \sigma\, dr \otimes dr + \sigma_i \left(dr \otimes d\theta^i + d\theta^i \otimes dr\right) + \sigma_{ij}\, d\theta^i \otimes d\theta^j,
\end{aligned}$$

where $\sigma, \sigma_i, \sigma_{ij}$ are infinitesimals of the wished order. □

We can rewrite formulæ (A.4), (A.5) in terms of (r, θ) as

$$u(r,\theta) = 1 - \frac{r^2}{2} \sum_{\alpha=1}^{n} \big[\lambda_\alpha^2 \, |\phi^\alpha|^2(\theta) \big] + w(r,\theta) \,, \tag{A.7}$$

$$|\mathrm{D}u|^2(r,\theta) = r^2 \sum_{\alpha=1}^{n} \big[\lambda_\alpha^4 \, |\phi^\alpha|^2(\theta) \big] + h(r,\theta) \,. \tag{A.8}$$

where $w(r,\theta) = o(r^2)$, $h(r,\theta) = o(r^2)$. Moreover, since we know from (A.4) that $\partial w / \partial x^\alpha = o(r)$ for any α, we have the following estimates on the order of the derivatives of w with respect to (r, θ)

$$\begin{aligned} \frac{\partial w}{\partial r}(r,\theta) &= \sum_{\alpha=1}^{n} \Big[\frac{\partial w}{\partial x^\alpha}(r,\theta) \, \frac{\partial x^\alpha}{\partial r}(r,\theta) \Big] \\ &= \sum_{\alpha=1}^{n} \Big[\frac{\partial w}{\partial x^\alpha}(r,\theta) \, \phi^\alpha(\theta) \Big] = o(r) \,, \qquad \text{as } r \to 0^+ \end{aligned} \tag{A.9}$$

$$\begin{aligned} \frac{\partial w}{\partial \theta^j}(r,\theta) &= \sum_{\alpha=1}^{n} \Big[\frac{\partial w}{\partial x^\alpha}(r,\theta) \, \frac{\partial x^\alpha}{\partial \theta^j}(r,\theta) \Big] \\ &= \sum_{\alpha=1}^{n} \Big[\frac{\partial w}{\partial x^\alpha}(r,\theta) \, r \, \frac{\partial \phi^\alpha}{\partial \theta^j}(\theta) \Big] = o(r^2) \,, \qquad \text{as } r \to 0^+ \end{aligned} \tag{A.10}$$

To estimate the limit in (A.2), we need to rewrite the set $\{u = t\} \cap B_d(y_0)$ and the density

$$\sqrt{\det(g_0|_{\{u=t\} \cap B_d(y_0)})}$$

as functions of our coordinates. In order to do so, it will prove useful to restrict our neighborhood $B_d(y_0)$ to a smaller domain where we have a better characterization of the level set $\{u = t\}$. In this regard, it is convenient, for any $\varepsilon > 0$, to define the set

$$C_\varepsilon = \Big\{ \theta = (\theta^1, \ldots, \theta^{n-1}) \in \mathbb{R}^{n-1} \, : \, \sum_{\alpha=1}^{n} \big[\lambda_\alpha^2 \, |\phi^\alpha|^2(\theta) \big] > \varepsilon \Big\} \,.$$

The following result shows that, for t small enough, the level set $\{u = t\}$, is a graph over C_ε.

Lemma A.4 *For any* $0 < \varepsilon < 1$, *there exists* $\eta = \eta(\varepsilon) > 0$ *such that*

(i) *the estimates* $|w|(r,\theta) < \frac{\varepsilon^2}{4} r^2$, $|\partial w/\partial r|(r,\theta) < \frac{\varepsilon}{2} r$, $|h|(r,\theta) < \varepsilon^2 r^2$ *holds on the whole* $B_\eta(y_0)$.
(ii) *it holds* $\frac{\partial u}{\partial r}(r,\theta) < 0$ *in* $(0,\eta) \times C_\varepsilon$.
(iii) *for every* $0 < \delta < \eta$, *there exists* $\tau = \tau(\delta,\varepsilon)$ *such that for any* $\tau < t < 1$, *there exists a smooth function* $r_t : C_\varepsilon \to (0,\delta)$ *such that*

$$\{u = t\} \cap B_\delta(y_0) \cap C_\varepsilon = \{(r_t(\theta), \theta) : \theta \in C_\varepsilon\}.$$

Proof Since the functions w, h in (A.7), (A.8) are $o(r^2)$, while $\partial w/\partial r$ is $o(r)$ thanks to (A.9), it is clear that statement (i) is true for some η small enough. Moreover, from expansion (A.7) we compute

$$\frac{\partial u}{\partial r}(r,\theta) = -r \sum_{\alpha=1}^{n} \left[\lambda_\alpha^2 \, |\phi^\alpha|^2(\theta) \right] + \frac{\partial w}{\partial r}(r,\theta) < -\varepsilon r + \frac{\varepsilon}{2} r = -\frac{\varepsilon}{2} r \, .$$

This proves point (ii). To prove (iii), fix $t \in (0,1)$ and consider the function $u(r,\theta) - t$. Since $u(r,\theta) \to 1^-$ as $r \to 0^+$, we have $u(r,\theta) - t > 0$ for small values of r.

On the other hand, from expansion (A.7) we find

$$\begin{aligned} u(\delta,\theta) - t = (1-t) - \frac{\delta^2}{2} \sum_{\alpha=1}^{n} \left[\lambda_\alpha^2 \, |\phi^\alpha|^2(\theta) \right] + w(\delta,\theta) &< (1-t) - \frac{\varepsilon}{2} \delta^2 \\ + w(\delta,\theta) &< (1-t) - \frac{\varepsilon}{4} \delta^2 \, , \end{aligned}$$

and the quantity on the right is negative for any $t > \tau = 1 - \frac{\varepsilon}{4}\delta^2$.

Therefore, fixed a $\theta \in C_\varepsilon$ the function $r \mapsto u(r,\theta) - t$ is positive for small values of r and negative for $r = \delta$. Moreover from point (ii) we have that $\frac{\partial u}{\partial r}(r,\theta) < 0$ for any $(r,\theta) \in (0,\delta) \times C_\varepsilon$, hence for any $\theta \in C_\varepsilon$, there exists one and only one value $0 < r_t(\theta) < \delta$ such that $(r_t(\theta), \theta) \in \{u = t\}$. The smoothness of the function $r_t(\theta)$ is a consequence of the Implicit Function Theorem applied to the function $u(r,\theta)$.

□

As anticipated, Lemma A.4 will now be used to estimate the density of the restriction of the metric g_0 on $\{u = t\} \cap ((0,\delta) \times C_\varepsilon)$.

Lemma A.5 *There exists* $0 < \delta < \eta(\varepsilon)$ *such that it holds*

$$\sqrt{\det(g_0{}_{|\{u=t\}})}\,(r_t(\theta), \theta) > (1-\varepsilon)\, r_t^{n-1}(\theta) \sqrt{\det(g_{\mathbb{S}^{n-1}})} \, ,$$

for every $\theta \in C_\varepsilon$, $\tau(\delta,\varepsilon) < t < 1$.

Proof Let r_t be the function introduced in Lemma A.4. Taking the total derivative of $u(r_t(\theta), \theta) = t$, we find, for any $\theta \in C_\varepsilon$

$$dr_t = -\Big[\frac{\partial u}{\partial r}(r_t(\theta), \theta)\Big]^{-1} \sum_{j=1}^{n-1} \frac{\partial u}{\partial \theta^j}(r_t(\theta), \theta)\, d\theta^j = -\, r_t(\theta) \sum_{j=1}^{n-1} \xi_j(\theta) d\theta^j\,,$$

where

$$\xi_j(r_t(\theta), \theta) = \frac{1}{r_t(\theta)} \frac{\frac{\partial u}{\partial \theta^j}}{\frac{\partial u}{\partial r}}(r_t(\theta), \theta)\,.$$

To ease the notation, in the rest of the proof we avoid to explicitate the dependence of the functions on the variables $r_t(\theta), \theta$. In order to compute the restriction of the metric on $\{u = t\} \cap ((0, \delta) \times C_\varepsilon)$, we substitute the term dr in formula (A.6) with the formula for dr_t computed above. We obtain

$$g_0{}_{|\{u=t\}} = r_t^2 \Big[\xi_i\, \xi_j (1+\sigma) + g_{ij}^{\mathbb{S}^{n-1}} - \frac{\sigma_i}{r_t}\xi_j - \frac{\sigma_j}{r_t}\xi_i + \sigma_{ij}\Big]\, d\theta^i \otimes d\theta^j\,.$$

Set $\xi = \sum_{j=1}^{n-1} \xi_j d\theta^j$. We have the following

$$\begin{aligned} \sqrt{\det\Big[\Big(\xi_i\, \xi_j + g_{ij}^{\mathbb{S}^{n-1}}\Big)\, d\theta^i \otimes d\theta^j\Big]} &= \sqrt{\det\Big(\xi \otimes \xi + g_{\mathbb{S}^{n-1}}\Big)} \\ &= \sqrt{(1 + |\xi|^2_{g_{\mathbb{S}^{n-1}}})\, \det(g_{\mathbb{S}^{n-1}})} \\ &\geq \sqrt{\det(g_{\mathbb{S}^{n-1}})}\,, \end{aligned} \tag{A.11}$$

where in the second equality we have used the Matrix Determinant Lemma.

On the other hand, since $\sigma_i = o(r^2)$ and $\sigma_{ij} = o(r^3)$, we deduce that

$$\sigma\, \xi_i\, \xi_j - \frac{\sigma_i}{r_t}\xi_j - \frac{\sigma_j}{r_t}\xi_i + \sigma_{ij} = o(r_t)\,,$$

hence

$$\sqrt{\det(g_0{}_{|\{u=t\}})} = (1+\omega)\, r_t^{n-1} \sqrt{\det\Big[\Big(\xi_i\, \xi_j + g_{ij}^{\mathbb{S}^{n-1}}\Big)\, d\theta^i \otimes d\theta^j\Big]}\,, \tag{A.12}$$

with $\omega = o(1)$ as $r \to 0^+$. In particular, we can choose δ small enough so that $|\omega| < \varepsilon$ on $(0, \delta) \times C_\varepsilon$. Combining Eqs. (A.11) and (A.12) we have the thesis. □

We also need an estimate of the integrand in (A.2), which is provided by the following lemma.

Lemma A.6 *We can choose* $0 < \delta < \eta(\varepsilon)$ *such that*

$$\frac{|\mathrm{D}u|^2}{1-u^2}(r,\theta) > (1-\varepsilon)\sum_{\alpha=1}^{n}\left[\lambda_\alpha^2\,|\phi^\alpha|^2(\theta)\right]$$

for every $(r,\theta) \in (0,\delta) \times C_\varepsilon$.

Proof To ease the notation, in this proof we avoid to explicitate the dependence of the functions on the coordinates r, θ. From expansions (A.7) and (A.8) we deduce

$$\begin{aligned}\frac{|\mathrm{D}u|^2}{2(1-u)} &= \frac{\sum_{\alpha=1}^n\left(\lambda_\alpha^4\,|\phi^\alpha|^2\right)+h}{\sum_{\alpha=1}^n\left(\lambda_\alpha^2\,|\phi^\alpha|^2\right)-2\,w}\\ &= \frac{\sum_{\alpha=1}^n\left(\lambda_\alpha^4\,|\phi^\alpha|^2\right)}{\sum_{\alpha=1}^n\left(\lambda_\alpha^2\,|\phi^\alpha|^2\right)}\left[1+\frac{h}{\sum_{\alpha=1}^n(\lambda_\alpha^4\,|\phi^\alpha|^2)}\right]\left[1-\frac{2\,w}{\sum_{\alpha=1}^n(\lambda_\alpha^2\,|\phi^\alpha|^2)}\right]^{-1}.\end{aligned}$$

Using the Cauchy-Schwarz Inequality we have

$$\sum_{\alpha=1}^n\left(\lambda_\alpha^2\,|\phi^\alpha|^2\right) \le \left[\sum_{\alpha=1}^n\left(\lambda_\alpha^4\,|\phi^\alpha|^2\right)\right]^{\frac{1}{2}}\cdot\left[\sum_{\alpha=1}^n|\phi^\alpha|^2\right]^{\frac{1}{2}} = \left[\sum_{\alpha=1}^n\left(\lambda_\alpha^4\,|\phi^\alpha|^2\right)\right]^{\frac{1}{2}}, \tag{A.13}$$

hence, recalling that $\sum_{\alpha=1}^n(\lambda_\alpha^2|\phi^\alpha|^2) > \varepsilon$ on C_ε, we have also $\sum_{\alpha=1}^n(\lambda_\alpha^4|\phi^\alpha|^2) > \varepsilon^2$. Therefore, from Lemma A.4-(i) we easily compute

$$\left|\left(1+\frac{h}{\sum_{\alpha=1}^n(\lambda_\alpha^4\,|\phi^\alpha|^2)}\right)\left(1-\frac{2\,w}{\sum_{\alpha=1}^n(\lambda_\alpha^2\,|\phi^\alpha|^2)}\right)^{-1}\right| > \frac{1-\delta^2}{1+\frac{\varepsilon\delta^2}{2}}.$$

In particular, we can choose δ small enough so that the right hand side of the inequality above is greater than $1-\varepsilon$. Hence, we get

$$\frac{|\mathrm{D}u|^2}{1-u^2} \ge \frac{|\mathrm{D}u|^2}{2(1-u)} \ge (1-\varepsilon)\frac{\sum_{\alpha=1}^n\left(\lambda_\alpha^4\,|\phi^\alpha|^2\right)}{\sum_{\alpha=1}^n\left(\lambda_\alpha^2\,|\phi^\alpha|^2\right)} \ge (1-\varepsilon)\sum_{\alpha=1}^n\left[\lambda_\alpha^2\,|\phi^\alpha|^2\right],$$

where in the first inequality we have used that $u \le 1$ on M and in the latter inequality we have used (A.13). □

Now we are finally able to prove our proposition.

Proof of Proposition A.2 For every $\varepsilon > 0$ and $0 < \delta \leq d$, we have the following estimate of the left hand side of condition (A.2)

$$\left(\frac{1}{1-t^2}\right)^{n-1} \int\limits_{\{u=t\}\cap B_d(y_0)} |\mathrm{D}u|^{n-1}\, \mathrm{d}\sigma \geq \left(\frac{1}{1-t^2}\right)^{n-1} \int\limits_{\{u=t\}\cap B_\delta(y_0)} |\mathrm{D}u|^{n-1}\, \mathrm{d}\sigma$$

$$= \int\limits_{\{u=t\}\cap B_\delta(y_0)} \left(\frac{|\mathrm{D}u|^2}{1-u^2}\right)^{\frac{n-1}{2}} \left(\frac{2}{1+u}\right)^{\frac{n-1}{2}} \left[\frac{1}{2(1-u)}\right]^{\frac{n-1}{2}} \mathrm{d}\sigma \tag{A.14}$$

Since $u \leq 1$, we have $2/(1+u) \geq 1$. Moreover, from (A.7) and Lemma A.4-(i), we obtain

$$[2(1-u)](r,\theta) = r^2 \sum_{\alpha=1}^{n} \left[\lambda_\alpha^2 |\phi^\alpha|^2(\theta)\right] \left(1 - \frac{w(r,\theta)}{r^2 \sum_{\alpha=1}^{n} \lambda_\alpha |\phi^\alpha|^2(\theta)}\right)$$

$$< (1+\varepsilon)\, r^2 \sum_{\alpha=1}^{n} \left[\lambda_\alpha^2 |\phi^\alpha|^2(\theta)\right].$$

Now fix a δ small enough so that Lemmas A.5, A.6 are in charge. Taking the limit of integrand (A.14) as $t \to 1^-$ we obtain the estimate

$$\liminf_{t\to 1^-} \left(\frac{1}{1-t^2}\right)^{n-1} \int\limits_{\{u=t\}\cap B_d(y_0)} |\mathrm{D}u|^{n-1}\, \mathrm{d}\sigma > \int\limits_{C_\varepsilon} (1-\varepsilon) \left(\frac{1-\varepsilon}{1+\varepsilon}\right)^{\frac{n-1}{2}} \sqrt{\det(g_{\mathbb{S}^{n-1}})}\, d\theta^1 \cdots d\theta^{n-1}$$

$$= \int\limits_{\mathbb{R}^{n-1}} \chi_{C_\varepsilon}(\theta) \frac{(1-\varepsilon)^{\frac{n+1}{2}}}{(1+\varepsilon)^{\frac{n-1}{2}}} \sqrt{\det(g_{\mathbb{S}^{n-1}})}\, d\theta^1 \cdots d\theta^{n-1}. \tag{A.15}$$

It is clear that the functions χ_{C_ε} converge to χ_{C_0} as $\varepsilon \to 0^+$, where

$$C_0 = \left\{\theta = (\theta^1, \ldots, \theta^{n-1}) \in \mathbb{R}^{n-1} : \sum_{\alpha=1}^{n} \left[\lambda_\alpha^2\, |\phi^\alpha|^2(\theta)\right] \neq 0\right\} \subseteq \mathbb{S}^{n-1}.$$

Therefore, taking the limit of (A.15) as $\varepsilon \to 0^+$ and using the Monotone Convergence Theorem, we find

$$\liminf_{t\to 1^-} \left(\frac{1}{1-t^2}\right)^{n-1} \int\limits_{\{u=t\}\cap B_d(y_0)} |\mathrm{D}u|^{n-1}\, \mathrm{d}\sigma \geq \int\limits_{\mathbb{S}^{n-1}} \chi_{C_0}\, \mathrm{d}\sigma_{\mathbb{S}^{n-1}}.$$

To end the proof, it is enough to show that the set $\mathbb{S}^{n-1} \setminus C_0$ is negligible. But this is clear. In fact, since $\sum_{\alpha=1}^{n} \lambda_\alpha^2 = n$, there exists at least one integer β such that $\lambda_\beta \neq 0$. Thus $\mathbb{S}^{n-1} \setminus C_0$ is contained in the hypersurface $\{\phi^\beta = 0\}$, hence its n-measure is zero. This proves inequality (A.2) and the thesis. □

This concludes the proof for the de Sitter case. In the anti-de Sitter case we can prove the following analogue of Theorem A.1.

Theorem A.7 *Let (M, g_0, u) be a conformally compact static solution of problem 1.17 satisfying Assumption 2. Then the set* MIN(u) *is discrete (and finite) and*

$$\liminf_{t \to 1^+} U_p(t) \geq |\mathrm{MIN}(u)|\, |\mathbb{S}^{n-1}|\,,$$

for every $p \leq n-1$.

The proof follows the exact same scheme as the de Sitter case, the only small modifications being in the proof of Lemma A.5 and in the computation (A.15), where we have used the fact that $u \leq 1$. This is not true anymore, however, since we are working around a minimum point, we can suppose $u < 1 + \kappa$, where κ is an infinitesimal quantity that can be chosen to be as small as necessary. Aside from this little expedient, the proof is virtually the same as the de Sitter case, thus we omit it.

We pass now to the proof of some other results that we have used in our work. The next lemma is useful in order to study the behavior of the static solutions of problem (1.17) near the conformal boundary.

Lemma A.8 *Let (M, g_0, u) be a conformally compact static solution to problem* (1.17)*. Suppose that $1/\sqrt{u^2-1}$ is a defining function, so that the metric $g = g_0/(u^2-1)$ extends to the conformal boundary ∂M. Then*

(i) $\lim_{x \to \bar{x}} (u^2 - 1 - |\mathrm{D}u|^2)$ *is well-definite and finite for every $\bar{x} \in \partial M$,*
(ii) ∂M *is a totally geodesic hypersurface in $(\overline{M}, g)$.*

Proof For the proof of this result, it is convenient to use the notations introduced in Sect. 3. Let φ be the function defined by (3.18). By hypothesis, M is the interior of a compact manifold $\overline{M}$ and the metric g is well defined on the whole $\overline{M}$. In particular, the scalar curvature R_g is a smooth finite function at ∂M. Therefore, from Eq. (3.24) we easily deduce that $\lim_{x \to \bar{x}} u^2(1 - |\nabla\varphi|_g^2)$ is well-definite and finite for every $\bar{x} \in \partial M$. Since

$$|\nabla\varphi|_g^2 = \frac{|\mathrm{D}u|^2}{u^2 - 1}\,,$$

this proves point (i).

To prove statement (ii), we first observe that, since $|\nabla\varphi|_g = 1$ at ∂M (as it follows immediately from point (i)), there exists $\delta > 0$ such that $|\nabla\varphi|_g \neq 0$ on the whole collar $\mathcal{U}_\delta = \{\varphi < \delta\}$. Therefore, proceeding as in Sect. 3.4, we find a set of coordinates $\{\varphi, \vartheta^1, \dots, \vartheta^{n-1}\}$ on $\mathcal{U}_\delta$, such that the metric g writes as

$$g = \frac{d\varphi \otimes d\varphi}{|\nabla\varphi|_g^2} + g_{ij}(\varphi, \theta^i, \dots, \theta^{n-1}) d\theta^i \otimes d\theta^j .$$

With respect to these coordinates, the second fundamental form of the boundary $\partial M = \{\varphi = 0\}$ is

$$\mathrm{h}_{ij}^{(g)} = \frac{\nabla^2_{ij}\varphi}{|\nabla\varphi|_g} = \nabla^2_{ij}\varphi , \qquad \text{for } i, j = 1, \dots, n-1 .$$

On the other hand, from the first equation of problem (3.23), we easily deduce that $\nabla^2\varphi = 0$ on ∂M. This concludes the proof of point (ii). □

Finally, in order to prove the integral identities in Sect. 5, we need an extension of the classical Divergence Theorem to the case of open domains whose boundary has a (not too big) nonsmooth portion. Note that [4, Theorem A.1] is not enough for our purposes, because hypothesis (ii) is not necessarily fulfilled. To avoid problems, we state the following generalization, due to De Giorgi and Federer.

Theorem A.9 ([16, 17, 19, 20]) *Let (M, g) be a n-dimensional Riemannian manifold, with $n \geq 2$, let $E \subset M$ be a bounded open subset of M with compact boundary ∂E of finite $(n-1)$-dimensional Hausdorff measure, and suppose that $\partial E = \Gamma \sqcup \Sigma$, where the subsets Γ and Σ have the following properties:*

(i) *For every $x \in \Gamma$, there exists an open neighborhood U_x of x in M such that $\Gamma \cap U_x$ is a smooth regular hypersurface.*
(ii) *The subset Σ is compact and $\mathscr{H}^{n-1}(\Sigma) = 0$.*

If X is a Lipschitz vector field defined in a neighborhood of $\overline{E}$ then the following identity holds true

$$\int_E \mathrm{div} X \, \mathrm{d}\mu = \int_\Gamma \langle X \,|\, \mathrm{n} \rangle \, \mathrm{d}\sigma, \tag{A.16}$$

where n *denotes the exterior unit normal vector field.*

Appendix B: Boucher-Gibbons-Horowitz Method

In this section we discuss an alternative approach to the study of the rigidity of the de Sitter and anti-de Sitter spacetime, which does not require the machinery of Sect. 3. Without the need of any assumptions, this method will allow to derive results that are comparable to Theorems 2.2, 2.6 (case $\Lambda > 0$) and Theorems 2.14, 2.18

(case $\Lambda < 0$). In the case $\Lambda > 0$, the computations that we are going to show are quite classical (see [12, 14]). However, to the author's knowledge, the analogous calculations in the case $\Lambda < 0$ are new.

As usual, we start with the case $\Lambda > 0$. Recalling the Bochner formula and the equations in (1.4) we compute

$$\begin{aligned}
\Delta|\mathrm{D}u|^2 &= 2\,|\mathrm{D}^2u|^2 + 2\,\mathrm{Ric}(\mathrm{D}u,\mathrm{D}u) + 2\langle \mathrm{D}\Delta u \,|\, \mathrm{D}u\rangle \\
&= 2\,|\mathrm{D}^2u|^2 + 2\Big[\frac{1}{u}\,\mathrm{D}^2u(\mathrm{D}u,\mathrm{D}u) + n\,|\mathrm{D}u|^2\Big] - 2n\,|\mathrm{D}u|^2 \\
&= 2\,|\mathrm{D}^2u|^2 + \frac{1}{u}\,\langle \mathrm{D}|\mathrm{D}u|^2 \,|\, \mathrm{D}u\rangle\,. \qquad \text{(B.1)}
\end{aligned}$$

Now, if we consider the field

$$Y = \mathrm{D}|\mathrm{D}u|^2 - \frac{2}{n}\,\Delta u\,\mathrm{D}u$$

we can compute its divergence using (B.1).

$$\begin{aligned}
\mathrm{div}(Y) &= \Delta|\mathrm{D}u|^2 - \frac{2}{n}\langle \mathrm{D}\Delta u \,|\, \mathrm{D}u\rangle - \frac{2}{n}(\Delta u)^2 \\
&= 2\Big[\,|\mathrm{D}^2u|^2 - \frac{(\Delta u)^2}{n}\Big] + \frac{1}{u}\,\langle \mathrm{D}|\mathrm{D}u|^2 \,|\, \mathrm{D}u\rangle + 2\,|\mathrm{D}u|^2\,.
\end{aligned}$$

More generally, for every nonzero $\mathscr{C}^1$ function $\alpha = \alpha(u)$:

$$\begin{aligned}
\frac{\mathrm{div}(\alpha\, Y)}{\alpha} &= \mathrm{div}(Y) + \frac{\dot\alpha}{\alpha}\,\langle Y \,|\, \mathrm{D}u\rangle \\
&= 2\Big[\,|\mathrm{D}^2u|^2 - \frac{(\Delta u)^2}{n}\Big] + \Big(\frac{\dot\alpha}{\alpha} + \frac{1}{u}\Big)\Big(\langle \mathrm{D}|\mathrm{D}u|^2 \,|\, \mathrm{D}u\rangle + 2\,u\,|\mathrm{D}u|^2\Big)\,.
\end{aligned}$$

where $\dot\alpha$ is the derivative of α with respect to u. The computation above suggests us to choose

$$\alpha(u) = \frac{1}{u}\,.$$

With this choice of α, we have

$$\mathrm{div}\Big(\frac{1}{u}\,Y\Big) = \frac{2}{u}\Big[\,|\mathrm{D}^2u|^2 - \frac{(\Delta u)^2}{n}\Big]. \qquad \text{(B.2)}$$

Proposition B.1 *Let (M, g_0, u) be a static solution to problem (1.4). Then, for every $t \in [0, 1)$ it holds*

$$\int_{\{u=t\}} \frac{1}{u} \left(|\mathrm{D}u|^2\, \mathrm{H} - \frac{n-1}{n}\, |\mathrm{D}u|\, \Delta u \right) \mathrm{d}\sigma \;=\; \int_{\{u>t\}} \frac{1}{u} \left[|\mathrm{D}^2 u|^2 - \frac{(\Delta u)^2}{n} \right] \mathrm{d}\mu \;\geq\; 0\,. \tag{B.3}$$

Moreover, if there exists $t_0 \in (0, 1)$ such that

$$\int_{\{u=t_0\}} \left(|\mathrm{D}u|^2\, \mathrm{H} \,-\, \frac{n-1}{n}\, |\mathrm{D}u|\, \Delta u \right) \mathrm{d}\sigma \;\leq\; 0\,, \tag{B.4}$$

then the triple (M, g_0, u) is isometric to the de Sitter solution.

Remark B.1 Recalling Remark 1, it is easy to realize that the integral on the left hand side of (B.3) is well defined also when t is a singular value of u.

Remark B.2 Note that the right hand side of inequality (B.3) is always nonnegative, as opposed to formula (5.8), where we needed to suppose Assumption 3 to achieve the same result. This is one of the reasons why this approach works without the need to suppose any assumption.

Proof of Proposition B.1 Suppose for the moment that $\{u = t\}$ is a regular level set. Integrating by parts identity (B.2), we obtain

$$\int_{\{u>t\}} \frac{2}{u} \left[|\mathrm{D}^2 u|^2 - \frac{(\Delta u)^2}{n} \right] \mathrm{d}\mu \;=\; \int_{\{u=t\}} \frac{1}{u}\, \langle Y \,|\, \mathrm{n} \rangle\, \mathrm{d}\sigma\,, \tag{B.5}$$

where $\mathrm{n} = -\mathrm{D}u/|\mathrm{D}u|$ is the outer g-unit normal of the set $\{u \geq t\}$ at its boundary. On the other hand, from the first formula in (3.27) it is easy to deduce that

$$\langle Y \,|\, \mathrm{D}u \rangle \;=\; 2 \left(\mathrm{D}^2 u(\mathrm{D}u, \mathrm{D}u) \;-\; \frac{|\mathrm{D}u|^2 \Delta u}{n} \right) \;=\; -2 \left(|\mathrm{D}u|^3\, \mathrm{H} - \frac{n-1}{n}\, |\mathrm{D}u|^2\, \Delta u \right).$$

Substituting in (B.5) proves formula (B.3) in the case where $\{u = t\}$ is a regular level set.

In the case where $t > 0$ is a singular value of u, we need to apply a slightly refined version of the Divergence Theorem, namely Theorem A.9 in the Appendix A, in order to perform the integration by parts which leads to identity (B.5). The rest of the proof is identical to what we have done for the regular case. We set

$$X = \frac{1}{u}\, Y \qquad \text{and} \qquad E = \{u > t\}\,.$$

so that $\partial E = \{u = t\}$.

As usual, we denote by $\mathrm{Crit}(u) = \{x \in M \mid \mathrm{D}u(x) = 0\}$ the set of the critical points of u, From [30] (see also [27, Theorem 6.3.3]), we know that there exists an open $(n-1)$-dimensional submanifold $N \subseteq \mathrm{Crit}(u)$ such that $\mathscr{H}^{n-1}(\mathrm{Crit}(u) \setminus N) = 0$. Set $\Sigma = \partial E \cap (\mathrm{Crit}(u) \setminus N)$ and $\Gamma = \partial E \setminus \Sigma$, so that ∂E can be written as the disjoint union of Σ and Γ. We have $\mathscr{H}^{n-1}(\Sigma) = 0$ by definition, while Γ is the union of the regular part of ∂E and of N, which are open $(n-1)$-submanifolds. Therefore the hypotheses of Theorem A.9 are met, and we can apply it to conclude that Eq. (B.5) holds true also when t is a singular value of u.

To prove the second part, we observe that from (B.3) and (B.4) one immediately gets $\mathrm{D}^2 u = (\Delta u/n)\, g_0$ in $\{u \geq t_0\}$. Since u is analytic, the same equality holds on the whole manifold M. Now we can use the results in [28] to conclude that (M, g_0, u) is the de Sitter solution. □

The proposition above is particularly interesting when applied at the boundary $\partial M = \{u = 0\}$.

Corollary B.2 *Let (M, g_0, u) be a static solution to problem* (1.4). *Then it holds*

$$\int_{\partial M} |\mathrm{D}u| \big[\, \mathrm{R}^{\partial M} - (n-1)(n-2)\, \big]\, \mathrm{d}\sigma \geq 0\,. \tag{B.6}$$

Moreover, if the equality holds then the triple (M, g_0, u) is isometric to the de Sitter solution.

Proof First we compute from the equations in (1.4) and formula (3.27), that

$$\frac{\mathrm{H}\,|\mathrm{D}u|}{u} = -\mathrm{Ric}(\nu, \nu)\,,$$

where $\nu = \mathrm{D}u/|\mathrm{D}u|$ as usual. In particular, we have $\mathrm{H} = 0$ on ∂M. Hence we can use the Gauss–Codazzi identity to find

$$\frac{\mathrm{H}\,|\mathrm{D}u|}{u} = \frac{\mathrm{R}^{\partial M} - \mathrm{R}}{2} = \frac{1}{2}\big[\, \mathrm{R}^{\partial M} - n(n-1)\, \big].$$

Substituting $t = 0$ in formula (B.3) and applying Proposition B.1, we have the thesis. □

Now we turn our attention to the case $\Lambda < 0$. Mimicking the computations done in the case $\Lambda > 0$, but using the equations in (1.17) instead of the ones in (1.4) we obtain

$$\mathrm{div}\Big(\frac{1}{u}\, Y \Big) = \frac{2}{u} \Big[\, |\mathrm{D}^2 u|^2 - \frac{(\Delta u)^2}{n}\, \Big]. \tag{B.7}$$

Incidentally, we notice that this equation coincides with the analogous formula (B.2) in the case $\Lambda > 0$. We are now ready to state the analogous of Proposition B.1.

Proposition B.3 *Let (M, g_0, u) be a static solution to problem (1.17). Then, for every $t \in (1, +\infty)$ it holds*

$$\int_{\{u=t\}} \frac{1}{u}\left(|\mathrm{D}u|^2\,\mathrm{H} - \frac{n-1}{n}\,|\mathrm{D}u|\,\Delta u\right)\mathrm{d}\sigma \;=\; -\int_{\{u<t\}} \frac{1}{u}\left[\,|\mathrm{D}^2 u|^2 - \frac{(\Delta u)^2}{n}\,\right]\mathrm{d}\mu \;\leq\; 0\,. \tag{B.8}$$

Moreover, if there exists $t_0 \in (1, +\infty)$ such that

$$\int_{\{u=t_0\}} \left(|\mathrm{D}u|^2\,\mathrm{H} - \frac{n-1}{n}\,|\mathrm{D}u|\,\Delta u\right)\mathrm{d}\sigma \;\geq\; 0\,, \tag{B.9}$$

then the triple (M, g_0, u) is isometric to the anti-de Sitter solution.

Remark B.3 Recalling Remark 1, it is easy to realize that the integral on the left hand side of (B.8) is well defined also when t is a singular value of u.

Proof of Proposition B.3 The proof is almost identical to the proof of Proposition B.3. The only change is that, when we apply the divergence theorem, we need the outer g-unit normal of the set $\{u \leq t\}$, that is $\mathrm{n} = \mathrm{D}u/|\mathrm{D}u|$ instead of $-\mathrm{D}u/|\mathrm{D}u|$. This is the reason of the different signs in formulæ (B.3), (B.8). □

Now suppose that the manifold M is conformally compact. We would like to use Proposition B.3 to study the behavior of a static solution at the conformal boundary ∂M. In order to simplify the computations and to emphasize the analogy with the case $\Lambda > 0$, it will prove useful to suppose that Assumption 2-bis holds. Therefore, from now on we suppose that $1/\sqrt{u^2 - 1}$ is a defining function, and that $\lim_{x \to \bar{x}}(u^2 - 1 - |\mathrm{D}u|^2) = 0$ for every $\bar{x} \in \partial M$. We are now ready to prove the analogous of Corollary B.2 in the case of a negative cosmological constant.

Corollary B.4 *Let (M, g_0, u) be a conformally compact static solution to problem (1.17) satisfying assumption 2-bis, and let $g = g_0/(u^2 - 1)$. Then it holds*

$$\int_{\partial M} \left[\,(n-1)(n-2) - \mathrm{R}_g^{\partial M}\,\right]\mathrm{d}\sigma_g \;\geq\; 0\,. \tag{B.10}$$

Moreover, if

$$\lim_{t \to +\infty}\; t^{n-1} \int_{\{u=t\}} \mathrm{Ric}_g(\nu_g, \nu_g)\,\mathrm{d}\sigma_g \;=\; 0\,, \tag{B.11}$$

where $\nu_g = \mathrm{D}u/|\mathrm{D}u|_g$, then the triple (M, g_0, u) is isometric to the anti-de Sitter solution.

Proof First we compute from the equations in (1.17) and formula (3.27), that

$$\frac{\mathrm{H}\,|\mathrm{D}u|}{u} \; = \; -\mathrm{Ric}(\nu,\nu)\,,$$

where $\nu = \mathrm{D}u/|\mathrm{D}u|$ as usual. Therefore, we can rewrite formula (B.8) as

$$\int\limits_{\{u=t\}} |\mathrm{D}u|\left[\,\mathrm{Ric}(\nu,\nu) + (n-1)\,\right]\mathrm{d}\sigma \;\geq\; 0\,. \tag{B.12}$$

Now we use Eq. (3.16) in order to rewrite the term in the square brackets in the following way

$$\left[\,(n-1)u^2-1\,\right]\left[\,\mathrm{Ric}(\nu,\nu)+(n-1)\,\right] \;=\; \mathrm{Ric}_g(\nu_g,\nu_g) - \left(\frac{(n-1)u^2+1}{u^2-1}\right)\left(u^2-1-|\mathrm{D}u|^2\right).$$

Now it is easy to obtain from inequality (B.12) the following formula

$$\int\limits_{\{u=t\}} \left(\frac{|\mathrm{D}u|}{\sqrt{u^2-1}}\right)\left[\,\mathrm{Ric}_g(\nu_g,\nu_g) - \left(\frac{(n-1)u^2+1}{u^2-1}\right)\left(u^2-1-|\mathrm{D}u|^2\right)\right]\mathrm{d}\sigma_g \;\geq\; 0\,.$$

Since $\lim_{x\to\bar{x}}(u^2-1-|\mathrm{D}u|^2) = 0$, in particular $|\mathrm{D}u|/\sqrt{u^2-1}$ goes to zero as $t\to+\infty$. Therefore, taking the limit as $t\to+\infty$ of the formula above, we obtain

$$\int\limits_{\partial M} \mathrm{Ric}_g(\nu_g,\nu_g)\,\mathrm{d}\sigma_g \;\geq\; 0\,, \tag{B.13}$$

where $\nu_g = \mathrm{D}u/|\mathrm{D}u|_g$. Since ∂M is a totally geodesic hypersurface by Lemma A.8-(ii), from the Gauss–Codazzi equation and formula (3.24) we obtain

$$2\,\mathrm{Ric}_g(\nu_g,\nu_g) \;=\; \mathrm{R}_g - \mathrm{R}_g^{\partial M} \;=\; (n-1)(n-2) - \mathrm{R}_g^{\partial M}\,.$$

Substituting in Eq. (B.13) we obtain formula (B.10).

To prove the rigidity statement, we observe that we can rewrite formula (B.11) as

$$\begin{aligned} 0 \;=\; &\lim_{t\to+\infty}\int\limits_{\{u=t\}} u^{n-1}\mathrm{Ric}_g(\nu_g,\nu_g)\,\mathrm{d}\sigma_g \\ =\; &\lim_{t\to+\infty}\int\limits_{\{u=t\}} u^{n-1}\left(\frac{|\mathrm{D}u|}{\sqrt{u^2-1}}\right) \\ &\times\left[(n-1)u^2+1\right]\left[\,\mathrm{Ric}(\nu,\nu)+(n-1)+1-\frac{|\mathrm{D}u|^2}{u^2-1}\right]\left[\frac{\mathrm{d}\sigma}{(u^2-1)^{\frac{n-1}{2}}}\right] \end{aligned}$$

$$= \lim_{t\to+\infty} \int_{\{u=t\}} (n-1)\Big[u\,|\mathrm{D}u|\big(\mathrm{Ric}(\nu,\nu)+(n-1)\big)+(u^2-1-|\mathrm{D}u|^2)\Big]\,\mathrm{d}\sigma$$

$$= \lim_{t\to+\infty} \int_{\{u=t\}} (n-1)\,u\,|\mathrm{D}u|\Big[\,\mathrm{Ric}(\nu,\nu)+(n-1)\,\Big]\,\mathrm{d}\sigma\,.$$

Now we recall that $u\,[\mathrm{Ric}(\nu,\nu)+(n-1)] = -\mathrm{H}|\mathrm{D}u| - (n-1)\Delta u/n$ and we conclude using Proposition B.3. □

Acknowledgements The authors would like to thank P. T. Chruściel for his interest in our work and for stimulating discussions during the preparation of the manuscript. The authors are members of the Gruppo Nazionale per l'Analisi Matematica, la Probabilità e le loro Applicazioni (GNAMPA) of the Istituto Nazionale di Alta Matematica (INdAM) and are partially founded by the GNAMPA Project "Principi di fattorizzazione, formule di monotonia e disuguaglianze geometriche".

References

1. V. Agostiniani, L. Mazzieri, Riemannian aspects of potential theory. J. Math. Pures Appl. **104**(3), 561–586 (2015)
2. V. Agostiniani, L. Mazzieri, Monotonicity formulas in potential theory (2016). https://arxiv.org/abs/1606.02489.
3. V. Agostiniani, L. Mazzieri, Comparing monotonicity formulas for electrostatic potentials and static metrics. Atti Accad. Naz. Lincei Rend. Lincei Mat. Appl. **28**(1), 7–20 (2017)
4. V. Agostiniani, L. Mazzieri, On the geometry of the level sets of bounded static potentials. Commun. Math. Phys. **355**(1), 261–301 (2017)
5. V. Agostiniani, S. Borghini, L. Mazzieri, On the torsion problem for domains with multiple boundary components (in preparation)
6. V. Agostiniani, M. Fogagnolo, L. Mazzieri, Sharp geometric inequalities for closed hypersurfaces in manifolds with nonnegative Ricci curvature. arXiv preprint arXiv:1812.05022 (2018)
7. L. Ambrosio, G. Da Prato, A. Mennucci, Introduction to measure theory and integration, in *Appunti. Scuola Normale Superiore di Pisa (Nuova Serie)* [Lecture Notes. Scuola Normale Superiore di Pisa (New Series)], vol. 10 (Edizioni della Normale, Pisa 2011)
8. L. Ambrozio, On static three-manifolds with positive scalar curvature. J. Differ. Geom. **107**(1), 1–45 (2017)
9. S. Borghini, L. Mazzieri, On the mass of static metrics with positive cosmological constant-II. 2017. ArXiv Preprint Server https://arxiv.org/abs/1711.07024
10. S. Borghini, L. Mazzieri, On the mass of static metrics with positive cosmological constant: I. Classical and Quantum Gravity **35**(12), 125001 (2018)
11. S. Borghini, G. Mascellani, L. Mazzieri, Some sphere theorems in linear potential theory. Trans. Am. Math. Soc. (2019). https://doi.org/10.1030/tran/7637
12. W. Boucher, G.W. Gibbons, G.T. Horowitz, Uniqueness theorem for anti-de Sitter spacetime. Phys. Rev. D (3) **30**(12), 2447–2451 (1984)
13. P.T. Chruściel, On analyticity of static vacuum metrics at non-degenerate horizons. Acta Phys. Polon. B **36**(1), 17–26 (2005)
14. P.T. Chruściel, Remarks on rigidity of the de sitter metric. http://homepage.univie.ac.at/piotr.chrusciel/papers/deSitter/deSitter2.pdf

15. P.T. Chruściel, M. Herzlich, The mass of asymptotically hyperbolic Riemannian manifolds. Pac. J. Math. **212**(2), 231–264 (2003)
16. E. De Giorgi, Complementi alla teoria della misura $(n-1)$-dimensionale in uno spazio n-dimensionale, in *Seminario di Matematica della Scuola Normale Superiore di Pisa, 1960–1961* (Editrice Tecnico Scientifica, Pisa, 1961)
17. E. De Giorgi, *Frontiere Orientate di Misura Minima* (Editr. Tecnico scientifica, 1961)
18. W. De Sitter, On the curvature of space. Proc. Kon. Ned. Akad. Wet **20**, 229–243 (1917)
19. H. Federer, The Gauss–Green theorem. Trans. Am. Math. Soc. **58**, 44–76 (1945)
20. H. Federer, A note on the Gauss-Green theorem. Proc. Am. Math. Soc. **9**, 447–451 (1958)
21. H. Federer, Geometric measure theory, in *Die Grundlehren der mathematischen Wissenschaften, Band 153* (Springer, New York, 1969)
22. M. Fogagnolo, L. Mazzieri, A. Pinamonti, Geometric aspects of p-capacitary potentials. Ann. Inst. H. Poincaré Anal. Non Linéaire. https://doi.org/10.1016/j.anihpc.2018.11.005
23. G.W. Gibbons, S.A. Hartnoll, C.N. Pope, Bohm and Einstein-Sasaki metrics, black holes, and cosmological event horizons. Phys. Rev. D (3) **67**(8), 084024 (2003)
24. O. Hijazi, S. Montiel, Uniqueness of the AdS spacetime among static vacua with prescribed null infinity. Adv. Theor. Math. Phys. **18**(1), 177–203 (2014)
25. O. Hijazi, S. Montiel, S. Raulot, Uniqueness of the de Sitter spacetime among static vacua with positive cosmological constant. Ann. Glob. Anal. Geom. **47**(2), 167–178 (2015)
26. F. Kottler, Über die physikalischen grundlagen der Einsteinschen gravitationstheorie. Ann. Phys. (Berlin) **361**(14), 401–462 (1918)
27. S.G. Krantz, H.R. Parks, A primer of real analytic functions, in *Birkhäuser Advanced Texts: Basler Lehrbücher*, 2nd edn. [Birkhäuser Advanced Texts: Basel Textbooks] (Birkhäuser, Boston, 2002)
28. J. Lafontaine, Sur la géométrie d'une généralisation de l'équation différentielle d'Obata. J. Math. Pures Appl. (9) **62**(1), 63–72 (1983)
29. L. Lindblom, Static uniform-density stars must be spherical in general relativity. J. Math. Phys. **29**(2), 436–439 (1988)
30. S. Łojasiewicz, *Introduction to Complex Analytic Geometry* (Birkhäuser, Basel, 1991). Translated from the Polish by Maciej Klimek
31. H. Nariai, On a new cosmological solution of Einstein's fieldequations of gravitation. Sci. Rep. Tohoku Univ. Ser. I **35**(1), 62–67 (1951)
32. M. Obata, Certain conditions for a Riemannian manifold to be isometric with a sphere. J. Math. Soc. Jpn. **14**, 333–340 (1962)
33. R. Penrose, Asymptotic properties of fields and space-times. Phys. Rev. Lett. **10**, 66–68 (1963)
34. J. Qing, On the uniqueness of AdS space-time in higher dimensions. Ann. Henri Poincaré **5**(2), 245–260 (2004)
35. K. Schwarzschild, On the gravitational field of a mass point according to Einstein's theory. Gen. Relativ. Gravit. **35**(5), 951–959 (2003). Translated from the original German article [Sitzungsber. Königl. Preussich. Akad. Wiss. Berlin Phys. Math. Kl. 1916, 189–196] by S. Antoci and A. Loinger
36. J. Serrin, Isolated singularities of solutions of quasi-linear equations. Acta Math. **113**, 219–240 (1965)
37. J. Souček, V. Souček, Morse-Sard theorem for real-analytic functions. Comment. Math. Univ. Carol. **13**, 45–51 (1972)
38. X. Wang, The mass of asymptotically hyperbolic manifolds. J. Differ. Geom. **57**(2), 273–299 (2001)
39. X. Wang, On the uniqueness of the AdS spacetime. Acta Math. Sin. (Engl. Ser.) **21**(4), 917–922 (2005)
40. H.F. Weinberger, Remark on the preceding paper of Serrin. Arch. Ration. Mech. Anal. **43**, 319–320 (1971)
41. H.M. Zum Hagen, On the analyticity of static vacuum solutions of Einstein's equations. Proc. Camb. Philos. Soc. **67**, 415–421 (1970)

Introduction to Controllability of Nonlinear Systems

Ugo Boscain and Mario Sigalotti

Abstract We present some basic facts about the controllability of nonlinear finite dimensional systems. We introduce the concepts of Lie bracket and of Lie algebra generated by a family of vector fields. We then prove the Krener theorem on local accessibility and the Chow-Rashevskii theorem on controllability of symmetric systems. We then introduce the theory of compatible vector fields and we apply it to study control-affine systems with a recurrent drift or satisfying the strong Lie bracket generating assumption. We conclude with a general discussion about the orbit theorem by Sussmann and Nagano.

Keywords Controllability · Chow theorem · Compatible vector fields · Orbit theorem

In this note we present some classical techniques to study the controllability of nonlinear systems. The discussion is kept as elementary as possible. Classical textbooks are [1–3].

Consider the nonlinear control system

$$\dot{x} = F(x, u(t)). \tag{1}$$

Here $x \in \mathbb{R}^n$ is the state of the system, $U \subset \mathbb{R}^m$ is the set of control values and $u(\cdot) : [0, \infty[\to U$ is the control. For the development of the theory it is not necessary to assume any structure on U. For example U could be a finite set of points, a polytope or the full $\mathbb{R}^m$.

U. Boscain (✉)
CNRS, Team Inria CAGE, Laboratoire Jacques-Louis Lions, Sorbonne Université, Paris, France
e-mail: ugo.boscain@upmc.fr

M. Sigalotti (✉)
Team Inria CAGE, Laboratoire Jacques-Louis Lions, Sorbonne Université, Paris, France
e-mail: mario.sigalotti@inria.fr

S. Dipierro (ed.), *Contemporary Research in Elliptic PDEs and Related Topics*, Springer INdAM Series 33, https://doi.org/10.1007/978-3-030-18921-1_4

We assume that F is a smooth[1] function of its arguments and that $u(\cdot)$ is regular enough in such a way that Eq. (1) with the initial condition $x(0) = x_0 \in \mathbb{R}^n$ has local existence and uniqueness of solutions. For instance we can assume that $u(\cdot)$ is a L^1_{loc} or a L^∞_{loc} function of the time. However, as it will be clear later, all conditions for controllability that we will get are actually sufficient conditions that are valid in the smaller class of piecewise constant controls.

Remark 1 For simplicity of notation in the following we also assume that for every $u(\cdot)$ a solution of (1) exists in $[0, \infty[$. However this hypothesis is not necessary for the validity of the theorems that we are going to prove.

Remark 2 In the discussion presented here we assume $x \in \mathbb{R}^n$. However almost nothing changes if we assume that x belongs to a smooth connected manifold M and that solutions of (1) exists in $[0, \infty[$. To extend the theory to smooth connected manifolds, if some non-trivial modification is necessary, we explicitly state it in the text.

Denote by $x(t; x_0, u(\cdot))$ the solution at time t of (1) starting from x_0 at $t = 0$ and corresponding to a control function $u(\cdot)$. We recall the definitions of the *reachable* (or *attainable*) sets starting from x_0:

- the *reachable set* from x_0 at time $\tau \geq 0$ is

$$\mathcal{A}(\tau, x_0) = \{x_1 \in \mathbb{R}^n \mid \exists\, u(\cdot) : [0, \tau] \to U,\ x(\tau; x_0, u(\cdot)) = x_1\};$$

- the *reachable set* from x_0 within time $\tau \geq 0$ is

$$\mathcal{A}(\leq \tau, x_0) = \cup_{t \in [0,\tau]} \mathcal{A}(t, x_0);$$

- the *reachable set* from x_0 is

$$\mathcal{A}(x_0) = \cup_{t \in [0,+\infty[} \mathcal{A}(t, x_0).$$

Given the control system (1), the purpose of the controllability theory is to characterize when these sets coincide with the entire state space.

Definition 3 The system (1) is said to be

- *controllable* if for every $x_0 \in \mathbb{R}^n$, $\mathcal{A}(x_0) = \mathbb{R}^n$;
- *small-time controllable* if for every $x_0 \in \mathbb{R}^n$ and $\tau > 0$, we have $\mathcal{A}(\leq \tau, x_0) = \mathbb{R}^n$;
- *small-time locally controllable at* x_0 if x_0 belongs to the interior of $\mathcal{A}(\leq \tau, x_0)$ for every $\tau > 0$.

[1] In this note by smooth we mean C^∞.

1 Control Systems as Families of Vector Fields

In the following it will be useful to think to the system (1) as a *family of vector fields*[2] parameterized by $u \in U$ (i.e., by constant controls). In other words we will often consider instead of the control system (1), the family of vector fields

$$\mathcal{F} = \{F(\cdot, v) \mid v \in U\}.$$

All vector fields of the family $\mathcal{F}$ are considered smooth and complete, i.e., for every $F \in \mathcal{F}$, $x_0 \in \mathbb{R}^n$, the equation $\dot{x} = F(x)$ with initial condition $x(0) = x_0$ admits a solution in $]-\infty, \infty[$.

1.1 Vec($\mathbb{R}^n$) *and Its Lie Algebra*

Let Vec($\mathbb{R}^n$) be the vector space of all smooth vector fields in $\mathbb{R}^n$. Given a compete vector field $f \in$Vec($\mathbb{R}^n$), let us indicate by e^{tf} its flow, i.e., the map that with x_0 associates the solution at time t to the Cauchy problem

$$\begin{cases} \dot{x} = f(x) \\ x(0) = x_0. \end{cases} \tag{2}$$

Although e^{tf} is not an exponential in the usual sense, this notation is useful thanks to the following properties that are a direct consequence of existence and uniqueness of solutions of (2) and of their differentiability w.r.t. x_0:

- for every $t, s \in \mathbb{R}$, $e^{(t+s)f} = e^{tf} \circ e^{sf}$;
- for every $t \in \mathbb{R}$, e^{tf} is a diffeomorphism. In particular, it is invertible and we have $(e^{tf})^{-1} = e^{-tf}$;
- for every $t \in \mathbb{R}$, $\left(\frac{d}{dt}e^{tf}\right)(x) = f(e^{tf}x)$. In particular, $\left(\frac{d}{dt}\big|_{t=0} e^{tf}\right)(x) = f(x)$.

A crucial object in studying the controllability of (1) is the Lie algebra generated by the vector fields of the corresponding family $\mathcal{F}$. Let us first define the Lie bracket between two vector fields f and g, as the vector field defined by

$$[f, g](x) = Dg(x)f(x) - Df(x)g(x). \tag{3}$$

[2]In an autonomous differential equation $\dot{x} = f(x)$, usually f it is called a vector field, since it is a map that with every position x associates a velocity vector $f(x)$.

Here, given a vector field $f = (f_1, \ldots, f_n)^T$, Df is the matrix of partial derivatives of the components of f, i.e.,

$$Df = \begin{pmatrix} \partial_1 f_1 & \ldots & \partial_n f_1 \\ \vdots & \ldots & \vdots \\ \partial_1 f_n & \ldots & \partial_n f_n \end{pmatrix}.$$

One immediately verifies the following properties of the Lie bracket:

- bilinearity: for every $\lambda_1, \lambda_2 \in \mathbb{R}$,

$$[f, \lambda_1 g_1 + \lambda_2 g_2] = \lambda_1 [f, g_1] + \lambda_2 [f, g_2],$$
$$[\lambda_1 f_1 + \lambda_2 f_2, g] = \lambda_1 [f_1, g] + \lambda_2 [f_2, g];$$

- antisymmetry: $[g, f] = -[f, g]$;
- Jacobi identity: $[f, [g, h]] + [h, [f, g]] + [g, [h, f]] = 0$.

A vector space V endowed with an operation $V \times V \to V$ that is bilinear, antisymmetric and satisfying the Jacobi identity is said to be a *Lie algebra*. It follows that $(\mathrm{Vec}(\mathbb{R}^n), [\cdot, \cdot])$ is a Lie algebra.

Remark 4 As a consequence of the antisymmetry of the Lie bracket, we have $[f, f] = 0$. Notice moreover that the value of $[f, g]$ at a point x does not depend only on the values of f and g at x, but also on their first order expansion at x. However, if at x_0 we have $f(x_0) = g(x_0) = 0$, it follows from the definition that $[f, g](x_0) = 0$.

Example 5 Given two linear vector fields Ax and Bx, where $A, B \in \mathbb{R}^{n\times n}$, we have that $[Ax, Bx] = BAx - ABx = -[A, B]x$.

The following lemma clarifies the geometric meaning of the Lie bracket: $[f, g]$ is a measure of the lack of commutation of the flows associated with f and g. We refer to Fig. 1.

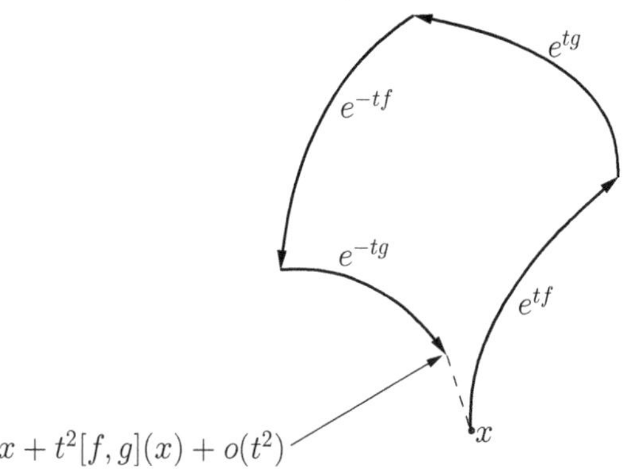

Fig. 1 Geometric meaning of the Lie bracket

Lemma 6 *For every* $x \in \mathbb{R}^n$,

$$e^{-tg} \circ e^{-tf} \circ e^{tg} \circ e^{tf}(x) = x + t^2[f, g](x) + O(t^3), \tag{4}$$

for t *that tends to zero.*

Proof It is enough to compute for each flow the Taylor expansion at order 3. We have

$$e^{tf}(x) = x + tf(x) + \frac{t^2}{2} Df(x)f(x) + O(t^3),$$

and

$$e^{tg} \circ e^{tf}(x) = x + t\big(f(x) + g(x)\big) + \frac{t^2}{2} Df(x)f(x) + t^2 Dg(x)f(x) + \frac{t^2}{2} Dg(x)g(x) + O(t^3).$$

Then

$$e^{-tf} \circ e^{tg} \circ e^{tf}(x) = x + tg(x) + t^2[f, g](x) + \frac{t^2}{2} Dg(x)g(x) + O(t^3).$$

At the next step the result follows. □

Remark 7 The previous lemma says in particular that if $[f, g](x) \notin \mathrm{Vect}(f(x), g(x))$, then it is possible, by alternating between the dynamics of f and g, to attain points that cannot be reached with the flow of linear combinations of f and g. This is the starting idea behind the conditions for controllability that we are going to study in this note. Notice however that in order to generate the Lie bracket $[f, g]$, one needs to be able to use, beside f and g, also $-f$ and $-g$, otherwise one is much more constrained in the possible movements. As a consequence, it will be easier to prove controllability results for symmetric systems (i.e., systems for which if $f \in \mathcal{F}$ then $-f \in \mathcal{F}$). See Sect. 3.

An important corollary of the previous lemma is the following.

Corollary 8 *The flows* e^{tf} *and* e^{tg} *(corresponding to the vector fields* f *and* g*) commute for every* $t \in \mathbb{R}$ *if and only if their Lie bracket* $[f, g](x) = 0$ *for every* $x \in \mathbb{R}^n$.

Proof The fact that commutation of the flows implies that the Lie bracket vanishes follows immediately from (4). Concerning the converse implication, let us consider the curve

$$\gamma(s) = e^{-\sqrt{s}g} \circ e^{-\sqrt{s}f} \circ e^{\sqrt{s}g} \circ e^{\sqrt{s}f}(x).$$

By Lemma 6, we have that γ is differentiable at each time s and

$$\left.\frac{d\gamma}{ds}\right|_{s=0} = [f, g](\gamma(0)).$$

If $[f, g]$ is identically equal to zero, γ turns out to be solution of

$$\frac{d\gamma}{ds} = 0, \quad \gamma(0) = 0.$$

It follows $\gamma(s) = 0$ for every s and $e^{\sqrt{s}g} \circ e^{\sqrt{s}f} = e^{\sqrt{s}g} \circ e^{\sqrt{s}f}$. Setting $t = \sqrt{s}$ the result follows. □

Remark 9 In a differentiable manifold, Definition 3 and formula (4) make sense only in coordinates. An intrinsic definition of Lie bracket is

$$[f, g](x) = \left.\frac{d}{ds}\right|_{s=0} e^{-\sqrt{s}g} \circ e^{-\sqrt{s}f} \circ e^{\sqrt{s}g} \circ e^{\sqrt{s}f}(x).$$

Definition 10 Let $\mathcal{F}$ be a family of vector fields. We call Lie($\mathcal{F}$) the smallest sub-algebra of Vec($\mathbb{R}^n$) containing $\mathcal{F}$. Namely, Lie($\mathcal{F}$) is the span of all vector fields of $\mathcal{F}$ and of their iterated Lie brackets of any order:

$$\mathrm{Lie}(\mathcal{F}) = \mathrm{span}\{f_1, [f_1, f_2], [f_1, [f_2, f_3]], [f_1, [f_2, [f_3, f_4]]], \ldots \mid f_1, f_2, \ldots \in \mathcal{F}\}.$$

Definition 11 We say that the family $\mathcal{F}$ is *Lie bracket generating at a point* x if the dimension of $\mathrm{Lie}_x(\mathcal{F}) := \{f(x) \mid f \in \mathrm{Lie}(\mathcal{F})\}$ is equal to n. We say that the family $\mathcal{F}$ is *Lie bracket generating* if this condition is verified for every $x \in \mathbb{R}^n$.

Remark 12 Notice that in general Lie($\mathcal{F}$) is an infinite-dimensional space, while $\mathrm{Lie}_x(\mathcal{F})$ is a subspace of $\mathbb{R}^n$.

Exercise 1 Let $\mathcal{F} = \{f_1, \ldots, f_m\}$ and let A be an invertible $m \times m$ matrix. Define $f_i' = \sum_j A_{ij} f_j$ and $\mathcal{F}' = \{f_1', \ldots, f_m'\}$. Prove that $\mathcal{F}$ is Lie bracket generating if and only if $\mathcal{F}'$ is.

1.2 Affine Control Systems

An affine control system is a system of the form

$$\dot{x} = f_0(x) + \sum_{i=1}^{m} u_i(t) f_i(x), \tag{5}$$

where $f_0, f_1, \dots, f_m$ belong to $\mathrm{Vec}(\mathbb{R}^n)$ and $u(\cdot) = (u_1(\cdot), \dots, u_m(\cdot)) : [0, \infty[\to U \subset \mathbb{R}^m$ is the control. In an affine control system it is also assumed that U contains a neighborhood of the origin in $\mathbb{R}^m$. The vector fields f_0 is called *drift*.

Exercise Let $\mathcal{F}$ be the family of vector fields associated with (5).

- Prove that if $\{f_0, f_1, \dots, f_m\}$ is Lie bracket generating then also $\mathcal{F}$ is.
- Prove that if $\{f_1, \dots, f_m\}$ is Lie bracket generating then also $\mathcal{F}$ is.

2 The Krener Theorem: Local Accessibility

The fact that a control system is Lie bracket generating does not permit in general to conclude that it is controllable. Consider for instance the control system on the plane

$$\begin{pmatrix} \dot{x}_1 \\ \dot{x}_2 \end{pmatrix} = \begin{pmatrix} 1 \\ 0 \end{pmatrix} + u(t) \begin{pmatrix} 0 \\ 1 \end{pmatrix},$$

where $u(\cdot) : [0, \infty[\to [-1, 1]$. It is Lie bracket generating (since the corresponding family $\mathcal{F}$ contains the vector fields $\{(1, 1), (1, -1)\}$, however starting from the origin one cannot reach any point whose first coordinate is negative. This is essentially due to the fact that the family $\mathcal{F}$ contains two vector fields but not their opposite (cf. Remark 7). The Lie bracket generating condition permits to say that a system is *locally accessible* in the following sense.

Theorem 13 (Krener) *If $\mathcal{F}$ is Lie bracket generating at x_0, then for every $\tau > 0$, x_0 belongs to the closure of the interior of $\mathcal{A}(\leq \tau, x_0)$.*

The conclusion of the Krener Theorem can be equivalently reformulated in the following way:

- for every $\tau > 0$, the set $\mathcal{A}(\leq \tau, x_0)$ has nonempty interior,
- x_0 is a density point of such an interior.

Krener's theorem says in particular that the trajectories starting from a point at which the system is Lie bracket generating can reach (in an arbitrarily small time) a set having nonempty interior. Figure 2 shows what one can expect/non-expect from $\mathcal{A}(\leq \tau, x_0)$, $\tau > 0$.

Proof of Theorem 13 First notice that if $\mathcal{F}$ is Lie bracket generating at a point, then the same property holds true in a neighborhood of that point. (If n vector fields are linearly independent at a point, then they are linearly independent in a neighborhood of that point.)

There exists $f \in \mathcal{F}$ such that $f(x_0) \neq 0$, otherwise $\mathrm{Lie}_{x_0}(\mathcal{F}) = \{0\}$. If $n = 1$ the conclusion follows.

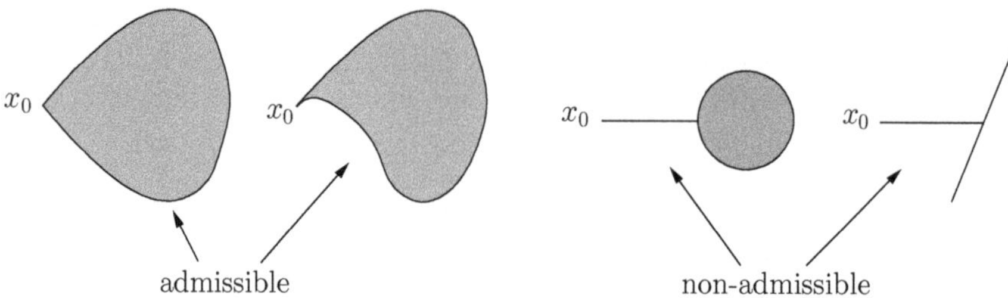

Fig. 2 Admissible and non-admissible reachable sets when the system is Lie bracket generating at x_0

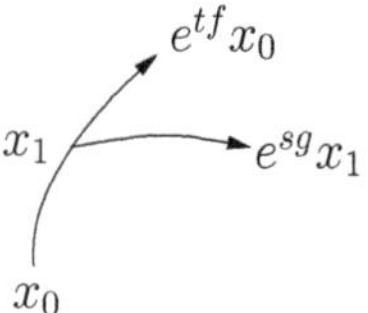

Fig. 3 Proof of Krener theorem

If $n > 1$ and all vector fields of $\mathcal{F}$ are tangent to the curve $t \mapsto e^{tf}(x_0)$, $0 < t < \varepsilon$, then from Lemma 6 it follows that $\text{Lie}_{e^{tf}(x_0)}(\mathcal{F})$ is also tangent to that curve and hence its dimension is less than 2. This contradicts the Lie bracket generating assumption. As a consequence, there exist $g \in \mathcal{F}$ and $0 < \bar{t} < \varepsilon$ such that f and g are linearly independent in a neighborhood of $x_1 = e^{\bar{t}f}(x_0)$ (see Fig. 3). Hence $(t, s) \mapsto e^{sg} \circ e^{tf}(x_0)$, $0 < s < \varepsilon'$, $\bar{t} - \varepsilon' < t < \bar{t} + \varepsilon'$ has as image a surface of dimension 2. If $n = 2$ the conlcusion follows.

Otherwise we iterate the same argument and we conclude by recurrence on n. □

Remark 14 Notice that for this proof we have only used piecewise constant controls.

Remark 15 From the proof of the Krener theorem it follows that $\mathcal{A}(\leq \tau, x_0)$, $\tau > 0$, contains an open set Ω having x_0 in its closure whose points can be reached by trajectories of the type $e^{t_n f_{i_n}} \circ \ldots \circ e^{t_1 f_{i_1}} x_0$ where $t_1, \ldots, t_n > 0$ and $f_{i_1}, \ldots, f_{i_n} \in \mathcal{F}$, i.e., by trajectories corresponding to piecewise constant controls made by n pieces. Notice that the vector fields $f_{i_1}, \ldots, f_{i_n}$ could be repeated. For instance if $\mathcal{F} = \{f_1, f_2\}$ is a Lie bracket generating family and we are in dimension 3, we could have for instance $f_{i_1} = f_1, \;\; f_{i_2} = f_2, \;\; f_{i_3} = f_1$.

3 Symmetric Systems

Definition 16 A family of vector fields $\mathcal{F}$ is said to be symmetric if $f \in \mathcal{F}$ implies $-f \in \mathcal{F}$.

When the family $\mathcal{F}$ is Lie bracket generating and symmetric one obtain that the system is controllable. This is the conclusion of the celebrated Chow–Rashevskii theorem.

Theorem 17 (Chow–Rashevskii) *If $\mathcal{F}$ is Lie bracket generating and symmetric, then for every $x_0 \in \mathbb{R}^n$ we have $\mathcal{A}(x_0) = \mathbb{R}^n$.*

Proof **Step 1** Fix $x_0 \in \mathbb{R}^n$ and let us show that $\mathcal{A}(x_0)$ contains a neighborhood of x_0. Since $\mathcal{F}$ is Lie bracket generating, $\mathcal{A}(x_0)$ contains a nonempty open set Ω whose points can be reached by trajectories corresponding to piecewise controls made by n pieces. Fix $t_1, \ldots, t_n > 0$ and $f_{i_1}, \ldots, f_{i_n} \in \mathcal{F}$ such that $\bar{x}_0 := e^{t_n f_{i_n}} \circ \ldots \circ e^{t_1 f_{i_1}} x_0 \in \Omega$. Since $\mathcal{F}$ is symmetric, $-f_{i_1}, \ldots, -f_{i_n} \in \mathcal{F}$ and we have that

$$e^{-t_1 f_{i_1}} \circ \ldots \circ e^{-t_n f_{i_n}} (\mathcal{A}(x_0)) \subset \mathcal{A}(x_0).$$

In particular $\mathcal{A}(x_0)$ contains the set

$$V = e^{-t_1 f_{i_1}} \circ \ldots \circ e^{-t_n f_{i_n}} (\Omega).$$

Now, since Ω is open and $e^{-t_1 f_{i_1}} \circ \ldots \circ e^{-t_n f_{i_n}}$ is a diffeomorphism, we have that V is open. Moreover V contains x_0 since $\bar{x}_0 \in \Omega$ and $e^{-t_1 f_{i_1}} \circ \ldots \circ e^{-t_n f_{i_n}} \bar{x}_0 = x_0$. It follows that, for every x_0, $\mathcal{A}(x_0)$ contains a neighborhood of x_0.

Step 2 Let us show that $\mathcal{A}(x_0)$ is open. If $x_1 \in \mathcal{A}(x_0)$, then $\mathcal{A}(x_1) \subset \mathcal{A}(x_0)$. It follows that $\mathrm{int}(\mathcal{A}(x_1)) \subset \mathrm{int}(\mathcal{A}(x_0))$. But from **Step 1** we have that $x_1 \in \mathrm{int}(\mathcal{A}(x_1))$. Hence $x_1 \in \mathrm{int}(\mathcal{A}(x_0))$.

Step 3 From the fact that $\mathcal{F}$ is symmetric, it follows that $x_1 \in \mathcal{A}(x_0)$ if and only if $x_0 \in \mathcal{A}(x_1)$. Let us consider the equivalence classes $\mathbb{R}^n / \sim$ where $\sim$ is the equivalence relation

$$x_1 \sim x_0 \text{ if and only if } x_1 \in \mathcal{A}(x_0).$$

Such equivalence classes are open and disjoint. Since $\mathbb{R}^n$ is connected, it follows that there is only one class. Hence, for every x_0 we have $\mathcal{A}(x_0) = \mathbb{R}^n$. □

Remark 18 Notice that in **Step 1** of the proof of Chow–Rashevskii theorem, since the times can be rendered arbitrarily small, we have proved that the system is small-time locally controllable in a neighborhood of every point x_0. Also, we have proved that every point of a neighborhood of x_0 can be reached with trajectories made by $2n$ pieces.

Remark 19 Notice that the Chow–Rashevskii theorem can be used in more general situations than those fixed by the hypotheses stated here. For instance

- to get controllability it is sufficient that the family $\mathcal{F}$ contains a symmetric family of Lie bracket generating vector fields;
- if one can prove that $\mathcal{F}$ is symmetric and Lie bracket generating in a connected open set Ω of $\mathbb{R}^n$ then one get the system is controllable in Ω.

4 Compatible Vector Fields

When a family of vector fields is Lie bracket generating but is not symmetric, in general it is not easy to understand if the system is controllable or not. A technique to study the controllability is the one of *compatible vector fields.*

Definition 20 A vector field g is said to be *compatible* with the family $\mathcal{F}$ if defining $\hat{\mathcal{F}} = \mathcal{F} \cup \{g\}$ we have the following: For every $x_0 \in \mathbb{R}^n$, the reachable set $\hat{\mathcal{A}}(x_0)$ of $\hat{\mathcal{F}}$ is contained in the closure of $\mathcal{A}(x_0)$.

The main result of the theory of compatible vector fields is the following.

Theorem 21 *If $\mathcal{F}$ is a Lie bracket generating family of vector fields, g is compatible with $\mathcal{F}$ and $\mathcal{F} \cup \{g\}$ is controllable, then $\mathcal{F}$ is controllable as well.*

This theorem should be used in the following way: one looks for a vector field g that added to the family $\mathcal{F}$ do not change the closure of the reachable set and such that it is easy to prove the controllability of the family $\mathcal{F} \cup \{g\}$.

The main ingredient to prove Theorem 21 is the following corollary of the Krener theorem.

Corollary 22 *If $\mathcal{F}$ is Lie bracket generating and $\mathcal{A}(x_0)$ is dense in $\mathbb{R}^n$ for some x_0, then $\mathcal{A}(x_0) = \mathbb{R}^n$.*

Proof Let $x_1 \in \mathbb{R}^n$ and consider the system

$$\dot{x} = -F(x, u(t)), \qquad x \in \mathbb{R}^n, \qquad u(\cdot) : [0, \infty[\to U \subset \mathbb{R}^m, \tag{6}$$

which is obtained from (1) by reversing the time. Let $\mathcal{F}^-$ be the family of vector fields associated with (6). Since $\mathcal{F}$ is Lie bracket generating, then $\mathcal{F}^-$ is Lie bracket generating as well.

Let $\mathcal{A}^-(x_1)$ be the reachable set for (6) starting from x_1. Thanks to Krener's theorem, $\mathcal{A}^-(x_1)$ contains a nonempty open set. In particular it has nonempty intersection with $\mathcal{A}(x_0)$ (being $\mathcal{A}(x_0)$ dense). See Fig. 4.

This means that $x_1 \in \mathcal{A}(x_0)$. Indeed from x_0 one can reach a point $\bar{x} \in \mathcal{A}(x_0) \cap \mathcal{A}^-(x_1)$ (since $\bar{x} \in \mathcal{A}(x_0)$) and from $\bar{x}$ one can reach x_1 (since $\bar{x}$ is reachable from x_1 for the system with reverted time).

Being x_1 arbitrary we have that $\mathcal{A}(x_0) = \mathbb{R}^n$. □

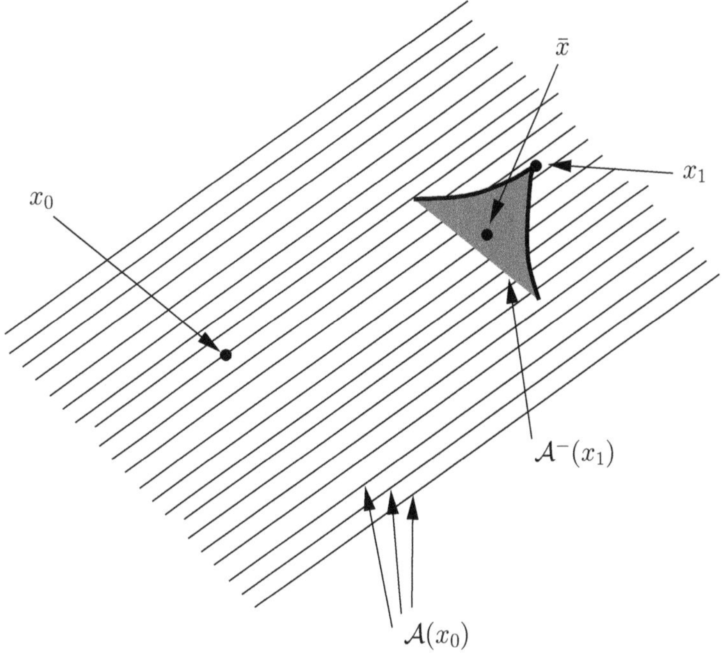

Fig. 4 Proof of Corollary 22

Proof of Theorem 21 Let $\mathcal{A}(x_0)$ be the reachable set from x_0 associated with $\mathcal{F}$ and $\hat{\mathcal{A}}(x_0)$ be the reachable set from x_0 associated with $\mathcal{F} \cup \{g\}$. For every $x_0 \in \mathbb{R}^n$ we have $\mathbb{R}^n = \hat{\mathcal{A}}(x_0) \subset \bar{\mathcal{A}}(x_0)$. Hence $\mathcal{A}(x_0)$ is dense. Since the system is Lie bracket generating, from Corollary 22 the conclusion follows. □

Next we present some important applications of the technique based on compatible vector fields.

4.1 *Affine Systems with Recurrent Drift*

In this section we apply the theory of compatible vector fields to affine control systems (cf. Sect. 1.2) that are Lie bracket generating and having a drift f_0 which is recurrent.

We refer to Fig. 5.

Definition 23 (Recurrent Vector Field) A vector field f is said to be *recurrent* if for every point $x_0 \in \mathbb{R}^n$, every neighborhood V of x_0 and every time $t > 0$, there exist $\bar{x}_0 \in V$ and $t^* > t$ such that $e^{t^* f}(\bar{x}_0) \in V$.

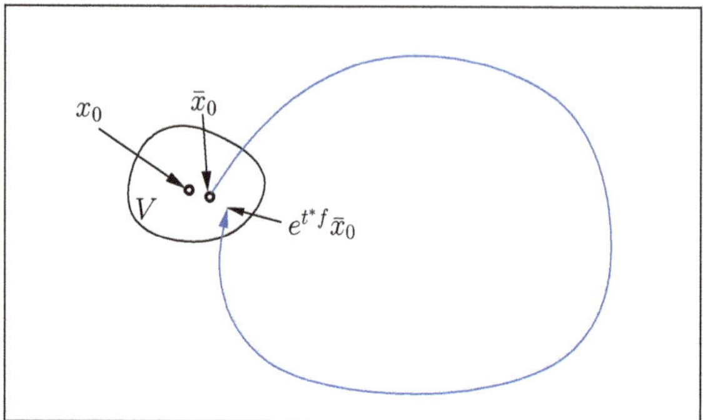

Fig. 5 Definition of recurrent vector field

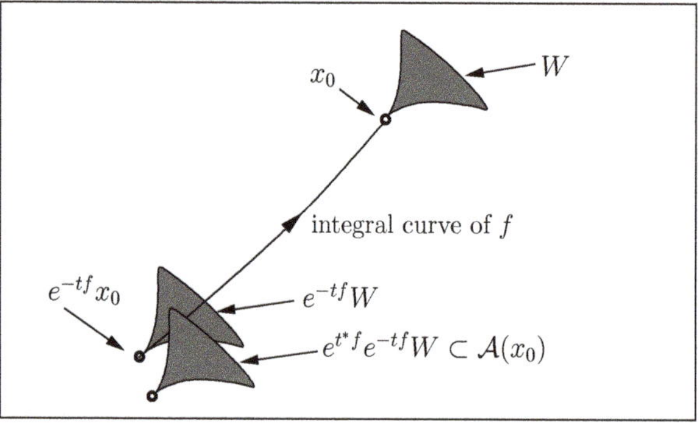

Fig. 6 Proof of Lemma 24

Notice that if the trajectories of f are periodic (possibly with period depending on the trajectory), then f is recurrent.

Lemma 24 *Let $\mathcal{F}$ be a Lie bracket generating family of vector fields and $f \in \mathcal{F}$. If f is recurrent then $-f$ is compatible with $\mathcal{F}$.*

Proof We have to prove that for every x_0 and for every $t > 0$, $e^{-tf}x_0$ can be obtained as limit of points belonging to the reachable set $\mathcal{A}(x_0)$.

We refer to Fig. 6. By Krener theorem (thanks to the fact that $\mathcal{F}$ is Lie-bracket generating), there exists an arbitrarily small open set $W \subset \mathcal{A}(x_0)$ such that $x_0 \in \overline{W}$.

Now since e^{-tf} is a diffeomorphism this implies that $e^{-tf}x_0 \in \overline{e^{-tf}W}$.

Since f is recurrent, there exists $t^* > t$ such that

$$e^{t^*f}e^{-tf}W \cap e^{-tf}W \neq \emptyset,$$

or equivalently

$$e^{(t^*-t)f}W \cap e^{-tf}W \neq \emptyset.$$

But since $t^* - t > 0$ and $W \subset \mathcal{A}(x_0)$ we have that $e^{(t^*-t)f} W \subset \mathcal{A}(x_0)$. It follows that

$$\mathcal{A}(x_0) \cap e^{-tf} W \neq \emptyset.$$

Hence in any neighborhood of $e^{-tf} x_0$ there are points of $\mathcal{A}(x_0)$. In other words $e^{-tf} x_0 \in \overline{\mathcal{A}(x_0)}$. □

As a consequence of the previous lemma we have the following.

Corollary 25 *Consider the control system*

$$\dot{x} = f_0(x) + \sum_{i=1}^{m} u_i(t) f_i(x), \qquad (u_1(\cdot), \dots, u_m(\cdot)) : [0, \infty[\to U \subset \mathbf{R}^m. \tag{7}$$

Assume that **(i)** 0 *belongs to the interior of* U, **(ii)** *the control system* (7) *is Lie bracket generating,* **(iii)** f_0 *is recurrent. Then the system is controllable.*

Proof Notice that $f_0 \in \mathcal{F}$ since 0 belongs to U. Lemma 24 states the equivalence between the controllability of (7) and that of

$$\dot{x} = \sum_{i=0}^{m} u_i(t) f_i(x), \qquad (u_0(\cdot), \dots, u_m(\cdot)) : [0, \infty[\to (\{-1\} \times \{0\}) \cup (\{1\} \times U).$$

The controllability of this system follows from the Chow–Rashevskii theorem after using Lemma 28 since $\{-1, 1\} \times U$ contains a symmetric set (Remark 19). □

Example (On a Manifold) Consider the control system on the sphere $S^2 = \{x \in \mathbb{R}^3 \mid x_1^2 + x_2^2 + x_3^2 = 1\}$ given by

$$\dot{x} = f_0(x) + u f_1(x), \qquad u(\cdot) : [0, \infty[\to (-1, 1), \qquad x \in S^2,$$

where

$$f_0(x) = \begin{pmatrix} -x_2 \\ x_1 \\ 0 \end{pmatrix}, \qquad f_1(x) = \begin{pmatrix} 0 \\ -x_3 \\ x_2 \end{pmatrix}.$$

The flows of f_0 and f_1 are rotations around the axes $(0, 0, 1)^T$ and $(1, 0, 0)^T$, respectively.

This system is controllable since

- the Lie bracket between f_0 and f_1 is given by

$$[f_0, f_1](x) = \begin{pmatrix} -x_3 \\ 0 \\ x_1 \end{pmatrix},$$

and hence the system is Lie bracket generating (for every $x \in S^2$, $\dim(\mathrm{span}\{f_0, f_1, f_2\}(x) = 2)$;
- the trajectories of f_0 are periodic and hence f_0 is recurrent.

4.2 *Affine Systems with Non-recurrent Drift*

When the drift is not recurrent one can still obtain that the system is controllable if the controls are unbounded and if it is not necessary to use the drift to get a Lie algebra of full dimension at every point. More precisely we have the following.

Proposition 26 (Strong Bracket Generating) *Consider the control system*

$$\dot{x} = f_0(x) + \sum_{i=1}^{m} u_i f_i(x), \qquad (u_1(\cdot), \ldots, u_m(\cdot)) : [0, \infty[\to \mathbb{R}^m. \tag{8}$$

If $\{f_1, \ldots, f_m\}$ is Lie bracket generating then (8) is controllable.

Remark 27 For an affine control system as (8), the condition that $\{f_1, \ldots, f_m\}$ is Lie bracket generating is called the *strong bracket generating condition.*

Proof First notice that as a consequence of Exercise 1.2, being $\{f_1, \ldots, f_m\}$ Lie bracket generating, then (8) is Lie bracket generating as well. Let $\mathcal{F}$ be the family of vector fields associated with (8). In the following we are going to prove that for every $(v_1, \ldots, v_m) \in \mathbb{R}^m$ the vector field $\sum_{i=1}^m v_i f_i$ is compatible with $\mathcal{F}$. Once this is done, the controllability of (8) follows, since the family $\mathcal{F} \cup \{\sum_{i=1}^m v_i f_i,\ v_i \in \mathbb{R}, i = 1, \ldots, m\}$ contains a symmetric and Lie bracket generating sub-family (Remark 19).

To show that $\sum_{i=1}^m v_i f_i$ is compatible with $\mathcal{F}$ for every $v_1, \ldots, v_m \in \mathbb{R}$, remark that

$$\sum_{i=1}^{m} v_i f_i = \lim_{n\to\infty} \frac{1}{n} \left(f_0 + \sum_{i=1}^{m} (n v_i) f_i \right) \tag{9}$$

and that $\frac{1}{n}(f_0 + \sum_{i=1}^m (n v_i) f_i)$ is compatible with $\mathcal{F}$ since a trajectory of (8) corresponding to controls $\tilde{u}_i(\cdot)$, $i = 1, \ldots, m$, is a time-reparameterisation of a trajectory of

$$\dot{x} = \frac{1}{n}\left(f_0(x) + \sum_{i=1}^{m} u_i f_i(x)\right)$$

corresponding to controls $n\tilde{u}_i(\cdot)$, $i = 1, \dots, m$.

The conclusion follows from the fact that if a vector field is the uniform limit on all compacts of a sequence of compatible vector fields, then it is compatible as well (from the continuity of solutions of ODEs with respect to the vector field). □

4.3 Convexification

A very useful criterium is the one that states that a convex combination of vector fields of $\mathcal{F}$ is compatible with $\mathcal{F}$. It formalize the intuition that if one commutes quickly between the dynamics of two vector fields, and one stays the same time on each dynamics, then the corresponding trajectory is close the trajectory of $\frac{f+g}{2}$ starting from the same point.

Lemma 28 *For every $\lambda_1, \dots, \lambda_k \geq 0$ and $f_1, \dots, f_k \in \mathcal{F}$, the vector field $\lambda_1 f_1 + \dots + \lambda_k f_k$ is compatible with $\mathcal{F}$.*

The proof of this Lemma is quite technical and it is based on the Gronwall inequality. See [1] for details.

From this lemma one can gets some useful corollaries of Theorem 17, Corollary 25, and Proposition 26.

Corollary 29 *If $\mathcal{F}$ is Lie bracket generating and its* convex hull in the space $\text{Vec}(\mathbb{R}^n)$ *is symmetric, then for every $x_0 \in \mathbb{R}^n$ we have $\mathcal{A}(x_0) = \mathbb{R}^n$.*

Corollary 30 *Consider the control system*

$$\dot{x} = f_0(x) + \sum_{i=1}^{m} u_i f_i(x), \qquad (u_1(\cdot), \dots, u_m(\cdot)) : [0, \infty[\to U \subset \mathbf{R}^m. \tag{10}$$

Assume that **(i)** 0 *belongs to the interior* of the convex hull of U, **(ii)** *the control system* (10) *is Lie bracket generating,* **(iii)** f_0 *is recurrent. Then the system is controllable.*

Corollary 31 *Consider the control system*

$$\dot{x} = f_0(x) + \sum_{i=1}^{m} u_i f_i(x), \qquad (u_1(\cdot), \dots, u_m(\cdot)) : [0, \infty[\to U \subset \mathbb{R}^m. \tag{11}$$

If the convex hull of U is $\mathbb{R}^m$ *and if* $\{f_1, \ldots, f_m\}$ *is Lie bracket generating then (11) is controllable.*

5 Orbits and Necessary Conditions for Controllability

We have seen in the previous sections several sufficient conditions for the controllability of a nonlinear system. In this section we discuss some necessary conditions that are consequence of a very deep theorem of geometric nature, the so-called *orbit theorem*. This theorem permits to conclude that, beside pathological cases, $\text{Lie}_x(\mathcal{F})$ measures precisely the dimension of the set of directions that can be used starting from a point.

We define the *orbit* of the family $\mathcal{F}$ starting from a point $x_0 \in \mathbb{R}^n$ as the set

$$\mathcal{O}(x_0) = \{e^{t_k f_k} \circ \cdots \circ e^{t_1 f_1}(x_0) \mid k \in \mathbb{N},\ t_1, \ldots, t_k \in \mathbb{R},\ f_1, \ldots, f_k \in \mathcal{F}\}.$$

Remark 32 $\mathcal{O}(x_0)$ can be interpreted as the reachable set, starting from x_0, of the family $-\mathcal{F} \cup \mathcal{F}$, using only piecewise constant controls.

We have the following result

Theorem 33 (Orbit Theorem) *For every* $x_0 \in \mathbb{R}^n$*, the set* $\mathcal{O}(x_0)$ *has the structure of an immersed sub-manifold of* $\mathbb{R}^n$*. In particular it has the same dimension at every point. Moreover if* $x \in \mathcal{O}(x_0)$ *then* $\text{Lie}_x(\mathcal{F}) \subseteq T_x\mathcal{O}(x_0)$*. The two spaces* $\text{Lie}_x(\mathcal{F})$ *and* $T_x\mathcal{O}(x_0)$ *coincide if one of the following two conditions is verified:*

- *every element of* $\mathcal{F}$ *is an analytic vector field;*
- *the dimension of* $\text{Lie}_x(\mathcal{F})$ *is constant with respect to* $x \in \mathcal{O}(x_0)$.

From the fact that $\text{Lie}_x(\mathcal{F}) \subseteq T_x\mathcal{O}(x_0)$ it follows that every element of $\mathcal{F}$ is tangent to $\mathcal{O}(x_0)$. As a consequence we have the following.

Lemma 34 *For every* $x_0 \in \mathbb{R}$ *we have that* $\mathcal{A}(x_0) \subseteq \mathcal{O}(x_0)$.

This result wouldn't be so obvious without the orbit theorem since $\mathcal{A}(x_0)$ is the set of points that one can reach using L^∞ controls (and not only piecewise constant ones).

The following corollary gives some consequence of the Orbit theorem on the controllability of nonlinear systems.

Corollary 35 *If* $\mathcal{F}$ *is not Lie bracket generating and either every vector field of* $\mathcal{F}$ *is analytic or the dimension of* $\text{Lie}_x(\mathcal{F})$ *is constant with respect to* $x \in \mathcal{O}(x_0)$*, then* $\mathcal{F}$ *is not controllable.*

Acknowledgements This work was supported by the ANR project "SRGI" ANR-15- CE40-0018, by the ANR project "Quaco" ANR-17-CE40-0007-01 and by a public grant as part of the Investissement d'avenir project, reference ANR-11-LABX-0056-LMH, LabEx LMH (in a joint call with Programme Gaspard Monge en Optimisation et Recherche Opérationnelle).

References

1. A.A. Agrachev, Y.L. Sachkov, *Control Theory from the Geometric Viewpoint. Encyclopaedia of Mathematical Sciences*, vol. 87 (Springer, Berlin, 2004). Control Theory and Optimization, II
2. F. Jean, *Control of Nonholonomic Systems: from Sub-Riemannian Geometry to Motion Planning.* Springer Briefs in Mathematics (Springer, Cham, 2014)
3. V. Jurdjevic, *Geometric Control Theory. Cambridge Studies in Advanced Mathematics*, vol. 52 (Cambridge University Press, Cambridge, 1997)

Introduction to Variational Methods for Viscous Ergodic Mean-Field Games with Local Coupling

Annalisa Cesaroni and Marco Cirant

Abstract We collect in these notes some results on the existence and uniqueness of classical solutions to viscous ergodic Mean-Field Game systems with local coupling. We present in particular some methods and ideas based on convex optimization techniques and elliptic regularity.

Keywords Ergodic Mean-Field Games · Elliptic systems · Variational methods

1 Introduction

In these notes we collect some basic results on second order stationary Mean-Field Games systems with local coupling with power growth. This material has been taught in a 5 h course by the first author at the IndAM Intensive Period **Contemporary Research in elliptic PDEs and related topics** held at the University of Bari from April to June 2017. We would like to thank the scientific organizer of the period, Serena Dipierro, for her kind invitation to give this course.

Mean-Field Games (MFG) is a recent theory that models the behaviour of a very large number of indistinguishable rational agents aiming at minimizing a common cost: each agent has to choose a strategy, in the form of a trajectory in a state space, which best fits his preferences, but is affected by the other agents through a global mean field effect. The theory was introduced in the seminal papers by Lasry, Lions [34–36] and by Huang, Caines, Malhamé [31]. For an introduction to the theory, we refer to the recent monographs [7, 30], to [37] and to the notes [5, 10, 27].

In the ergodic framework, the idea of mean-field Nash equilibrium, and its characterization in terms of a system of PDEs can be derived as follows. For

A. Cesaroni (✉)
Dipartimento di Scienze Statistiche, Università di Padova, Padova, Italy
e-mail: annalisa.cesaroni@unipd.it

M. Cirant
Dipartimento di Matematica Tullio Levi Civita, Università di Padova, Padova, Italy
e-mail: cirant@math.unipd.it

S. Dipierro (ed.), *Contemporary Research in Elliptic PDEs and Related Topics*, Springer INdAM Series 33, https://doi.org/10.1007/978-3-030-18921-1_5

simplicity, we will consider a periodic setting: suppose that the population of players has density $m \in C(Q)$, where Q is the N-dimensional torus $\mathbb{R}^N/\mathbb{Z}^N$. The dynamics of a typical agent is given by the controlled stochastic differential equation

$$dX_s = -v_s ds + \sqrt{2} dB_s, \tag{1}$$

where v_s is the control and B_s is a Brownian motion, and the cost, of long-time average form, is given by

$$\lim_{T\to\infty} \frac{1}{T}\mathbb{E}\int_0^T [L(v_s) + f(X_s, m(X_s))]ds.$$

The (convex) Lagrangian function L is associated to the cost of moving with velocity v_s, while the term $f(X_s, m(X_s))$ is the cost of being at position X_s; the dependance of f with respect to the density m describes the interaction between the individual and the overall population.

Every agent seeks to optimize this cost. A classical result in control theory states that the optimal control for this problem is given in feedback form by

$$v_s = -\nabla H(\nabla u(X_s)),$$

where the Hamiltonian H is the Legendre transform of L, i.e. $H(p) = L^*(p) = \sup_{v\in\mathbb{R}^N}[p \cdot v - L(v)]$, and u is the solution of the stationary Hamilton-Jacobi-Bellman (HJB) equation

$$-\Delta u(x) + H(\nabla u(x)) + \lambda = f(x, m(x)) \quad \text{on } Q.$$

This fact can be obtained by means of the Ito formula and the definition of H (see, e.g. [6]). Note that in the HJB equation the constant $\lambda \in \mathbb{R}$ is itself an unknown.

A crucial feature of the stochastic differential equation (1) is that, when the velocity v_s is given in feedback form. i.e. $v_s = b(X_s)$, where b is some autonomous drift, then the law of X_s becomes stable in the long time regime. In other words,

$$\mathscr{L}(X_s) \to \bar{m}, \quad \text{as } s \to \infty,$$

where $\bar{m}$ is the so-called invariant measure. It can be proven that (see [6, 32] for further details) $\bar{m}$ is the unique solution of

$$-\Delta\bar{m} + \mathrm{div}(\bar{m}b) = 0 \quad \text{on } Q, \qquad \int_Q \bar{m} = 1. \tag{2}$$

Hence, if all the agents play optimally, all their invariant measures $\bar{m}$ will coincide, and solve the Kolmogorov (Fokker-Planck) Eq. (2) with drift $b(x) = -\nabla H(\nabla u(x))$. Therefore, the overall population will be distributed (as $s \to \infty$)

with density $\bar{m}$. In an equilibrium situation,

$$\bar{m} = m,$$

independently on the initial state X_0. This heuristically leads to the stationary MFG system

$$\begin{cases} -\Delta u + H(\nabla u) + \lambda = f(x, m(x)) \\ -\Delta m - \operatorname{div}(m \nabla H(\nabla u)) = 0 & \text{on } Q, \\ \int_Q m = 1, \end{cases} \tag{3}$$

that is associated to equilibria of the game (see also [2]). These notes will be focused on the existence of solutions to this system of PDEs; we will discuss in particular some methods based on convex optimization and elliptic regularity (for further details and results see [5, 11, 13, 16, 19, 28, 29, 38]).

We stress that we are assuming the coupling f to be a local function of $m(x)$. In other words, each player is affected at time s just by the value of $m(X_s)$. This setting is opposed to the "non-local" case, where f can be a functional defined on the space of probability measures. For this reason, it is important in the local case to obtain solutions of (3) such that m is at least continuous on Q, for the optimization problem to be meaningful. This requires assumptions on the data f, H. Classical regularity of solutions to (3) has been considered in several works (see, e.g., [21, 30, 39], and reference therein), but it is in some cases still an open problem.

We mention that (3) has also a deep derivation from Nash equilibria of games with N-players, in the limit $N \to \infty$. This striking result has been obtained in the seminal works [34–36], and recently in [15] for some time-dependent problems. We also refer to [25] for additional details.

Finally, ideas presented here can be adapted to time-dependent problems, i.e. finite horizon MFG systems, which consists of a backward HJB equation coupled with a forward Fokker-Planck equation (see, e.g, [12, 22, 30] and references therein) and also to fractional MFG, that is MFG systems driven by fractional laplacian (see [17]). The treatment of MFG with more than one population is more delicate, and a variational formulation is available only in certain situations, see [14, 19, 23].

These notes are organized as follows. In Sect. 2 we provide our standing assumptions and notations throughout the paper. In Sects. 3 and 4 we discuss some existence and regularity results for the stationary Fokker-Planck and Hamilton-Jacobi-Bellman equations respectively. In Sect. 5 we derive the main result on existence of solutions to (3), while in Sect. 6 we prove that solutions are unique under the additional assumption that f is monotone increasing in the m variable.

2 Standing Assumptions

As explained in the introduction, the problem we aim to solve is finding a constant $\lambda \in \mathbb{R}$ for which (3) has a solution (u, m). A (classical) solution to the system (3) is a triple $(u, \lambda, m) \in C^{2,\theta}(Q) \times \mathbb{R} \times W^{1,p}(Q)$, for all $\theta \in (0, 1)$ and for all $p > 1$.

We assume that $H : \mathbb{R}^N \to \mathbb{R}$ is strictly convex, that $H \in C^2(\mathbb{R}^N \setminus \{0\})$ and that there exist some $C_H > 0$, $K > 0$ and $\gamma > 1$ such that, for all $p \in \mathbb{R}^N$,

$$\begin{aligned} &C_H|p|^\gamma - C_H^{-1} \le H(p) \le C_H(|p|^\gamma + 1), \\ &\nabla H(p) \cdot p - H(p) \ge C_H|p|^\gamma - K \text{ and } \quad |\nabla H(p)| \le C_H|p|^{\gamma-1}. \end{aligned} \tag{4}$$

The coupling term f is local, that is $f : \mathbb{R}^N \times [0, +\infty) \to \mathbb{R}$ is locally Lipschitz continuous in both variables, and $\mathbb{Z}^N$-periodic in x, that is $f(x + z, m) = f(x, m)$ for all $z \in \mathbb{Z}^N$, all $x \in \mathbb{R}^N$ and all $m \in [0, +\infty)$. We assume that there exist $C > 0$ and $K > 0$ such that

$$-Cm^{q-1} - K \le f(x, m) \le Cm^{q-1} + K, \quad \text{with } 1 < q < 1 + \frac{1}{N}\frac{\gamma}{\gamma - 1}. \tag{5}$$

Notations For every $p \ge 1$, $p' = \frac{p}{p-1}$ will be the usual conjugate exponent of p. Q is the N-dimensional torus $Q := \mathbb{R}^N/\mathbb{Z}^N$ and we identify functions on Q with their periodic extension to $\mathbb{R}^N$. Finally, $C, C_1, C_2, K \dots$ denote (positive) constants we need not to specify.

3 Steady States of the Fokker-Planck Equation

We provide here some results on existence, uniqueness and regularity of steady state solutions to the Fokker-Planck equations in the periodic setting. These results are classical and well known, we refer to [6], nevertheless in some cases we add also some sketch of proof for readers' convenience.

From classical elliptic regularity, we recall the following result.

Proposition 1 *Let $p > 1$, $\Omega \subset \mathbb{R}^N$ be a bounded open set, and $m \in L^p(\Omega)$ be such that*

$$\left| \int_\Omega m\, \Delta\varphi\, dx \right| \le K \|\nabla\varphi\|_{L^{p'}(\Omega)} \quad \textit{for all } \varphi \in C_0^\infty(\Omega) \tag{6}$$

for some $K > 0$. *Then,* $m \in W^{1,p}(\Omega')$ *for every* $\Omega' \subset\subset \Omega$ *and there exists* $C > 0$ *depending only on p, and* Ω' *such that*

$$\|\nabla m\|_{L^p(\Omega')} \leq C(K + \|m\|_{L^p(\Omega)}).$$

Proof For the proof we refer to [1, Theorem 6.1].

Note that in the periodic setting, namely if $m \in L^p(Q)$ and (6) holds with $\Omega = Q$, then the conclusion of Proposition 1 holds with $\Omega' = \Omega = Q$.

Using this result we prove the following a priori estimate.

Lemma 1 *Let* $w \in L^p(Q; \mathbb{R}^N)$. *Then there exists a unique solution* $m \in W^{1,p}(Q)$ *to the problem*

$$-\Delta m = div\, w, \qquad with \int_Q m\, dx = 1. \tag{7}$$

Moreover there exists $C > 0$, *depending on p, such that*

$$\|m\|_{W^{1,p}(Q)} \leq C(\|w\|_{L^p(Q)} + \|m\|_{L^p(Q)}). \tag{8}$$

Proof Assume first that w is smooth, let m be the unique smooth solution[1] to (7), and let $v \in C^\infty(Q)$ be a test function. Multiplying (7) by v and integrating by parts, we get

$$\int_Q m(-\Delta v)\, dx = \int_Q w \cdot \nabla v\, dx \leq \|w\|_{L^p(Q)}\, \|\nabla v\|_{L^{p'}(Q)}.$$

We conclude by Proposition 1 that $\|m\|_{W^{1,p}(Q)} \leq C(\|w\|_{L^p(Q)} + \|m\|_{L^p(Q)})$.

The result in the general case then follows by approximating w with smooth vector fields.

Finally we consider steady state solutions to the periodic Fokker-Planck equation.

Proposition 2 *Let* $b \in L^\infty(Q; \mathbb{R}^N)$. *Then, there exists a unique solution* $m \in W^{1,p}(Q)$, *for all* $p > 1$, *to the problem*

$$-\Delta m + div(bm) = 0, \tag{9}$$

[1]This solution m can be found for example by standard methods in calculus of variations, i.e. by minimizing the convex functional $m \mapsto \frac{1}{2}\int_Q |\nabla m|^2 - (\text{div} w)m$ subject to the constraint $\int_Q m = 1$; regularity of the minimizing weak solution is classical by uniform ellipticity.

with $\int_Q m\,dx = 1$, *and*

$$\|m\|_{W^{1,p}(Q)} \le C,$$

where $C > 0$ *depends only on* N, p *and* $\|b\|_{L^\infty(Q;\mathbb{R}^N)}$. *In particular, we have that* $m \in C^\theta(Q)$, *for every* $\theta \in (0,1)$.

Furthermore, we get that there exists a constant $C = C(N,b) > 0$ *such that*

$$0 < C \le m(x) \le C^{-1}, \quad \textit{for any } x \in Q.$$

Proof Assume b to be smooth, the general case will follow by an approximation argument.

Step 1: Existence and Uniqueness of a Solution The existence result follows by the Fredholm alternative.

More precisely, for K large enough, by Lax-Milgram Theorem, the equation

$$-\Delta v - b \cdot \nabla v + Kv = \psi$$

has a unique solution $u \in W^{2,2}Q)$, for any fixed $\psi \in L^2(Q)$. Therefore, the mapping $\mathcal{G}_K$, defined by $v = \mathcal{G}_K\psi$, is a compact mapping of $L^2(Q)$ into itself.

Now, Eq. (9) may be rewritten as

$$(I - K\mathcal{G}_K^*)m = 0. \tag{10}$$

By the Fredholm alternative, the number of linearly independent solutions of (10) is the same as that of the adjoint problem, that is

$$(I - K\mathcal{G}_K)v = 0,$$

that corresponds to

$$-\Delta v - b \cdot \nabla v = 0. \tag{11}$$

Any $v \in W^{2,2}(Q)$ solving (11) is in $C^2(Q)$ (due to classical elliptic regularity theory, see [26]), and then it must be constant by the Strong Maximum Principle (see [26]). We conclude that there exists m solving (9) (in the distributional sense), and such $m \in L^2(Q)$ is unique up to a multiplicative constant. Moreover $m \in W^{1,2}(Q)$ and $\|m\|_{W^{1,2}(Q)} \le C\|b\|_{L^\infty}$, see [6, Thm II.4.3]).

Step 2: Positivity Fix a nonnegative periodic continuous initial datum z_0, and consider the following Cauchy problem,

$$\begin{cases} \partial_t z - \Delta z - b \cdot \nabla z = 0 & \text{in } \mathbb{R}^N \times (0,\infty), \\ z(\cdot,0) = z_0(\cdot). \end{cases} \tag{12}$$

So, by Comparison Principle, the solution z of (12) satisfies $0 \le \min z_0 \le z(x,t) \le \max z_0$ for all t. By [4, Theorem 4.1], $z(\cdot,t) - \Lambda t - \bar{z}(\cdot)$ converges uniformly to zero as $t \to +\infty$, where the couple $(\bar{\lambda}, \bar{z})$ solves the stationary problem

$$-\Delta \bar{z} - b \cdot \nabla \bar{z} = \bar{\lambda} \quad \text{in } Q. \tag{13}$$

Note that $(\bar{\lambda}, \bar{z})$ solving (13) must satisfy $\bar{\lambda} = 0$, so that $\bar{z}$ is identically constant on Q; hence $z(\cdot,t) \to \bar{z}$ uniformly on Q as $t \to +\infty$. Moreover $\min z_0 \le \bar{z} \le \max z_0$.

By multiplying the equation in (12) by m, the equation in (9) by z, and integrating by parts on Q, we obtain that, for all $t > 1$,

$$\int_Q \partial_t z(x,t) m(x) dx = 0,$$

so, since $\int_Q m\, dx = 1$, as $t \to +\infty$, we get

$$\int_Q z_0(x) m(x) dx = \int_Q z(x,t) m(x) dx \to \int_Q \bar{z} m(x) dx = \bar{z}. \tag{14}$$

Therefore we conclude that for every periodic continuous function z_0,

$$0 \le \min z_0 \le \int_Q z_0(x) m(x) dx = \bar{z} \le \max z_0.$$

In particular this implies that $m \ge 0$. So, by Harnack inequality (see [26]), we get that $m > 0$.

Step 3: Boundedness and Regularity The regularity and boundedness follow by an iterative argument, starting from the initial regularity estimate in $W^{1,2}$ and applying Lemma 1 and Sobolev embeddings. We refer for more details for example to [3, Lemma 2.3].

4 Hamilton Jacobi Equations with Coercive Hamiltonian

We collect some well known results on the Lipschitz continuity of viscosity solutions to Hamilton-Jacobi equations and on the solution to the ergodic problem.

We consider the following Hamilton-Jacobi equation

$$-\Delta u + H(\nabla u) + \lambda = f(x), \qquad x \in \mathbb{R}^N. \tag{15}$$

We assume that $f \in C(\mathbb{R}^N)$, and that f is $\mathbb{Z}^N$-periodic, so we look for periodic solutions to (15).

Theorem 1 *There exists a unique constant* $\lambda \in \mathbb{R}$ *such that* (15) *has a periodic solution* $u \in C^2(\mathbb{R}^N)$ *and*

$$\lambda = \sup\{c \in \mathbb{R} \ s.t. \ \exists u \in C^2(\mathbb{R}^N) \ s.t. \ -\Delta u + H(\nabla u) + c \leq f(x)\}. \tag{16}$$

Moreover, there exists a constant $K > 0$*, depending on* $\|f\|_{L^\infty(\mathbb{R}^N)}$ *and* $|\lambda|$ *such that*

$$\|\nabla u\|_{L^\infty} \leq K.$$

Finally, u *is the unique Lipschitz viscosity solution to* (15) *up to addition of constants.*

Proof This result is very well known, see e.g. [4]. For completeness we provide a brief sketch of the proof. We will assume f to be smooth, the general case will follow by an approximation argument, once we show that the estimates are only depending on $\|f\|_{L^\infty}$.

Step 1: Ergodic Approximation Let $\delta > 0$ and u_δ be the continuous periodic solution to

$$\delta u_\delta - \Delta u_\delta + H(\nabla u_\delta) = f(x), \qquad x \in \mathbb{R}^N. \tag{17}$$

Note that (17) admits a unique periodic viscosity solution, and that by comparison $\|\delta u_\delta\|_{L^\infty(\mathbb{R}^N)} \leq \|f\|_{L^\infty(\mathbb{R}^N)} + |H(0)|$. Moreover, recalling that we assumed f to be smooth, we can apply classical Bernstein method, see [4], to get that $\|\nabla u_\delta\|_{L^\infty} \leq C$, for some $C > 0$ depending on the C^1 norm of f. Therefore, by classical elliptic regularity theory, since $\Delta u_\delta \in L^\infty$, we have that $u_\delta \in C^{1,\alpha}$ for every $\alpha \in (0, 1)$. Then by bootstrap, we conclude that $u_\delta \in C^2$.

Step 2: A Priori Gradient Bounds By assumption (4), we get that

$$|-\Delta u_\delta + C_H|\nabla u_\delta|^\gamma| \leq K. \tag{18}$$

Therefore, by [33, Thm A.1], we get that for every $r \in (1, +\infty)$, there exists C depending on C_H, r, N, γ, K such that $\|\nabla u_\delta\|_{L^r(Q)} \leq C$. So, this implies also that $\|\nabla u_\delta\|_{L^\infty} \leq C$ by a constant depending on K, γ, C_H, N. Indeed if r is large enough, then $-\Delta u_\delta$ is bounded in $L^q(Q)$ for some $q > N$, and the statement follows by elliptic regularity theory and Sobolev embeddings.

So, there exists a constant $K > 0$, depending on $\|f\|_{L^\infty(\mathbb{R}^N)}$ (and in principle by $\|\delta u_\delta\|_{L^\infty(\mathbb{R}^N)}$ also, that is itself controlled by $\|f\|_{L^\infty(\mathbb{R}^N)} + |H(0)|$, see Step 1), such that

$$\|\nabla u_\delta\|_{L^\infty} \leq K. \tag{19}$$

Step 3: Solution of the Ergodic Problem We define $v_\delta = u_\delta - u_\delta(0)$. Then $\|\nabla v_\delta\|_{L^\infty} \leq K$ and moreover $\|v_\delta\|_{L^\infty} \leq K$. Therefore by Ascoli-Arzelá theorem, we can extract a subsequence converging uniformly to v. Moreover $\delta v_\delta \to \lambda$, where λ is a constant. Note that v_δ is a solution to

$$\delta v_\delta - \Delta v_\delta + H(\nabla v_\delta) = f(x) - \delta v_\delta(0), \qquad x \in \mathbb{R}^N.$$

Therefore v, by stability of viscosity solution with respect to uniform convergence is a solution to (15).

Step 4: Uniqueness Assume there exist two solutions to (15), (v_1, λ_1), (v_2, λ_2). Assume that $\lambda_1 > \lambda_2$. Up to addition of constants we can assume that $v_1 \leq v_2$. Then $\lambda_1 t + v_1$ is a solution to

$$\begin{cases} v_t - \Delta v + H(\nabla v) = f(x) & x \in \mathbb{R}^N, t > 0 \\ v(0, x) = v_1(x). \end{cases}$$

On the other hand $v_2 + \lambda_2 t$ is a subsolution to the same problem, since $v_1 \leq v_2$. Therefore by comparison principle $v_1(x) + \lambda_1 t \leq v_2(x) + \lambda_2 t$, for every $t \geq 0$. Recalling that v_1, v_2 are bounded, this is in contradiction with $\lambda_1 > \lambda_2$, for t sufficiently large.

So, λ in (15) is unique. Moreover, a classical argument based on the strong maximum principle, shows that v is unique up to addition of constants (see [4]).

Remark 1 We can make the gradient bound in Step 2 of the previous proof more precise by a suitable rescaling argument: we show that there exists a constant C depending only on γ, N, C_H such that if (18) holds, then

$$\|\nabla u_\delta\|_{L^\infty} \leq C(1 + K)^{1/\gamma}.$$

Indeed, arguing as in Step 2 of the proof, if (18) holds with $K = 1$, then $\|\nabla u_\delta\|_{L^\infty} \leq C_1$ for some positive C_1 depending only on γ, N, C_H. For general $K > 0$, let $\kappa := (1 + K)^{-1/\gamma'}$, and

$$v(y) := \kappa^{\frac{2-\gamma}{\gamma-1}} u_\delta(\kappa y) \quad \text{on } \mathbb{R}^N.$$

Then, v satisfies on $\mathbb{R}^N$

$$| - \Delta v + C_H |\nabla v|^\gamma| \leq \kappa^{\gamma'} K = \frac{K}{1 + K} \leq 1.$$

Therefore, $\|\nabla v\|_{L^\infty} \leq C_1$, that implies $\|\nabla u_\delta\|_{L^\infty} = \kappa^{-\frac{1}{\gamma-1}} \|\nabla v\|_{L^\infty} \leq C_1(1 + K)^{1/\gamma}$.

5 Existence of Solutions to the MFG System

In this section we prove existence of classical solutions to (3). This result has been proved with a somehow different method in [21], and also in [39] under the additional assumption that f is bounded from below.

Theorem 2 *Under the assumptions* (4) *and* (5) *there exists a classical solution* (u, λ, m) *to the MFG system* (3)*. Moreover,* $m > 0$.

The proof is obtained using variational techniques and a regularization procedure.

Remark 2 It is often useful to consider (3) in the quadratic case, namely when $H(p) = |p|^2/2$. Then, the so-called Hopf-Cole transformation reduces the number of unknowns, namely setting $v^2 := m = \frac{e^{-u/2}}{\int_Q e^{-u/2}}$, then v solves

$$-\Delta v + f(x, v^2)v = \lambda v \quad \text{in } Q, \qquad \int_Q v^2 = 1. \tag{20}$$

Information on this equation can be used to deduce some features of (3) in general. For example, (20) has a variational structure (with L^2-constraint); this is true, as we will see in the sequel, also for (3). On the other hand, if $H(p) \neq |p|^2/2$ a similar transformation is available only in particular cases (see [20]).

The Energy Associated to the System We denote by $\tilde{L}$ the Legendre transform of H, i.e.

$$\tilde{L}(v) := \sup_{p \in \mathbb{R}^N} [p \cdot v - H(p)], \qquad \text{for any } v \in \mathbb{R}^N.$$

The assumptions on H guarantee the following (see [18, Proposition 2.1], and also [8, 24]).

Proposition 3 *There exist* $C_H, C_L > 0$ *such that for all* $p, b \in \mathbb{R}^N$,

i) $\tilde{L} \in C^2(\mathbb{R}^N \setminus \{0\})$ *and it is strictly convex,*
ii) $0 \leq C_L|q|^{\gamma'} \leq \tilde{L}(q) \leq C_L^{-1}(|q|^{\gamma'} + 1)$, *where* $\gamma' = \frac{\gamma}{\gamma-1}$ *is the conjugate exponent of* γ,
iii) $\nabla\tilde{L}(q) \cdot q - \tilde{L}(q) \geq C_L|q|^{\gamma'} - C_L^{-1}$,
iv) $C_L|q|^{\gamma'-1} - C_L^{-1} \leq |\nabla\tilde{L}(q)| \leq C_L^{-1}(|q|^{\gamma'-1} + 1)$.
v) $C_H|p|^{\gamma-1} - C_H^{-1} \leq |\nabla H(p)| \leq C_H^{-1}(|p|^{\gamma-1} + 1)$.

We let

$$\mathcal{K} := \Big\{(m, w) \in L^1(Q) \cap L^q(Q) \times L^1(Q) \text{ s.t.}$$
$$\int_Q m(-\Delta\varphi)\,dx = \int_Q w \cdot \nabla\varphi\,dx \quad \forall\varphi \in C_0^\infty(Q), \tag{21}$$
$$\int_Q m\,dx = 1, \quad m \geq 0 \text{ a.e.}\Big\}.$$

We associate to the mean field game (3) the following energy

$$\mathcal{E}(m, w) := \begin{cases} \displaystyle\int_Q mL\left(-\frac{w}{m}\right) + F(x, m)\,dx & \text{if } (m, w) \in \mathcal{K}, \\ +\infty & \text{otherwise,} \end{cases} \tag{22}$$

where[2]

$$L\left(-\frac{w}{m}\right) := \begin{cases} \tilde{L}\left(-\frac{w}{m}\right) & \text{if } m > 0, \\ 0 & \text{if } m = 0,\, w = 0, \\ +\infty & \text{otherwise} \end{cases}$$
$$\text{and} \qquad F(x, m) := \begin{cases} \displaystyle\int_0^m f(x, n)\,dn & \text{if } m \geq 0, \\ +\infty & \text{if } m < 0. \end{cases} \tag{23}$$

Proposition 4 *The map*

$$(m, w) \to mL\left(-\frac{w}{m}\right)$$

is convex, and is strictly convex if restricted to $m > 0$. Moreover,

$$mH(p) = \sup_w \left[-p \cdot w - mL\left(-\frac{w}{m}\right)\right]. \tag{24}$$

Finally

$$C_L \frac{|w|^{\gamma'}}{m^{\gamma'-1}} - C_L^{-1} m \leq mL\left(-\frac{w}{m}\right) \leq C_L^{-1} \frac{|w|^{\gamma'}}{m^{\gamma'-1}} + C_L^{-1} m. \tag{25}$$

[2]Note that $L(\cdot)$ here coincides with L as in the introduction. Indeed, $\tilde{L} = H^* = (L^*)^* = L$.

Proof Note that, since $\tilde{L}$ is the Legendre transform of H, for every $m \geq 0$ we have

$$mL\left(-\frac{w}{m}\right) = m \sup_p \left(p \cdot \frac{-w}{m} - H(p)\right) = \sup_p \left(-p \cdot w - mH(p)\right).$$

So $mL\left(-\frac{w}{m}\right)$ is the supremum of a family of linear functions in (m, w), therefore is convex. A similar argument gives (24). As for the strict convexity, observe that since H is strictly convex, also $\tilde{L}$ is strictly convex. Then it is easy to check the strict convexity of $mL(-w/m)$ where $m > 0$. The estimate (25) comes from Proposition 3.

Now, we provide a priori estimates for couples $(m, w) \in \mathcal{K}$ with finite energy.

Proposition 5 *Let*

$$\frac{1}{r} = \frac{1}{\gamma'} + \frac{1}{\gamma q}.$$

Assume that $(m, w) \in \mathcal{K}$ *is such that there exists* $K > 0$ *with*

$$E := \int_Q \frac{|w|^{\gamma'}}{m^{\gamma'-1}} dx \leq K.$$

Then, there exist $\delta > 0$ *and* $C > 0$ *such that*

$$\|m\|_{W^{1,r}(Q)} \leq C \left[\int_Q \frac{|w|^{\gamma'}}{m^{\gamma'-1}} dx + 1\right] \leq C(K+1) \tag{26}$$

$$\|m\|_{L^q(Q)}^{q(1+\delta)} \leq C \left[\int_Q \frac{|w|^{\gamma'}}{m^{\gamma'-1}} dx + 1\right] \leq C(K+1). \tag{27}$$

Proof By (21), we see that

$$\int_Q m(-\Delta\phi)\, dx = \int_Q w \cdot \nabla\phi\, dx \leq \int_Q \left(\frac{|w|^{\gamma'}}{m^{\gamma'-1}}\right)^{\frac{1}{\gamma'}} m^{\frac{1}{\gamma}} |\nabla\phi|\, dx$$

$$\leq \left(\int_Q \frac{|w|^{\gamma'}}{m^{\gamma'-1}}\, dx\right)^{\frac{1}{\gamma}} \|m\|_{L^q(Q)}^{\frac{1}{\gamma}} \|\nabla\phi\|_{L^{r'}(Q)} \leq E^{\frac{1}{\gamma'}} \|m\|_{L^q(Q)}^{\frac{1}{\gamma}} \|\nabla\phi\|_{L^{r'}(Q)},$$

for any $\phi \in C_0^\infty(Q)$. Here above we used the notation $r' = \frac{r}{r-1}$.

Therefore, by Proposition 1 we get that

$$\|\nabla m\|_{L^r(Q)} \leq C\left(E^{\frac{1}{\gamma'}} \|m\|_{L^q(Q)}^{\frac{1}{\gamma}} + \|m\|_{L^r(Q)}\right).$$

Moreover, by interpolation, we get that $\|m\|_{L^r(Q)} \leq \|m\|_{L^q(Q)}^{\frac{1}{\gamma}} \|m\|_{L^1(Q)}^{\frac{1}{\gamma'}} = \|m\|_{L^q(Q)}^{\frac{1}{\gamma}}$. So, we conclude that

$$\|m\|_{W^{1,r}(Q)} \leq C(E^{\frac{1}{\gamma'}} + 1)\|m\|_{L^q(Q)}^{\frac{1}{\gamma}}. \tag{28}$$

Let $r^\star$ be such that $\frac{1}{r^\star} = \frac{1}{r} - \frac{1}{N}$ if $r < N$, and $r^\star = +\infty$ if $r \geq N$. Notice that by (5), $q < r^\star$. Therefore by Sobolev embedding, there exists C such that $\|m\|_{W^{1,r}(Q)} \geq C\|m\|_{L^q(Q)}$, and so, substituting in (28) we get that

$$\|m\|_{L^q(Q)} \leq C(E+1)$$

which in turns gives (26), again substituting in (28).

To obtain (27), we need to use also interpolation. Note that since $q < r^\star$, by interpolation we get

$$\|m\|_{L^q(Q)} \leq \|m\|_{L^1(Q)}^{1-\theta}\|m\|_{L^{r^\star}(Q)}^{\theta} \leq C(1+E^{\frac{\theta}{\gamma'}})\|m\|_{L^q(Q)}^{\frac{\theta}{\gamma}},$$

where θ is such that

$$\frac{1}{q} = 1 - \theta + \frac{\theta}{r^\star}.$$

We then obtain that

$$\|m\|_{L^q(Q)}^{q(1+\delta)} \leq C(1+E),$$

where

$$\delta = \frac{1}{q-1}\left(\frac{\gamma'+N}{N} - q\right) > 0$$

by (5).

Using the previous estimates, we deduce the existence of a minimizer of the energy in the class $\mathscr{K}$.

Theorem 3 *There exists* $(m, w) \in \mathscr{K}$ *such that*

$$\mathscr{E}(m, w) = \min_{(m,w)\in\mathscr{K}} \mathscr{E}.$$

Proof First of all observe that, by Proposition 5 and (25), there exists $C > 0$ such that, for every $(m, w) \in \mathcal{K}$,

$$\mathcal{E}(m, w) \geq C\|m\|_{L^q(Q)}^{(1+\delta)q} - C + \int_Q F(x, m)dx\,.$$

From this, recalling assumption (5) and the definition of F in (23), we conclude that there exists a constant K, depending on q, such that

$$\mathcal{E}(m, w) \geq C\|m\|_{L^q(Q)}^{(1+\delta)q} - C'\|m\|_{L^q(Q)}^q - C' \geq K.$$

Let $e := \inf_{(m,w)\in\mathcal{K}} \mathcal{E}(m, w)$. We fix a minimizing sequence (m_n, w_n). Therefore $\mathcal{E}(m_n, w_n) \leq e + 1$, for every n sufficiently large. Therefore, again by assumption (5), (25) and Proposition 5, we get

$$\int_Q \frac{|w_n|^{\gamma'}}{m_n^{\gamma'-1}} dx \leq C_L^{-1}(e + 1 - \int F(x, m_n)dx)) \leq C_L^{-1}(e + 1 - C + C\|m_n\|_{L^q(Q)}^q)$$

$$\leq C_L^{-1}\left(e + 1 + C' + K\left(\int_Q \frac{|w_n|^{\gamma'}}{m_n^{\gamma'-1}} dx + 1\right)^{\frac{1}{1+\delta}}\right).$$

This implies in particular that $\left(\int_Q \frac{|w_n|^{\gamma'}}{m_n^{\gamma'-1}} dx\right)$ is equibounded in n.

By Proposition 5 this implies that $\|m_n\|_{W^{1,r}(Q)} \leq C$, which in turn implies, recalling that $q < r^\star$ and using Sobolev embeddings, that up to a subsequence

$$m_n \to m \quad \text{strongly in } L^q(Q), \qquad m_n \rightharpoonup m \quad \text{weakly in } W^{1,r}(Q).$$

This implies in particular that $m_n \to m$ in L^1 and $\int_Q m dx = 1$.

Note that by Hölder inequality

$$\int_Q |w_n|^{\frac{\gamma' q}{\gamma'+q-1}} dx \leq \left(\int_Q \frac{|w_n|^{\gamma'}}{m_n^{\gamma'-1}} dx\right)^{\frac{q}{\gamma'+q-1}} \|m_n\|_{L^q(Q)}^{\frac{\gamma'-1}{q(\gamma'+q-1)}}. \tag{29}$$

This gives that w_n is equibounded in $L^{\frac{\gamma' q}{\gamma'+q-1}}(Q)$ and so, up to subsequences,

$$w_n \to w \quad \text{weakly in } L^{\frac{\gamma' q}{\gamma'+q-1}}(Q).$$

Note that the convergences are strong enough to assure that $(m, w) \in \mathcal{K}$. We conclude by the lower semicontinuity of the kinetic part of the functional (i.e. $\int mL(-w/m)$) and by the strong convergence in $L^q(Q)$ of m_n.

Regularization Procedure In order to pass from minimizers to classical solutions to (3), we need first to regularize the problem and then pass to the limit in the approximation. We consider the following approximation of the system (3).

$$\begin{cases} -\Delta u + H(\nabla u) + \lambda = f_\varepsilon[m](x), \\ -\Delta m - \operatorname{div}(m \nabla H(\nabla u)) = 0, \\ \int_Q m\, dx = 1, \end{cases} \tag{30}$$

where

$$f_\varepsilon[m](x) = f(\cdot, m \star \chi_\varepsilon) \star \chi_\varepsilon(x) = \int_Q \chi_\varepsilon(x-y) f\left(y, \int_Q m(z)\chi_\varepsilon(y-z)dz\right) dy$$

and χ_ε, for $\varepsilon > 0$, is a sequence of standard symmetric mollifiers approximating the unit.

We observe that $f_\varepsilon[m](x)$ is the L^2-gradient of a C^1 potential $F_\varepsilon : \mathbb{P}(Q) \to \mathbb{R}$, where $\mathbb{P}(Q)$ is the set of probability measures on Q (see [9]), defined as follows

$$F_\varepsilon[m] := \int_Q F(x, m \star \chi_\varepsilon(x))dx, \tag{31}$$

where F is given in (23). In particular there holds

$$F_\varepsilon[m'] - F_\varepsilon[m] = \int_0^1 \int_Q f_\varepsilon[(1-t)m + tm'](x)(m'-m)(x)dx \tag{32}$$

for all $m, m' \in L^1(Q)$ with $\int_Q m = \int_Q m' = 1$.

Note that by the properties of mollifiers and Jensen inequality, (5) still holds with constants independent of ε: there exists $C > 0$ such that

$$-C \int_Q m^q dx - K \le F_\varepsilon[m] \le C \int_Q m^q dx + K \tag{33}$$

for all functions $m \in L^1(Q)$, $m \ge 0$, $\int_Q m = 1$.

We associate to this approximate system (30) the energy

$$\mathscr{E}_\varepsilon(m, w) := \begin{cases} \displaystyle\int_Q mL\left(-\frac{w}{m}\right) dx + F_\varepsilon[m] & \text{if } (m, w) \in \mathscr{K}, \\ +\infty & \text{otherwise.} \end{cases} \tag{34}$$

Proposition 6 *For every $\varepsilon > 0$ there exists $(m_\varepsilon, w_\varepsilon) \in \mathscr{K}$ such that*

$$\mathscr{E}_\varepsilon(m_\varepsilon, w_\varepsilon) = \inf_{(m,w)\in\mathscr{K}} \mathscr{E}_\varepsilon(m, w).$$

Moreover, there exists $C > 0$ independent of ε such that

$$\|m_\varepsilon\|_{L^q(Q)} \le C \qquad \text{and} \qquad \|m_\varepsilon\|_{W^{1,r}(Q)} \le C \tag{35}$$

where r is as in Proposition 5.

Proof The proof follows exactly the same argument of Theorem 3, due to the fact that F_ε satisfies (33) and so we can use the estimates in Proposition 5. So, we get for every $\varepsilon > 0$ a minimizer.

Observe moreover that $\inf_{(m,w)\in\mathcal{K}} \mathcal{E}_\varepsilon(m,w) \le \mathcal{E}(1,0) = F_\varepsilon[1] \le C$, by (33). Therefore, following the same argument as in the proof of Theorem 3, we get that $\int_Q m_\varepsilon L\left(-\frac{w_\varepsilon}{m_\varepsilon}\right) dx \le C$, for some C independent of ε. So, applying again Proposition 5, we conclude.

In order to construct a solution to the Mean Field Game system (30), we associate to the energy in (34) a dual problem, using standard arguments in convex analysis. First of all, following [9], we pass to a convex problem.

Given a minimizer $(m_\varepsilon, w_\varepsilon)$ as obtained in Proposition 6 we introduce the following functional

$$J_\varepsilon(m,w) := \int_Q mL\left(-\frac{w}{m}\right) + f_\varepsilon[m_\varepsilon](x)m\,dx. \tag{36}$$

We claim that for $(m,w) \in \mathcal{K}$ we have that

$$\int_Q mL\left(-\frac{w}{m}\right) dx - \int_Q m_\varepsilon L\left(-\frac{w_\varepsilon}{m_\varepsilon}\right) \ge -\int_Q f_\varepsilon[m_\varepsilon](x)(m - m_\varepsilon)dx.$$

This can be proved as in [9, Proposition 3.1], using the convexity of L and the regularity of F. The idea is to consider, for every $\lambda \in (0,1)$, $m_\lambda := \lambda m + (1-\lambda)m_\varepsilon$ and the same definition for w_λ, and to observe that by minimality

$$\int_Q m_\lambda L\left(-\frac{w_\lambda}{m_\lambda}\right) dx - \int_Q m_\varepsilon L\left(-\frac{w_\varepsilon}{m_\varepsilon}\right) \ge -F[m_\lambda] + F[m_\varepsilon].$$

Then, using the convexity to estimate the left hand side and (32) on the right hand side, and finally sending $\lambda \to 0$, we get that

$$\min_{(m,w)\in\mathcal{K}} J_\varepsilon(m,w) = J_\varepsilon(m_\varepsilon, w_\varepsilon).$$

We now construct a solution to (30).

Theorem 4 *Let $(m_\varepsilon, w_\varepsilon)$ be a minimizer of J_ε as given by Proposition 6.*

Then $m_\varepsilon \in W^{1,p}(Q)$ *for all* $p > 1$, *and there exist* $\lambda_\varepsilon \in \mathbb{R}$ *and* $u_\varepsilon \in C^{2,\theta}(Q)$, *for all* $\theta \in (0,1)$ *such that* $(u_\varepsilon, \lambda_{,\varepsilon}, m_\varepsilon)$ *is a classical solution to the MFG system* (30). *Moreover* $w_\varepsilon = -m_\varepsilon \nabla H(\nabla u_\varepsilon)$.

Finally there exists C *independent of* ε *such that*

$$|\lambda_\varepsilon| \le C. \tag{37}$$

Proof The functional (36) is convex, so we can write the dual problem as follows, by standard arguments in convex analysis. First of all we write the following functional

$$\mathscr{A}(m,w,u,c) := \int_Q \left[mL\left(-\frac{w}{m}\right) + f_\varepsilon[m_\varepsilon](x)m + m\Delta u + \nabla u \cdot w - cm \right] dx + c.$$

We recall Proposition 6, in particular that $m_\varepsilon \in W^{1,r}(Q)$. From this, using the same argument as in (29) in the proof of Theorem 3, we have that $w_\varepsilon \in L^{\frac{\gamma' q}{\gamma'+q-1}}$. It is easy to observe that

$$J_\varepsilon(m_\varepsilon, w_\varepsilon) = \inf_{\{(m,w)\in(W^{1,r}(Q))\times L^{\frac{\gamma' q}{\gamma'+q-1}}(Q),\ \int_Q m=1, m\ge 0\}} \ \sup_{(u,c)\in C^2(Q)\times\mathbb{R}} \mathscr{A}(m,w,u,c), \tag{38}$$

where r is as in Proposition 5. So the infimum is actually a minimum.

Note that $\mathscr{A}(\cdot,\cdot,u,c)$ is convex and weakly lower semicontinuous, and $\mathscr{A}(m,w,\cdot,\cdot)$ is linear (so in particular concave). Hence we can use the min-max Theorem, see in particular [8, Thm 2.3.7], to interchange minimum and supremum, that is

$$\begin{aligned} &\min_{\{(m,w)\in W^{1,r}\times L^{\frac{\gamma' q}{\gamma'+q-1}},\ \int m=1,\ m\ge 0\}} \ \sup_{(u,c)\in C^2\times\mathbb{R}} \mathscr{A}(m,w,u,c) \\ &= \sup_{(u,c)\in C^2\times\mathbb{R}} \ \min_{\{(m,w)\in W^{1,r}\times L^{\frac{\gamma' q}{\gamma'+q-1}},\ \int m=1,\ m\ge 0\}} \mathscr{A}(m,w,u,c). \end{aligned} \tag{39}$$

Finally, thanks to the Rockafellar's Interchange Theorem [40] between infimum and integral (based on measurable selection arguments, and lower semicontinuity of the functional) we get, using the fact that H is the Legendre transform of L, that

$$\begin{aligned} &\min_{\{(m,w)\in W^{1,r}\times L^{\frac{\gamma' q}{\gamma'+q-1}},\ \int m=1, m\ge 0\}} \mathscr{A}(m,w,u,c) \\ &= \int_Q \min_{m\ge 0, w} \left[mL\left(-\frac{w}{m}\right) + f_\varepsilon[m_\varepsilon](x)m + m\Delta u + \nabla u \cdot w - cm \right] dx + c \\ &= \int_Q \min_{m\ge 0} m\left[-H(\nabla u) + \Delta u + f_\varepsilon[m_\varepsilon](x) - c\right] dx + c. \end{aligned}$$

Note that

$$\min_{m\geq 0} m\,[\Delta u - H(\nabla u) + f_\varepsilon[m_\varepsilon](x) - c] = \begin{cases} 0, & \text{if } \Delta u - H(\nabla u) + f_\varepsilon[m_\varepsilon](x) - c \geq 0, \\ -\infty, & \text{if } \Delta u - H(\nabla u) + f_\varepsilon[m_\varepsilon](x) - c < 0. \end{cases}$$

Therefore, from (38), (39) and (40) we get that

$$\begin{aligned} J_\varepsilon(m_\varepsilon, w_\varepsilon) &= \sup_{(u,c)\in C^2\times\mathbb{R}} \int_Q \min_{m\geq 0} m\,[-H(\nabla u) + \Delta u + f_\varepsilon[m_\varepsilon](x) - c]\,dx + c \\ &= \sup\left\{c \in \mathbb{R}\;|\exists u \in C^2, \text{ s.t. } -\Delta u + H(\nabla u) + c \leq f_\varepsilon[m_\varepsilon](x)\right\}. \end{aligned} \tag{40}$$

Since $f_\varepsilon[m_\varepsilon](x)$ is a smooth function, due to Theorem 1, such supremum is actually a maximum: there exist $\lambda_\varepsilon \in \mathbb{R}$ and a periodic function $u_\varepsilon \in C^{2,\theta}(Q)$, for every $\theta \in (0, 1)$ which is unique up to additive constants and solves

$$-\Delta u_\varepsilon + H(\nabla u_\varepsilon) + \lambda_\varepsilon = f_\varepsilon[m_\varepsilon](x). \tag{41}$$

So, equality (38) reads

$$\lambda_\varepsilon = J_\varepsilon(m_\varepsilon, w_\varepsilon) = \int_Q m_\varepsilon \left[L\left(-\frac{w_\varepsilon}{m_\varepsilon}\right) + f_\varepsilon[m_\varepsilon](x)\right] dx. \tag{42}$$

Therefore, recalling that $\int_Q m_\varepsilon = 1$ and using both (41) and (21) with test function u_ε, we obtain that

$$\begin{aligned} 0 &= \int_Q m_\varepsilon \left[L\left(-\frac{w_\varepsilon}{m_\varepsilon}\right) + f_\varepsilon[m_\varepsilon](x) - \lambda_\varepsilon\right] dx \\ &= \int_Q m_\varepsilon \left[L\left(-\frac{w_\varepsilon}{m_\varepsilon}\right) - \Delta u_\varepsilon + H(\nabla u_\varepsilon)\right] dx \\ &= \int_Q m_\varepsilon \left[L\left(-\frac{w_\varepsilon}{m_\varepsilon}\right) + \nabla u_\varepsilon \cdot \frac{w_\varepsilon}{m_\varepsilon} + H(\nabla u_\varepsilon)\right] dx. \end{aligned}$$

Using the fact that H is the Legendre transform of L and (24), we thus conclude that

$$\frac{w_\varepsilon}{m_\varepsilon} = -\nabla H(\nabla u_\varepsilon),$$

where $m_\varepsilon \neq 0$. Moreover, by the definition of L, we get that $w_\varepsilon = 0$ where $m_\varepsilon = 0$.

In particular, recalling (21), we find that m_ε is a solution of

$$-\Delta m_\varepsilon - \operatorname{div}(m_\varepsilon \nabla H(\nabla u_\varepsilon)) = 0, \qquad \text{with} \quad \int_Q m_\varepsilon = 1.$$

Since $\nabla H(\nabla u_\varepsilon) \in L^\infty(Q)$, by Theorem 1, we get by Proposition 2 that $m_\varepsilon \in W^{1,p}(Q)$ for every $p > 1$ and $m_\varepsilon > 0$. This implies that $(u_\varepsilon, \lambda_\varepsilon, m_\varepsilon)$ is a classical solution to (30).

Finally to prove the uniform bound on λ_ε, we use (42). Using (25) and (5) we get that there exists C, K independent of ε such that

$$\lambda_\varepsilon = J_\varepsilon(m_\varepsilon, w_\varepsilon) \geq -C\|m_\varepsilon\|^q_{L^q(Q)} - K,$$

from which we conclude recalling (35) that $\lambda_\varepsilon \geq -C$ for some constant independent of ε. Moreover by minimality and recalling (5) and the definition of f_ε, we get

$$\lambda_\varepsilon = J_\varepsilon(m_\varepsilon, w_\varepsilon) \leq J_\varepsilon(1, 0) = \int_Q f_\varepsilon[m_\varepsilon](x)dx \leq C\int_Q m_\varepsilon^{q-1}dx - C,$$

from which we conclude by recalling (35) and the fact that $\int_Q m_\varepsilon dx = 1$, that $\lambda_\varepsilon \leq C$, for some C independent of ε.

Passage to the Limit Now we can to pass to the limit as $\varepsilon \to 0$ in the system (30) in order to get a solution to (3). To do so, we need an a-priori estimate in L^∞ for the function m_ε; we provide here two slightly different proofs of such an estimate, both based on rescalings and elliptic regularity.

Proposition 7 *Let $(u_\varepsilon, \lambda_\varepsilon, m_\varepsilon)$ be a classical solution to the mean field game* (30), *as constructed in Theorem 4.*

Then there exists a constant C independent of ε such that

$$\|m_\varepsilon\|_{L^\infty(Q)} \leq C.$$

Proof We argue by contradiction, so we assume that

$$\max_Q m_\varepsilon = m_\varepsilon(x_\varepsilon) := M_\varepsilon \to +\infty.$$

Let

$$\mu_\varepsilon := M_\varepsilon^{-\beta} \qquad \beta = (q-1)\frac{\gamma - 1}{\gamma} > 0.$$

So, observe that $\mu_\varepsilon \to 0$ as $\varepsilon \to 0$.

We define the following rescaling

$$\begin{cases} v_\varepsilon(x) = \mu_\varepsilon^{\frac{2-\gamma}{\gamma-1}} u_\varepsilon(\mu_\varepsilon x + x_\varepsilon) \\ n_\varepsilon(x) = M_\varepsilon^{-1} m_\varepsilon(\mu_\varepsilon x + x_\varepsilon). \end{cases}$$

Note that $n_\varepsilon(0) = 1$ and $0 \le n_\varepsilon(x) \le 1$. We define

$$H_\varepsilon(q) = \mu_\varepsilon^{\frac{\gamma}{\gamma-1}} H(\mu_\varepsilon^{\frac{1}{1-\gamma}} q) \qquad \nabla H_\varepsilon(q) = \mu_\varepsilon \nabla H(\mu_\varepsilon^{\frac{1}{1-\gamma}} q).$$

Recalling (4) we get that

$$C_H|q|^\gamma - C_H^{-1} \le H_\varepsilon(q) \le C_H^{-1}(|q|^\gamma + 1) \qquad |\nabla H_\varepsilon(q))| \le C_H|q|^{\gamma-1}. \tag{43}$$

Moreover, we define

$$\tilde{f}_\varepsilon(x) = \mu_\varepsilon^{\frac{\gamma}{\gamma-1}} f[m_\varepsilon](x_\varepsilon + \mu_\varepsilon x).$$

Recalling that $0 \le m_\varepsilon \le M_\varepsilon$, (5) and the definition of f_ε we get that for all x

$$|\tilde{f}_\varepsilon(x)| \le \mu_\varepsilon^{\frac{\gamma}{\gamma-1}}(C + CM_\varepsilon^{q-1}) \le C + CM_\varepsilon^{q-1-\beta\frac{\gamma}{\gamma-1}} \le 2C \tag{44}$$

where we used the fact that $\mu_\varepsilon = M_\varepsilon^{-\beta}$ with $\beta = (q-1)\frac{\gamma-1}{\gamma}$. Finally, we let

$$\tilde{\lambda}_\varepsilon = \mu_\varepsilon^{\frac{\gamma}{\gamma-1}} \lambda_\varepsilon$$

and we observe that, by (37),

$$|\tilde{\lambda}_\varepsilon| \le C. \tag{45}$$

An easy computation shows that by rescaling, recalling that $\mu_\varepsilon = M_\varepsilon^{-\beta}$,

$$\begin{cases} \Delta v_\varepsilon(x) = \mu_\varepsilon^{\frac{\gamma}{\gamma-1}} \Delta u_\varepsilon(x\mu_\varepsilon + x_\varepsilon) \\ H_\varepsilon(\nabla v_\varepsilon(x)) = \mu_\varepsilon^{\frac{\gamma}{\gamma-1}} H(\nabla u_\varepsilon(x\mu_\varepsilon + x_\varepsilon)) \\ \Delta n_\varepsilon(x) = \mu_\varepsilon^{\frac{1}{\beta}+2} \Delta m_\varepsilon(x\mu_\varepsilon + x_\varepsilon) \\ \nabla H_\varepsilon(\nabla v_\varepsilon(x)) = \mu_\varepsilon \nabla H(\nabla u_\varepsilon(x\mu_\varepsilon + x_\varepsilon)) \\ \operatorname{div}(n_\varepsilon \nabla H_\varepsilon(\nabla v_\varepsilon(x))) = \mu_\varepsilon^{2+\frac{1}{\beta}} \operatorname{div}(m_\varepsilon(x\mu_\varepsilon + x_\varepsilon)\nabla H(\nabla u_\varepsilon(x\mu_\varepsilon + x_\varepsilon))). \end{cases}$$

Therefore, recalling that $(u_\varepsilon, \lambda_\varepsilon, m_\varepsilon)$ solves (30), we get that $(v_\varepsilon, n_\varepsilon)$ is a solution to

$$\begin{cases} -\Delta v_\varepsilon + H_\varepsilon(\nabla v_\varepsilon) + \tilde{\lambda}_\varepsilon = \tilde{f}_\varepsilon(x), \\ -\Delta n_\varepsilon - \operatorname{div}(n_\varepsilon \nabla H_\varepsilon(\nabla v_\varepsilon)) = 0. \end{cases}$$

Since by (37) and (44) both $\tilde{\lambda}_\varepsilon$ and $\tilde{f}_\varepsilon(x)$ are uniformly bounded in ε, by Theorem 1 we get that there exists a constant C independent of ε such that $\|\nabla v_\varepsilon\|_\infty \le$

C. So, using (43), have the existence of another C independent of ε such that $\|\nabla H_\varepsilon(\nabla v_\varepsilon)\|_\infty \le C$. This implies by Proposition 2 that n_ε is equibounded in every space $W^{1,p}(Q)$, which by Sobolev embedding, implies that n_ε is equibounded in ε in C^θ for every $\theta \in (0,1)$. Recall that $n_\varepsilon(0) = 1$. So, using equiboundness in C^θ for some θ, we obtain that there exist $\delta > 0$, $0 < R < 1$ independent of ε such that $\int_{B(0,R)} n_\varepsilon^q(x)dx \ge \delta > 0$. Therefore we get, recalling the rescaling,

$$0 < \delta \le \int_{B(0,R)} n_\varepsilon^q(x)dx \le \|n_\varepsilon\|_{L^q(Q)}^q = M_\varepsilon^{-q}\mu_\varepsilon^{-N}\|m_\varepsilon\|_{L^q(Q)}^q = M_\varepsilon^{-q+\beta N}\|m_\varepsilon\|_{L^q(Q)}^q.$$

Recalling the definition of β we get that, using assumption (5)

$$-q+\beta N = q\left[\frac{N(\gamma-1)}{\gamma} - 1\right] - \frac{N(\gamma-1)}{\gamma} < -\frac{\gamma}{N(\gamma-1)} < 0.$$

We recall that by (35) there exists a constant C such that $\|m_\varepsilon\|_{L^q(Q)} \le C$, so

$$0 < \delta \le M_\varepsilon^{-\frac{\gamma}{N(\gamma-1)}}\|m_\varepsilon\|_{L^q(Q)}^q \le M_\varepsilon^{-\frac{\gamma}{N(\gamma-1)}} C \to 0$$

which gives a contradiction.

Alternative Proof of Proposition 7 We first choose $p > N$ such that

$$q - 1 < \frac{1}{p}\frac{\gamma}{\gamma-1} < \frac{1}{N}\frac{\gamma}{\gamma-1}.$$

This can be done because (5) is in force. By the same hypothesis and properties of the convolution,

$$\|f_\varepsilon[m]\|_{L^\infty(Q)} \le \|f(\cdot, m\star\chi_\varepsilon)\|_{L^\infty(Q)} \le C\|m\star\chi_\varepsilon\|_{L^\infty(Q)}^{q-1} + K \le C_1(\|m\|_{L^\infty(Q)}^{q-1}+1).$$

By (37) and (4) we then infer that on Q

$$|-\Delta u_\varepsilon + C_H|\nabla u_\varepsilon|^\gamma| \le C_2(\|m\|_{L^\infty(Q)}^{q-1} + 1).$$

This gives the existence of C_3 such that $\|\nabla u_\varepsilon\|_{L^\infty(Q)} \le C_3(\|m_\varepsilon\|_{L^\infty(Q)}^{(q-1)/\gamma} + 1)$ by Remark 1.

We now turn to the Kolmogorov equation in (30). By (4) and Lemma 1,

$$\|m_\varepsilon\|_{W^{1,p}(Q)} \le C_4(\|\nabla H(\nabla u_\varepsilon)m_\varepsilon\|_{L^p(Q)} + \|m_\varepsilon\|_{L^p(Q)}) \le C_5(\|\nabla u_\varepsilon\|_{L^\infty(Q)}^{\gamma-1} + 1)\|m_\varepsilon\|_{L^p(Q)}.$$

By the previous L^∞ estimate on ∇u_ε and interpolation of the L^p norm of m between L^1 and L^∞ we get

$$\|m_\varepsilon\|_{W^{1,p}(Q)} \le C_6(\|m_\varepsilon\|_{L^\infty(Q)}^{(q-1)(\gamma-1)/\gamma} + 1)\|m_\varepsilon\|_{L^1(Q)}^{1/p}\|m_\varepsilon\|_{L^\infty(Q)}^{1-1/p}.$$

Recall that $\|m_\varepsilon\|_{L^1} = 1$; then, since $p > N$, by Sobolev embeddings (see, e.g., [26]) we obtain that

$$\|m_\varepsilon\|_{L^\infty(Q)} \leq C_7(\|m_\varepsilon\|_{L^\infty(Q)}^{(q-1)(\gamma-1)/\gamma} + 1)\|m_\varepsilon\|_{L^\infty(Q)}^{1-1/p},$$

and get the desired conclusion because $(q-1)(\gamma-1)/\gamma + 1 - 1/p < 1$ by the initial choice of p.

Note that the previous proof of Proposition 7 is based on a rescaling in the x variable; though such a rescaling is not appearing here, it is implicitly used also in this version of the proof, as it is involved in the gradient estimate in Remark 1.

We are now ready to prove the final result on existence of solutions to the MFG system (3).

Proof of Theorem 2 Let $(u_\varepsilon, \lambda_\varepsilon, m_\varepsilon)$ be a classical solution to (30) as in Theorem 4. We claim that, up to subsequences, $\lambda_\varepsilon \to \lambda$, $u_\varepsilon \to u$ uniformly in C^2, and $m_\varepsilon \to m$ in $W^{1,p}$ for every p and (u, λ, m) is a classical solution to (3).

By (37), up to subsequences we get that $\lambda_\varepsilon \to \lambda$. Now, using the L^∞ estimates on m_ε obtained in Proposition 7, and recalling the definition of f_ε and condition (5), we conclude that $\|f_\varepsilon[m_\varepsilon]\|_\infty \leq C$ for some C independent of ε. So, by Theorem 1, we get that $\|\nabla u_\varepsilon\|_\infty \leq K$ for some K uniform in ε. This implies, using elliptic regularity in the HJB equation, that for every $p > 1$, there exists a constant (just depending on p) such that $\|u\|_{W^{2,p}(Q)} \leq C$. So, by Sobolev embedding this gives also that u_ε are equibounded in $C^{1,\alpha}(Q)$ for every $\alpha \in (0, 1)$.

Using the equation for m_ε, and recalling that $\|\nabla H(\nabla u_\varepsilon)\|_\infty \leq C$ for some C independent of ε, we get by Proposition 2 $\|m\|_{W^{1,p}(Q)} \leq C$ uniformly, for every $p > 1$, which gives also uniform bounds of m_ε in $C^\theta(Q)$ for every $\theta \in (0, 1)$. Moreover, up to subsequences, we get that $m_\varepsilon \to m$ in $W^{1,p}(Q)$ for every p (and also uniformly).

Therefore $f_\varepsilon[m_\varepsilon](x)$ is equibounded in C^θ for every θ, and this in turn gives, by using the equation for u_ε again, that u_ε are equibounded in $C^{2,\theta}$. By Ascoli-Arzelá, we can extract a converging subsequence $u_\varepsilon \to u$ in C^2.

Note that the convergences are sufficiently strong to pass to the limit in the equations, so, we conclude that (u, λ, m) is a classical solution to (3). Finally, the fact that $m > 0$ comes for Proposition 2.

6 Uniqueness of Solutions to the MFG System

In this section we consider the case in which the potential term $f(x, m)$ is nondecreasing with respect to m. We start by showing that in this case solutions to (3) are actually minimizers of the energy in (22).

Proposition 8 *Assume that $f(x, \cdot)$ is nondecreasing. Then if (u, λ, m) is a solution to* (3), *$(m, -m\nabla H(\nabla u))$ is a minimizer of* (22).

Proof Let (u, λ, m) be a solution to (3). Then by Proposition 2, $m > 0$. Define $w := -m\nabla H(\nabla u)$. Then, recalling the regularity of solutions to (3), we have that $w \in L^\infty(Q)$. In particular, we obtain that $(m, w) \in \mathcal{K}$.

Moreover, if we multiply the first equation in (3) by m, the second by u, integrate in Q and subtract one from the other, we get

$$0 = \lambda + \int_Q [-mL(\nabla H(\nabla u)) - mf(x, m)]\, dx. \tag{46}$$

Observe that if $f(x, \cdot)$ is nondecreasing, then $F(x, \cdot)$ is a convex function. Let $(m', w') \in \mathcal{K}$, and we aim to prove that

$$\mathcal{E}(m, -m\nabla H(\nabla u)) \leq \mathcal{E}(m', w'). \tag{47}$$

For this, first of all observe that, by the convexity of F, we get

$$\begin{aligned}\mathcal{E}(m', w') &= \int_Q m'L\left(-\frac{w'}{m'}\right) + F(x, m')\, dx \\ &\geq \int_Q m'L\left(-\frac{w'}{m'}\right) + F(x, m) + f(x, m)(m' - m)\, dx.\end{aligned} \tag{48}$$

Using the fact that u is a solution to the first equation in (3), the duality between H and L, in particular by (24), and the fact that (m', w') satisfies (7), we obtain

$$\begin{aligned}\int_Q m'L\left(-\frac{w'}{m'}\right) + f(x, m)m'\, dx &= \int_Q m'\left[L\left(-\frac{w'}{m'}\right) - \Delta u + H(\nabla u) + \lambda\right] dx \\ &\geq \int_Q \left[-m'\Delta u - \nabla u \cdot w' + \lambda m'\right] dx \\ &= \lambda.\end{aligned} \tag{49}$$

By putting together (48) and (49), and recalling (46), we obtain

$$\begin{aligned}\mathcal{E}(m', w') &\geq \lambda + \int_Q [F(x, m) - f(x, m)m]\, dx \\ &= \int_Q mL\left(\nabla H(\nabla u)\right)) + F(x, m)\, dx \\ &= \mathcal{E}(m, -m\nabla H(\nabla u)),\end{aligned}$$

which gives (47), and completes the proof of Proposition 8.

Under the same assumptions, we have uniqueness of solutions to MFG systems.

Theorem 5 *Assume that the map $m \mapsto f(x, m)$ is nondecreasing for all $x \in Q$. Then the system in* (3) *admits a unique solution (u, λ, m), where u is defined up to addition of constants.*

Proof Note that under the assumptions of Theorem 5, the energy (22) is convex. Moreover, since $mL(-w/m)$ is strictly convex on the set where $m > 0$ (see Proposition 4), we have that if (m, w) and (m', w') are two minimizers with $m > 0, m' > 0$, then $w = w'$.

Suppose by contradiction that there exist two different solutions (u_1, λ_1, m_1) and (u_2, λ_2, m_2). Recall that by Proposition 2, $m_1, m_2 > 0$. By Proposition 8, we get that $(m_1, -m_1 \nabla H(\nabla u_1))$ and $(m_2, -m_2 \nabla H(\nabla u_2))$ are both minimizers. Then $m_1 \nabla H(\nabla u_1) = m_2 \nabla H(\nabla u_2)$. Using the second equation in (3) and recalling Lemma 1, we get that $m_1 = m_2$. This implies, by Theorem 1, that $\lambda_1 = \lambda_2$ and that $u_1 = u_2 + C$ for some constant C. The proof of Theorem 5 is thus complete.

Acknowledgements This work has been supported by the INdAM Intensive Period "Contemporary Research in elliptic PDEs and related topics". The authors are partially supported by the Fondazione CaRiPaRo Project "Nonlinear Partial Differential Equations: Asymptotic Problems and Mean-Field Games".

References

1. S. Agmon, The L^p approach to the Dirichlet problem. I. Regularity theorems. Ann. Scuola Norm. Sup. Pisa (3) **13**, 405–448 (1959)
2. M. Bardi, E. Feleqi, The derivation of ergodic mean field game equations for several populations of players. Dyn. Games Appl. **3**(4), 523–536 (2013)
3. M. Bardi, E. Feleqi, Nonlinear elliptic systems and mean field games. NoDEA Nonlinear Differ. Equ. Appl. **23**, 23–44 (2016)
4. G. Barles, P. E. Souganidis, Space-time periodic solutions and long-time behavior of solutions to quasi-linear parabolic equations. SIAM J. Math. Anal. **32**(6), 1311–1323 (2001)
5. J.-D. Benamou, G. Carlier, F. Santambrogio, *Variational Mean Field Games*, ed. by Bellomo, Degond, Tadmor. Active Particles, vol. 1 (Springer, Berlin, 2017)
6. A. Bensoussan, *Perturbation Methods in Optimal Control* (John Wiley & Sons, Hoboken, 1988)
7. A. Bensoussan, J. Y. Frehse, *Mean Field Games and Mean Field Type Control* (Springer, Berlin, 2013)
8. J. Borwein, J. Vanderwerff, *Convex Functions: Constructions, Characterizations and Counterexamples* (Cambridge University Press, Cambridge, 2010)
9. A. Briani, P. Cardaliaguet, Stable solutions in potential mean field game systems. Nonlinear Differ. Equ. Appl. **25**, 1 (2018)
10. P. Cardaliaguet, Notes on mean-field games (2011). Available at https://www.ceremade.dauphine.fr/ cardaliaguet/MFG20130420.pdf
11. P. Cardaliaguet, P.-J. Graber, Mean field games systems of first order. ESAIM Control Optim. Calc. Var. **21**, 690–722 (2015)
12. P. Cardaliaguet, P.J. Graber, A. Porretta, D. Tonon, Second order mean field games with degenerate diffusion and local coupling. NoDEA Nonlinear Differ. Equ. Appl. **22**(5), 1287–1317 (2015)

13. P. Cardaliaguet, A.R. Mészáros, F. Santambrogio, First order mean field games with density constraints: pressure equals price. SIAM J. Control. Optim. **54**(5), 2672–2709 (2016)
14. P. Cardaliaguet, A. Porretta, D. Tonon, A segregation problem in multi-population mean field games, in *Advances in Dynamic and Mean Field Games. ISDG 2016. Annals of the International Society of Dynamic Games* ed. by J. Apaloo, B. Viscolani, vol. 15 (Birkhäuser, Basel, 2017), pp. 49–70
15. P. Cardaliaguet, F. Delarue, J.-M. Lasry, P.-L. Lions, The master equation and the convergence problem in mean field games. Preprint arXiv:1509.02505
16. A. Cesaroni, M. Cirant, Concentration of ground states in stationary mean-field games systems. Anal. PDE **12**(3), 737–787 (2019)
17. A. Cesaroni, M. Cirant, S. Dipierro, M. Novaga, E. Valdinoci, On stationary fractional mean field games. J. Math. Pures Appl. **122**, 1–22 (2019)
18. M. Cirant, On the solvability of some ergodic control problems in $\mathbb{R}^d$. SIAM J. Control Optim. **52**(6), 4001–4026 (2014)
19. M. Cirant, Multi-population mean field games systems with Neumann boundary conditions. J. Math. Pures Appl. (9) **103**(5), 1294–1315 (2015)
20. M. Cirant, A generalization of the Hopf-Cole transformation for stationary mean field games systems. C.R. Math. **353**(9), 807–811 (2015)
21. M. Cirant, Stationary focusing mean-field games. Commun. Part. Diff. Eq. **41**(8), 1324–1346 (2016)
22. M. Cirant, D. Tonon, Time-dependent focusing mean-field games: the sub-critical case. J. Dyn. Diff. Equat. **31**(1), 49–79 (2019)
23. M. Cirant, G. Verzini, Bifurcation and segregation in quadratic two-populations mean field games systems. ESAIM Control Optim. Calc. Var. **23**, 1145–1177 (2017)
24. I. Ekeland, R. Temam, *Convex Analysis and Variational Problems* (North-Holland Publishing Co., Amsterdam-Oxford, 1976)
25. E. Feleqi, The derivation of ergodic mean field game equations for several populations of players. Dyn. Games Appl. **3**(4), 523–536 (2013)
26. D. Gilbarg, N.S. Trudinger, *Elliptic Partial Differential Equations of Second Order*, Classics in Mathematics. (Springer-Verlag, Berlin, 2001)
27. D.A. Gomes, J. Saude, Mean field games models–a brief survey. Dyn. Games Appl. **4**(2), 110–154 (2014)
28. D.A. Gomes, S. Patrizi, V. Voskanyan, On the existence of classical solutions for stationary extended mean field games. Nonlinear Anal. **99**, 49–79 (2014)
29. D.A. Gomes, L. Nurbekyan, M. Prazeres, One-dimensional stationary mean-field games with local coupling. Dyn. Games Appl. **8**(2), 315–351 (2018)
30. D.A. Gomes, E.A. Pimentel, V. Voskanyan, *Regularity Theory for Mean-Field Game Systems* (Springer, Berlin, 2016)
31. M. Huang, R. Malhamé, P. Caines, Large population stochastic dynamic games: closed-loop McKean-Vlasov systems and the nash certainty equivalence principle. Commun. Inf. Syst. **6**(3), 221–251 (2006)
32. R. Khasminskii, Stochastic stability of differential equations, in *Stochastic Modelling and Applied Probability*, vol. 66 (Springer, Berlin, 2012)
33. J.-M. Lasry, P.-L. Lions, Nonlinear elliptic equations with singular boundary conditions and stochastic control with state constraints. Math. Ann. **283**(4), 583–630 (1989)
34. J.-M. Lasry, P.-L. Lions, Jeux à champ moyen. II. Horizon fini et contrôle optimal. C. R. Math. Acad. Sci. Paris **343**(10), 679–684 (2006)
35. J.-M. Lasry, P.-L. Lions, Jeux à champ moyen. I. Le cas stationnaire. C. R. Math. Acad. Sci. Paris **343**(9), 619–625 (2006)
36. J.-M. Lasry, P.-L. Lions, Mean field games. Japan. J. Math. **2**, 229–260 (2007)
37. P.-L. Lions, *Cours au Collége de France*, www.college-de-france.fr
38. A.R. Mészáros, F.J. Silva, A variational approach to second order mean field games with density constraints: the stationary case. J. Math. Pures Appl. (9) **104**(6), 1135–1159 (2015)

39. E. Pimentel, V. Voskanyan, Regularity for second-order stationary mean-field games. Indiana Univ. Math. J. **66**, 1–22 (2017)
40. R.T. Rockafellar, Integral functionals, normal integrands and measurable selections. Lect. Notes Math. **543**, 157–207 (1976)

Flatness Results for Nonlocal Phase Transitions

Eleonora Cinti

Abstract We consider a nonlocal version of the Allen-Cahn equation, which models phase transitions problems. In the classical setting, the connection between the Allen-Cahn equation and the classification of entire minimal surfaces is well known and motivates a celebrated conjecture by E. De Giorgi on the one-dimensional symmetry of bounded monotone solutions to the (classical) Allen-Cahn equation up to dimension 8. In this work, we present some recent results in the study of the nonlocal analogue of this phase transition problem. In particular we describe the results obtained in several contributions where the classification of certain entire bounded solutions to the fractional Allen-Cahn equation has been obtained. Moreover we describe the connection between the fractional Allen-Cahn equation and the fractional perimeter functional, and we present also some results in the classifications of nonlocal minimal surfaces.

Keywords Fractional Laplacian · Symmetry properties · Nonlocal minimal surfaces

1 Introduction

In this paper we present some recent results concerning the classification of certain solutions to the fractional Allen-Cahn equation

$$(-\Delta)^s u = u - u^3 \quad \text{in } \mathbb{R}^n, \tag{1.1}$$

where s is a real parameter in $(0, 1)$. More precisely, we are interested in the analogue, for problem (1.1), of a well known conjecture by E. De Giorgi for solutions of the classical Allen-Cahn equation.

E. Cinti (✉)
Dipartimento di Matematica, Università di Bologna, Bologna, Italy
e-mail: eleonora.cinti5@unibo.it

S. Dipierro (ed.), *Contemporary Research in Elliptic PDEs and Related Topics*, Springer INdAM Series 33, https://doi.org/10.1007/978-3-030-18921-1_6

In 1978, De Giorgi conjectured that the level sets of every bounded solution of

$$-\Delta u = u - u^3 \quad \text{in } \mathbb{R}^n, \tag{1.2}$$

which is monotone in one direction, must be hyperplanes at least if $n \leqslant 8$. That is, such solutions depend only on one Euclidean variable.

The original motivation for this conjecture was given by a classical result in the Calculus of Variations due to Modica and Mortola [35], who proved that, after a suitable rescaling, the energy functional associated to (1.2), Γ-converges to the perimeter functional (see Sect. 2 for more details). Moreover, the classification of area-minimizing surfaces was known: any area-minimizing set in the whole $\mathbb{R}^n$ is necessarily flat if $n \leqslant 7$. The dimension 7 is optimal, indeed in $\mathbb{R}^8$ there exists an area-minimizing singular cone, the Simons cone, defined in the following way:

$$\mathcal{C} := \{(x_1, \dots, x_8) \in \mathbb{R}^8 \mid x_1^2 + \cdots + x_4^2 = x_5^2 + \cdots + x_8^2\}.$$

A related result concerns the classification of minimal graphs (the so-called Bernstein problem): any area-minimizing graph in $\mathbb{R}^n$ is necessarily a hyperplane if $n \leqslant 8$.

Coming back to the Allen-Cahn equation, by the Modica-Mortola result, one knows that the level sets of certain solutions to $-\Delta u = u - u^3$ are asymptotically area-minimizing surfaces. Moreover, if we assume the solution to be monotone in some direction, we have that the level sets are graphs. Hence, by the previous result on the classification of entire minimal graph, we know that the level sets of bounded monotone solutions to the Allen-Cahn equation are asymptotically flat. The De Giorgi conjecture asserts that they are indeed flat, not only asymptotically. The fact that a function u has level sets which are parallel hyperplanes, means that u depends only on one Euclidean direction (the direction perpendicular to all these hyperplanes). When this happens, we say that u has one-dimensional symmetry, or is 1-D, for short.

We consider now the fractional version of the Allen-Cahn equation (1.1) and we are interested in the validity of the analogue of the De Giorgi conjecture. First of all, a natural question is whether a Modica-Mortola type result is valid for the energy functional associated to (1.1) and whether there is a natural connection with an area-minimizing problem. The answer to this question was given by Savin and Valdinoci in [42]: they proved that, after a suitable rescaling the energy associated to (1.1) Γ-converges to the classical perimeter functional if $1/2 \leqslant s < 1$ and to the so-called *nonlocal perimeter* if $0 < s < 1/2$. Hence, when $1/2 \leqslant s < 1$, one expects the analogue of the De Giorgi conjecture to be true up to dimension $n = 8$ as in the classical setting. While, when $0 < s < 1/2$, the level sets of solutions to (1.1) looks, at large scales, like *nonlocal minimal surfaces*.

The nonlocal (or fractional) s-perimeter functional was introduced by Caffarelli et al. in [18] (see formula (4.1) in Sect. 4) and the classification of minimizers for this functional is still widely open. In [43], Savin and Valdinoci proved that any s-minimal set in $\mathbb{R}^2$ is necessarily an half-plane. Moreover in [30], Figalli and Valdinoci addressed the nonlocal analogue of the Bernstein problem and they obtained flatness of s-minimal graphs in $\mathbb{R}^3$. These are the only known results about the classification for s-minimal surfaces, except for some asymptotic results that are valid only for s sufficiently close to $1/2$ (see Sect. 4 for all the precise results).

This lack of information in large dimensions for the geometric problem, is reflected on the PDE side, where the De Giorgi conjecture for s below $1/2$ is still open in dimensions $n > 3$. We recall here the main references for the fractional De Giorgi conjecture: it has been proven in dimension $n = 2$ and for any $s \in (0, 1)$ in [10, 43], in dimension $n = 3$ for $s \in [1/2, 1)$ in [7, 8], in dimension $n = 3$ for $s \in (0, 1/2)$ in [25] and in the forthcoming paper [13], in dimensions $4 \leqslant n \leqslant 8$ for $1/2 \leqslant s < 1$ (under an additional assumption on the limits at infinity of the solution) in [40, 41].

We comment now on the proof of the De Giorgi conjecture for the fractional problem (1.1). As in the classical setting, two different approaches have been used to deal with the low or high dimensional case. Indeed, for the classical Allen-Cahn equation, the proof of the conjecture in dimensions $n = 2, 3$ is a purely PDE proof, which relies on some energy estimates and a Liouville-type argument, but never uses the classification for area-minimizing surfaces (see [2, 3, 33]). Instead, for $4 \leqslant n \leqslant 8$, the fact that the only area-minimizing surfaces in the whole $\mathbb{R}^n$ are hyperplanes if $n \leqslant 7$ plays a crucial role. The proof of the conjecture in dimensions larger than 3 was given by Savin in [39] who, using the so-called improvement of flatness, proved that if the level sets of certain solutions are asymptotically flat, then the solution needs to be one-dimensional. In [39] all the ingredients needed in the proofs (energy estimates, density estimates, improvement of flatness) require the solution to be a minimizer for the associated energy functional. A first result in Savin's paper is, in fact, the validity of the De Giorgi conjecture for minimizers in dimensions $n \leqslant 7$. On the geometric side, this statement corresponds to the fact that any area-minimizing surface in the whole $\mathbb{R}^n$ is flat if $n \leqslant 7$. The original conjecture by E. De Giorgi was for bounded monotone solutions (which in general are not minimizers without further assumption). A second result in [39] asserts that if u is a bounded monotone solution for the classical Allen-Cahn equation in $\mathbb{R}^n$ (e.g. $u_{x_n} > 0$), such that $\lim_{x_n \to \pm\infty} u(x) = \pm 1$, then u is 1-D for $n \leqslant 8$. This statement corresponds, on the geometric side, to the fact that any area-minimizing graph is flat up to dimension $n = 8$. We stress that the additional assumption on the limits at infinity are needed to ensure that the solution is a minimizer. The conjecture in dimension $4 \leqslant n \leqslant 8$ for monotone solutions without the limits assumption is still open.

Concerning the fractional case, when $n = 2$ for any $0 < s < 1$, and when $n = 3$ for $1/2 \leqslant s < 1$ the pure PDE proof, which uses the ideas developed in [3] for the classical conjecture in the low-dimensional case, still works (see [7, 8, 10, 43]). While for treating the case $n = 3$ and $0 < s < 1/2$ (see [25, 26]), and the case $4 \leqslant n \leqslant 8$ with $1/2 < s < 1$ (see [40, 41]) one needs to use the idea of Savin based on an improvement of flatness result. As said above, in this approach, the classification for nonlocal minimal surfaces is crucial, that is why when $0 < s < 1/2$ and $n > 3$ the conjecture is still open.

We conclude this section, commenting on the class of solutions for which one expects 1-D symmetry to hold true. As already mentioned, the original conjecture was for monotone solutions, which corresponds to having area-minimizing graphs on the geometric side. For these solutions the conjecture is true up to dimension 8 for $s \in [1/2, 1]$ with the additional assumptions on the limits at infinity (as we will see in Sect. 3, when $n = 3$ this additional assumption is not needed). On the other hand the problem has a variational structure and it is natural to ask the same question for minimizers of the energy: in this version the conjecture is true up to dimension 7 for $s \in [1/2, 1]$. Another class of solutions for which one expects the conjecture to hold true is the one of *stable* solutions (here stability is in the variational sense, that is one requires the second variation of the energy functional to be nonnegative). For stable solutions, even the conjecture for the classical Allen-Cahn equation is still open in all dimensions $n > 2$. This lack of information for the PDE is reflected at the geometric level: it is still an open question whether *stable* minimal surfaces are necessarily hyperplanes in dimension $3 < n \leqslant 7$ (see Sect. 4 for the precise references). We stress that, instead, stable minimal *cones* are completely classified: they are hyperplanes in dimensions $n \leqslant 7$. In $\mathbb{R}^8$ the Simons cone is an example of stable singular cone. One would expect that this classification of stable cones would imply an analogue classification for any stable surfaces and, on the PDE side, the 1-D symmetry of stable solutions. One of the main obstruction in classifying stable objects is given by the lack of energy estimates. Surprisingly, in the nonlocal setting, some techniques have been recently developed to attack the study of stable objects and some results (such as energy and BV estimates) have been obtained. The analogue results in the local setting are still unknown and the study of stable solutions to the Allen-Cahn equation (and of stable classical minimal surfaces) is still widely open. We will address the classification of stable objects for both the Allen-Cahn equation and the theory of minimal surfaces, in the last section.

The paper is organized as follows:

- In Sect. 2, we describe the connections between the fractional Allen-Cahn equation and the theory of local/nonlocal minimal surfaces. The main results of this section have been obtained in [42, 45];
- Section 3 deals with the De Giorgi conjecture for the fractional Allen-Cahn equation. In particular we address the low dimensional case, giving a sketch of the proofs of the results contained in [7, 8].
- In Sect. 4, we describe some recent results concerning the classification of nonlocal minimal surfaces contained in [12, 20, 43];
- In Sect. 5 we present some very recent results on the classification of stable objects, both for the fractional Allen-Cahn equation and for the fractional perimeter and we conclude with some related open questions.

2 Γ-Convergence Results for Nonlocal Phase Transitions

The Classical Setting We start by describing a classical model for phase transitions and the rigorous mathematical results which explains the connection between the Allen-Cahn equation and the theory of minimal surfaces. For this part, we refer to [1] and references therein.

Let us consider a container, represented by a bounded and regular subset Ω of $\mathbb{R}^3$, which is filled with two phases of the same fluid. The configuration of the system is described by a function u. There are two different models for the phase transition phenomenon, depending whether the transition between the two phases is given by a separating interface or is a continuous transition which occurs in a thin layer.

In the first model, usually called *sharp-interface model*, the configuration function u only takes two values, e.g. $+1$ and -1, which correspond to the two pure phases. The classical theory of phase transitions, assume that at equilibrium the two fluids arrange themselves in order to minimize the area of the separating interface, that is the measure of the jump set of u. Hence, in this model, the energy of the system is a pure interface energy given by

$$F(u) = \sigma \mathcal{H}^2(S_u), \tag{2.1}$$

where S_u denotes the jump set of u, $\mathcal{H}^2$ the two-dimensional Hausdorff measure, and σ is a parameter representing the surface tension between the two phases.

Imposing a volume constraint, the space of all admissible configurations is given by $A = \{u : \Omega \to \{-1, 1\} : \int_\Omega u = V\}$, where $-|\Omega| < V < |\Omega|$ and the configuration of the system at equilibrium is obtained by minimizing F over A.

The second model, often called the *diffusive model*, was proposed by J.W. Cahn and J.E. Hilliard and allows a fine mixture of the two phases, which corresponds to the fact that the configuration function u can take values in the whole interval $[-1, 1]$. Now, the space of all admissible configurations is given by $A = \{u : \Omega \to [-1, 1] : \int_\Omega u = V\}$ and the energy has the following form:

$$\mathcal{E}_\varepsilon(u, \Omega) = \int_\Omega \left(\frac{\varepsilon^2}{2} |\nabla u|^2 + W(u) \right) dx, \tag{2.2}$$

where $\varepsilon > 0$ is a small parameter and W is a continuous function which vanishes only at -1 and 1 and is positive elsewhere (usually called a double-well potential). We observe that the Dirichlet term and the potential one are in competition, indeed $W(u)$ forces the configurations to take values close to -1 and 1 and hence favourites the separation of the two phases, while the first term in the energy penalizes the spatial inhomogeneity of u. For small ε the potential term prevails, and the minimum of $\mathcal{E}_\varepsilon$ is attained by a function u_ε which takes values close to -1 and 1 and the transition between these two phases happens in a thin layer of thickness ε.

The Euler-Lagrange equation for the energy (2.2) is given by the (rescaled) Allen-Cahn equation

$$\varepsilon^2 \Delta u = W'(u).$$

A rigorous mathematical justification of the connection between the sharp-interface and the diffuse models was given by Modica and Mortola in [35]. They proved that, when $\varepsilon \to 0$, the rescaled functional $\varepsilon^{-1}\mathcal{E}_\varepsilon$ Γ-converges to F defined by (2.1), and hence minimizers of $\mathcal{E}_\varepsilon$ converges to minimizers of F (this Γ-convergence result holds in any dimension n). The right setting for functions u_0 obtained as limits of minimizers u_ε of $\mathcal{E}_\varepsilon$ is the one of BV functions and the limit functional is the perimeter in the sense of De Giorgi of the sublevelsets of u_0 (which agrees with the $(n-1)$-dimensional Hausdorff measure for smooth objects). The Modica-Mortola theorem establishes convergence, in the L^1-sense, for sequences of minimizers u_ε to a BV function taking values in $\{-1, 1\}$, whose jump set is an area-minimizing surface. Later, in [14] Caffarelli and Cordoba proved that actually the convergence of minimizers is not only in L^1 but in the Hausdorff distance sense.

The Nonlocal Setting We pass now to describe what happens when one replaces the standard Dirichlet energy with a nonlocal term which takes into account long range interactions. For a bounded subset Ω of $\mathbb{R}^n$, we consider an energy functional of the form

$$\mathcal{E}^s(u, \Omega) = \frac{1}{4} \iint_{(\mathbb{R}^n \times \mathbb{R}^n)\setminus(\Omega^c \times \Omega^c)} \frac{|u(x) - u(\bar{x})|^2}{|x - \bar{x}|^{n+2s}}\, dx\, d\bar{x} + \int_\Omega W(u)\, dx, \tag{2.3}$$

where Ω^c denotes the complement of Ω.

Observe that the set of integration in the Dirichlet term is given by $(\mathbb{R}^n \times \mathbb{R}^n) \setminus (\Omega^c \times \Omega^c)$. This term represents to contribution of the H^s-seminorm of u in Ω and takes into account the interactions between all possible couple of points x, $\bar{x}$ except the ones for which both x and $\bar{x}$ do not belong to Ω. The reason for this choice is that the energy in the whole space $\mathbb{R}^n \times \mathbb{R}^n$ could not be finite, and the notion of minimality that we consider is the one with fixed "boundary" data, that is competitors must agree with the minimizer u in the complement of Ω, according to the following definition.

Definition 2.1 We say that a function u is a *minimizer* for the energy $\mathcal{E}^s$ if

$$\mathcal{E}^s(u, \Omega) \leqslant \mathcal{E}^s(w, \Omega), \quad \text{for any } w \text{ such that } u \equiv w \text{ in } \Omega^c.$$

In [42], Savin and Valdinoci proved a Γ-convergence result for the (suitably rescaled) functional $\mathcal{E}^s$, that is the analogue of the Modica-Mortola theorem in the nonlocal setting. Interestingly, the Γ-limit of $\mathcal{E}^s$ is different depending whether s is below or above $1/2$. Before stating the main result in [42], we introduce all the necessary ingredients.

In the sequel, W will denote a potential with a double-well structure, i.e. we assume that

$$W : [-1, 1] \to [0, +\infty), \quad W \in C^2([-1, 1]), \quad W > W(\pm 1) = 0 \text{ in } (-1, 1)$$

$$W'(\pm 1) = 0 \quad \text{and} \quad W''(\pm 1) > 0.$$

The class of admissible functions is given by

$$X = \{u \in L^\infty(\mathbb{R}^n) \mid ||u||_\infty \leqslant 1\}.$$

We set

$$\mathcal{E}^s_\varepsilon(u, \Omega) = \frac{1}{4}\varepsilon^{2s} \iint_{\mathbb{R}^n \times \mathbb{R}^n \setminus (\Omega^c \times \Omega^c)} \frac{|u(x) - u(\bar{x})|^2}{|x - \bar{x}|^{n+2s}} \, dx \, d\bar{x} + \int_\Omega W(u) \, dx,$$

and we consider the functional

$$\mathcal{F}^s_\varepsilon(u, \Omega) = \begin{cases} \varepsilon^{-2s}\mathcal{E}^s_\varepsilon(u, \Omega) & \text{if } 0 < s < 1/2 \\ |\varepsilon \log \varepsilon|^{-1}\mathcal{E}^s_\varepsilon(u, \Omega) & \text{if } s = 1/2 \\ \varepsilon^{-1}\mathcal{E}^s_\varepsilon(u, \Omega) & \text{if } 1/2 < s < 1. \end{cases}$$

We can now state the main result in [42].

Theorem 2.2 (Theorem 1.4 in [42]) *Let $s \in (0, 1)$ and Ω be a bounded Lipschitz domain of $\mathbb{R}^n$.*

Then,

$$\mathcal{F}^s_\varepsilon(u, \Omega) \xrightarrow{\Gamma} \mathcal{F}(u, \Omega) \quad as \quad \varepsilon \to 0,$$

where $\mathcal{F}(u, \Omega)$ is defined as follows:

$$\text{if } 0 < s < 1/2 \qquad \mathcal{F}(u, \Omega) = \begin{cases} \operatorname{Per}_s(E) & \text{if } u_{|\Omega} = \chi_E - \chi_{E^c} \text{ for some } E \subset \Omega \\ +\infty & \text{otherwise,} \end{cases}$$

$$\text{if } 1/2 \leqslant s < 1 \qquad \mathcal{F}(u, \Omega) = \begin{cases} c_* \operatorname{Per}(E) & \text{if } u_{|\Omega} = \chi_E - \chi_{E^c} \text{ for some } E \subset \Omega \\ +\infty & \text{otherwise,} \end{cases}$$

and c_* is a constant depending on n, s and W.

The s-perimeter Per_s will be precisely defined in Sect. 4 below. This Γ-convergence theorem, together with a compactness result (see Theorem 1.5 in [42]), implies that if u_ε is a sequence of minimizers for $\mathcal{F}_\varepsilon^s$ such that $\mathcal{F}_\varepsilon^s$ are uniformly bounded as $\varepsilon \to 0$, then there exists a subsequence, that we still call u_ε, which converges in L^1 to a function $u_0 = \chi_E - \chi_{E^c}$ where E is a minimizer for the fractional perimeter Per_s in Ω if $0 < s < 1/2$ and a minimizer for the classical perimeter Per in Ω if $1/2 \leqslant s < 1$.

As for the case of the classical phase transition model, one can prove that the convergence is not just in L^1 but in a stronger sense. In [45], Savin and Valdinoci proved some density estimates for minimizers of $\mathcal{E}^s$ which gives a bound on the measure of the volume occupied by the level sets of a minimizer in a ball. As a consequence of the density estimates, we have that level sets of minimizers of $\mathcal{E}_\varepsilon^s$ converge locally uniformly as $\varepsilon \to$ to a nonlocal s-minimal surface when $0 < s < 1/2$, and to a classical minimal surface when $1/2 \leqslant s < 1$.

As already explained in Sect. 1, this convergence results motivate the analogue of the De Giorgi conjecture for certain solutions (monotone solutions and minimizers) to the fractional Allen-Cahn equation, which is the Euler-Lagrange equation of the energy functional $\mathcal{E}_\varepsilon^s$. More precisely, since the Γ-limit of $\mathcal{E}_\varepsilon^s$ when $1/2 \leqslant s < 1$ is exactly the same as in the local case, one expects to have one-dimensional symmetry of bounded monotone solutions up to dimension $n = 8$. This has been proven in [7, 8, 10, 43] in the low-dimensional case $n = 2, 3$, and in [40, 41], under the additional assumption on the limits at infinity, for $4 \leqslant n \leqslant 8$. Instead, when $0 < s < 1/2$, the level sets of minimizers for $\mathcal{E}^s$ are asymptotically nonlocal minimal surfaces, and not much is known yet on their classification. For this range of s, the conjecture is known to be true only in dimensions $n = 2, 3$ (see [10, 25, 43]), while it is still open in dimensions $n > 3$.

3 The De Giorgi Conjecture for the Fractional Laplacian

In this section we describe the main ideas in the proof of the one-dimensional symmetry for minimizers and bounded monotone solutions to

$$(-\Delta)^s u = f(u) \quad \text{in } \mathbb{R}^n, \tag{3.1}$$

in the low-dimensional case, that is for $n = 2$ with $0 < s < 1$, and for $n = 3$ with $1/2 \leqslant s < 1$. These results are contained in [7–9, 11, 46]. In all these works, in order to prove the De Giorgi conjecture for solutions to the nonlocal equation (3.1), the authors considered the, so-called, Caffarelli-Silvestre extension (that we recall in the next subsection) and work with solutions to a local problem in the half-space $\mathbb{R}^{n+1}_+$. We emphasize that a new proof of the conjecture in dimension $n = 2$ and for any $0 < s < 1$ has been found by Bucur and Valdinoci in [6], without making use of the extension and working "downstairs". This proof is based on some techniques introduced in [20] and it works only in dimension $n = 2$.

Here, we present the proofs in [7, 8] that cover both cases $n = 2$ with $0 < s < 1$ and $n = 3$ with $1/2 \leqslant s < 1$. As already explained in Sect. 1, in this setting the same approach used to prove the original De Giorgi conjecture in dimension $n = 3$ in [2, 3] based on some sharp energy estimates and a Liouville-type argument, works. We stress that this approach allows to consider a general nonlinearity f, not necessarily associated to a double well potential.

We give now the precise statement of this result.

Theorem 3.1 (See [7, 8, 10, 43]) *Let f be any $C^{1,\gamma}$ nonlinearity with $\gamma > \max\{0, 1-2s\}$ and u be either a bounded minimizer or a bounded solution which is monotone in some direction, of*

$$(-\Delta)^s u = f(u) \quad \text{in } \mathbb{R}^n.$$

Then, if $n = 2$ and $0 < s < 1$ or $n = 3$ and $1/2 \leqslant s < 1$, u depends only on one variable, or equivalently, the level sets of u are flat.

Here below, we describe the main steps of the proof of Theorem 3.1, emphasizing which are the main difficulties in the nonlocal setting. The notion of minimizer that we use is given precisely in Definition 3.2 below.

The proof uses the Caffarelli-Silvestre extension and is based on the three following main ingredients:

- stability of solutions;
- a Liouville-type result;
- energy estimates.

In the following subsections we recall briefly all these ingredients.

3.1 *The Caffarelli-Silvestre Extension and the Notion of Minimality*

In [15], Caffarelli and Silvestre gave an equivalent formulation for nonlocal problems involving the fractional Laplacian in $\mathbb{R}^n$, introducing a new local problem in the positive half-space $\mathbb{R}^{n+1}_+$. More precisely, they established that a bounded function u is a solution of (3.1) if and only if v is a solution of

$$\begin{cases} \operatorname{div}(y^{1-2s}\nabla v) = 0 & \text{in } \mathbb{R}^{n+1}_+ \\ -d_s \lim_{y\to 0} y^{1-2s}\partial_y v = f(v(x,0)) & \text{in } \mathbb{R}^n, \end{cases} \tag{3.2}$$

where v is the bounded extension in the positive half-space of u, that is $v(x, 0) = 0$ in $\mathbb{R}^n$, and d_s is a positive constant depending only on s. Here, we denote by $(x, y) = (x_1, \dots, x_n, y)$ a point in $\mathbb{R}^{n+1}_+ = \mathbb{R}^n \times \mathbb{R}^+$.

In the sequel we will call the extension v satisfying the first equation in (3.2), the *s-extension* of u.

We set $a := 1 - 2s$, so that $-1 < a < 1$ for any $0 < s < 1$. We recall that the first equation in (3.2) is an equation in divergence form with a weight that belongs to the Muckenhoupt class A_2 and hence is a "good" weight, in the sense that we have a Poincaré inequality, Harnack inequality and Hölder regularity of weak solutions for this kind of equations by the theory of Fabes, Kenig and Serapioni, developed in [28]. We observe moreover that, depending whether s is above or below $1/2$, the equation becomes degenerate or singular.

Problem (3.2) has a variational structure and therefore it is natural to consider the associated energy functional and the related notion of minimizer. Let B_R denote the ball in $\mathbb{R}^n$ centered at 0 and of radius R and let $C_R = B_R \times (0, R)$ denote the cylinder of radius R and height R in the positive half-space $\mathbb{R}^{n+1}_+$. We consider a localized energy functional (since the energy in the whole space is not finite in general) in cylinders, which has the following form:

$$E^s(v, C_R) = \frac{1}{2} \int_{C_R} y^a |\nabla v|^2 \, dx \, dy + \int_{B_R} W(v(x, 0)) \, dx,$$

where the potential W is such that $W' = -f$.

We can now give the notion of minimizer for problem (3.2).

Definition 3.2 We say that a bounded $C^1(\mathbb{R}^{n+1}_+)$ function v is a minimizer for (3.2) if

$$E^s(v, C_R) \leqslant E^s(w, C_R)$$

for every $R > 0$ and for every bounded competitor w such that $v \equiv w$ on $\partial C_R \cap \{y > 0\}$.

We say that a bounded $C^1(\mathbb{R}^n)$ function u is a minimizer for (3.1) if its s-extension v is a minimizer for (3.2).

3.2 *Stability of Solutions*

We recall the definition of *stable solution* for (3.1).

Definition 3.3 We say that a bounded solution v of (3.2) is *stable* if

$$\int_{\mathbb{R}^{n+1}_+} y^a |\nabla \xi|^2 \, dx \, dy - \int_{\mathbb{R}^n \times \{y=0\}} f'(u) \xi^2 \, dx \geqslant 0$$

for every function $\xi \in C^1_0(\overline{\mathbb{R}^{n+1}_+})$.

We say that a bounded function u is a *stable* solution for (3.1) if its s-extension v is a stable solution for (3.2).

We observe that if u is a minimizer for (3.1) then, in particular, it is a stable solution. Moreover, as established in Lemma 6.1 in [10], we have a characterization of stability in terms of existence of a positive solution for the linearized problem.

Let $H^1_{loc}(\overline{\mathbb{R}^{n+1}_+}, y^a)$ denote the following weighted Sobolev space:

$$H^1_{\text{loc}}(\overline{\mathbb{R}^{n+1}_+}, y^a) = \{\sigma : \overline{\mathbb{R}^{n+1}_+} \to \mathbb{R} \mid y^a(\sigma^2 + |\nabla\sigma|^2) \in L^1_{\text{loc}}(\overline{\mathbb{R}^{n+1}_+})\}.$$

One can prove that a solution u to (3.1) is stable if and only if there exists a positive Hölder continuous function $\varphi \in H^1_{loc}(\overline{\mathbb{R}^{n+1}_+}, y^a)$ with $\varphi > 0$ in $\overline{\mathbb{R}^{n+1}_+}$, satisfying

$$\begin{cases} \operatorname{div}(y^a \nabla\varphi) = 0 & \text{in } \mathbb{R}^{n+1}_+ \\ -y^a \partial_y \varphi = f'(u)\varphi & \text{on } \{y = 0\}. \end{cases} \tag{3.3}$$

Suppose that u is monotone in some direction, e.g. $\partial_{x_n} u > 0$ then, as an application of the maximum principle, one can easily see that its s-extension v satisfies $\partial_{x_n} v > 0$ in the whole $\mathbb{R}^{n+1}_+$. By using the previous characterization of stability, we deduce that v is a stable solution to (3.2) since its derivative in the x_n direction is a positive solution to the linearized problem (3.3).

3.3 A Liouville-Type Result

A second ingredient in the proof of the De Giorgi conjecture is the following Liouville-type lemma.

Lemma 3.4 (Theorem 6.1 in [8] and Theorem 4.10 in [10]) *Let φ be a positive function in $L^\infty_{loc}(\mathbb{R}^{n+1}_+)$, $\sigma \in H^1_{loc}(\overline{\mathbb{R}^{n+1}}_+, y^a)$ such that:*

$$\begin{cases} -\sigma \operatorname{div}(y^a \varphi^2 \nabla\sigma) \leqslant 0 & \textit{in } \mathbb{R}^{n+1}_+ \\ -y^a \sigma\, \partial_y \sigma \leqslant 0 & \textit{on } \mathbb{R}^n \times \{y = 0\} \end{cases}$$

in the weak sense. If in addition:

$$\int_{C_R} y^a (\varphi\sigma)^2 \leqslant C R^2 \log R \tag{3.4}$$

holds for every $R > 1$, then σ is constant.

For the proof of this lemma under the stronger assumption that the quantity in (3.4) grows at most like R^2 (instead of $R^2 \log R$) it is enough to multiply the equation by a cutoff function and integrate by parts. For allowing the logarithmic term one needs a refinement of this argument found in [36].

3.4 *Sketch of the Proof of the De Giorgi Conjecture in Low Dimensions*

We can now describe the main ideas in the proof of the one-dimensional symmetry of minimizers and monotone solutions for $n = 2$ with $0 < s < 1$ and $n = 3$ with $1/2 \leqslant s < 1$.

In order to prove that the solution u to (3.1) is one-dimensional, we will show that its s-extension v depends only on y and on one direction in $\mathbb{R}^n$.

Suppose that u is a stable solution to (3.1) and v is its s-extension, that is a stable solution for problem (3.2). By the characterization of stability, we know that there exists some positive function φ satisfying (3.3) (if in particular v is monotone in the x_n direction, one can take $\varphi = v_{x_n}$).

For any $i = 1, \ldots, n$, we define the functions

$$\sigma_i := \frac{v_{x_i}}{\varphi}.$$

An easy computation shows that $\varphi^2 \nabla \sigma_i = \varphi \nabla v_{x_i} - v_{x_i} \nabla \varphi$ and using that both v_{x_i} and φ satisfy the linearized problem (3.3), we deduce

$$\operatorname{div}(y^a \varphi^2 \nabla \sigma_i) = 0 \quad \text{in } \mathbb{R}^{n+1}_+. \tag{3.5}$$

Moreover, using again that v_{x_i} and φ satisfy the same linearized problem (in particular they have the same Neumann condition on $\{y = 0\}$), we have

$$y^a \sigma_i \partial_y \sigma_i = y^a \frac{v_{x_i}}{\varphi^2} v_{x_i y} - y^a \frac{v_{x_i}^2}{\varphi^2} \frac{\varphi_y}{\varphi} = 0 \qquad \text{on } \mathbb{R}^n \times \{y = 0\}. \tag{3.6}$$

Suppose now that the following estimate for the Dirichlet energy of v holds:

$$\int_{C_R} y^a |\nabla v|^2 \, dx \, dy \leqslant C R^2 \log R.$$

Then, we can apply Lemma 3.4 with $\sigma = \sigma_i$. Indeed (3.3) is satisfied by (3.5) and (3.6), moreover,

$$\int_{C_R} y^a (\varphi \sigma_i)^2 \, dx \, dy = \int_{C_R} y^a |v_{x_i}|^2 \, dx \, dy \leqslant \int_{C_R} y^a |\nabla v|^2 \, dx \, dy \leqslant C R^2 \log R.$$

Hence we deduce that σ_i is constant for any $i = 1, \ldots, n$. This concludes the proof observing that if $c_1 = \cdots = c_n = 0$ then v is constant. Otherwise we have $c_i v_{x_j} = c_j v_{x_i} = 0$ for every $i \neq j$ and we deduce that v depends only on y and on the variable parallel to the vector $(c_1, \ldots, c_n)$.

Hence, the crucial missing ingredient to conclude the proof is the following estimate for the Dirichlet energy

$$\int_{C_R} y^a |\nabla v|^2 \, dx \, dy \leqslant C R^2 \log R. \tag{3.7}$$

3.5 Energy Estimates

By the previous discussion, in order to conclude the proof of Theorem 3.1, it remains to prove that both minimizers and bounded monotone solutions (which are in particular stable solutions) satisfy estimate (3.7), in $\mathbb{R}^2$ with $0 < s < 1$ and in $\mathbb{R}^3$ with $1/2 \leqslant s < 1$. This is the aim of this subsection.

We start by stating the energy estimate for minimizers contained in [7, 8], which holds in any dimension n.

Theorem 3.5 (Theorem 1.2 in [7]) *Let $f \in C^{1,\gamma}(\mathbb{R})$, with $\gamma > \max\{0, -a\}$, and let v be a bounded minimizer for problem* (3.2).

Then, the following estimates hold

$$E^s(v, C_R) \leqslant \begin{cases} C R^{n-2s} & \text{if } 0 < s < 1/2 \\ C R^{n-1} \log R & \text{if } s = 1/2 \\ C R^{n-1} & \text{if } 1/2 < s < 1, \end{cases} \tag{3.8}$$

for any $R \geqslant 2$.

In dimension $n = 3$, the same energy estimate holds also for bounded monotone solutions, according to the following result

Theorem 3.6 (Theorem 1.4 in [7]) *Let $f \in C^{1,\gamma}(\mathbb{R})$, with $\gamma > \max\{0, -a\}$, and let v be a bounded solution of (3.2) with $n = 3$ such that its trace $u(x) = v(x, 0)$ is monotone in some direction.*

Then, the following estimates hold

$$E^s(v, C_R) \leqslant \begin{cases} C R^{3-2s} & \text{if } 0 < s < 1/2 \\ C R^{2} \log R & \text{if } s = 1/2 \\ C R^{2} & \text{if } 1/2 < s < 1, \end{cases} \tag{3.9}$$

for any $R \geqslant 2$.

As a consequence of Theorems 3.5 and 3.6, we have that the required estimate (3.7) is satisfied by minimizers and bounded monotone solutions in dimension $n = 2$ for any $s \in (0, 1)$ and in dimension $n = 3$ for $s \in [1/2, 1)$.

We stress that the main difficulty in the proof of the fractional De Giorgi conjecture in low dimensions, relies precisely in establishing the sharp energy estimates for minimizers, since all the other ingredients (the characterization of stability and the Liouville-type result described above are not difficult adaptations of the analogous local results to the nonlocal setting). The energy estimates for minimizers have also been proven by Savin and Valdinoci in [42], without making use of the extension and working with the nonlocal energy functional associated to (3.1). Here, we present the approach via extension of [7], since, as already explained at the beginning of this section, the extension is needed in order to prove the De Giorgi conjecture for the fractional Laplacian in dimension $n = 3$ and for $1/2 \leqslant s < 1$.

Before giving an idea of the proof of Theorem 3.5, we recall how one can get the sharp energy estimate for minimizers of the classical Allen-Cahn equation $-\Delta u = u - u^3$, whose associated energy functional is given by

$$\mathcal{E}(u, B_R) = \int_{B_R} \left(\frac{1}{2} |\nabla u|^2 + W(u) \right) dx.$$

To give a bound for the energy of a minimizer u, we argue using a comparison argument, that is we construct a suitable competitor w, which agrees with u on ∂B_R and whose energy is bounded above by CR^{n-1} (we recall that in the classical setting the energy of minimizers grows like R^{n-1}, that is exactly the same growth as the case $1/2 < s < 1$ in Theorem 3.5).

Given a cut-off function $\eta(x) = \eta(|x|)$ compactly supported in B_R and identically equal to 1 in B_{R-1}, we define the competitor $w = \eta + (1 - \eta)u$ in the whole $\mathbb{R}^n$, so that

$$w = \begin{cases} 1 & \text{in } B_{R-1} \\ u & \text{on } \partial B_R. \end{cases}$$

With this choice of w, it is easy to verify that

$$\begin{aligned} \mathcal{E}(u, B_R) \leqslant \mathcal{E}(w, B_R) &= \int_{B_R} \left(\frac{1}{2} |\nabla w|^2 + W(w) \right) dx \\ &= \int_{B_R \setminus B_{R-1}} \left(\frac{1}{2} |\nabla w|^2 + W(w) \right) dx \leqslant CR^{n-1}, \end{aligned} \tag{3.10}$$

where in the first inequality we have used the minimality of u and in the last inequality we have used that $|\nabla u| \in L^\infty(\mathbb{R}^n)$ by standard elliptic estimates, and that the measure of the annulus $B_R \setminus B_{R-1}$ in $\mathbb{R}^n$ is estimated by CR^{n-1}.

We consider now the case of the fractional Laplacian. In this case, we need to construct a suitable competitor w for the minimizers v of E^s in C_R, which fulfills the energy estimates of Theorem 3.5. We recall that, due to the Neumann condition in problem (3.2), now the competitor w has to agree with v on $\partial C_R \cap \{y > 0\}$ but it is free on the bottom of the cylinder $\partial C_R \times \{y = 0\}$. This fact will play a crucial role in the construction of w. On the other hand, let us emphasize that now the cylinder C_R is an $(n+1)$-dimensional object, and we hope for an estimate for the energy that grows at most like $R^{n-1} \log R$.

Let us describe now how we build the competitor w. We start by defining a function $\bar{w}$ on ∂C_R. Then, we will define w as a suitable extension of $\bar{w}$ inside C_R. First of all, in order to use a comparison argument, $\bar{w}$ needs to agree with v on $\partial C_R \cap \{y > 0\}$. Secondly, since the potential energy appears only as a boundary term on the bottom of the cylinder C_R, and in this part of the boundary w is free, we define $\bar{w}$ as done for the local case, that is in such a way that it agrees with $v(x, 0)$ on $B_R \times \{y = 0\}$ and is identically 1 in $B_{R-1} \times \{y = 0\}$. Resuming, $\bar{w}$ is defined on the whole ∂C_R and satisfies

$$\bar{w} = \begin{cases} 1 & \text{in } B_{R-1} \times \{y = 0\} \\ v & \text{on } \partial C_R \cap \{y > 0\}. \end{cases}$$

It remains now to extend $\bar{w}$ to a function w defined on the whole cylinder C_R. Since we want w to have the least possible Dirichlet energy, we choose w to be the solution of the Dirichlet problem

$$\begin{cases} \operatorname{div}(y^a \nabla w) = 0 & \text{in } C_R \\ w = \bar{w} & \text{on } \partial C_R. \end{cases} \tag{3.11}$$

With this choice of w, we have

$$\begin{aligned} E^s(v, C_R) \leqslant E^s(w, C_R) &= \int_{C_R} y^a |\nabla w|^2 \, dx \, dy + \int_{B_R} W(w) \, dx \\ &\leqslant \int_{C_R} y^a |\nabla w|^2 \, dx \, dy + C R^{n-1}. \end{aligned}$$

The final step of the proof of the energy estimate for minimizers consists in giving an estimate for the Dirichlet energy of w. This is achieved in [7, 8] using some suitable trace inequalities and optimal gradient bounds for the solutions of (3.2) (for details see Theorems 1.7 and 1.9 in [7]).

To conclude, we comment on the proof of the energy estimate for monotone solutions. In [7, 8], the authors follow the idea of [2] which is based on the following result: bounded monotone solutions are minimizers in the restricted class of functions

$$\mathcal{A}_v := \{ \lim_{x_n \to -\infty} v \leqslant w \leqslant \lim_{x_n \to +\infty} v \}.$$

This result can be proven by a sliding argument and the use of the maximum principle (see proof of Proposition 6.1 in [7]). Once one has this minimality property of monotone solutions, it is enough to show that the competitor w constructed in the proof of Theorem 3.5 belongs to the class $\mathcal{A}_v$. For this last step we refer to Lemma 6.1 and the proof of Theorem 1.4 in [7].

4 Classification for Nonlocal Minimal Surfaces

We start by recalling the notion of fractional perimeter, which was introduced in [18].

Let $s \in (0, 1/2)$ and let Ω be a bounded domain in $\mathbb{R}^n$. We define the fractional s-perimeter of a measurable set $E \subset \mathbb{R}^n$ relative to Ω as

$$\mathrm{Per}_s(E, \Omega) := \int_{E\cap\Omega} \int_{E^c} \frac{1}{|x - \bar{x}|^{n+2s}} \, dx \, d\bar{x} + \int_{E\setminus\Omega} \int_{\Omega\setminus E} \frac{1}{|x - \bar{x}|^{n+2s}} \, dx \, d\bar{x}, \tag{4.1}$$

where E^c denotes the complement of E in $\mathbb{R}^n$.

Observe that here we use the notation Per_s for the perimeter associated with the kernel $|z|^{-n-2s}$, with $0 < s < 1/2$. In many references the order $2s$ is replaced by s, that is one writes Per_s for the perimeter associated to the power $|z|^{-n-s}$ and in this notation s belongs to $(0, 1)$. Here, we use the first notations for consistency with the notations used for the fractional Laplacian $(-\Delta)^s$.

The choice of the set of integration in the definition of the fractional perimeter is the natural one, similarly as for the Dirichlet term in the energy $\mathcal{E}^s$ defined in (2.3), in order to avoid infinite contributions coming from the complement of Ω and it does not change the variational structure of the functional once we have fixed the set E outside of Ω. More precisely, similarly to Definition 2.1, we give the following definition.

Definition 4.1 We say that a set E is a *minimizer* for the s-perimeter in Ω if

$$\mathrm{Per}_s(E, \Omega) \leqslant \mathrm{Per}_s(F, \Omega), \qquad \text{for all } F \text{ such that } E \setminus \Omega = F \setminus \Omega.$$

Moreover, we say that E is a minimizer for the s-perimeter in $\mathbb{R}^n$, if E is a minimizer in B_R for all $R > 0$.

The (boundaries of) minimizers of the s-perimeter are usually called *nonlocal minimal* (or s-minimal) *surfaces*.

While the classical perimeter (in the De Giorgi sense) of a set E relative to Ω is the BV-seminorm of the characteristic function χ_E in Ω, the s-perimeter is the H^s (or $W^{2s,1}$) seminorm of the characteristic function χ_E in Ω (we remind that the characteristic function of a set belongs to H^s only if $0 < s < 1/2$). Hence, a nonlocal minimal surface is the boundary of a set E, whose characteristic function minimizes the H^s seminorm, among all sets which coincide with E in the complement of Ω.

Another motivation for referring to Per_s as a fractional *perimeter* comes from the asymptotics of this nonlocal functional as $s \to 1/2$. Indeed it is known that the s-perimeter (multiplied by the factor $1/2 - s$) tends to the classical perimeter as $s \to 1/2$, up to a dimensional constant. This fact has been established in several contributions where different notions of convergence are considered (see [22] for the precise limit in the class of BV functions, [16, 17] for a geometric approach to prove regularity and [4] for a Γ-convergence result). The limit as $s \to 0$ was studied in [24], where the authors proved that it is related to the Lebesgue measure of the sets $E \cap \Omega$ and $\Omega \setminus E$.

Making the first variation of the nonlocal perimeter functional, one can introduce the notion of *nonlocal mean curvature*. The nonlocal mean curvature of a set E at a point $x \in \partial E$ is defined as follows

$$H_E^s(x) := \int_{\mathbb{R}^n} \frac{\chi_{E^c}(y) - \chi_E(y)}{|x-y|^{n+2s}}\, dy.$$

Hence, a necessary condition for a set E to be an s-minimal surface is that $H_E^s = 0$ (see Theorem 5.1 in [18]). The first example of a surface with zero nonlocal mean curvature is a half-space. Other examples of sets with vanishing nonlocal mean curvature have been studied in the recent contributions [19, 23]. In [23], the nonlocal analogue of catenoids are constructed, but they differ from the standard catenoids since they approach a singular cone at infinity instead of having a logarithmic growth. These surfaces are constructed using perturbative methods, by performing small perturbation along the normal vector to ∂E. Instead in [19] it is proven that the standard helicoids are surfaces with zero nonlocal mean curvature.

We pass now to describe the main results in the study of regularity of nonlocal minimal surfaces. In [18], Caffarelli, Roquejoffre, and Savin established some results that are fundamental tools in the study of regularity, such as density estimates, the improvement of flatness for minimizers, a monotonicity formula, a blow up and a dimension reduction argument. Nevertheless, the study of regularity for minimizers of the fractional perimeter is still widely open. In this section we recall the main results related to the classification of entire s-minimal surfaces and to the study of regularity, and we describe the main open questions in the field.

4.1 The Classical Setting

We start by recalling the main results in the theory of *classical area minimizing surfaces*:

(a) Every minimal cone in $\mathbb{R}^n$ is a hyperplane, whenever $n < 8$;
(b) In $\mathbb{R}^8$ the Simons cone defined as

$$\mathcal{C} := \{x \in \mathbb{R}^8 \mid x_1^2 + \cdots + x_4^2 = x_5^2 + \cdots + x_8^2\}$$

is a minimizer for the perimeter functional;
(c) If E is a minimizer of the perimeter functional in the whole $\mathbb{R}^n$, then E is a half-space, whenever $n < 8$;
(d) If E is a minimizer of the perimeter functional and ∂E is a graph, then E is a half-space, whenever $n < 9$;
(e) Any area-minimizing surface is smooth outside of a singular set Σ of Hausdorff dimension $n - 8$.

The classification of minimal cones (point a) is one of the basic tools in both the classification of entire minimal surfaces (that is surfaces that are minimizer of the perimeter functional in the whole $\mathbb{R}^n$) and in the study of regularity for minimizers of the perimeter in a bounded set Ω. Indeed, the classification of minimal cones leads, on one side, to the classification of *any* entire area minimizing surfaces (point c) via a blow-down argument. On the other hand the nonexistence of singular minimal cones in space dimension $n \leqslant 7$ implies, via a blow up and a dimension reduction argument, that any minimal surface is $C^{1,\alpha}$ outside of a singular set of Hausdorff dimension $n - 8$ (point e). Moreover, again the classification of minimal cones leads to the classification of entire minimal graphs (the so called Bernstein problem). Note that the critical dimension for a graph to be flat is one more than the one for a general set (point d). The main ingredients in the proof of these results are given by density estimates, perimeter estimates, improvement of flatness for minimizers and a monotonicity formula.

4.2 The Nonlocal Setting

We describe now, more in details, what is known in the nonlocal framework and which are the main open questions in the field.

The study of regularity for nonlocal minimal surfaces was started in [18], where density and perimeter estimates, the improvement of flatness and a monotonicity formula were established. With these tools, the authors could reduce the study of regularity for nonlocal minimal surfaces to the classification of nonlocal minimal cones. More precisely they proved that, if the blow up, around the origin, of an s-minimal set E is flat, then ∂E is $C^{1,\alpha}$ in a neighborhood of the origin (see

Theorem 9.4 in [18]). As a consequence of a dimension reduction argument they proved $C^{1,\alpha}$ regularity outside of a singular set of Hausdorff dimension at most $n-2$ (see Theorem 10.4 in [18]). The bound $n-2$ on the dimension of the singular set was not optimal due to the fact that in [18] the classification of nonlocal minimal cones was not known, not even in $\mathbb{R}^2$. Basically, they had all the needed ingredients to pass from (a) to (e) in the above scheme, but the starting point (a) was missing.

Later, in [43] Savin and Valdinoci proved that in $\mathbb{R}^2$ an s-minimal cone is necessarily a half-plane. As a consequence they could improve the bound on the Hausdorff dimension of the singular set from $n-2$ to $n-3$ and via a blow-down argument they obtained the classification of any s-minimal surface in $\mathbb{R}^2$. Moreover, in [5] Barrios, Figalli, and Valdinoci shows that if E is an s-minimal set such that $\partial E \in C^{1,\alpha}$, then ∂E is in fact C^∞. This is a consequence of a more general regularity result for solutions to integro-differential equations via a bootstrap argument. In [30], Figalli and Valdinoci address the fractional version of the Bernstein problems and they prove that, if there are not s-minimal singular cones in $\mathbb{R}^n$, then the only entire s-minimal graphs in $\mathbb{R}^{n+1}$ are the hyperplanes (they show how to pass from point (a) to point (d) in the nonlocal analogue of the previous scheme).

Resuming all these results, we have the following statement:

Theorem 4.2

(1) *Every s-minimal cone in $\mathbb{R}^2$ is a hyperplane [43];*
(2) *If E is a minimizer of the s-perimeter in the whole $\mathbb{R}^2$, then E is a half-plane [43];*
(3) *If E is a minimizer of the s-perimeter in $\mathbb{R}^n$ and ∂E is a graph, then E is a half-space, whenever $n \leqslant 3$ [30];*
(4) *If E is a minimizer of the s-perimeter, then ∂E is C^∞ outside of a singular set Σ of Hausdorff dimension $n-3$ [5, 18, 43].*

In addition, when s is close to $1/2$ Caffarelli and Valdinoci proved that all the regularity results that hold in the classical setting are inherited, by a compactness argument, by s-nonlocal minimal surfaces (see [16, 17]).

Theorem 4.3 ([17]) *There exists $\varepsilon_0 \in (0, 1/2)$ such that if $s \geqslant 1/2 - \varepsilon_0$, then any s-minimal surfaces is C^∞ outside of a singular set Σ of Hausdorff dimension $n-8$.*

Finally, in the very recent contribution [12], Cabré, Serra and the author proved flatness for nonlocal s-minimal cones in $\mathbb{R}^3$ for s close to $1/2$ (see Theorem 5.3 of the next section). We emphasize that in this last result, the proof is not by compactness perturbing from $s = 1/2$ and it gives a quantifiable value for the required closeness of s to $1/2$. This last result holds not only for cones that are minimizers for the s-perimeter, but for the more general class of *stable* cones. We will describe this result more in details in the next section, were we address the classification for stable objects.

We focus now on the classification of s-minimal cones in $\mathbb{R}^2$ proven in [43] (i.e. point (1) in Theorem 4.2). The idea of the proof of this result relies in considering perturbations of the minimizer E, that are translations of E inside a ball $B_{R/2}$ and that coincide with E outside the double ball B_R. The authors work with the extended problem (the Caffarelli-Silvestre extension but, in this setting, for functions that take only values ± 1 on the boundary of the positive half-space) and compare the energy of (the extension of) these competitors with the energy of (the extension of) E itself. A computation shows that this difference in energy is controlled from above with R^{n-2s-2}. Hence, when $n = 2$, the difference in energy between E and the competitors can be made arbitrarily small as $R \to \infty$. On the other hand, if E was not a half-plane, they showed that it could be modified in order to decrease its energy by a small but fixed amount and this leads to a contradiction. We emphasize here that this argument works only in dimension $n = 2$ since a crucial fact that is needed is that the estimate R^{n-2s-2} goes to 0 as $R \to \infty$, and this holds true only when $n = 2$. We emphasize that the factor R^{n-2s} comes from an optimal bound for the perimeter of minimizers. Indeed, by a comparison argument one can show that if E is a minimizers for the s-perimeter in B_R, then

$$\mathrm{Per}_s(E, B_R) \leqslant C R^{n-2s},$$

and this bound is optimal.

These ideas were recently used in [20] to prove a quantitative version of this two-dimensional flatness result. By point (1) in Theorem 4.2, we know that if E is a minimizer for the nonlocal perimeter in the whole $\mathbb{R}^2$ (that is, it is a minimizer in balls B_R of radius R *for any* $R > 1$), then E is a half-plane.

Suppose now that E is a minimizer for Per_s in a ball B_R for *some* R large enough. Can we deduce that E is "close" to be a half-plane in B_1? Moreover, can we give an estimate on this closeness depending on R? The following result, contained in [20], gives an answer to these questions.

Theorem 4.4 (Theorem 1.3 in [20]) *Let $n = 2$. Let $R \geqslant 2$ and E be a minimizer for the s-perimeter in the ball $B_R \subset \mathbb{R}^2$.*

Then, there exists a half-plane $\mathfrak{h}$ such that

$$|(E \Delta \mathfrak{h}) \cap B_1| \leqslant C R^{-s}. \tag{4.2}$$

Moreover, after a rotation, we have that $E \cap B_1$ is the subgraph of a measurable function $g : (-1, 1) \to (-1, 1)$ with $\mathrm{osc}\, g \leqslant C R^{-s}$ outside a "bad" set $\mathcal{B} \subset (-1, 1)$ with measure $C R^{-s}$.

The above result can be seen as a quantitative version of the flatness result of Savin and Valdinoci. It says that if E is a minimizer for Per_s in B_R, with R large but fixed, then E is close, in the L^1-sense, to be a half-plane in B_1. The second part of the statement gives an even more precise information: outside of a bad set $\mathcal{B}$ of small measure, $E \cap B_1$ coincides with the subgraph of a function g which has small oscillation. Again, the smallness of both the bad set and the oscillation of g is given explicitly in terms of R.

The proof of Theorem 4.4 follows the main ideas contained in the proof of flatness of s-minimal cones in $\mathbb{R}^2$ in [43]. We consider again perturbations of the minimizer E obtained by small translations in some fixed direction and we try to refine the arguments in [43] in order to get some quantitative estimates. Differently from [43], we do not use the Caffarelli-Silvestre extension. This will allow us to obtain a statement analogous to the one of Theorem 4.4 for more general notions of nonlocal perimeter (such as the anisotropic fractional perimeter). Here below, we explain the main steps in the proof of Theorem 4.4.

Sketch of the Proof of Theorem 4.4 We start by defining two (small) perturbations of the minimizer E. Let φ_R be a smooth function such that

$$\varphi_R(x) = \begin{cases} 1 & \text{for } |x| < R/2 \\ 0 & \text{for } |x| > R. \end{cases}$$

For $\boldsymbol{v} \in S^{n-1} := \{x \in \mathbb{R}^n \, : \, |x| = 1\}$ and $t \in [0, 1]$ we define

$$\Psi_{R,+}(x) := x + t\varphi_R(x)\boldsymbol{v} \quad \text{and} \quad \Psi_{R,-}(x) := x - t\varphi_R(x)\boldsymbol{v}. \tag{4.3}$$

We set $u = \chi_E$ and define the new functions

$$u_R^{\pm}(x) := u\big(\Psi_{R,\pm}^{-1}(x)\big). \tag{4.4}$$

In set notations, we are considering the sets E_R^+ and E_R^- defined as

$$E_R^{\pm} = \{x \, : \, u_R^{\pm}(x) = 1\}. \tag{4.5}$$

We recall the following crucial energy estimate for minimizers, obtained via a comparison argument: if E is a minimizer for the s-perimeter in B_R, then

$$\mathrm{Per}_s(E, B_R) \leqslant C R^{n-2s}. \tag{4.6}$$

We divide the proof in three steps:

- **Step 1**: *Estimating the difference* $\mathrm{Per}_s(E_R^{\pm}, B_R) - \mathrm{Per}_s(E, B_R)$ (see Lemma 2.1 in [20]): using the change of variable formula and after some computations one can prove that

$$\mathrm{Per}_s(E_R^+, B_R) + \mathrm{Per}_s(E_R^-, B_R) - 2\mathrm{Per}_{s,B_R}(E) \leqslant Ct^2 \frac{\mathrm{Per}_s(E, B_R)}{R^2}. \tag{4.7}$$

Using the estimate (4.6) and the fact that we are in dimension $n = 2$, we get

$$\mathrm{Per}_s(E_R^+, B_R) + \mathrm{Per}_s(E_R^-, B_R) - 2\mathrm{Per}_s(E, B_R) \leqslant Ct^2 R^{-2s}.$$

Observe that here the fact that the we are working in dimension 2 is crucial in order to get a bound that goes to 0 as $R \to \infty$. As described above, in [43] this fact leads to a contradiction if E was not flat. Here we refine this argument, by keeping the above estimate R^{-2s} in order to get a quantitative estimate (depending on R) on how E differs from being a half-plane.

- **Step 2**: *a purely nonlocal Lemma*:

Lemma 4.5 (Lemma 2.2 in [20]) *Let E, $F \subset \mathbb{R}^2$. Assume that E is a minimizer for* Per_s *in B_R and that F coincides with E outside B_R, that is, $E \setminus B_R = F \setminus B_R$. Assume moreover that*

$$\mathrm{Per}_s(F, B_R) \leqslant \mathrm{Per}_s(E, B_R) + \delta. \tag{4.8}$$

Then,

$$2\int_{F\setminus E}\int_{E\setminus F} \frac{1}{|x-\bar{x}|^{2+2s}} dx d\bar{x} \leqslant \delta.$$

Applying this Lemma to $F = E_R^+$ (and similarly to E_R^-), we deduce that

$$\int_{E_R^+\setminus E}\int_{E\setminus E_R^+} \frac{1}{|x-\bar{x}|^{2+2s}} dx d\bar{x} \leqslant Ct^2 R^{-2s}.$$

Therefore, in B_1 we have that for any $\boldsymbol{v} \in S^1$ and any $t \in (0, 1)$:

$$\big|\{(E + t\boldsymbol{v}) \setminus E\} \cap B_1\big| \cdot \big|\{E \setminus (E + t\boldsymbol{v})\} \cap B_1\big| \leqslant Ct^2 R^{-2s}$$

and thus

$$\min\Big\{\big|\{(E + t\boldsymbol{v}) \setminus E\} \cap B_1\big|, \big|\{E \setminus (E + t\boldsymbol{v})\} \cap B_1\big|\Big\} \leqslant CtR^{-s}. \tag{4.9}$$

In this step, the nonlocal character of the s-perimeter is crucial and allows to pass from an estimate in the difference of the s-perimeter between the minimizer E and the competitors $E_R^\pm$ to an estimate on the volume of their symmetric difference.

Setting $u := \chi_E$, estimate (4.9) can be written as

$$\min\Big\{\int_{B_1} \big(u(x + t\boldsymbol{v}) - u(x)\big)_+ dx\,,\ \int_{B_1} \big(u(x + t\boldsymbol{v}) - u(x)\big)_- dx\Big\} \leqslant CtR^{-s}. \tag{4.10}$$

- **Step 3**: *Some geometric lemmas and conclusion*. Dividing (4.10) by t and taking the limit as $t \to 0$, we deduce that for any $\boldsymbol{v} \in S^1$, the following holds:

$$\min\{(\nabla u \cdot \boldsymbol{v})_+(B_1), (\nabla u \cdot \boldsymbol{v})_-(B_1)\} \leqslant CR^{-s}, \tag{4.11}$$

where $\nabla u \cdot \boldsymbol{v}$ denotes the distributional derivative in the direction $\boldsymbol{v}$ of u. This last part of the proof is more technical and needs several geometric lemmas (for the details, we refer to Lemma 2.5 and to all lemmas and propositions of Section 4 in [20]). The main underlying idea is the following: If we set $\Phi_{\pm}(\boldsymbol{v}) := (\nabla u \cdot \boldsymbol{v})_{\pm}(B_1)$, by (4.11) we have that

$$\min\{\Phi_+(\boldsymbol{v}), \Phi_-(\boldsymbol{v})\} \leqslant CR^{-s}, \quad \text{for any } \boldsymbol{v} \in S^1.$$

Moreover, since

$$\Phi_+(\boldsymbol{v}) = \Phi_-(-\boldsymbol{v}),$$

by a continuity argument we have that there exists $\boldsymbol{v}^* \in S^1$ such that

$$\max\left\{(\nabla u \cdot \boldsymbol{v}^*)_+(B_1), (\nabla u \cdot \boldsymbol{v}^*)_-(B_1)\right\} \leqslant CR^{-s}.$$

Hence, except for a bad set $\mathfrak{B}$ of measure less than CR^{-s} the function $u = \chi_E$ restricted to all lines parallel to $\boldsymbol{v}^*$ will be at the same time monotone nondecreasing and non-increasing; i.e., constant. Since we also have that u is also monotone along most (for large R) lines perpendicular to $\boldsymbol{v}^*$, the only possibility is that the set $E = \{u = 1\}$ is equal to a half plane up to the bad set $\mathfrak{B}$ with $|\mathfrak{B}| \leqslant CR^{-s}$. The rigorous proof for this fact is contained in Section 4 of [20].

We emphasize that in [43], the authors first prove a flatness result for minimizing cones, and then they deduce, by a blow-down argument, flatness for any s-minimal set in $\mathbb{R}^2$. In this blow-down procedure the monotonicity formula is needed and unfortunately such a formula is available only for the energy functional of the extended problem (see [18]). Instead, in the proof of Theorem 4.4, we consider E to be *any* set which minimizes the s-perimeter, not necessarily a cone and, as a consequence of the quantitative estimate (4.2) after letting $R \to \infty$, we deduce that if E is a minimizer in the whole $\mathbb{R}^2$, then E is an half-plane. Hence, we give an alternative proof of the classification result in [43], without using the Caffarelli-Silvestre extension and without needing a monotonicity formula. For this reason, we can generalize our Theorem 4.4 and hence the classification of nonlocal minimal surfaces in $\mathbb{R}^2$ to more general notions of nonlocal perimeter, such as the anisotropic fractional perimeter (see [20]).

The techniques developed in [20] and, more precisely, the estimate (4.11) implies also some estimates for the s-perimeter and the classical perimeter of an s-minimal set E. More interestingly, these estimates holds true in the more general class of *stable sets*. We are going to state and comment on these results on stable sets in the next section.

5 What About Stable Objects?

In this section we present some very recent results in the study of *stable* solutions to the fractional Allen-Cahn equation and of *stable* nonlocal minimal surfaces. In both cases the notion of stability that we use is the variational one, that is the nonnegativity of the second variation of the associated energy functional. Surprisingly, some results recently established for stable objects in the nonlocal setting, are still unknown in the local setting. The nonlocality of the energy functional (for the Allen-Cahn equation or for the nonlocal perimeter) helps in giving sharp estimates that are crucial for classifying stable solutions. In order to explain which are the main difficulties in this setting and to compare the local and nonlocal framework, we start by recalling what is known for classical stable minimal surfaces.

5.1 The Classical Setting

Stable minimal cones (for the classical perimeter) are completely classified: they are hyperplanes in space dimensions $n \leqslant 7$. In $\mathbb{R}^8$, the Simons cone is an example of stable cone which is singular. The classification that we have presented in the previous section for classical minimal surfaces holds true for stable cones. In order to pass from the classification of stable cones to the classification of *any* stable surface in the whole $\mathbb{R}^n$, one would like to perform a blow-down procedure using the monotonicity formula. A crucial tool needed for using a blow-down argument would be an optimal estimate for the perimeter of stable sets. It is well known that any minimizer of the classical perimeter in a ball B_R satisfies the estimate

$$\text{Per}(E, B_R) \leqslant C R^{n-1}. \tag{5.1}$$

Unfortunately, an estimate like (5.1) is not known to hold for stable sets, unless we are in dimension $n = 3$ and we require some topological assumption on the set E (see Theorem 5.1 below). While for proving an energy estimate for minimizers it is enough to construct a suitable competitor, which has to agree with E outside B_R but can be modified arbitrarily in B_R, and that satisfies the needed estimate, for proving such an estimate for stable sets we are allowed to consider only competitors which are *small perturbations* of the given set E.

We recall here below the perimeter estimate for classical stable sets, which was proven by Pogorelov [38], and Colding and Minicozzi [21]—see also [27, 32] [34, Theorem 2] and [47, Lemma 34].

Theorem 5.1 ([21, 38]) *Let D be a simply connected, immersed, stable minimal disk of geodesic radius r_0 on a minimal (two-dimensional) surface $\Sigma \subset \mathbb{R}^3$, then*

$$\pi r_0^2 \leqslant \text{Area}\,(D) \leqslant \frac{4}{3}\pi r_0^2.$$

In dimension $n > 3$ the perimeter estimate for stable sets is still completely open. As explained above, having a universal bound for the classical perimeter of embedded minimal surfaces in every dimension $n > 3$ would be a decisive step towards proving the following well-known and long standing conjecture: *The only stable embedded minimal (hyper)surfaces in $\mathbb{R}^n$ are hyperplanes as long as the dimension of the ambient space is less than or equal to 7*. On the other hand, without a universal perimeter bound, the sequence of blow-downs could have perimeters converging to ∞.

In a similar way, one can ask whether the De Giorgi conjecture on one-dimensional symmetry for solutions to the Allen-Cahn equation, holds in the more general class of *stable solutions*. This is known only in dimension $n = 2$ and it is still open in higher dimensions. We have already seen in Sect. 3 that stability plays a crucial role in the proof of the conjecture, but again another fundamental ingredient was given by the energy estimate (3.10). Also in this case the optimal estimate is known to hold only for minimizers (and for monotone solutions in dimension 3) and it is completely open for stable sets. Nevertheless, when $n = 2$ one can prove the conjecture for stable solutions because, in order to apply the Liouville-type argument described in Sect. 3, an estimate of the form

$$\mathcal{E}(u, B_R) \leqslant CR^2,$$

is enough. In $\mathbb{R}^2$ this (not sharp) estimate holds true since the measure of B_R is of order R^2 (and $|\nabla u|$ is bounded by standard elliptic estimates).

One important open question in the classification of solutions to the classical Allen-Cahn equation is, then, the following:

Open Question Is it true that any bounded *stable* solution of $-\Delta u = u - u^2$ in $\mathbb{R}^n$ is one-dimensional for $3 \leqslant n \leqslant 7$?

One would expect a positive answer to this question for all dimensions $3 \leqslant n \leqslant 7$, in the same way one expects a positive answer to the conjecture for stable minimal surfaces stated above. Starting from dimension $n = 8$, instead there are examples of stable solutions to the Allen-Cahn equations which are not one-dimensional. This was established by Pacard and Wei in [37].

5.2 The Nonlocal Setting

Surprisingly, when dealing with stable sets for the nonlocal perimeter (or the nonlocal Allen-Cahn equation) some of the above open problems received a positive answer, at least in some particular cases.

In this section we describe some recent results for stable objects obtained in [12, 20] (see also [44]). Since the notion of stability that we consider is the one of nonnegativity for the second variation of the s-perimeter functional, we recall here the expression for $\partial^2 \mathrm{Per}_s$, given in [23, 31].

Let $E \subset \mathbb{R}^n$ be such that ∂E is C^2 away from 0. We denote by H^{n-1} the $(n-1)$-dimensional Hasudorff measure. E is a stationary set for Per_s (i.e $H_E^S \equiv 0$), then, the second variation of the s-perimeter is given by

$$\int_{\partial E} c_{s,\partial E}^2(x)|\zeta(x)|^2\, dH^{n-1}(x) - \iint_{\partial E\times\partial E} \frac{\left|\zeta(x)-\zeta(\bar{x})\right|^2}{|x-\bar{x}|^{n+2s}}\, dH^{n-1}(x)\, dH^{n-1}(\bar{x}),$$

where

$$c_{s,\partial E}^2(x) := \int_{\partial E} \frac{\left|\nu_E(x)-\nu_E(\bar{x})\right|^2}{|x-\bar{x}|^{n+2s}}\, dH^{n-1}(\bar{x}),$$

$\nu_E(x)$ denotes the outward normal vector to ∂E at $x \in \partial E$ and $\zeta \in C_0^2(\mathbb{R}^n \setminus \{0\})$.

In Section 3 of [12], we deal with different possible notions of stability, that, in the case of smooth sets E, are equivalent to require that the expression above is nonnegative for any $\zeta \in C_c^2(\mathbb{R}^n \setminus \{0\})$.

As anticipated in the previous section, the techniques developed in [20] allow to prove some perimeter and energy estimates for nonlocal stable sets.

Theorem 5.2 (Theorem 1.1 in [20]) *Let $s \in (0, 1/2)$, $R > 0$ and E be a stable set in the ball B_{2R} for the nonlocal s-perimeter functional. Then,*

$$\mathrm{Per}(E, B_R) \leqslant CR^{n-1},$$

and

$$\mathrm{Per}_{2s}(E, B_R) \leqslant CR^{n-2s}.$$

As a consequence of Theorem 5.2, in [20] we obtained that any nonlocal stable set in the whole $\mathbb{R}^2$ is a half-plane (by using the same argument that we sketch in the proof of Theorem 4.4 in the previous section).

Analogue estimates for classical stable surfaces are not known when $n > 2$, and even comparing our result with the two-dimensional result of Theorem 5.1 above, we stress that here we do not need ∂E to be simply connected. In fact, an estimate exactly like ours can not hold for classical stable minimal surfaces since a large number of parallel planes is always a classical stable minimal surface with arbitrarily large perimeter in B_1.

Moreover, we believe that our result in Theorem 5.2 can be used to reduce the classification of stable s-minimal surfaces in the whole $\mathbb{R}^n$ to the classification of stable cones, by means of a blow-down argument and using a monotonicity formula. Somehow the difficulties in the local/nonlocal setting are interchanged: in the local setting we have the complete classifications of stable cones but it is not known yet how to pass from the classification of cones to the classification of any stable surfaces (due to the lack of perimeter estimates). On the other hand, in the nonlocal setting, we have the energy estimates for stable sets, but the classification of stable s-minimal cones is still widely open.

Concerning the classification of stable s-minimal cones, in [12], Cabré, Serra and the author, proved the following Theorem, which is the first result in the three-dimensional case.

Theorem 5.3 (Theorem 1.2 in [12]) *There exists $s_* \in (0, 1/2)$ such that for every $s \in (s_*, 1/2)$ the following statement holds.*

Let $\Sigma \subset \mathbb{R}^3$ be a cone with nonempty boundary of class C^2 away from 0. Assume that Σ is a stable set for the s-perimeter. Then, Σ is a half-space.

In the proof of Theorem 5.3, the estimate of Theorem 5.2 plays a crucial role, together with several other ingredients, such as the fractional Hardy inequality and some geometric lemmas. We stress that our result is not a perturbative result from $s = 1/2$ which can be obtained by some sort of compactness argument. In fact, a careful inspection of our proof gives an explicit (computable) value for s_*, something impossible when using compactness arguments.

We conclude with some considerations and open problems on the classification of stable solutions for the fractional Allen-Cahn equation. As in the classical setting, when the dimension $n = 2$, one can prove that any bounded stable solution to (3.1) is one-dimensional for any $0 < s < 1$, using the same approach described in Sect. 3. Indeed, in this case a not optimal energy estimate is enough to apply the Liouville-type argument. What about $n > 3$? In the very recent contribution [29], Figalli and Serra proved that when $n = 3$ and $s = 1/2$, any stable bounded solutions to (3.1) is one-dimensional. Again, surprisingly, in the nonlocal case (even if only for the half-Laplacian) something that is not known for the local case, has been established. To conclude, we announce that the forthcoming paper [13] will contain a careful study of stable solutions to the fractional Allen-Cahn equation in the case $0 < s < 1/2$, including energy estimates, density estimates, convergence of blow-down and some new classification results.

Acknowledgements The author is supported by MINECO grants MTM2014-52402-C3-1-P and MTM2017-84214-C2-1-P, and is part of the Catalan research group 2014 SGR 1083.

References

1. G. Alberti, Variational models for phase transitions, an approach via Γ-convergence, in *Calculus of Variations and Partial Differential Equations (Pisa, 1996)* (Springer, Berlin, 2000), pp. 95–114
2. G. Alberti, A. Ambrosio, X. Cabré, On a long-standing conjecture of E. De Giorgi: symmetry in 3D for general nonlinearities and a local minimality property. Acta Appl. Math. **65**, 9–33 (2001)
3. A. Ambrosio, X. Cabré, Entire solutions of semilinear elliptic equations in $\mathbb{R}^3$ and a conjecture of De Giorgi. J. Am. Math. Soc. **13**, 725–739 (2000)
4. L. Ambrosio, G. De Philippis, L. Martinazzi, Gamma-convergence of nonlocal perimeter functionals. Manuscripta Math. **134**(3–4), 377–403 (2011)
5. B. Barrios, A. Figalli, E. Valdinoci, Bootstrap regularity for integro-differential operators and its application to nonlocal minimal surfaces. Ann. Sc. Norm. Super. Pisa Cl. Sci. (5) **13**(3), 609–639 (2014)
6. C. Bucur, E. Valdinoci, *Nonlocal Diffusion and Applications*. Lecture Notes of the Unione Matematica Italiana, vol. 20 (Springer, Berlin, 2016), xii+155 pp.
7. X. Cabré, E. Cinti, Energy estimates and 1-D symmetry for nonlinear equations involving the half-Laplacian. Discrete Contin. Dyn. Syst. **28**(3), 1179–1206 (2010)
8. X. Cabré, E. Cinti, Sharp energy estimates for nonlinear fractional diffusion equations. Calc. Var. Partial Differ. Equ. **49**(1–2), 233–269 (2014)
9. X. Cabré, Y. Sire, Nonlinear equations for fractional Laplacians, I: regularity, maximum principles, and Hamiltonian estimates. Ann. Inst. H. Poincaré Anal. Non Linéaire **31**(1), 23–53 (2014)
10. X. Cabré, Y. Sire, Nonlinear equations for fractional Laplacians II: existence, uniqueness, and qualitative properties of solutions. Trans. Am. Math. Soc. **367**(2), 911–941 (2015)
11. X. Cabré, J. Sola-Morales, Layer solutions in a half-space for boundary reactions. Commun. Pure Appl. Math. **58**(12), 1678–1732 (2005)
12. X. Cabré, E. Cinti, J. Serra, Stable s-minimal cones in $\mathbb{R}^3$ are flat for $s \sim 1$. J. Reine Angew. Math. (to appear, 2017). Available at https://arxiv.org/abs/1710.08722
13. X. Cabré, E. Cinti, J. Serra, Stable solutions to the fractional Allen-Cahn equation. Preprint
14. L. Caffarelli, A. Cordoba, Uniform convergence of a singular perturbation problem. Commun. Pure Appl. Math. **48**, 1–12 (1995)
15. L. Caffarelli, L. Silvestre, An extension problem related to the fractional Laplacian. Commun. Partial Differ. Equ. **32**(7–9), 1245–1260 (2007)
16. L. Caffarelli, E. Valdinoci, Uniform estimates and limiting arguments for nonlocal minimal surfaces. Calc. Var. Partial Differ. Equ. **41**(1–2), 203–240 (2011)
17. L. Caffarelli, E. Valdinoci, Regularity properties of nonlocal minimal surfaces via limiting arguments. Adv. Math. **248**, 843–871 (2013)
18. L. Caffarelli, J.-M. Roquejoffre, O. Savin, Nonlocal minimal surfaces. Commun. Pure Appl. Math. **63**, 1111–1144 (2010)
19. E. Cinti, J. Dávila, M. del Pino, Solutions of the fractional Allen-Cahn equation which are invariant under screw motion. J. Lond. Math. Soc. **94**, 295–313 (2016)
20. E. Cinti, J. Serra, E. Valdinoci, Quantitative flatness results and BV-estimates for stable nonlocal minimal surfaces. J. Differ. Geom. (to appear)
21. T.H. Colding, W.P. Minicozzi II, Estimates for parametric elliptic integrands. Int. Math. Res. Not. **2002**(6), 291–297 (2002)
22. J. Dávila, On an open question about functions of bounded variation. Calc. Var. Partial Differ. Equ. **15**(4), 519–527 (2002)
23. J. Dávila, M. del Pino, J. Wei, Nonlocal s-minimal surfaces and Lawson cones. J. Differ. Geom. **109**(1), 111–175 (2018)
24. S. Dipierro, A. Figalli, G. Palatucci, E. Valdinoci, Asymptotics of the s-perimeter as $s \downarrow 0$. Discrete Contin. Dyn. Syst. **33**(7), 2777–2790 (2013)

25. S. Dipierro, A. Farina, E. Valdinoci, A three-dimensional symmetry result for a phase transition equation in the genuinely nonlocal regime (2017). Preprint. Available at https://arxiv.org/abs/1705.00320
26. S. Dipierro, J. Serra, E. Valdinoci, Improvement of flatness for nonlocal phase transitions. Preprint (2017). Available at https://arxiv.org/abs/1611.10105
27. M. do Carmo, C.K. Peng, Stable complete minimal surfaces in $\mathbb{R}^3$ are planes. Bull. AMS **1**, 903–906 (1979)
28. E.B. Fabes, C.E. Kenig, R.P. Serapioni, The local regularity of solutions of degenerate elliptic equations. Commun. Partial Differ. Equ. **7**, 77–116 (1982)
29. A. Figalli, J. Serra, On stable solutions for boundary reactions: a De Giorgi-type result in dimension 4+1 (2017). Preprint. Available at https://arxiv.org/abs/1705.02781
30. A. Figalli, E. Valdinoci, Regularity and Bernstein-type results for nonlocal minimal surfaces. J. Reine Angew. Math. **729**, 263–273 (2017)
31. A. Figalli, N. Fusco, F. Maggi, V. Millot, M. Morini, Isoperimetry and stability properties of balls with respect to nonlocal energies. Commun. Math. Phys. **336**(1), 441–507 (2015)
32. D. Fischer-Colbrie, R. Schoen, The structure of complete stable minimal surfaces in 3-manifolds of nonnegative scalar curvature. Commun. Pure Appl. Math. **33**, 199–211 (1980)
33. N. Ghoussoub, C. Gui, On a conjecture of De Giorgi and some related problems. Math. Ann. **311**, 481–491 (1998)
34. H.W. Meeks III, Proofs of some classical theorems in minimal surface theory. Indiana Univ. Math. J. **54**(4), 1031–1045 (2005)
35. L. Modica, S. Mortola, Un esempio di Γ^--convergenza (Italian). Boll. Un. Mat. Ital. B (5) **14**(1), 285–299 (1977)
36. L. Moschini, New Liouville theorems for linear second order degenerate elliptic equations in divergence form. Ann. I. H. Poincaré **22**, 11–23 (2005)
37. F. Pacard, J. Wei, Stable solutions of the Allen-Cahn equation in dimension $n = 8$ and minimal cones. J. Funct. Anal. **264**, 1131–1167 (2013)
38. A.V. Pogorelov, On the stability of minimal surfaces. Sov. Math. Dokl. **24**, 274–276 (1981)
39. O. Savin, Regularity of flat level sets in phase transitions. Ann. Math. (2) **169**, 41–78 (2009)
40. O. Savin, Rigidity of minimizers in nonlocal phase transitions. Anal. PDE **11**, 1881–1900 (2018)
41. O. Savin, Rigidity of minimizers in nonlocal phase transitions II. Available at https://arxiv.org/abs/1802.01710
42. O. Savin, E. Valdinoci, Γ-Convergence for nonlocal phase transitions. Ann. Inst. H. Poincaré Anal. Non Lineaire **29**(4), 479–500 (2012)
43. O. Savin, E. Valdinoci, Regularity of nonlocal minimal cones in dimension 2. Calc. Var. Partial Differ. Equ. **48**(1–2), 33–39 (2013)
44. O. Savin, E. Valdinoci, Some monotonicity results for minimizers in the calculus of variations. J. Funct. Anal. **264**(10), 2469–2496 (2013)
45. O. Savin, E. Valdinoci, Density estimates for a variational model driven by the Gagliardo norm. J. Math. Pures Appl. **101**, 1–26 (2014)
46. Y. Sire, E. Valdinoci, Fractional Laplacian phase transitions and boundary reactions: a geometric inequality and a symmetry result. J. Funct. Anal. **256**(6), 1842–1864 (2009)
47. B. White, Lectures on Minimal Surface Theory (2013). arXiv:1308.3325

Fractional De Giorgi Classes and Applications to Nonlocal Regularity Theory

Matteo Cozzi

Abstract We present some recent results obtained by the author on the regularity of solutions to nonlocal variational problems. In particular, we review the notion of fractional De Giorgi class, explain its role in nonlocal regularity theory, and propose some open questions in the subject.

Keywords Fractional De Giorgi classes · Nonlocal Caccioppoli inequality · Hölder continuity · Harnack inequality · Nonlocal functionals · Nonlinear integral operators

2010 Mathematics Subject Classification 49N60, 35B45, 35B50, 35B65, 35R11, 47G20

1 Introduction

De Giorgi classes are a powerful tool in the regularity theory of Partial Differential Equations and Calculus of Variations. By definition, their elements are functions that belong to a Sobolev space and satisfy Caccioppoli inequalities at all of their levels. Their introduction can be dated back to the fundamental work of De Giorgi [10], where he devised them to prove the Hölder continuity of solutions to second order equations in divergence form with bounded measurable coefficients. Later on, Giaquinta and Giusti [15] discovered that De Giorgi classes could also be utilized to prove Hölder estimates for minimizers of non-differentiable functionals, one of the

This note is mostly based on a talk given by the author at a conference held in Bari on May 29–30, 2017, as part of the INdAM intensive period "Contemporary research in elliptic PDEs and related topics".

M. Cozzi (✉)
University of Bath, Department of Mathematical Sciences, Bath, UK
e-mail: m.cozzi@bath.ac.uk

S. Dipierro (ed.), *Contemporary Research in Elliptic PDEs and Related Topics*, Springer INdAM Series 33, https://doi.org/10.1007/978-3-030-18921-1_7

first general regularity results that did not make use of the Euler-Lagrange equation. A couple of years later, DiBenedetto and Trudinger [14] showed that De Giorgi classes are not only responsible for continuity properties, but also lead to Harnack inequalities. See the classical books [17, 21], and the more recent [16] for additional information.

The aim of this work is to review and further enrich the theory developed by the author in [8], about fractional notions of De Giorgi classes and their applications to the regularity properties of solutions to nonlocal variational problems.

We consider the class $\widetilde{\mathrm{DG}}^{s,p}$ and its subclass $\mathrm{DG}^{s,p}$, both made up by functions that are contained in a Sobolev space of fractional order and satisfy a family of nonlocal Caccioppoli-type estimates. The inequality defining $\widetilde{\mathrm{DG}}^{s,p}$ has a somewhat similar structure to that of standard De Giorgi classes and it is by now fairly understood, thanks to a number of contributions available in the literature, such as [1, 12, 20]. On the other hand, the inequality that corresponds to the class $\mathrm{DG}^{s,p}$ is stronger and incorporates a purely nonlocal term that has no classical counterpart. To the best of our knowledge, this last inequality has been previously considered only by Caffarelli et al. [7] in a nonlocal parabolic context.

Throughout Sect. 2 we state several results pertaining to these classes. In particular, we establish that:

(a) the elements of the class $\widetilde{\mathrm{DG}}^{s,p}$ (and, therefore, of its smaller subset $\mathrm{DG}^{s,p}$) are locally bounded functions—see Theorem 2.2;
(b) the functions of $\mathrm{DG}^{s,p}$ are locally uniformly Hölder continuous—see Theorem 2.5;
(c) Harnack-type inequalities are true for functions that belong to $\mathrm{DG}^{s,p}$ and that are non-negative—see Theorems 2.9 and 2.10.

Furthermore, in Appendix A we show by means of an explicit example that, for some choices of the parameters s and p, the results of points (b) and (c) cannot be extended to the larger class $\widetilde{\mathrm{DG}}^{s,p}$.

Sections 3 and 4 are devoted to applications in the regularity theory for nonlocal variational problems. In Sect. 3 we deal with minimizers of energy functionals obtained as the sum of a possibly non-differentiable potential and of an interaction term comparable to the Gagliardo seminorm of a fractional Sobolev space. By showing that these extrema are contained in the fractional De Giorgi class $\mathrm{DG}^{s,p}$, we deduce their Hölder continuity and the validity of Harnack inequalities, thanks to the statements of Sect. 2. In Sect. 4 we approach in a similar way the regularity of solutions to equations driven by singular integral operators, such as fractional Laplacians and other nonlinear variations. These results complement and extend several available contributions, as for instance [11, 12, 18, 25].

2 Fractional De Giorgi Classes

We begin by introducing the larger set $\widetilde{\mathrm{DG}}^{s,p}$, which we will sometimes call *weak fractional De Giorgi class*. To do this, we first need to fix some terminology.

Unless otherwise stated, throughout the whole paper $n \geqslant 1$ is an integer indicating the dimension of the Euclidean space under consideration, $s \in (0, 1)$ is a parameter representing a fractional order of differentiability, and $p > 1$ is an integrability exponent. Also, Ω always denotes a bounded open subset of the space $\mathbb{R}^n$.

With the symbol $W^{s,p}(\Omega)$ we denote the fractional Sobolev space composed by those functions u that lie in the Lebesgue space $L^p(\Omega)$ and have finite Gagliardo seminorm

$$[u]_{W^{s,p}(\Omega)} := \left(\int_\Omega \int_\Omega \frac{|u(x) - u(y)|^p}{|x - y|^{n+sp}} \, dxdy \right)^{\frac{1}{p}}.$$

As it is customary, we endow $W^{s,p}(\Omega)$ with the norm $\| \cdot \|_{W^{s,p}(\Omega)}$ defined by the identity $\|u\|^p_{W^{s,p}(\Omega)} := \|u\|^p_{L^p(\Omega)} + [u]^p_{W^{s,p}(\Omega)}$ and we simply write $H^s(\Omega) := W^{s,2}(\Omega)$ when $p = 2$.

Another functional space that we will often use is the weighted Lebesgue space $L^{p-1}_s(\mathbb{R}^n)$ made up by all measurable functions $u : \mathbb{R}^n \to \mathbb{R}$ for which

$$\int_{\mathbb{R}^n} \frac{|u(x)|^{p-1}}{(1 + |x|)^{n+sp}} \, dx < +\infty.$$

For $u \in L^{p-1}_s(\mathbb{R}^n)$, the quantities

$$\mathrm{Tail}_{s,p}(u; x_0, R) := \left(R^{sp} \int_{\mathbb{R}^n \setminus B_R(x_0)} \frac{|u(x)|^{p-1}}{|x - x_0|^{n+sp}} \, dx \right)^{\frac{1}{p-1}} \tag{2.1}$$

and $\overline{\mathrm{Tail}}_{s,p}(u; x_0, R) := R^{-\frac{sp}{p-1}} \mathrm{Tail}_{s,p}(u; x_0, R)$ are finite for every point $x_0 \in \mathbb{R}^n$ and every radius $R > 0$. The tail term (2.1)—introduced in [11, 12]—is conveniently used to describe the behavior of u far away from x_0. When x_0 is the origin of $\mathbb{R}^n$, we just write $B_R := B_R(0)$, $\mathrm{Tail}_{s,p}(u; R) := \mathrm{Tail}_{s,p}(u; 0, R)$, and $\overline{\mathrm{Tail}}_{s,p}(u; R) := \overline{\mathrm{Tail}}_{s,p}(u; 0, R)$.

For $k \in \mathbb{R}$, we indicate the super- and sublevel sets of a function $u : \mathbb{R}^n \to \mathbb{R}$ respectively with $A^+(k)$ and $A^-(k)$. In symbols,

$$A^+(k) := \{u > k\} \quad \text{and} \quad A^-(k) := \{u < k\}.$$

We also write $A^\pm(k, x_0, R) := A^\pm(k) \cap B_R(x_0)$ for their intersections with the ball $B_R(x_0)$ and, as before, $A^\pm(k, R) := A^\pm(k, 0, R)$.

Finally, $v_+ := \max\{v, 0\}$ and $v_- := (-v)_+ = \max\{-v, 0\}$ indicate respectively the positive and negative parts of a function v.

With this in hand, we can now state the definition of *weak fractional De Giorgi class*.

Definition 2.1 (Weak Fractional De Giorgi Class $\widetilde{\mathrm{DG}}^{s,p}$) Let $d, \lambda \geqslant 0$ and $H \geqslant 1$. A function $u \in L_s^{p-1}(\mathbb{R}^n)$ with $u|_\Omega \in W^{s,p}(\Omega)$ belongs to $\widetilde{\mathrm{DG}}_\pm^{s,p}(\Omega; d, H, \lambda)$ if

$$
\begin{aligned}
&[(u-k)_\pm]^p_{W^{s,p}(B_r(x_0))} \\
&\quad \leqslant H \Bigg\{ R^\lambda d^p |A^\pm(k, x_0, R)| + \frac{R^{(1-s)p}}{(R-r)^p} \|(u-k)_\pm\|^p_{L^p(B_R(x_0))} \\
&\qquad + \frac{R^{n+sp}}{(R-r)^{n+sp}} \|(u-k)_\pm\|_{L^1(B_R(x_0))} \overline{\mathrm{Tail}}_{s,p}((u-k)_\pm; x_0, r)^{p-1} \Bigg\}
\end{aligned}
\tag{2.2}
$$

holds for every point $x_0 \in \Omega$, radii $0 < r < R < \mathrm{dist}(x_0, \partial\Omega)$, and level $k \in \mathbb{R}$. In addition, $u \in \widetilde{\mathrm{DG}}^{s,p}(\Omega; d, H, \lambda)$ if and only if $u \in \widetilde{\mathrm{DG}}_+^{s,p}(\Omega; d, H, \lambda) \cap \widetilde{\mathrm{DG}}_-^{s,p}(\Omega; d, H, \lambda)$.

According to (2.2), functions in $\widetilde{\mathrm{DG}}^{s,p}$ satisfy a fractional and nonlocal version of the usual Caccioppoli inequality at all levels k. Broader definitions can be considered, along the lines of those of [8, Section 6]. Here, we preferred to keep things as simple as possible, in order to favor readability over generality. Of course, we could take into account an even simpler definition, by removing the last line of (2.2) and thus neglecting the presence of tail terms. This choice would certainly be more elegant, as then the class $\widetilde{\mathrm{DG}}^{s,p}(\Omega; d, H, \lambda)$ would be a subset of the Sobolev space $W^{s,p}(\Omega)$. However, this definition would be too restrictive in light of our applications in Sects. 3 and 4, which ultimately motivate the structure of (2.2).

As for their classical counterparts, prototypical examples of functions belonging to weak fractional De Giorgi classes are the solutions of elliptic equations. While for standard De Giorgi classes, these equations are second order PDEs, the ones that are naturally associated to $\widetilde{\mathrm{DG}}^{s,p}$ are fractional order equations driven by singular integral operators, such as the fractional Laplacian and nonlinear variations. This connection has been already observed by many authors—see, e.g., [1, 12, 20].

As it has been partially anticipated in the introduction, classical De Giorgi classes were introduced for their importance in the regularity theory for second order equations, as they encode virtually all the information concerning the basic regularity properties enjoyed by the solutions of such equations—namely, local boundedness, Hölder continuity, and the validity of Harnack inequalities. The first goal of this section is to discuss whether these properties continue to hold for the fractional class $\widetilde{\mathrm{DG}}^{s,p}$.

As a first observation, we show that the elements of $\widetilde{\mathrm{DG}}^{s,p}$ are locally bounded functions. Of course, when $n < sp$ their boundedness (and Hölder continuity) is guaranteed by the Morrey-type embedding $W^{s,p} \hookrightarrow C^{s-n/p}$ (see, e.g., [13, Theorem 8.2]). Hence, at least for what concerns regularity, we can restrict ourselves to dealing with the case of $n \geqslant sp$. For the sake of a simpler exposition, we will in fact suppose throughout the whole section that $n > sp$. We stress that the critical case $n = sp$—which is excluded here—only poses few additional technical difficulties and can be treated similarly—see [8, Section 6].

Theorem 2.2 (Local Boundedness of Functions in $\widetilde{\mathrm{DG}}^{s,p}$) *Let $u \in \widetilde{\mathrm{DG}}^{s,p}(\Omega; d, H, \lambda)$ for some $d, \lambda \geqslant 0$ and $H \geqslant 1$. Then, $u \in L^\infty_{\mathrm{loc}}(\Omega)$ and there exists a constant $C \geqslant 1$, depending only on n, s, p, and H, such that*

$$\|u\|_{L^\infty(B_R(x_0))} \leqslant C \left\{ \left(\fint_{B_{2R}(x_0)} |u(x)|^p \, dx \right)^{\frac{1}{p}} + \mathrm{Tail}_{s,p}(u; x_0, R) + R^{\frac{\lambda+sp}{p}} d \right\}$$

for every $x_0 \in \Omega$ and $0 < R < \mathrm{dist}\,(x_0, \partial\Omega)/2$.

We observe that a different version of Theorem 2.2, valid for variants of weak fractional De Giorgi classes that do not include the presence of a tail term on the right-hand side of (2.2) and for $p = 1$, is contained in [22].

The estimate of Theorem 2.2 follows from analogous one-sided bounds for the elements of $\widetilde{\mathrm{DG}}^{s,p}_+$ and $\widetilde{\mathrm{DG}}^{s,p}_-$. By symmetry, it suffices to prove the following result.

Proposition 2.3 *Let $u \in \widetilde{\mathrm{DG}}^{s,p}_+(\Omega; d, H, \lambda)$ for some $d, \lambda \geqslant 0$ and $H \geqslant 1$. Then, there exists a constant $C \geqslant 1$, depending only on n, s, p, and H, such that*

$$\sup_{B_R(x_0)} u \leqslant C \left\{ \left(\fint_{B_{2R}(x_0)} u_+(x)^p \, dx \right)^{\frac{1}{p}} + \mathrm{Tail}_{s,p}(u_+; x_0, R) + R^{\frac{\lambda+sp}{p}} d \right\} \tag{2.3}$$

for every $x_0 \in \Omega$ and $0 < R < \mathrm{dist}\,(x_0, \partial\Omega)/2$.

Proof Our argument is a simple variation of the one that leads to, say, [17, Theorem 7.2].

Up to a translation, we may assume that x_0 is the origin. Let two radii $R \leqslant \rho < \tau \leqslant 2R$ be fixed, take $k \geqslant 0$, and set $w_k := (u - k)_+$. Using Hölder and fractional Sobolev inequalities, it is not hard to infer that

$$\|w_k\|^p_{L^p(B_\rho)} \leqslant C |A^+(k, \rho)|^{sp/n} \left([w_k]^p_{W^{s,p}(B_{(\tau+\rho)/2})} + \frac{\tau^{(1-s)p}}{(\tau - \rho)^p} \|w_k\|^p_{L^p(B_\tau)} \right),$$

for some constant $C \geqslant 1$ depending only on n, s, and p. This and (2.2) give

$$\|w_k\|^p_{L^p(B_\rho)} \leqslant C|A^+(k,\rho)|^{sp/n}\left\{\tau^\lambda d^p|A^+(k,\tau)| + \frac{\tau^{(1-s)p}}{(\tau-\rho)^p}\|w_k\|^p_{L^p(B_\tau)} + \frac{\tau^{n+sp}}{(\tau-\rho)^{n+sp}}\|w_k\|_{L^1(B_\tau)}\overline{\mathrm{Tail}}_{s,p}(w_k;R)^{p-1}\right\}, \tag{2.4}$$

where C may now depend on H as well.

Letting $0 \leqslant h < k$, it is easy to see that

$$|A^+(k,r)| \leqslant \frac{\|w_h\|^p_{L^p(B_r)}}{(k-h)^p}, \quad \|w_k\|^p_{L^p(B_r)} \leqslant \|w_h\|^p_{L^p(B_r)}, \quad \|w_k\|_{L^1(B_r)} \leqslant \frac{\|w_h\|^p_{L^p(B_r)}}{(k-h)^{p-1}},$$

and

$$\overline{\mathrm{Tail}}_{s,p}(w_k;r)^{p-1} \leqslant \overline{\mathrm{Tail}}_{s,p}(w_0;r)^{p-1} = r^{-sp}\,\mathrm{Tail}_{s,p}(u_+;r)^{p-1}$$

for every $r > 0$. Accordingly, (2.4) yields the estimate

$$\varphi(k,\rho) \leqslant \frac{C\tau^{-sp}}{(k-h)^{\frac{sp^2}{n}}}\left\{\frac{\tau^{\lambda+sp}d^p}{(k-h)^p} + \frac{\tau^p}{(\tau-\rho)^p} + \frac{\tau^{n+sp}\,\mathrm{Tail}_{s,p}(u_+;R)^{p-1}}{(\tau-\rho)^{n+sp}(k-h)^{p-1}}\right\}\varphi(h,\tau)^{1+\frac{sp}{n}}$$

for the quantities $\varphi(\ell,r) := \|w_\ell\|^p_{L^p(B_r)}$.

Consider now the sequences $\{k_i\}$ and $\{\rho_i\}$, respectively defined by $k_i := M(1-2^{-i})$ and $\rho_i := (1+2^{-i})R$ for all integers $i \geqslant 0$ and for some $M > 0$ to be chosen later. Set $\varphi_i := \varphi(k_i,\rho_i)$. By evaluating the last inequality along these two sequences, we obtain

$$\varphi_{i+1} \leqslant \frac{C\,2^{(n+3p)i}}{M^{sp^2/n}R^{sp}}\left\{\frac{R^{\lambda+sp}d^p}{M^p} + 1 + \frac{\mathrm{Tail}_{s,p}(u_+;R)^{p-1}}{M^{p-1}}\right\}\varphi_i^{1+\frac{sp}{n}}.$$

By choosing $M \geqslant M_1 := \mathrm{Tail}_{s,p}(u_+;R) + R^{(\lambda+sp)/p}d$, we are finally led to the estimate

$$\varphi_{i+1} \leqslant \frac{C\,2^{(n+3p)i}}{M^{sp^2/n}R^{sp}}\varphi_i^{1+\frac{sp}{n}}.$$

Thanks to a standard numerical lemma (e.g., [17, Lemma 7.1]), we conclude that φ_i converges to 0, provided M is greater than the constant $M_2 := C'R^{-n/p}\|u_+\|_{L^p(B_{2R})}$ with $C' \geqslant 1$ large enough, in dependence of n, s, p, and H only. This gives (2.3). □

Following the theory of classical De Giorgi classes, the natural next step would be to understand whether the functions of $\widetilde{\mathrm{DG}}^{s,p}$ are Hölder continuous. It turns out that this is not the case, at least when $sp < 1$. This is a consequence of an explicit one-dimensional example that we will present in Appendix A.

Question 1 Is it true that functions in $\widetilde{\mathrm{DG}}^{s,p}$ are Hölder continuous, when $sp \geqslant 1$?

In order to extend the Hölder regularity estimates that are true for classical De Giorgi classes, we are thus forced in general to consider a strict subset of $\widetilde{\mathrm{DG}}^{s,p}$. To this aim, we propose the following definition.

Definition 2.4 (Fractional De Giorgi Class $\mathrm{DG}^{s,p}$) Let $d, \lambda \geqslant 0$ and $H \geqslant 1$. A function $u \in L_s^{p-1}(\mathbb{R}^n)$ with $u|_\Omega \in W^{s,p}(\Omega)$ belongs to $u \in \mathrm{DG}_\pm^{s,p}(\Omega; d, H, \lambda)$ if

$$\begin{aligned} &[(u-k)_\pm]^p_{W^{s,p}(B_r(x_0))} + \int_{B_r(x_0)} (u(x)-k)_\pm \left\{ \int_{\mathbb{R}^n} \frac{(u(y)-k)_\mp^{p-1}}{|x-y|^{n+sp}}\, dy \right\} dx \\ &\quad \leqslant H \left\{ R^\lambda d^p |A^\pm(k, x_0, R)| + \frac{R^{(1-s)p}}{(R-r)^p} \|(u-k)_\pm\|^p_{L^p(B_R(x_0))} \right. \\ &\qquad \left. + \frac{R^{n+sp}}{(R-r)^{n+sp}} \|(u-k)_\pm\|_{L^1(B_R(x_0))} \overline{\mathrm{Tail}}_{s,p}((u-k)_\pm; x_0, r)^{p-1} \right\} \end{aligned} \tag{2.5}$$

holds for every point $x_0 \in \Omega$, radii $0 < r < R < \operatorname{dist}(x_0, \partial\Omega)$, and level $k \in \mathbb{R}$. We then set $\mathrm{DG}^{s,p}(\Omega; d, H, \lambda) := \mathrm{DG}_+^{s,p}(\Omega; d, H, \lambda) \cap \mathrm{DG}_-^{s,p}(\Omega; d, H, \lambda)$.

We will call $\mathrm{DG}^{s,p}$ a *strong fractional De Giorgi class* or, simply, a *fractional De Giorgi class*. It is clear that $\mathrm{DG}^{s,p}$ is a subset of $\widetilde{\mathrm{DG}}^{s,p}$. The difference between the two classes lies in the fact that the elements of $\mathrm{DG}^{s,p}$ satisfy the stronger Caccioppoli-type inequality (2.5), which improves (2.2) via the presence of an additional summand on its left-hand side. We remark that the specific structure of this term can be partially altered without totally spoiling the results that will follow in the remainder of the section. For instance, if one replaces it with the smaller (and, perhaps, more natural) quantity

$$\int_{B_r(x_0)} \int_{B_r(x_0)} \frac{(u(x)-k)_\pm (u(y)-k)_\mp^{p-1}}{|x-y|^{n+sp}}\, dx dy,$$

all future statements will still hold, apart from the Harnack inequality of Theorem 2.9.

Though more artificial than (2.2), inequality (2.5) is still satisfied by solutions of problems involving energies and operators of fractional order, as we will see in Sects. 3 and 4. In addition, it turns out that definition (2.5) is strong enough to guarantee the Hölder continuity of the functions that satisfy it. This has been first realized by Caffarelli et al. [7] for a similar inequality in the context of nonlocal parabolic equations.

Here is our Hölder regularity result for functions in $\mathrm{DG}^{s,p}$.

Theorem 2.5 (Hölder Continuity of Functions in $\mathrm{DG}^{s,p}$**)** *Let $u \in \mathrm{DG}^{s,p}(\Omega; d, H, \lambda)$ for some $d, \lambda \geqslant 0$ and $H \geqslant 1$. Then, $u \in C^{\alpha}_{\mathrm{loc}}(\Omega)$ for some $\alpha \in (0,1)$ and there exists a constant $C \geqslant 1$ such that*

$$[u]_{C^{\alpha}(B_R(x_0))} \leqslant \frac{C}{R^{\alpha}} \left(\|u\|_{L^{\infty}(B_{2R}(x_0))} + \mathrm{Tail}_{s,p}(u; x_0, 2R) + R^{\frac{\lambda+sp}{p}} d \right)$$

for every $x_0 \in \Omega$ and $0 < R < \mathrm{dist}\,(x_0, \partial\Omega)\,/2$. The constants α and C depend only on n, s, p, H, and λ.

Theorem 2.5 can be proved via an inductive argument based on subsequent applications of a suitable *growth lemma* at smaller and smaller scales. This method goes back to De Giorgi [10] and our implementation of it in this framework follows rather closely the approaches of Silvestre [25], Kassmann [18, 19], and Caffarelli and Vasseur [4]. We omit further details, that can be found in the proof of [8, Theorem 6.4].

The statement of the growth lemma is as follows.

Lemma 2.6 *Let $u \in \mathrm{DG}^{s,p}_{-}(B_{4R}; d, H, \lambda)$ for some $d, \lambda \geqslant 0$, $H \geqslant 1$, and $R > 0$. Assume that*

$$u \geqslant 0 \quad \text{in } B_{4R}$$

and

$$|B_{2R} \cap \{u \geqslant 1\}| \geqslant \frac{1}{2} |B_{2R}| \, .$$

There exists a constant $\delta \in (0, 1/8]$, depending only on n, s, p, H, and λ, such that, if

$$R^{\frac{\lambda+sp}{p}} d + \mathrm{Tail}_{s,p}(u_{-}; 4R) \leqslant \delta,$$

then

$$u \geqslant \delta \quad \text{in } B_R.$$

We split the proof of Lemma 2.6 into two sublemmata. Interestingly, the first one only relies on the weaker Caccioppoli inequality (2.2) and is therefore valid for all functions in $\widetilde{\mathrm{DG}}^{s,p}_{-}$.

Lemma 2.7 *Let $u \in \widetilde{\mathrm{DG}}^{s,p}_{-}(B_4; d, H, \lambda)$ for some $d, \lambda \geqslant 0$ and $H \geqslant 1$. There exists a constant $\tau \in (0, 2^{-n-1}]$, depending only on n, s, p, H, and λ, such that* *if* $u \geqslant 0$ *in* B_2,

$$|B_2 \cap \{u < 2\delta\}| \leqslant \tau \, |B_2| \,, \tag{2.6}$$

and

$$d + \mathrm{Tail}_{s,p}(u_-; 2) \leqslant \delta, \tag{2.7}$$

for some $\delta \in (0, 1/2]$, *then*

$$u \geqslant \delta \quad in \ B_1. \tag{2.8}$$

Proof Let $\delta \leqslant h < k \leqslant 2\delta$ and $1 \leqslant \rho < r \leqslant 2$ be fixed, and $\tau \in (0, 2^{-n-1}]$ to be later taken small. Setting $z_k := (u-k)_-$, we first observe that, by (2.6) and the fact that $\tau \leqslant 2^{-n-1}$, it holds

$$\big|B_\rho \cap \{z_k = 0\}\big| = \big|B_\rho \setminus \{u < k\}\big| \geqslant |B_\rho| - \big|B_\rho \cap \{u < 2\delta\}\big| \geqslant |B_\rho| - \tau |B_2| \geqslant |B_\rho|/2.$$

By this, we may apply the fractional Sobolev inequality for functions that vanish over a set with positive density (see, e.g., [8, Corollary 4.9]) and get that

$$\begin{aligned}(k-h)|A^-(h,\rho)|^{\frac{2n-s}{2n}} &\leqslant \left(\int_{B_\rho} z_k(x)^{\frac{2n}{2n-s}}\,dx\right)^{\frac{2n-s}{2n}} \leqslant C\int_{A^-(k,\rho)}\int_{B_\rho} \frac{|z_k(x)-z_k(y)|}{|x-y|^{n+\frac{s}{2}}}\,dxdy \\ &\leqslant C|A^-(k,\rho)|^{\frac{p-1}{p}}[z_k]_{W^{s,p}(B_\rho)},\end{aligned}$$

for some constant $C \geqslant 1$ depending only on n, s, and p. Note that the last estimate follows by Hölder's inequality—see, e.g., [8, Lemma 4.6] for the detailed computation. Taking advantage of (2.2), we further obtain that

$$\begin{aligned}&(k-h)^p|A^-(h,\rho)|^{\frac{(2n-s)p}{2n}} \\ &\qquad \leqslant C|A^-(k,\rho)|^{p-1}\left\{d^p|A^-(k,r)| + \frac{\|z_k\|^p_{L^p(B_r)}}{(r-\rho)^p} + \frac{\|z_k\|_{L^1(B_r)}\overline{\mathrm{Tail}}_{s,p}(z_k;\rho)^{p-1}}{(r-\rho)^{n+sp}}\right\},\end{aligned}$$

where C may now also depend on H and λ. Now, thanks to assumption (2.7), the non-negativity of u in B_2, and the fact that $\delta \leqslant k$, from the previous inequality we easily deduce that

$$|A^-(h,\rho)|^{\frac{2n-s}{2n}} \leqslant \frac{C\,k}{(r-\rho)^{\frac{n+p}{p}}(k-h)}\,|A^-(k,r)|.$$

By evaluating this inequality along two sequences of radii $\{\rho_i\}$ and of levels $\{k_i\}$—exponentially decreasing from 2 to 1 and from 2δ to δ, respectively—and arguing as in the last part of the proof of Proposition 2.3, we are led to the conclusion (2.8), provided τ is chosen sufficiently small. □

The second step in the proof of Lemma 2.6 is represented by the next result. Unlike Lemma 2.7, it heavily relies on the presence of the second term on the left-hand side of (2.5) and, therefore, it only holds true for functions in the smaller class $\mathrm{DG}_-^{s,p}$.

Lemma 2.8 *Let $u \in \mathrm{DG}_-^{s,p}(B_4; d, H, \lambda)$ for some $d, \lambda \geqslant 0$, $H \geqslant 1$. For every $\tau \in (0, 1)$, there exists $\delta \in (0, 1/8]$, depending only on n, s, p, H, λ, and τ, such that if $u \geqslant 0$ in B_4,*

$$|B_2 \cap \{u \geqslant 1\}| \geqslant \frac{1}{2} |B_2| , \tag{2.9}$$

and

$$d + \mathrm{Tail}_{s,p}(u_-; 4) \leqslant \delta,$$

then

$$|B_2 \cap \{u < 2\delta\}| \leqslant \tau |B_2| . \tag{2.10}$$

The lemma tells that if u is a non-negative function belonging to $\mathrm{DG}_-^{s,p}$ (with, say, $d = 0$) and for which $\{u \geqslant 1\}$ has positive measure in B_2, then the measure of the sublevel set $B_2 \cap \{u < 2\delta\}$ goes to zero as $\delta \searrow 0$.

For classical De Giorgi classes, this result is usually proved by estimating the mass that is lost by passing from the level $\{u < \delta_2\}$ to $\{u < \delta_1\}$, with $0 < \delta_1 < \delta_2$ small. A crucial role in this argument is played by an isoperimetric-type inequality valid for level sets of functions in $W^{1,p}$ and established by De Giorgi in [10]. This inequality gives a quantification of the fact that classical Sobolev functions cannot have jump discontinuities.

As will be discussed more extensively at the end of this section, inequalities like De Giorgi's may not hold true in fractional Sobolev spaces—indeed, step functions may belong to $W^{s,p}$ when $sp \leqslant 1$. Estimate (2.5), when used to bound the second term on its left-hand side, provides an alternative inequality to De Giorgi's, no longer holding for all functions of $W^{s,p}$ but only for those that lie in $\mathrm{DG}^{s,p}$. Proposition A.1 in Appendix A shows that, instead, estimate (2.2) alone does not lead in general to a similar conclusion.

Here below we make this argument rigorous and establish Lemma 2.8 through it.

Proof of Lemma 2.8 We apply (2.5) with $x_0 = 0$, $r = 2$, $R = 3$, and $k = 4\delta$, for some $\delta \in (0, 1/8]$ to be determined. By arguing as in the last part of the proof of Lemma 2.7, it is easy to see that the right-hand side of (2.5) can be controlled from above by $C\delta^p$, for some constant $C \geqslant 1$ depending only on n, s, p, H, and λ. On the other hand, its left-hand side—and, in particular, its second summand—is larger than

$$\int_{B_2} (4\delta - u(x))_+ \left\{ \int_{B_2} \frac{(u(y) - 4\delta)_+^{p-1}}{|x-y|^{n+sp}} \, dy \right\} dx \geqslant c\, \delta \, |B_2 \cap \{u < 2\delta\}| \, |B_2 \cap \{u \geqslant 1\}| \,,$$

for some constant $c \in (0, 1)$ depending only on n, s, and p. See the end of the proof of [8, Lemma 6.3] for more details. By combining these two facts and recalling hypothesis (2.9), we deduce that $|B_2 \cap \{u < 2\delta\}| \leqslant C\delta^{p-1}$, from which (2.10) readily follows, provided δ is small enough. □

Since, by scaling, we may reduce ourselves to the case of $R = 1$, it is clear that the joint application of Lemmata 2.7 and 2.8 leads to Lemma 2.6.

The growth lemma is the key ingredient of another important result valid for the elements of the class $\mathrm{DG}^{s,p}$: the Harnack inequality.

Theorem 2.9 (Harnack Inequality for $\mathrm{DG}^{s,p}$**)** *Let* $u \in \mathrm{DG}^{s,p}(\Omega; d, H, \lambda)$ *for some constants* $d, \lambda \geqslant 0$ *and* $H \geqslant 1$*. There exists a constant* $C \geqslant 1$*, depending on* n*,* s*,* p*,* λ*, and* H*, such that, if* $u \geqslant 0$ *in* Ω*, then*

$$\sup_{B_R(x_0)} u + \mathrm{Tail}_{s,p}(u_+; x_0, R) \leqslant C \left(\inf_{B_R(x_0)} u + \mathrm{Tail}_{s,p}(u_-; x_0, R) + R^{\frac{\lambda+sp}{p}} d \right) \tag{2.11}$$

for every $x_0 \in \Omega$ *and* $0 < R < \mathrm{dist}(x_0, \partial\Omega)/2$*.*

Notice the presence of tail terms on both sides of the inequality. The one on the right accounts for the possible negativity of u outside of Ω and cannot be removed, as it was noticed by Kassmann [19] for s-harmonic functions. Conversely, the one on the left makes the inequality stronger. To the best of our knowledge, the possibility of including such a term was first realized by Ros-Oton and Serra [23] in the case of the *weak* Harnack inequality (see the forthcoming Theorem 2.10) for supersolutions of fully nonlinear nonlocal equations. We also mention the recent [3], by Cabré and the author of this note, where the presence of this extra term is crucially exploited to obtain a gradient bound for nonlocal minimal graphs.

As for the Hölder continuity result, Theorem 2.9 does not hold for the elements of the larger class $\widetilde{\mathrm{DG}}^{s,p}$ when $sp < 1$, in view of Proposition A.1.

To obtain Theorem 2.9, we first establish the aforementioned weak Harnack inequality for the class $\mathrm{DG}_-^{s,p}$.

Theorem 2.10 (Weak Harnack Inequality for $\mathrm{DG}_-^{s,p}$**)** *Let* $u \in \mathrm{DG}_-^{s,p}(\Omega; d, H, \lambda)$ *for some* $d, \lambda \geqslant 0$ *and* $H \geqslant 1$*. There exist an exponent* $\varepsilon > 0$ *and a constant* $C \geqslant 1$*, both depending only on* n*,* s*,* p*,* λ*, and* H*, such that, if* $u \geqslant 0$ *in* Ω*, then*

$$\left(\fint_{B_R(x_0)} u(x)^\varepsilon \, dx \right)^{\frac{1}{\varepsilon}} \leqslant C \left(\inf_{B_R(x_0)} u + \mathrm{Tail}_{s,p}(u_-; x_0, R) + R^{\frac{\lambda+sp}{p}} d \right) \tag{2.12}$$

for every $x_0 \in \Omega$ *and* $0 < R < \mathrm{dist}(x_0, \partial\Omega)/2$*.*

For the sake of conciseness, we do not include here the proof of Theorem 2.10, which essentially relies on a scaled version of Lemma 2.6 along with a Krylov-Safonov-type covering lemma. This argument is similar to the one developed in [14, Section 3] for classical De Giorgi classes and can be found in [8, Subsection 6.4].

Question 2 Is it possible to establish a weak Harnack inequality for functions in $\mathrm{DG}_-^{s,p}$ identical in structure to those of [23, Theorem 2.2] and [3, Theorem 1.6]? That is, does (2.12) hold with $\varepsilon = 1$ and with the additional term $\mathrm{Tail}_{s,p}(u_+; x_0, R)$ added on the left-hand side, such as in (2.11)?

Next is the following result, that reduces (2.11) to the verification of the corresponding inequality for the essential supremum of u in $B_R(x_0)$ only.

Lemma 2.11 *Let $u \in \mathrm{DG}_-^{s,p}(B_R; d, H, \lambda)$ for some $d, \lambda \geqslant 0$, $H \geqslant 1$, and $R > 0$. There is a constant $C \geqslant 1$, depending only on n, s, p, and H, such that, if $u \geqslant 0$ in B_R, then*

$$\mathrm{Tail}_{s,p}(u_+; R) \leqslant C\left(\sup_{B_R} u + \mathrm{Tail}_{s,p}(u_-; R) + R^{\frac{\lambda+sp}{p}} d\right).$$

Proof It suffices to apply inequality (2.5) with $x_0 = 0$, $r = R/2$, and $k = 2M$, where we set $M := \sup_{B_R} u + R^{(\lambda+sp)/p} d$. On the one hand, it is not hard to see that

$$\int_{B_{R/2}} (u(x) - 2M)_- \left\{ \int_{\mathbb{R}^n} \frac{(u(y) - 2M)_+^{p-1}}{|x-y|^{n+sp}} \, dy \right\} dx$$

$$\geqslant \frac{MR^n}{C} \int_{\mathbb{R}^n \setminus B_R} \frac{(u(y) - 2M)_+^{p-1}}{|y|^{n+sp}} \, dy \geqslant R^{n-sp} \left\{ \frac{M}{C} \mathrm{Tail}_{s,p}(u_+; R)^{p-1} - CM^p \right\},$$

for some constant $C \geqslant 1$ depending only on n, s, and p. Note that the last inequality in the above formula is immediate for $p = 2$. For a general $p > 1$, one may deduce it using a numerical inequality such as the one provided by [8, Lemma 4.4]—see the beginning of the proof of [8, Theorem 6.9] for more details. On the other hand, the right-hand side of (2.5) is controlled by $CR^{n-sp}\left\{M^p + M\,\mathrm{Tail}_{s,p}(u_-; R)^{p-1}\right\}$, with C now depending on H as well. The lemma then plainly follows by comparing these two expressions. □

The full Harnack inequality of Theorem 2.9 follows by putting together Theorem 2.10, (a slightly improved version of) Proposition 2.3, and Lemma 2.11. Indeed, arguing as in [8, Subsection 6.4] and, in particular, the proof of Theorem 6.9 there, one gets that $\sup_{B_R(x_0)} u$ can be controlled by the right-hand side of (2.11). As anticipated before, the analogous bound for the tail term on its left-hand side can be deduced using Lemma 2.11.

We conclude the section with a comment on the stability of the results that we just presented in the limit as $s \nearrow 1$.

Essentially all the estimates that we obtained can be made uniform with respect to this limit—that is, the constants that govern them can be chosen to be independent of s, for s bounded away from zero—, provided a couple of changes in the definitions of fractional De Giorgi classes are carried out: one needs to replace H with $H/(1-s)$ in both (2.2) and (2.5), and to correct the definition of the tail term by adding the factor $(1-s)$ in front of the integral that appears, within round brackets, on the right-hand side of (2.1). After these modifications, all results are uniform as $s \nearrow 1$ and coherent with those that are known for classical De Giorgi classes. See [8] for the precise statements.

Such uniformity can be achieved mostly by keeping track of the dependence in s of all the constants involved in the various results. Everything goes through with little effort besides one point: the behavior of the constant δ in Lemma 2.8. As can be easily checked, the proof of Lemma 2.8 is based exclusively on the estimate for the second term on the left-hand side of (2.5), a purely nonlocal quantity that, when multiplied by $(1-s)$, vanishes in the limit as $s \nearrow 1$. As a result, the proof of Lemma 2.8 is not uniform in s as it is. To make it uniform, one can interpolate such proof with an argument closer in spirit to one that leads to the growth lemma for classical De Giorgi classes, such as [17, Lemma 7.5].

A key element of the proof of this classical result is an isoperimetric-type inequality for the level sets of functions in $W^{1,p}$ due to De Giorgi [10]—see, e.g., [5] for its statement when $p = 2$ and [8, Lemma 5.2] for the general case. Next is a partial extension of this inequality to the fractional Sobolev space $W^{s,p}$, when s is close to 1.

Proposition 2.12 *Let $n \geqslant 2$, $M > 0$, and $\gamma \in (0,1)$. There exist two constants $\bar{s} \in (0,1)$ and $C > 0$ such that the inequality*

$$\Big\{|B_1 \cap \{u \leqslant 0\}||B_1 \cap \{u \geqslant 1\}|\Big\}^{\frac{n-1}{n}} \leqslant C(1-s)^{1/p}[u]_{W^{s,p}(B_1)}|B_1 \cap \{0 < u < 1\}|^{\frac{p-1}{p}}$$

holds true for every $s \in [\bar{s}, 1)$ and every function $u \in W^{s,p}(B_1)$ satisfying

$$\begin{gathered} \|u\|^p_{L^p(B_1)} + (1-s)[u]^p_{W^{s,p}(B_1)} \leqslant M, \\ |B_1 \cap \{u \leqslant 0\}| \geqslant \gamma |B_1| \quad \text{and} \quad |B_1 \cap \{u \geqslant 1\}| \geqslant \gamma |B_1|. \end{gathered} \tag{2.13}$$

The constant C depends only on n and p, while $\bar{s}$ also depends on M and γ.

The proof of Proposition 2.12 presented in [8, Section 5] is by contradiction and based on a compactness argument that relies on the aforementioned De Giorgi's isoperimetric inequality in the Sobolev space $W^{1,p}$. As a result, the optimal value of $\bar{s}$ is unknown, as well as its possible independence from M and γ. However, it necessarily holds that $\bar{s} \geqslant 1/p$, due to the fact that $\chi_E \in W^{s,p}(B_1)$ for every $s \in (0, 1/p)$ and every smooth set $E \subset B_1$.

Question 3 Is it possible to obtain an inequality similar to the one of Proposition 2.12 for every function of the space $W^{s,p}(B_1)$, every $s \in [1/p, 1)$, and without assuming (2.13)?

3 Applications to Minimizers of Nonlocal Functionals

In this section we present the main application of fractional De Giorgi classes, which ultimately motivates their introduction: the Hölder regularity of minimizers of possibly non-differentiable nonlocal functionals.

Let $K : \mathbb{R}^n \times \mathbb{R}^n \to \mathbb{R}$ be a non-negative measurable function satisfying

$$K(x, y) = K(y, x) \quad \text{for a.e. } x, y \in \mathbb{R}^n \tag{3.1}$$

and

$$\frac{1}{\Lambda |x - y|^{n+sp}} \leqslant K(x, y) \leqslant \frac{\Lambda}{|x - y|^{n+sp}} \quad \text{for a.e. } x, y \in \mathbb{R}^n, \tag{3.2}$$

for some constants $s \in (0, 1)$, $p > 1$, and $\Lambda \geqslant 1$. Let $F : \Omega \times \mathbb{R} \to \mathbb{R}$ be a Carathéodory function and assume that

$$|F(x, u)| \leqslant F_0 \quad \text{for a.e. } x \in \Omega \text{ and every } u \in \mathbb{R}, \tag{3.3}$$

for some constant $F_0 \geqslant 0$. Associated to these two functions, we consider the energy functional $\mathcal{E}$, defined on every measurable function $u : \mathbb{R}^n \to \mathbb{R}$ by

$$\mathcal{E}(u; \Omega) := \frac{1}{2p} \iint_{\mathscr{C}_\Omega} |u(x) - u(y)|^p \, K(x, y) \, dx dy + \int_\Omega F(x, u(x)) \, dx,$$

where $\mathscr{C}_\Omega := \mathbb{R}^{2n} \setminus (\mathbb{R}^n \setminus \Omega)^2$. More general kernels K and unbounded potentials F (with subcritical growth in u) can also be considered. For simplicity of exposition, here we restrict ourselves to those that are allowed by hypotheses (3.1)–(3.3). We refer the interested reader to [8] for a broader setting.

The functional $\mathcal{E}$ has been recently considered by several authors, since it allows to model nonlinear phenomena that occur in the presence of long-range interactions. Here, we are particularly interested in the case when F is not differentiable (and, perhaps, not even continuous) in the variable u. Examples of such potentials have been considered for instance in [6], with $F(u) = \chi_{(0,+\infty)}(u)$, and in [9], with $F(u)$ comparable to $|1 - u^2|^d$ for some $d > 0$.

Notice that, under the sole assumption (3.3), the functional $\mathcal{E}$ is not differentiable and therefore the regularity properties of its minimizers cannot be inferred from a Euler-Lagrange equation. Instead, we will deduce such properties directly from the minimizing inequality, as done in [15] in the case of a local functional.

We now specify the notion of minimizers that we take into account. To this aim, we will say that a function $u : \mathbb{R}^n \to \mathbb{R}$ belongs to $\mathbb{W}^{s,p}(\Omega)$ if $u|_\Omega \in L^p(\Omega)$ and

$$\iint_{\mathscr{C}_\Omega} \frac{|u(x) - u(y)|^p}{|x - y|^{n+sp}} \, dx dy < +\infty.$$

By (3.2) and (3.3), this is equivalent to ask that $u|_\Omega \in L^p(\Omega)$ and $\mathcal{E}(u; \Omega) < +\infty$.

Definition 3.1 A function $u \in \mathbb{W}^{s,p}(\Omega)$ is a *superminimizer* of $\mathcal{E}$ in Ω if

$$\mathcal{E}(u; \Omega) \leqslant \mathcal{E}(u + \varphi; \Omega) \tag{3.4}$$

for every non-negative measurable function $\varphi : \mathbb{R}^n \to \mathbb{R}$ supported inside Ω. Similarly, u is a *subminimizer* of $\mathcal{E}$ in Ω if (3.4) holds true for every non-positive such φ. Finally, u is a *minimizer* of $\mathcal{E}$ in Ω if (3.4) holds for every measurable $\varphi : \mathbb{R}^n \to \mathbb{R}$ supported inside Ω.

It is not hard to check that u is a minimizer if and only if it is at the same time a sub- and a superminimizer.

In the following result, we establish that minimizers of the energy functional $\mathcal{E}$ belong to a fractional De Giorgi class.

Theorem 3.2 *Let $u \in L_s^{p-1}(\mathbb{R}^n) \cap \mathbb{W}^{s,p}(\Omega)$. There exists a constant $H \geqslant 1$, depending only on n, s, p, and Λ, such that:*

(a) if u is a superminimizer of $\mathcal{E}$ in Ω, then $u \in \mathrm{DG}_-^{s,p}(\Omega; F_0^{1/p}, H, 0)$;
(b) if u is a subminimizer of $\mathcal{E}$ in Ω, then $u \in \mathrm{DG}_+^{s,p}(\Omega; F_0^{1/p}, H, 0)$;
(c) if u is a minimizer of $\mathcal{E}$ in Ω, then $u \in \mathrm{DG}^{s,p}(\Omega; F_0^{1/p}, H, 0)$.

For the sake of simplicity, we present the proof of Theorem 3.2 only for $p = 2$. With little additional technical effort, the argument can be easily extended to the case of a general $p > 1$, as shown in the proof of [8, Proposition 7.5].

Proof of Theorem 3.2 for $p = 2$ We only deal with point (a), since (b) is completely analogous. Clearly, (c) immediately follows from (a) and (b).

Let $x_0 \in \Omega$ and $0 < r \leqslant \rho < \tau \leqslant R < \mathrm{dist}(x_0, \partial\Omega)$. Up to a translation, we may suppose that $x_0 = 0$. Let $\eta \in C^\infty(\mathbb{R}^n)$ be a cutoff function satisfying $0 \leqslant \eta \leqslant 1$ in $\mathbb{R}^n$, $\mathrm{supp}(\eta) \subseteq B_{(\tau+\rho)/2}$, $\eta = 1$ in B_ρ, and $|\nabla \eta| \leqslant 4/(\tau - \rho)$ in $\mathbb{R}^n$.

For any fixed $k \in \mathbb{R}$, let $w_\pm := (u - k)_\pm$, $\varphi := \eta w_-$, and choose $v := u + \varphi$ as a competitor for u in (3.4). It holds

$$\iint_{\mathscr{C}_{B_\tau}} \Xi(x, y) \, d\mu(x, y) \leqslant 4 \int_{B_\tau} \Big\{ F(x, v(x)) - F(x, u(x)) \Big\} \, dx, \tag{3.5}$$

where $\Xi(x, y) := |u(x) - u(y)|^2 - |v(x) - v(y)|^2$ and $d\mu(x, y) := K(x, y) \, dx dy$.

Now, on the one hand, by (3.3) we have that

$$F(x, v(x)) - F(x, u(x)) = F(x, u(x) + \eta(x) w_-(x)) - F(x, u(x)) \leqslant 2F_0 \, \chi_{A^-(k,\rho)}(x)$$

for every $x \in B_\rho$. By this, we easily obtain an upper bound for the right-hand side of (3.5):

$$\int_{B_\tau} \Big\{ F(x, v(x)) - F(x, u(x)) \Big\} \, dx \leqslant 2F_0 \left| A^-(k, \tau) \right|. \tag{3.6}$$

On the other hand, we estimate the left-hand side of (3.5) as follows. Using the definition of w_- along with Young's inequality, we get that, for every $(x, y) \in A^-(k) \times A^-(k)$,

$$\begin{aligned} \Xi(x, y) &= |w_-(x) - w_-(y)|^2 - |(1 - \eta(x)) \, (w_-(x) - w_-(y)) - (\eta(x) - \eta(y)) \, w_-(y)|^2 \\ &\geqslant \left(1 - 2(1 - \eta(x))^2\right) |w_-(x) - w_-(y)|^2 - 2 \, |\eta(x) - \eta(y)|^2 \, w_-(y)^2. \end{aligned}$$

In particular, when $x \in A^-(k) \setminus B_\tau$ and $y \in A^-(k)$, it also holds

$$\begin{aligned} \Xi(x, y) &= |w_-(x) - w_-(y)|^2 - |(w_-(x) - w_-(y)) + \eta(y) w_-(y)|^2 \\ &\geqslant -2\eta(y) w_-(x) w_-(y). \end{aligned}$$

For $(x, y) \in A^-(k) \times \left(\mathbb{R}^n \setminus A^-(k)\right)$ we have

$$\begin{aligned} \Xi(x, y) &= \eta(x) w_-(x) \big(w_-(x) + 2w_+(y) + (1 - \eta(x)) w_-(x)\big) \\ &\geqslant \eta(x) \left(|w_-(x) - w_-(y)|^2 + 2w_-(x) w_+(y) \right). \end{aligned}$$

By these inequalities, the fact that $\Xi(x, y) = 0$ for $x, y \in \mathbb{R}^n \setminus A^-(k)$, hypotheses (3.1)–(3.2) on the kernel K, and the properties of the cutoff η, it is not hard to conclude that

$$\begin{aligned} \iint_{\mathscr{C}_{B_\tau}} \Xi(x, y) \, d\mu(x, y) \geqslant \frac{1}{C} & \left\{ [w_-]_{H^s(B_\rho)}^2 + \int_{B_\rho} w_-(x) \left\{ \int_{\mathbb{R}^n} \frac{w_+(y)}{|x - y|^{n+2s}} \, dy \right\} dx \right\} \\ & - C \left\{ \iint_{B_\tau^2 \setminus B_\rho^2} \frac{|w_-(x) - w_-(y)|^2}{|x - y|^{n+2s}} \, dx dy \right. \\ & \left. + \frac{R^{2-2s}}{(\tau - \rho)^2} \|w_-\|_{L^2(B_R)}^2 + \frac{R^{n+2s}}{(\tau - \rho)^{n+2s}} \|w_-\|_{L^1(B_R)} \overline{\mathrm{Tail}}_{s,2}(w_-; r) \right\} \end{aligned}$$

for some constant $C \geqslant 1$ depending only on n, s, p, and Λ.

By putting together the last estimate, (3.6), and (3.5), and applying Widman's hole-filling technique (with respect to the term $[w_-]^2_{H^s(B_\rho)}$), we obtain

$$[w_-]^2_{H^s(B_\rho)} + \int_{B_\rho} w_-(x) \left\{ \int_{\mathbb{R}^n} \frac{w_+(y)}{|x-y|^{n+2s}} \, dy \right\} dx \leqslant \gamma \left\{ [w_-]^2_{H^s(B_\tau)} \right.$$
$$\left. + F_0 \left| A^-(k,R) \right| + \frac{R^{2-2s}}{(\tau-\rho)^2} \| w_- \|^2_{L^2(B_R)} + \frac{R^{n+2s}}{(\tau-\rho)^{n+2s}} \| w_- \|_{L^1(B_R)} \overline{\mathrm{Tail}}_{s,2}(w_-; r) \right\}$$

for some constant $\gamma \in (0,1)$ depending only on n, s, p, and Λ. From this and a simple iteration lemma (see, e.g., [8, Lemma 4.11]) it follows that $u \in \mathrm{DG}^{s,p}_-(\Omega; F_0^{1/p}, H, 0)$ for some $H \geqslant 1$ depending only on n, s, p, and Λ. □

Notice that an important role in the above proof is played by the so-called hole-filling technique of Widman [26], that we applied to the Gagliardo seminorm of the space H^s. The same trick was used in [15] to deduce Caccioppoli inequalities for minimizers of energies with gradient structure. For integro-differential equations, it has been recently employed in [24].

By combining Theorem 3.2 with Theorems 2.2 and 2.5 of Sect. 2, we deduce the Hölder continuity of minimizers of $\mathcal{E}$.

Corollary 3.3 (Hölder Continuity of Minimizers) *Let $u \in L^{p-1}_s(\mathbb{R}^n) \cap \mathbb{W}^{s,p}(\Omega)$ be a minimizer of $\mathcal{E}$ in Ω. Then, $u \in C^\alpha_{\mathrm{loc}}(\Omega)$ for some exponent $\alpha \in (0,1)$ and there exists a constant $C \geqslant 1$ such that*

$$\|u\|_{L^\infty(B_R(x_0))} + R^\alpha [u]_{C^\alpha(B_R(x_0))} \leqslant C \left(\frac{\|u\|_{L^p(B_{2R}(x_0))}}{R^{n/p}} + \mathrm{Tail}_{s,p}(u; x_0, R) + R^s F_0^{1/p} \right)$$

for every point $x_0 \in \Omega$ and radius $0 < R < \mathrm{dist}(x_0, \partial\Omega)/2$. The constants α and C depend only on n, s, p, and Λ.

Similarly, by Theorems 2.9, 2.10, and 3.2, non-negative minimizers of $\mathcal{E}$ satisfies the following Harnack-type inequalities.

Corollary 3.4 (Harnack Inequalities for Minimizers) *Let $u \in L^{p-1}_s(\mathbb{R}^n) \cap \mathbb{W}^{s,p}(\Omega)$ with $u \geqslant 0$ in Ω. The following statements hold true:*

(a) if u is a superminimizer of $\mathcal{E}$ in Ω, then there exist an exponent $\varepsilon > 0$ and a constant $C \geqslant 1$, both depending only on n, s, p, and Λ, such that

$$\left(\fint_{B_R(x_0)} u(x)^\varepsilon \, dx \right)^{\frac{1}{\varepsilon}} \leqslant C \left(\inf_{B_R(x_0)} u + \mathrm{Tail}_{s,p}(u_-; x_0, R) + R^s F_0^{1/p} \right)$$

for every $x_0 \in \Omega$ and $0 < R < \mathrm{dist}(x_0, \partial\Omega)/2$;

(b) if u is a minimizer of $\mathcal{E}$ in Ω, then there exists a constant $C \geqslant 1$, only depending on n, s, p, and Λ, such that

$$\sup_{B_R(x_0)} u + \mathrm{Tail}_{s,p}(u_+; x_0, R) \leqslant C \left(\inf_{B_R(x_0)} u + \mathrm{Tail}_{s,p}(u_-; x_0, R) + R^s F_0^{1/p} \right)$$

for every $x_0 \in \Omega$ and $0 < R < \mathrm{dist}(x_0, \partial\Omega)/2$.

4 Applications to Solutions of Nonlocal Equations

Another application of fractional De Giorgi classes is represented by the regularity results that will be discussed in this section, concerning weak solutions of equations driven by nonlocal operators.

Let K be a kernel satisfying (3.1) and (3.2), for some $s \in (0, 1)$, $p > 1$, and $\Lambda \geqslant 1$. We introduce the nonlinear and nonlocal operator $\mathcal{L} = \mathcal{L}_{K,p}$ as formally defined on a measurable function $u : \mathbb{R}^n \to \mathbb{R}$ and at a point $x \in \mathbb{R}^n$ by

$$\begin{aligned} \mathcal{L}u(x) &:= \mathrm{P.V.} \int_{\mathbb{R}^n} |u(x) - u(y)|^{p-2}(u(x) - u(y))K(x, y)\, dy \\ &= \lim_{\delta \searrow 0} \int_{\mathbb{R}^n \setminus B_\delta(x)} |u(x) - u(y)|^{p-2}(u(x) - u(y))K(x, y)\, dy. \end{aligned}$$

Also, let $f \in L^\infty(\Omega)$ and $f_0 \geqslant \|f\|_{L^\infty(\Omega)}$ be given.

Throughout the section, we will consider (sub-/super-)solutions of the equation

$$\mathcal{L}u = f \quad \text{in } \Omega, \tag{4.1}$$

defined in the following weak sense.

Definition 4.1 A function $u \in \mathbb{W}^{s,p}(\Omega)$ is a *weak supersolution* of (4.1) if

$$\frac{1}{2} \int_{\mathbb{R}^n} \int_{\mathbb{R}^n} |u(x) - u(y)|^{p-2}(u(x) - u(y))(\varphi(x) - \varphi(y))K(x, y)\, dxdy \leqslant \int_{\mathbb{R}^n} f(x)\varphi(x)\, dx$$

for every non-negative function $\varphi \in W^{s,p}(\mathbb{R}^n)$ supported inside Ω. Conversely, u is a *weak subsolution* of (4.1) if the reverse inequality holds for every such φ. Finally, u is a *weak solution* of (4.1) if

$$\frac{1}{2} \int_{\mathbb{R}^n} \int_{\mathbb{R}^n} |u(x) - u(y)|^{p-2}(u(x) - u(y))(\varphi(x) - \varphi(y))K(x, y)\, dxdy = \int_{\mathbb{R}^n} f(x)\varphi(x)\, dx$$

for every $\varphi \in W^{s,p}(\mathbb{R}^n)$ supported inside Ω.

Note that, if $F = F(x,u)$ is a differentiable function in the variable u, the minimizers of the energy $\mathcal{E}$ considered in Sect. 3 are weak solutions of (4.1), with $f = -F_u(\cdot, u)$. As for those minimizers, solutions of equations driven by the operator $\mathcal{L}$ are contained in a fractional De Giorgi class. This is the content of the next result.

Theorem 4.2 *Let $u \in L_s^{p-1}(\mathbb{R}^n) \cap \mathbb{W}^{s,p}(\Omega)$. There exists a constant $H \geqslant 1$, depending only on n, s, p, and Λ, such that:*

(a) if u is a weak supersolution of (4.1), then $u \in \mathrm{DG}_-^{s,p}(\Omega; f_0^{1/(p-1)}, H, sp/(p-1))$;
(b) if u is a weak subsolution of (4.1), then $u \in \mathrm{DG}_+^{s,p}(\Omega; f_0^{1/(p-1)}, H, sp/(p-1))$;
(c) if u is a weak solution of (4.1), then $u \in \mathrm{DG}^{s,p}(\Omega; f_0^{1/(p-1)}, H, sp/(p-1))$;

Theorem 4.2 can be proved through a strategy analogous to Theorem 3.2. We omit the details for brevity and refer the interested reader to [8, Section 8].

By putting together this result with Theorems 2.2 and 2.5, we are able to deduce the Hölder regularity of weak solutions to (4.1).

Corollary 4.3 (Hölder Continuity of Solutions) *Let $u \in L_s^{p-1}(\mathbb{R}^n) \cap \mathbb{W}^{s,p}(\Omega)$ be a weak solution of (4.1). Then, $u \in C^{\alpha}_{\mathrm{loc}}(\Omega)$ for some exponent $\alpha \in (0,1)$ and there exists a constant $C \geqslant 1$ such that*

$$\|u\|_{L^\infty(B_R(x_0))} + R^\alpha [u]_{C^\alpha(B_R(x_0))} \leqslant C \left(\frac{\|u\|_{L^p(B_{2R}(x_0))}}{R^{n/p}} + \mathrm{Tail}_{s,p}(u; x_0, R) + R^{\frac{sp}{p-1}} f_0^{1/(p-1)} \right)$$

for every $x_0 \in \Omega$ and every $0 < R < \mathrm{dist}(x_0, \partial\Omega)/2$. The constants α and C only depend on n, s, p, and Λ.

When $p = 2$, the C^α character of solutions to (4.1) is well-known—see, e.g., Silvestre [25] and Kassmann [18]. For a general $p > 1$, such regularity has been obtained by Di Castro et al. [12] in the case of $\mathcal{L}$-harmonic functions, i.e., when $f \equiv 0$. To the best of our knowledge, Corollary 4.3—appeared in [8] as Theorem 2.4—is the first result establishing Hölder estimates for solutions of (4.1) when $p \neq 2$ and in the presence of a non-vanishing right-hand side f. See also the very recent [2] for almost sharp results when $p \geqslant 2$ and $s < (p-1)/p$.

Taking advantage of Theorems 2.9 and 2.10, we also have the following Harnack inequalities.

Corollary 4.4 (Harnack Inequalities for Solutions) *Let $u \in L_s^{p-1}(\mathbb{R}^n) \cap \mathbb{W}^{s,p}(\Omega)$ with $u \geqslant 0$ in Ω. The following statements hold true:*

(a) *if u is a weak supersolution of* (4.1)*, then there exist an exponent $\varepsilon > 0$ and a constant $C \geqslant 1$, both depending only on n, s, p, and Λ, such that*

$$\left(\fint_{B_R(x_0)} u(x)^\varepsilon \, dx \right)^{\frac{1}{\varepsilon}} \leqslant C \left(\inf_{B_R(x_0)} u + \mathrm{Tail}_{s,p}(u_-; x_0, R) + R^{\frac{sp}{p-1}} f_0^{1/(p-1)} \right)$$

for every $x_0 \in \Omega$ and $0 < R < \mathrm{dist}(x_0, \partial\Omega)/2$;

(b) *if u is a weak solution of* (4.1)*, then there exists a constant $C \geqslant 1$, only depending on n, s, p, and Λ, such that*

$$\sup_{B_R(x_0)} u + \mathrm{Tail}_{s,p}(u_+; x_0, R) \leqslant C \left(\inf_{B_R(x_0)} u + \mathrm{Tail}_{s,p}(u_-; x_0, R) + R^{\frac{sp}{p-1}} f_0^{1/(p-1)} \right)$$

for every $x_0 \in \Omega$ and $0 < R < \mathrm{dist}(x_0, \partial\Omega)/2$.

Similar Harnack inequalities appeared in [19], for $p = 2$, and in [11], with a general $p > 1$ but with $f = 0$.

Appendix A: An Explicit Example

It is easy to see that the characteristic function of a sufficiently smooth subset E of $\mathbb{R}^n$ is contained in the fractional Sobolev space $W^{s,p}$, provided $sp < 1$. In this appendix we show that, in dimension $n = 1$ and under this assumption on s and p, a step function may also belong to a *weak* fractional De Giorgi class $\widetilde{\mathrm{DG}}^{s,p}$—but never to a *strong* class $\mathrm{DG}^{s,p}$. From this, it follows that the C^α estimates of Theorem 2.5 and the Harnack inequality of Theorem 2.9—both valid for the elements of the smaller class $\mathrm{DG}^{s,p}$—cannot be extended to $\widetilde{\mathrm{DG}}^{s,p}$.

Proposition A.1 *Let $n = 1$ and $sp < 1$. Then,*

$$\chi_{(0,+\infty)} \in \widetilde{\mathrm{DG}}^{s,p}((-1, 1); 0, H, 0) \tag{A.1}$$

for some constant $H \geqslant 1$. Furthermore,

$$\chi_{(0,+\infty)} \notin \mathrm{DG}_-^{s,p}((-1, 1); d, H, \lambda) \tag{A.2}$$

for every $d, \lambda \geqslant 0$ and $H \geqslant 1$.

Proof We begin by showing that (A.1) holds true. We only check that $u := \chi_{(0,+\infty)}$ belongs to the class $\widetilde{\mathrm{DG}}_-^{s,p}$, as the verification of its inclusion in $\widetilde{\mathrm{DG}}_+^{s,p}$ is analogous.

Fix any $x_0 \in (-1, 1)$, $0 < r < R \leq 1 - |x_0|$, and $k \in \mathbb{R}$. In order to check the validity of the inequality defining $\widetilde{\mathrm{DG}}_-^{s,p}$, we clearly can restrict ourselves to considering the case of $k > 0$, since otherwise $(u - k)_- \equiv 0$. For shortness, we only deal with $k \in (0,1]$, the case $k > 1$ being similar. We first estimate from above the left-hand side of (2.2):

$$\begin{aligned}
[(u-k)_-]^p_{W^{s,p}((x_0-r,x_0+r))} &= \int_{x_0-r}^{x_0+r}\int_{x_0-r}^{x_0+r} \frac{|(u(x)-k)_- - (u(y)-k)_-|^p}{|x-y|^{1+sp}}\,dxdy \\
&= 2k^p \chi_{(|x_0|,+\infty)}(r) \int_{x_0-r}^{0}\int_{0}^{x_0+r} \frac{dxdy}{|x-y|^{1+sp}} \\
&\leqslant \frac{2(r-|x_0|)_+^{1-sp} k^p}{sp(1-sp)}.
\end{aligned} \tag{A.3}$$

In view of this, it suffices to estimate from below the right-hand side of (2.2) when $r > |x_0|$. In this case, also $R > |x_0|$ and therefore such right-hand side is larger than

$$H\frac{R^{(1-s)p}}{(R-r)^p}\|(u-k)_-\|^p_{L^p(x_0-R,x_0+R)} = H\frac{R^{(1-s)p}}{(R-r)^p}k^p\int_{x_0-R}^{0} dx \geqslant H(R-|x_0|)^{1-sp}k^p.$$

As $R > r$, the latter quantity controls the one appearing on the last line of (A.3), provided H is sufficiently large (in dependence of s and p only). Consequently, u belongs to the class $\widetilde{\mathrm{DG}}_-^{s,p}((-1,1);0,H,0)$.

We now turn our attention to (A.2). We point out that, arguing by contradiction, its validity could be inferred from Theorem 2.10. Nevertheless, we present here a proof of it based on a direct computation, for we show that inequality (2.5) does not hold when $x_0 = 0$ and $R = 2r$, with $k, r > 0$ suitably small. Indeed, under these assumptions the left-hand side of (2.5) is larger than

$$\begin{aligned}
\int_{-r}^{r}(u(x)-k)_-\left\{\int_{\mathbb{R}}\frac{(u(y)-k)_+^{p-1}}{|x-y|^{1+sp}}\,dy\right\}dx &\geqslant \int_{-r}^{0}k\left\{\int_{0}^{x+r}\frac{(1-k)^{p-1}}{(y-x)^{1+sp}}\,dy\right\}dx \\
&= \frac{r^{1-sp}k(1-k)^{p-1}}{1-sp}.
\end{aligned}$$

On the other hand, it is easy to check that the right-hand side of (2.5) is bounded above by $CH(r^{1+\lambda}d^p + r^{1-sp}k^p)$, for some constant $C \geqslant 1$ depending only on s and p. By taking r and k smaller and smaller (but positive), it follows that the latter quantity cannot control the one displayed above, no matter how large H is. Hence, (A.2) holds true. □

Acknowledgements The author wishes to thank Serena Dipierro, the Università degli Studi di Bari, and INdAM for their kind invitation, warm hospitality, and financial support. The author also thanks the anonymous referee for her/his keen comments on a previous version of this note. The author is supported by the "María de Maeztu" MINECO grant MDM-2014-0445, by the MINECO grant MTM2017-84214-C2-1-P, and by a Royal Society Newton International Fellowship.

References

1. L. Brasco, E. Parini, The second eigenvalue of the fractional p-Laplacian. Adv. Calc. Var. **9**(4), 323–355 (2016)
2. L. Brasco, E. Lindgren, A. Schikorra, Higher Hölder regularity for the fractional p-Laplacian in the superquadratic case. Adv. Math. **338**, 782–846 (2018)
3. X. Cabré, M. Cozzi, A gradient estimate for nonlocal minimal graphs. Duke Math. J. **168**(5), 775–848 (2019)
4. L.A. Caffarelli, A. Vasseur, Drift diffusion equations with fractional diffusion and the quasi-geostrophic equation. Ann. Math. (2) **171**(3), 1903–1930 (2010)
5. L.A. Caffarelli, A. Vasseur, The De Giorgi method for nonlocal fluid dynamics, in *Nonlinear Partial Differential Equations*. Advanced Courses in Mathematics. CRM Barcelona (Birkhäuser/Springer Basel AG, Basel, 2012), pp. 1–38
6. L.A. Caffarelli, J.-M. Roquejoffre, Y. Sire, Variational problems for free boundaries for the fractional Laplacian. J. Eur. Math. Soc. **12**(5), 1151–1179 (2010)
7. L.A. Caffarelli, C.H. Chan, A. Vasseur, Regularity theory for parabolic nonlinear integral operators. J. Am. Math. Soc. **24**(3), 849–869 (2011)
8. M. Cozzi, Regularity results and Harnack inequalities for minimizers and solutions of nonlocal problems: a unified approach via fractional De Giorgi classes. J. Funct. Anal. **272**(11), 4762–4837 (2017)
9. M. Cozzi, E. Valdinoci, Plane-like minimizers for a non-local Ginzburg-Landau-type energy in a periodic medium. J. Éc. Polytech. Math. **4**, 337–388 (2017)
10. E. De Giorgi, Sulla differenziabilità e l'analiticità delle estremali degli integrali multipli regolari. Mem. Accad. Sci. Torino. Cl. Sci. Fis. Mat. Nat. (3) **3**, 25–43 (1957)
11. A. Di Castro, T. Kuusi, G. Palatucci, Nonlocal Harnack inequalities. J. Funct. Anal. **267**(6), 1807–1836 (2014)
12. A. Di Castro, T. Kuusi, G. Palatucci, Local behavior of fractional p-minimizers. Ann. Inst. H. Poincaré Anal. Non Linéaire **33**(5), 1279–1299 (2016)
13. E. Di Nezza, G. Palatucci, E. Valdinoci, Hitchhiker's guide to the fractional Sobolev spaces. Bull. Sci. Math. **136**(5), 521–573 (2012)
14. E. DiBenedetto, N.S. Trudinger, Harnack inequalities for quasiminima of variational integrals. Ann. Inst. H. Poincaré Anal. Non Linéaire **1**(4), 295–308 (1984)
15. M. Giaquinta, E. Giusti, On the regularity of the minima of variational integrals. Acta Math. **148**, 31–46 (1982)
16. M. Giaquinta, L. Martinazzi, *An Introduction to the Regularity Theory for Elliptic Systems, Harmonic Maps and Minimal Graphs*. Appunti, Scuola Normale Superiore di Pisa (Nuova Serie) [Lecture Notes. Scuola Normale Superiore di Pisa (New Series)], vol. 11 (Edizioni della Normale, Pisa, 2012)
17. E. Giusti, *Direct Methods in the Calculus of Variations* (World Scientific Publishing Co., Inc., River Edge, 2003)
18. M. Kassmann, A priori estimates for integro-differential operators with measurable kernels. Calc. Var. Partial Differ. Equ. **34**(1), 1–21 (2009)
19. M. Kassmann, Harnack inequalities and Hölder regularity estimates for nonlocal operators revisited (2011). Preprint
20. T. Kuusi, G. Mingione, Y. Sire, Nonlocal self-improving properties. Anal. PDE **8**(8), 57–114 (2015)
21. O.A. Ladyzhenskaya, N.N. Ural'tseva, *Linear and Quasilinear Elliptic Equations* (Academic, New York, 1968). Translated from the Russian by Scripta Technica, Inc., Translation editor: Leon Ehrenpreis
22. G. Mingione, Gradient potential estimates. J. Eur. Math. Soc. **13**(2), 459–486 (2011)
23. X. Ros-Oton, J. Serra, The boundary Harnack principle for nonlocal elliptic operators in non-divergence form. Potential Anal. (to appear). https://doi.org/10.1007/s11118-018-9713-7

24. A. Schikorra, Integro-differential harmonic maps into spheres. Commun. Partial Differ. Equ. **40**(3), 506–539 (2015)
25. L. Silvestre, Hölder estimates for solutions of integro-differential equations like the fractional Laplace. Indiana Univ. Math. J. **55**(3), 1155–1174 (2006)
26. K.-O. Widman, Hölder continuity of solutions of elliptic systems. Manuscripta Math. **5**, 299–308 (1971)

Harnack and Pointwise Estimates for Degenerate or Singular Parabolic Equations

Fatma Gamze Düzgün, Sunra Mosconi, and Vincenzo Vespri

Abstract In this paper we give both a historical and technical overview of the theory of Harnack inequalities for nonlinear parabolic equations in divergence form. We start reviewing the elliptic case with some of its variants and geometrical consequences. The linear parabolic Harnack inequality of Moser is discussed extensively, together with its link to two-sided kernel estimates and to the Li-Yau differential Harnack inequality. Then we overview the more recent developments of the theory for nonlinear degenerate/singular equations, highlighting the differences with the quadratic case and introducing the so-called *intrinsic* Harnack inequalities. Finally, we provide complete proofs of the Harnack inequalities in some paramount case to introduce the reader to the *expansion of positivity* method.

Keywords Degenerate and singular parabolic equations · Pointwise estimates · Harnack estimates · Weak solutions · Intrinsic geometry

2010 Mathematics Subject Classification 35K67, 35K92, 35K20

1 Introduction

Generally speaking, given a class $\mathcal{C}$ of nonnegative functions defined on a set Ω, a Harnack inequality is a pointwise control of the form $u(x) \leq C\, u(y)$ for all $u \in \mathcal{C}$ (with a constant independent of u) where the inequality holds for $x \in X \subseteq \Omega$ and

---

F. G. Düzgün
Department of Mathematics, Hacettepe University, Ankara, Turkey
e-mail: gamzeduz@hacettepe.edu.tr

S. Mosconi (✉)
Dipartimento di Matematica e Informatica, Università degli Studi di Catania, Catania, Italy
e-mail: mosconi@dmi.unict.it

V. Vespri
Dipartimento di Matematica e Informatica "U. Dini", Università di Firenze, Firenze, Italy
e-mail: vespri@math.unifi.it

S. Dipierro (ed.), *Contemporary Research in Elliptic PDEs and Related Topics*, Springer INdAM Series 33, https://doi.org/10.1007/978-3-030-18921-1_8

$y \in Y \subseteq \Omega$, (X, Y) belonging to a certain family $\mathcal{F}$ determined by $\mathcal{C}$. Thus it takes the form

$$\exists\, C = C(\mathcal{C}, \mathcal{F}) \quad \text{such that} \quad \sup_X u \le C \inf_Y u \qquad \forall\, (X, Y) \in \mathcal{F},\ u \in \mathcal{C}. \tag{1.1}$$

Given $\mathcal{C}$, one is ideally interested in maximal families $\mathcal{F}$. In this respect, certain properties of maximal families are immediate, e.g., if $X' \subseteq X$ and $(X, Y) \in \mathcal{F}$, then $(X', Y) \in \mathcal{F}$. The so-called *Harnack chain* argument consists in the elementary observation that if both (X, Y) and (Y', Z) belong to $\mathcal{F}$ and $y_0 \in Y \cap Y' \neq \emptyset$, then

$$\sup_X u \le C \inf_Y u \le C\, u(y_0) \le C \sup_{Y'} u \le C^2 \inf_Z u,$$

hence we can add all such couples (X, Z) to $\mathcal{F}$ by considering the constant C^2. Other properties of $\mathcal{F}$ follow from the structure of $\mathcal{C}$: if, for instance, $\mathcal{C}$ is invariant by a suitable semi-group $\{\Phi_\lambda\}_{\lambda>0}$ of domain transformations (meaning that $u \in \mathcal{C} \;\Rightarrow\; u \circ \Phi_\lambda \in \mathcal{C}$ for all $\lambda > 0$), then $\mathcal{F}$ should also exhibits this invariance.

Formally, to a larger class $\mathcal{C}$ corresponds a smaller family $\mathcal{F}$ and the more powerful Harnack inequalities aim at "maximize" the two sets at once. Typically, $\mathcal{C}$ is the set of nonnegative solutions to certain classes of PDE in an ambient metric space Ω and $\mathcal{F}$ should at least cluster near each point of Ω (i.e. $\forall P \in \Omega, r > 0$ there exists $(X, Y) \in \mathcal{F}$ such that both X and Y lie in the ball of center P and radius r). Another example is the class of ratios of nonnegative harmonic functions vanishing on the same set, giving rise to the so-called *boundary Harnack inequalities*. Given $\mathcal{C}$, searching for a suitable maximal family $\mathcal{F}$ such that (1.1) holds, informally takes the name of *finding the right form* of the Harnack inequality in $\mathcal{C}$. Rich examples of such instance arise in the theory of hypoelliptic PDE's.

Historically, the first of such pointwise control was proved by Harnack in 1887 for the class $\mathcal{C}$ of nonnegative harmonic functions in a domain $\Omega \subseteq \mathbb{R}^2$, with $\mathcal{F}$ being made of couples of identical balls well contained in Ω. Since then, extensions and variants of the Harnack inequality grew steadily in the mathematical literature, with plentiful applications in PDE and differential geometry. Correspondingly, its proof in the various settings has been obtained through many different points of view. To mention a few: the original potential theoretic approach, the measure-theoretical approach of Moser [70], the probabilistic one of Krylov-Safonov [58] and the differential approach of Li and Yau [65].

Many very good books and surveys on the Harnack inequality already exist (see e.g. [54]) and we are thus forced to justify the novelty of this one. Our main focus will be the quest for the right form $\mathcal{F}$ of various Harnack inequalities and, to this end, we will mainly deal with parabolic ones, which naturally exhibit a richer structure. Even restricting the theme to the parabolic setting requires a further choice, as the theory naturally splits into two large branches: one can either consider *divergence form* (or variational) equations, whose basic linear example is $u_t = \operatorname{div}(A(x)\, Du)$, or equations in *non-divergence form* (or non-variational), such as $u_t = A(x) \cdot D^2 u$. While some attempts to build a unified approach to the Harnack inequality has been made (see [36]), structural differences seem unavoidable. Moreover, both examples have nonlinear counterparts and the corresponding theories rapidly diverge. We will deal with parabolic nonlinear equations in divergence form, referring to the surveys [49, 57] for the non-divergence theory.

Rather than simply collecting known result to describe the state of the art, we aim at giving both a historical and technical overview on the subject, with emphasis on the different proofs and approaches to the subject.

The first part, consisting in Sects. 2–4, will focus on the various form of (1.1), mentioning some applications and giving from time to time proofs of well-known facts which we found somehow hard to track in the literature. In particular, we will deal with the elliptic case in Sect. 2, with the linear parabolic Harnack inequality in Sect. 3 and with the singular and degenerate parabolic setting in Sect. 4. Here we will describe the so-called *intrinsic Harnack inequalities*, by which we mean a generalization of (1.1) where the sets X and Y also depend on u (or, equivalently, (1.1) holds in a restricted class $\mathcal{C}$ determined by non-homogeneous scalings).

The second part consists of the final and longest section, which is devoted to detailed proofs of the most relevant Harnack inequalities for equations in divergence form. Our aim is to obtain the elliptic and parabolic Harnack inequalities in a unified way, following the measure-theoretical approach of De Giorgi to regularity and departing from Moser's one. This roadmap has been explored before (see [67] for an axiomatic treatment), but we push it further to gather what we believe are the most simple proofs of the Harnack inequalities up to date. Credits to the main ideas and techniques should be given to the original De Giorgi paper [18], the book of Landis [63] and the work of Di Benedetto and collaborators gathered in the monograph [31]. We will focus on model problems rather than on generality in the hope to make the proofs more transparent and attract non-experts to this fascinating research field.

2 Elliptic Harnack Inequality

2.1 Original Harnack

In 1887, the german mathematician C.G. Axel von Harnack proved the following result in [47].

Theorem 2.1 *Let u be a nonnegative harmonic function in $B_R(x_0) \subseteq \mathbb{R}^2$. Then for all $x \in B_r(x_0) \subset B_R(x_0)$ it holds*

$$\frac{R-r}{R+r}\, u(x_0) \le u(x) \le \frac{R+r}{R-r}\, u(x_0).$$

The estimate can be generalized to any dimension $N \ge 1$ through the Poisson representation formula, resulting in

$$\left(\frac{R}{R+r}\right)^{N-2} \frac{R-r}{R+r}\, u(x_0) \le u(x) \le \left(\frac{R}{R-r}\right)^{N-2} \frac{R+r}{R-r}\, u(x_0), \tag{2.1}$$

and the constants can be seen to be optimal by looking at the solutions u_n of the Dirichlet problem on the ball B_R with boundary data $\varphi_n \to \delta_{x_0}$, $|x_0| = R$. However, the modern version of the Harnack inequality for harmonic functions is the following special case of the previous one.

Theorem 2.2 *Let $N \ge 1$. Then there exists a constant $C = C(N) > 1$, such that if u is a nonnegative, harmonic function in $B_{2r}(x_0)$, then*

$$\sup_{B_r(x_0)} u \le C \inf_{B_r(x_0)} u. \tag{2.2}$$

The proof of this latter form of the Harnack inequality is an easy consequence of the mean value theorem. For the early historical developments related to the first Harnack inequality we refer to the survey [54].

The Harnack inequality has several deep and powerful consequences. On the local side, Harnack himself in [47] derived from it a precisely quantified oscillation estimate. Due to the ubiquity of this argument we recall its elementary proof. Let $x_0 = 0$ and

$$M_r(u) = \sup_{B_r} u, \qquad m_r(u) = \inf_{B_r} u, \quad \operatorname{osc}(u, B_r) = M_r(u) - m_r(u).$$

Both $M_{2r}(u) - u$ and $u - m_{2r}(u)$ are nonnegative and harmonic in B_{2r}, so (2.2) holds for them, resulting in

$$M_{2r}(u) - m_r(u) \le C(M_{2r}(u) - M_r(u)), \quad M_r(u) - m_{2r}(u) \le C(m_r(u) - m_{2r}(u)),$$

which added together give

$$M_{2r}(u) - m_{2r}(u) + M_r(u) - m_r(u) \le C\big(M_{2r}(u) - m_{2r}(u) - (M_r(u) - m_r(u))\big).$$

Rearranging, we obtain

$$\mathrm{osc}(u, B_r) \leq \frac{C-1}{C+1}\mathrm{osc}(u, B_{2r}),$$

which is the claimed quantitive estimate of decrease in oscillation.

Removable singularity results can also be obtained through the Harnack inequality, as well as two classical convergence criterions for sequences of harmonic functions. At the global level, it implies Liouville and Picard type theorems. For example, Liouville's theorem asserts that any globally defined harmonic function bounded from below must be constant, as can be clearly seen by applying (2.2) to $u - \inf_{\mathbb{R}^N} u$ and letting $r \to +\infty$.

2.2 Modern Developments

In his celebrated paper [18], De Giorgi introduced the measure theoretical approach to regularity, proving the local Hölder continuity of weak solutions of linear elliptic equations in divergence form

$$L(u) := \sum_{i,j=1}^{N} D_i(a_{ij}(x)D_j u) = 0 \tag{2.3}$$

with merely measurable, symmetric coefficients satisfying the ellipticity condition

$$\lambda|\xi|^2 \leq \sum_{i,j=1}^{N} a_{ij}(x)\xi_i\xi_j \leq \Lambda|\xi|^2, \qquad 0 < \lambda \leq \Lambda < +\infty. \tag{2.4}$$

The modern regularity theory descending from his ideas is a vast field and the relevant literature is huge. We refer to [69] for a general overview and bibliographic references; the monograph [43] contains the regularity theory of quasi-minima, while for systems one should see [61] and the literature therein.

Regarding the Harnack inequality, Moser extended in his fundamental work [70] its validity to solutions of (2.3).

Theorem 2.3 *Suppose $u \geq 0$ solves* (2.3) *in a ball $B_{2r}(x_0)$ where* (2.4) *holds. Then there exists a constant $C > 1$ depending only on N and the* ellipticity ratio Λ/λ *such that*

$$\sup_{B_r(x_0)} u \leq C \inf_{B_r(x_0)} u.$$

Moser's proof is also measure-theoretical, stemming from the De Giorgi approach but introducing pioneering new ideas. It relied on the John-Nirenberg Lemma [51] and certainly contributed to its diffusion in the mathematical community. Such a level of generality allowed to apply essentially the same technique for the general quasilinear equation

$$\mathrm{div} A(x, u, Du) = 0. \tag{2.5}$$

Indeed, in [81, 86], the same statement of the Harnack inequality has been proved for (2.5) instead of the linear equation (2.3), provided A satisfies for some $p > 1$ and $\Lambda \geq \lambda > 0$ the ellipticity condition

$$\begin{cases} A(x, s, z) \cdot z \geq \lambda |z|^p \\ |A(x, s, z)| \leq \Lambda |z|^{p-1} \end{cases} \qquad x \in B_{2r}(x_0), s \in \mathbb{R}, z \in \mathbb{R}^N. \tag{2.6}$$

The power of the measure-theoretical approach was then fully exploited in [25], where the Harnack inequality has been deduced without any reference to an elliptic equation, proving that it is a consequence of very general energy estimates of Caccioppoli type, encoded in what are the nowadays called *De Giorgi classes*. For a comprehensive treatment of the latters see [23].

2.3 Moser's Proof and Weak Harnack Inequalities

Moser's proof of the Harnack inequality is splitted in two steps:

(I) $L^p - L^\infty$ *bound*:
Let u be a nonnegative subsolution of (2.3) in B_{2r}, i.e., u obeys $-L(u) \leq 0$ weakly (supersolutions being defined through the opposite inequality). For any $p > 0$ it holds

$$\sup_{B_r} u \leq C \left(\fint_{B_{2r}} |u|^p \, dx \right)^{\frac{1}{p}} \tag{2.7}$$

for some constant $C = C(N, \Lambda/\lambda, p)$. If on the other hand u is a positive supersolution, then u^{-1} is a positive subsolution, and (2.7) can be rewritten as

$$\inf_{B_r} u \geq C^{-1} \left(\fint_{B_{2r}} u^{-p} \, dx \right)^{-\frac{1}{p}}.$$

(II) *Crossover Lemma*. The Harnack inequality then follows if one has

$$\fint_{B_r} u^{\bar{p}}\, dx \fint_{B_r} u^{-\bar{p}}\, dx \le C(N) \tag{2.8}$$

for some (small) $\bar{p} = \bar{p}(N, \Lambda, \lambda) > 0$. This is the most delicate part of Moser's approach, and is dealt with the so-called *logarithmic estimate*. The idea is to prove a universal bound on $\log u$, as suggested by the Harnack inequality itself. To this end, consider a ball $B_{2\rho}(x_0) \subseteq B_{2r}$ and test the equation with $u^{-1}\eta^2$, η being a cutoff function in $C_c^\infty(B_{2\rho}(x_0))$. This yields

$$\lambda \int_{B_{2\rho}(x_0)} |Du|^2 u^{-2} \eta^2\, dx \le 2\Lambda \int_{B_{2\rho}(x_0)} |Du|\, u^{-1}\, |\eta|\, |D\eta|\, dx$$

with λ, Λ given in (2.4). Apply Young inequality on the right and note that we can assume $|D\eta| \le c\,\rho^{-1}$ to get

$$\fint_{B_\rho(x_0)} |D \log u|^2\, dx \le C(\Lambda/\lambda)\, \rho^{-2} \tag{2.9}$$

as long as $\eta = 1$ in $B_\rho(x_0)$. The Poincaré inequality then implies

$$\fint_{B_\rho(x_0)} \Big(\log u - \fint_{B_\rho(x_0)} \log u\, dx\Big)^2 dx \le C(N, \Lambda/\lambda), \qquad \text{for all } B_{2\rho}(x_0) \subseteq B_{2r},$$

which means that $\log u \in BMO(B_{2r})$. Then John-Nirenberg's Lemma ensures

$$\fint_{B_r} e^{\bar{p}\,|w|}\, dx \le c, \qquad w = \log u - m, \qquad m = \fint_{B_r} \log u\, dx$$

for some small $\bar{p} = \bar{p}(N, \Lambda) > 0$ and $c = c(N)$, and inequality (2.8) follows by multiplying

$$\fint_{B_r} u^{\bar{p}}\, dx = e^{\bar{p}\,m} \fint_{B_r} e^{\bar{p}\,w}\, dx \le c\, e^{\bar{p}\,m} \quad \text{and} \quad \fint_{B_r} u^{-\bar{p}}\, dx = e^{-\bar{p}\,m} \fint_{B_r} e^{-\bar{p}\,w}\, dx \le c\, e^{-\bar{p}\,m}.$$

In particular, Moser's proof shows that a weaker form of Harnack inequality holds for the larger class of non-negative supersolutions to (2.3) in B_{2r}. Namely, the following *weak Harnack inequality* holds

$$\Big(\fint_{B_{2r}} u^p\, dx\Big)^{\frac{1}{p}} \le C \inf_{B_r} u, \qquad \text{for any } p \in \Big]0, \frac{N}{N-2}\Big[$$

for some constant $C = C(N, \Lambda/\lambda, p)$. The range of exponents in the weak Harnack inequality is optimal, as the fundamental solution for the Laplacian shows. Notice that the $L^\infty - L^p$ bound also implies an L^p-Liouville theorem, as letting $r \to +\infty$ in (2.7) shows that 0 is the only nonnegative solution globally in $L^p(\mathbb{R}^N)$. On the other hand, the previous weak Harnack inequality gives a lower asymptotic estimate for positive $L^p_{\rm loc}(\mathbb{R}^N)$ supersolutions of the form $\inf_{B_r} u \gtrsim r^{-N/p}$ for $r \to +\infty$. From the local point of view, the weak form of the Harnack inequality is also sufficient for the Hölder regularity and for strong comparison principles.

A different and detailed proof of the elliptic Harnack inequality via the expansion of positivity technique will be given in Sect. 5.1.

2.4 Harnack Inequality on Minimal Surfaces

After considering the Harnack inequality for nonlinear operator, a very fruitful framework was to consider its validity for linear elliptic operators defined on *nonlinear* ambient spaces, such as Riemannian manifolds. One of the first examples of this approach was the Bombieri–De Giorgi–Miranda gradient bound [10] for solutions of the *minimal surface equation*

$$\operatorname{div}\left(\frac{Du}{\sqrt{1+|Du|^2}}\right) = 0. \tag{2.10}$$

The approach of [10], later simplified in [88], consisted in showing that $w = \log\sqrt{1+|Du|^2}$ is a subsolution of the Laplace-Beltrami operator naturally defined on the graph of u considered as a Riemannian manifold. Since a Sobolev-Poincaré inequality can be proved for minimal graphs (see [68] for a refinement to smooth minimal submanifolds), the Moser iteration yields an $L^\infty - L^1$ bound on w which is the core of the proof.

Another realm of application of the Harnack inequality are Bernstein theorem, i.e. Liouville type theorem for the minimal surface equation (2.10). More precisely Bernstein's theorem asserts that *any entire solution to* (2.10) in $\mathbb{R}^2$ is affine. This statement is known to be true in all dimension $N \le 7$ and false from $N = 8$ onwards. One of the first applications in [70] of Moser's (Euclidean) Harnack inequality was to show that if in addition u has bounded gradient, the Bernstein statement holds true in any dimension. Indeed, one can differentiate (2.10) with respect to x_i, giving a nonlinear equation which however can be seen as linear in u_{x_i} with freezed coefficients. It turns out that if $|Du|$ is bounded then the coefficients are elliptic and the Liouville property gives the conclusion.

The approach of [10] was pushed forward in [9], where a pure Harnack inequality was shown for general linear operators on minimal graphs. Taking advantage of their Harnack inequality, Bombieri and Giusti proved that if $N-1$ derivatives of a solution to (2.10) are bounded, then also the N-th one is bounded, thus ensuring the

Bernstein statement in any dimension thanks to the Moser result. See also [35] for a direct proof of this fact using the Harnack inequality on minimal graph alone.

For other applications of the Harnack inequality on minimal graphs, see [16].

2.5 *Differential Harnack Inequality*

A natural way to look at the Harnack estimate $u(x) \leq C\, u(y)$ is to rewrite it as

$$\log u(x) - \log u(y) \leq \log C = C', \qquad \text{for all } x, y \in B_r$$

as long as $u > 0$ in B_{2r}. If one considers smooth functions (such as solutions to smooth elliptic equations), a way to prove the latter would be to look at it as a gradient bound on $\log u$. More concretely, it is a classical fact that harmonic functions in $B_{2r}(x_0)$ satisfy the gradient estimate

$$|Du(x_0)| \leq C(N)\frac{\sup_{B_r(x_0)} |u|}{r},$$

therefore Harnack's inequality implies that

$$u \geq 0 \text{ in } B_r(x_0) \quad \Rightarrow \quad |Du(x_0)| \leq C(N)\frac{u(x_0)}{r}.$$

This can be rewritten in the following form:

Theorem 2.4 (Differential Harnack Inequality) *Let $u > 0$ be harmonic in $B_r(x_0) \subseteq \mathbb{R}^N$. Then*

$$|D \log u(x_0)| \leq \frac{C(N)}{r}. \tag{2.11}$$

Inequality (2.11) can be seen as the pointwise version of the integral estimate (2.9) and as such it can be integrated back along segments, to give the original Harnack inequality. The differential form (2.11) of the Harnack inequality clearly requires much more regularity than the Moser's one, however, it was proved to hold in the Riemannian setting for the Laplace-Beltrami equation in the ground-breaking works [16, 94], under the assumption of non-negative Ricci curvature for the manifold. To appreciate the result, notice that all proofs of the Harnack inequality known at the time required a global Sobolev inequality, which is known to be false in general under the Ric ≥ 0 assumption alone.

The elliptic Harnack inequality in the Riemannian setting proved in [94] (and, even more importantly, its parabolic version proved soon after in [65]) again implies the Liouville property for semi-bounded harmonic functions and it was one of the pillars on which modern geometric analysis grew. See for example the survey article

[64] for recent results on the relationship between Liouville-type theorems and geometric aspects of the underlying manifold. The book [73] gives an in-depth exposition of the technique of differential Harnack inequalities in the framework of Ricci flow, culminating in Perelman differential Harnack inequality.

2.6 *Beyond Smooth Manifolds*

Clearly, the differential approach to the Harnack inequality is restricted to the Laplace-Beltrami operator, due to its smoothness and its close relationship with Ricci curvature given by the Bochner identity

$$\Delta u = 0 \quad \Rightarrow \quad \Delta \frac{|Du|^2}{2} = |D^2 u|^2 + \mathrm{Ric}(Du, Du).$$

It was only after the works [44, 78] that a different approach to Moser's Harnack inequality on manifolds was found.[1] Essentially, it was realized that in order to obtain the Harnack inequality on a Riemannian manifold (M, g) with corresponding volume form m and geodesic distance, two ingredients suffice:

– *Doubling condition:* $\quad m\big(B_{2r}(x_0)\big) \le C m\big(B_r(x_0)\big)$

– *Poincaré inequality:*
$$\int_{B_r(x_0)} \Big|u - \fint_{B_r(x_0)} u\, dm\Big|^2 dm \le C \int_{B_r(x_0)} |Du|^2\, dm \tag{2.12}$$

for any $x_0 \in M$ and $r > 0$. These two properties hold in any Riemannian manifold with nonnegative Ricci curvature, thus giving a Moser-theoretic approach to the Harnack inequality in this framework. What's more relevant here is that Doubling and Poincaré are stable with respect to quasi-isometries (i.e. bilipschitz homeomorphisms) and thus can hold in non-smooth manifolds, manifolds where $\mathrm{Ric} \ge 0$ does not hold (since curvature is not preserved through quasi-isometries), and/or for merely measurable coefficients elliptic operators. It is worth mentioning that Doubling and Poincaré were also shown in [17] to be sufficient conditions for the solution of Yau's conjecture on the finite-dimensionality of the space of harmonic functions of polynomial growth.

It was a long standing problem to give geometric conditions which are actually *equivalent* to the validity of the elliptic Harnack inequality, and thus to establish the stability of the latter with respect to quasi (or even rough) isometries. This

[1]Actually, to a *parabolic version* of the Harnack inequality, which readily implies the elliptic one. For further details see the discussion on the parabolic Harnack inequality below and for a nice historical overview on the subject see [80, Section 5.5].

problem has recently been settled in [6], to which we refer the interested reader for bibliographic reference and discussion.

3 Parabolic Harnack Inequality

3.1 Original Parabolic Harnack

Looking at the fundamental solution for the heat equation

$$u_t - \Delta u = 0,$$

one finds out that there is no hope to prove a straightforward generalization of the Harnack inequality (2.2). In the stationary case, ellipticity is preserved by spatial homotheties and translations, thus the corresponding Harnack inequality turns out to be scale and translation invariant. For the heat equation, the natural scaling $(x, t) \mapsto (\lambda x, \lambda^2 t)$ preserves the equation and one expects a parabolic Harnack inequality to obey this invariance. In order to guess its form it is useful to look at the special caloric function $w(x, t) = t^{-1/2} e^{-x^2/t}$ defined on $\mathbb{R}\times\,]0, +\infty[$. Given two times $t_1, t_2 > 0$ and $\xi \geq 0$, one easily computes

$$\sup_{x\in B_r(\xi)} w(x, t_1) = t_1^{-\frac{1}{2}} e^{-\frac{(\xi-r)_+^2}{t_1}}, \qquad \inf_{x\in B_r(\xi)} w(x, t_2) = t_2^{-\frac{1}{2}} e^{-\frac{(\xi+r)^2}{t_2}}.$$

In order for the latters to be comparable for all large r and $\xi = 0$, it must hold $t_1 \simeq t_2 \simeq r^2$. Moreover, $t_1 = t_2 = r^2$ won't do when we choose $\xi = kr$ with $k \to +\infty$, so that the control must happen at different times. Even if $t \mapsto t^{-1/2}$ is decreasing while $t \mapsto e^{-x^2/t}$ is increasing, a growth rate argument suggests that, in order for a Harnack inequality to hold, one must require $t_1 < t_2$. Indeed, setting

$$\tilde{w}(x, t) = \begin{cases} w(x, t) & \text{if } x > 0,\ t > 0 \\ 0 & \text{if } x > 0,\ t \leq 0 \end{cases}$$

also gives a solution on the half-space $]0, +\infty[\, \times \mathbb{R}$, vanishing for $t \leq 0$. Thus we see that the supremum at a certain time can only be controlled by the infimum at *later* times.

The explicit parabolic form of the Harnack inequality was found and proved independently by Pini and Hadamard in [45, 75] and reads as follows.

Theorem 3.1 *Let $u \geq 0$ be a solution of the heat equation in $B_{2\rho}(x_0) \times [t_0 - 4\rho^2, t_0 + 4\rho^2]$. Then there exists a constant $C(N)$, N being the dimension, such that*

$$\sup_{B_\rho(x_0)} u(\cdot, t_0 - \rho^2) \leq C(N) \inf_{B_\rho(x_0)} u(\cdot, t_0 + \rho^2). \tag{3.1}$$

As expected, this form of the Harnack's inequality complies with the scaling of the equation and introduces the notion of *waiting time* for a pointwise control to hold. It represents a quantitative bound from below on how much the positivity of $u(x_0, t_0)$ (physically, the temperature of a body at a certain point) propagates forward in time: in order to have such a bound in a whole ball of radius r we have to wait a time proportional to r^2.

Another way of expressing this propagation for a nonnegative solution on $B_{2\sqrt{T}}(x_0) \times [0, 4T]$ is the following, which, up to numerical factors is equivalent to (3.1),

$$C \inf_{P_T^+(x_0)} u \geq u(x_0, 2T) \geq C^{-1} \sup_{P_T^-(x_0)} u, \tag{3.2}$$

where $P_T^{\pm}(x_0)$ are the part of the forward (resp. backward) space-time paraboloid with vertex $(x_0, 2T)$ in $B_{\sqrt{T}}(x_0) \times [T, 3T]$ (see Fig. 1):

$$P_T^+(x_0) = \{(x,t) : T - t_0 \geq t - t_0 \geq |x - x_0|^2\}, \quad P_T^-(x_0) = \{(x,t) : t_0 - T \geq t_0 - t \geq |x - x_0|^2\}.$$

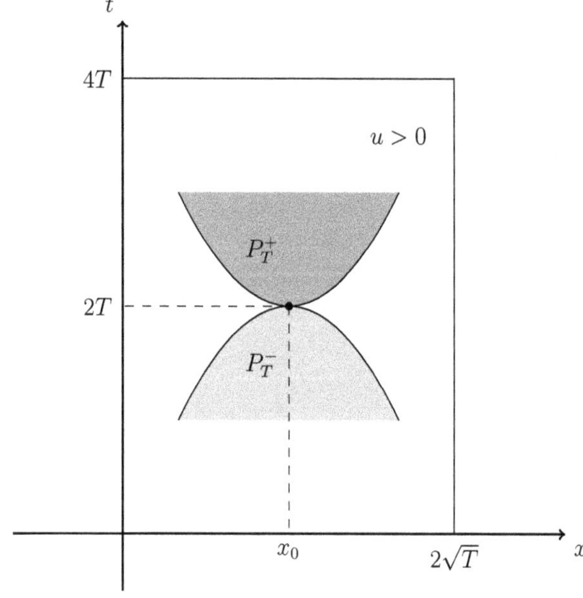

Fig. 1 Representation of (3.2): assuming $u > 0$ in the boxed region, the dark grey area is P_T^+ where u is bounded below by $u(x_0, 2T)$, while the light grey one is P_+^T where u is bounded above by $u(x_0, 2T)$

A consequence of the parabolic Harnack inequality is the following form of the strong maximum principle. We sketch a proof here since this argument will play a rôle in the discussion of the Harnack inequality for nonlinear equations.

Corollary 3.2 (Parabolic Strong Minimum Principle) *Let $u \geq 0$ be a solution of the heat equation in $\Omega \times [0, T]$, where Ω is connected, and suppose $u(x_0, t_0) = 0$ for some $x_0 \in \Omega$ and $t_0 \in]0, T[$. Then $u \equiv 0$ in $\Omega \times [0, t_0]$.*

Proof (Sketch) Pick $(x_1, t_1) \in \Omega\times]0, t_0[$ and join it to (x_0, t_0) with a smooth curve $\gamma : [0, 1] \to \Omega\times]0, t_0]$ such that γ' always has positive t-component. By compactness there is $\delta > 0$ and a small forward parabolic sector $P^+_\varepsilon = \{\varepsilon \geq t \geq |x|^2\}$ such that: *1)* $\gamma(\sigma) \in \gamma(\tau) + P^+_\varepsilon$ for all $\sigma \in [\tau, \tau + \delta]$ and *2)* the Harnack inequality holds in the form (3.2) for all $s \in [0, 1]$, i.e.

$$u(\gamma(s)) \leq \inf_{\gamma(s)+P^+_\varepsilon} u.$$

These two properties and $u(\gamma(1)) = 0$ readily imply $u \circ \gamma \equiv 0$. □

3.2 *The Linear Case with Coefficients*

In the seminal paper [74] on the Hölder regularity of solutions to parabolic equations with measurable coefficients, Nash already mentioned the possibility to obtain a parabolic Harnack inequality through his techniques. However, the first one to actually prove it was again Moser, who in [71] extended the Harnack inequality to linear parabolic equations of the form

$$u_t = \sum_{j,i=1}^{N} D_i(a_{ij}(x, t)D_j u). \tag{3.3}$$

Theorem 3.3 (Moser) *Let u be a positive weak solution of* (3.3) *in $B_{2r} \times [0, T]$, where a_{ij} are measurable and satisfy the ellipticity condition* (2.4) *for all $t \in [0, T]$. For any $0 < t_1^- < t_2^- < t_1^+ < t_2^+ < T$ define (see Fig. 2)*

$$C_- := B_r \times [t_1^-, t_2^-], \quad C_+ := B_r \times [t_1^+, t_2^+].$$

Then it holds

$$\sup_{C_-} u \leq C(N, \Lambda, \lambda, t^{\pm}_{1,2}) \inf_{C_+} u, \tag{3.4}$$

with a constant which is bounded as long as $t_1^+ - t_2^-$ is bounded away from 0.

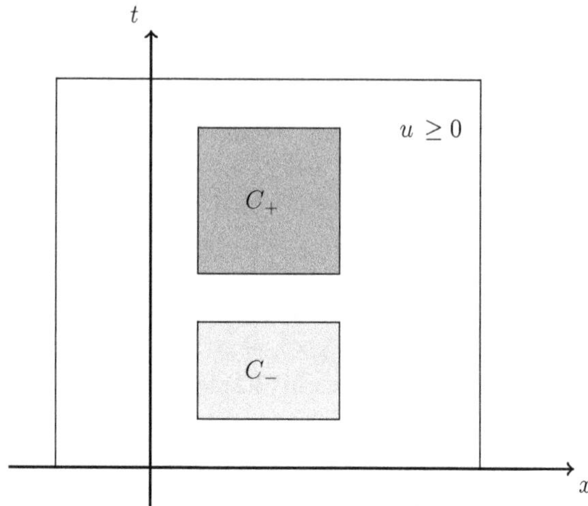

Fig. 2 The cylinders C_+ and C_- where the Harnack inequality is stated

Using the natural scaling of the equation, the previous form the parabolic Harnack inequality can be reduced to (3.1).

As in the elliptic case, the first step of Moser's proof consisted in the $L^p - L^\infty$ estimates for subsolutions, obtained by testing the equation with recursively higher powers of the solution. This leads to

$$\sup_{Q(\rho)} u^p \leq \frac{C(N, p, \Lambda, \lambda)}{(r-\rho)^{N+2}} \iint_{Q(r)} u^p \, dx \, dt, \qquad r > \rho, \quad p > 0 \tag{3.5}$$

where $Q(r)$ are parabolic cylinders having top boundary at the same fixed time t_0, say $Q(r) = B_r \times [t_0 - r^2, t_0]$. Since if u is a positive solution, u^{-1} is a positive subsolution, (3.5) holds true also for negative powers p, yielding a bound from below for u in terms of integrals of u^p. Similarly to the elliptic case, in order to obtain the parabolic Harnack inequality, Moser proceeded to prove a crossover lemma which reads as

$$\int_{-1}^{0} \int_{B_1} u^{p_0} \, dx \int_{1}^{2} \int_{B_1} u^{-p_0} \, dx \leq C, \tag{3.6}$$

for some C and a small $p_0 > 0$ depending on N and the ellipticity constants. This proved to be much harder than in the elliptic case, mainly because the integrals are taken on the two different and distant sets and no appropriate John-Nirenberg inequality dealing with this situation was known at the time. Moser himself proved such a parabolic version of the John-Nirenberg lemma yielding (3.6), but the proof was so involved that he was forced to an erratum 3 years later. In [72] he gave a different proof avoiding it, following an approach of Bombieri and Giusti [9]. For this to work, he refined his $L^p - L^\infty$ estimates (3.5), showing that they hold with constants independent from p, at least for sufficiently small values of $|p|$. As we will see, this was necessary for the Bombieri-Giusti argument to carry over. Despite the parabolic John-Nirenberg Lemma has later been given a simpler proof in [33], the *abstract John-Nirenberg Lemma* technique of [9] is nowadays the standard tool to prove parabolic Harnack inequalities, see e.g. [56, 80]. On the other hand, Nash's program was later established in [34].

We next sketch the proof in [72]. The starting point is a logarithmic estimate, obtained by multiplying the equation by $u^{-1}\eta^2$, with $\eta \in C^\infty_c(B_3)$, $\eta \geq 0$ and $\eta \equiv 1$ on B_2 and integrate in space only. Proceeding as in the elliptic case we obtain the differential inequality

$$\frac{d}{dt}\fint_{B_3} \eta^2(x) \log u(x,t)\, dx + c \fint_{B_3} |D \log u(x,t)|^2\, \eta^2(x)\, dx \leq C.$$

Under mild concavity assumptions on η, a weighted Poincaré inequality holds true with respect to the measure $d\mu = \eta^2(x)\, dx$, so that we infer

$$\frac{d}{dt}\fint_{B_3} \log u(x,t)\, d\mu + c \fint_{B_3} \Big(\log u(x,t) - \fint_{B_3} \log u(x,t)\, d\mu\Big)^2 d\mu \leq C.$$

By letting

$$v(x,t) = \log u(x,t) - C\, t, \qquad M(t) = \fint_{B_3} v(x,t)\, d\mu$$

the previous inequality can be rewritten as

$$\frac{d}{dt} M(t) + c \fint_{B_3} \big(v(x,t) - M(t)\big)^2 d\mu \leq 0,$$

so that $M(t)$ is decreasing. Next, for $\lambda > 0$ and $t \in [0,4]$, restrict the integral over $\{x \in B_2 : v(x,t) \geq M(0) + \lambda\}$ where, by monotonicity, $v(x,t) - M(t) \geq M(0) - M(t) + \lambda \geq \lambda$, to get

$$\frac{d}{dt} M(t) + c\,(M(0) - M(t) + \lambda)^2\, |B_2 \cap \{v(x,t) \geq M(0) + \lambda\}| \leq 0,$$

(notice that $d\mu = dx$ on B_2). Dividing by $(M(0) - M(t) + \lambda)^2$, integrating in $t \in [0,4]$ and recalling that $M(t) \leq M(0)$ we deduce

$$|Q_+(2) \cap \{v \geq M(0) + \lambda\}| \leq C/\lambda, \qquad Q_+(2) := B_2 \times [0,4].$$

Similarly, for any $t \in [-4,0]$, on $\{x \in B_2 : v(x,t) \leq M(0) - \lambda\}$ it holds $M(t) - v \geq M(t) - M(0) + \lambda \geq \lambda$ being M decreasing and proceeding as before we get

$$|Q_-(2) \cap \{v \leq M(0) - \lambda\}| \leq C/\lambda, \qquad Q_-(2) := B_2 \times [-4,0].$$

Recalling the definition of v, the last two displays imply the weak-L^1 estimate

$$|Q_+(2) \cap \{\log u \geq M(0) + \lambda\}| \leq C/\lambda, \quad |Q_-(2) \cap \{\log u \leq M(0) - \lambda\}| \leq C/\lambda, \tag{3.7}$$

where $M(0)$ is a weighted mean of $\log u$. To proceed, we let

$$w = u\, e^{-M(0)}, \qquad Q_+(r) = B_r \times [4 - r^2, 4], \qquad \varphi(r) = \sup_{Q_+(r)} \log w$$

for $r \in [1, 2]$. Since $Q_+(r) \subseteq Q_+(2)$, for all $\lambda > 0$,

$$|Q_+(r) \cap \{\log w \geq \lambda\}| \leq C/\lambda.$$

We will prove a universal bound on $\varphi(r)$ so we may suppose that $\varphi(r)$ is large. Estimate the integral of w^p on $Q_+(r)$ splitting it according to $\log w \leq \varphi(r)/2$ or $\log w > \varphi(r)/2$, to get

$$\begin{aligned}
\iint_{Q_+(r)} w^p \, dx\, dt &= \iint_{Q_+(r)} e^{p \log w} \, dx\, dt \\
&\leq e^{p\varphi(r)} |Q_+(r) \cap \{\log w \geq \varphi(r)/2\}| + |Q_+(r)|\, e^{\frac{p}{2}\varphi(r)} \\
&\leq \frac{2\,C}{\varphi(r)}\, e^{p\varphi(r)} + c_N\, e^{\frac{p}{2}\varphi(r)}.
\end{aligned}$$

Choose now $p = p(r)$ such that

$$\frac{2\,C}{\varphi(r)}\, e^{p\varphi(r)} = c_N\, e^{\frac{p}{2}\varphi(r)} \quad \Leftrightarrow \quad p = \frac{2}{\varphi(r)} \log(c\,\varphi(r)), \quad c := c_N/(2C)$$

(where $\varphi(r)$ is so large that p is positive and sufficiently small), so that

$$\iint_{Q_+(r)} w^p \, dx\, dt \leq 2c_N\, e^{\frac{p}{2}\varphi(r)}.$$

We use (3.5) (with constant independent of p for small p), obtaining for a larger C,

$$\begin{aligned}
\varphi(\rho) &\leq \frac{1}{p} \log\left(\frac{C\, e^{\frac{p}{2}\varphi(r)}}{(r-\rho)^{N+2}}\right) = \frac{\varphi(r)}{2} + \frac{1}{p} \log\left(\frac{C}{(r-\rho)^{N+2}}\right) \\
&= \frac{\varphi(r)}{2}\left(1 + \frac{\log(C/(r-\rho)^{N+2})}{\log(c\,\varphi(r))}\right)
\end{aligned}$$

Therefore, either the second term in the parenthesis is greater than $1/2$, which is equivalent to

$$\varphi(r) \leq \frac{C^2}{c\,(r-\rho)^{2(N+2)}}$$

or the opposite is true, giving $\varphi(\rho) \le \frac{3}{4}\varphi(r)$. All in all we obtained

$$\varphi(\rho) \le \frac{3}{4}\varphi(r) + \frac{C}{(r-\rho)^{2N+4}}.$$

The latter can be iterated on an infinite sequence of radii $1 = r_0 \le r_n \le r_{n+1} \le \dots \le 2$ with, say, $r_{n+1} - r_n \simeq (n+1)^{-2}$, to get

$$\varphi(1) \le C \sum_{n=0}^{\infty} (3/4)^n \, n^{4(N+2)},$$

which implies $\sup_{Q_+(1)} u\, e^{-M(0)} \le C$ for some C depending on N, Λ and λ. Thanks to the second estimate in (3.7), a completely similar argument holds true for $w = u^{-1}\, e^{M(0)}$ on the cylinders $Q_-(r) = B_r \times [-r^2, 0]$, yielding $\sup_{Q_-(1)} u^{-1}\, e^{M(0)} \le C$, i.e. $\inf_{Q_-(1)} u\, e^{M(0)} \ge C^{-1}$. Therefore we obtained

$$\frac{\sup_{Q_+(1)} u}{\inf_{Q_-(1)} u} = \frac{\sup_{Q_+(1)} u\, e^{M(0)}}{\inf_{Q_-(1)} u\, e^{M(0)}} \le C^2.$$

3.3 First Consequences

As in the elliptic case, the parabolic Harnack inequality provides an oscillation estimate giving the Hölder continuity of solutions to (3.3) subjected to (2.4). Moreover, (3.4) readily yields a strong minimum principle like the one in Corollary 3.2 for nonnegative solutions of (3.3).

On the other hand, Liouville theorems in the parabolic setting are more subtle and don't immediately follow from the parabolic version of the Harnack inequality. In fact, the Liouville property is false in general since, for example, the function $u(x,t) = e^{x+t}$ is clearly a nontrivial positive eternal (i.e., defined on $\mathbb{R}^N \times \mathbb{R}$) solution of the heat equation. A two sided bound is needed, and a fruitful setting where to state Liouville properties in the one of *ancient* solutions, i.e. those defined on an unbounded interval $]-\infty, T_0[$. An example is the following.

Theorem 3.4 (Widder) *Let $u > 0$ solve the heat equation in $\mathbb{R}^N \times]-\infty, T_0[$. Suppose for some $t_0 < T_0$ it holds*

$$u(x, t_0) \le C e^{o(|x|)}, \qquad \text{for } |x| \to +\infty.$$

Then, u is constant.

The latter has been proved for $N = 1$ in [93], and we sketch the proof in the general case.

Proof By the Widder representation for ancient solutions (see [66]) it holds

$$u(x,t) = \int_{\mathbb{R}^N} e^{x\cdot\xi + t|\xi|^2}\, d\mu(\xi) \tag{3.8}$$

for some nonnegative Borel measure μ. Let $\nu := e^{t_0|\xi|^2}\mu$ and observe that Hölder's inequality with respect to the measure ν implies that for all $s \in]0,1[$

$$\begin{aligned} u(sx+(1-s)y, t_0) &= \int_{\mathbb{R}^N} e^{(sx+(1-s)y)\cdot\xi}\, d\nu(\xi) \\ &\le \left(\int e^{x\cdot\xi}\, d\nu(\xi)\right)^s \left(\int e^{y\cdot\xi}\, d\nu(\xi)\right)^{1-s} = u^s(x,t_0)\, u^{1-s}(y,t_0) \end{aligned}$$

i.e., $x \mapsto \log u(x,t_0)$ is convex. As $\log u(x,t_0) = o(|x|)$ by assumption, it follows that $x \mapsto u(x,t_0)$ is constant. Differentiating under the integral sign the Widder representation, we obtain

$$0 = P(D_x)u(x,t_0)|_{x=0} = \int_{\mathbb{R}^N} P(\xi)\, d\nu(\xi)$$

for any polynomial P such that $P(0) = 0$. By a classical Fourier transform argument, this implies that $\nu = c\,\delta_0$ and thus $u(x,t) \equiv c$ due to the representation (3.8). □

Compare with [83] where it is proved that under the growth condition $0 \le u \le Ce^{o(|x|+\sqrt{|t|})}$ for $t \le T_0$, there are no ancient non-constant solutions to the heat equation on a complete Riemannian manifold with Ric ≥ 0.

Using Moser's Harnack inequality, Aronsson proved in [1] a *two sided* bound on the fundamental solution of (3.3) with symmetric coefficients, which reads

$$\frac{1}{C(t-s)^{N/2}} e^{-C\frac{|x-y|^2}{t-s}} \le \Gamma(x,t;y,s) \le \frac{C}{(t-s)^{N/2}} e^{-\frac{1}{C}\frac{|x-y|^2}{t-s}} \tag{3.9}$$

for some $C = C(N,\Lambda,\lambda)$ and $t > s > 0$, where the fundamental solution (or *heat kernel*) solves, for any fixed $(y,s) \in \mathbb{R}^N \times \mathbb{R}_+$

$$\begin{cases} \partial_t \Gamma = \sum_{j,i=1}^N D_{x_i}(a_{ij}(x,t)D_{x_j}\Gamma) & \text{in } \mathbb{R}^N \times]s,+\infty[, \\ \Gamma(\cdot,t;y,s) \rightharpoonup^* \delta_y, & \text{as } t \downarrow s, \text{ in the measure sense.} \end{cases}$$

In [34], the previous kernel estimate was proved through Nash's approach, and was shown to be equivalent to the parabolic Harnack inequality.

A *global* Harnack inequality also follows from (3.9), whose proof we will now sketch. If $u \geq 0$ is a solution to (3.3) on $\mathbb{R}^N \times \mathbb{R}_+$ and $t > s > \tau \geq 0$, then using the representation

$$u(x,t) = \int_{\mathbb{R}^N} \Gamma(x,t;\xi,\tau)u(\xi,\tau)\,d\xi, \qquad t > \tau,$$

and the analogous one for (y, s), we get

$$\begin{aligned} u(x,t) &= \int_{\mathbb{R}^N} \Gamma(x,t;\xi,\tau)\,\Gamma^{-1}(y,s;\xi,\tau)\,\Gamma(y,s;\xi,\tau)\,u(\xi,\tau)\,d\xi \\ &\geq \frac{u(y,s)}{C^2}\left(\frac{s-\tau}{t-\tau}\right)^{\frac{N}{2}} \inf_{\xi} e^{\frac{1}{C}\frac{|y-\xi|^2}{s-\tau} - C\frac{|x-\xi|^2}{t-\tau}}, \end{aligned}$$

where $\tau \geq 0$ is a free parameter. Recalling that

$$\inf_{\xi} a\,|y-\xi|^2 - b\,|x-\xi|^2 = \frac{a\,b}{b-a}|x-y|^2, \qquad a > b \geq 0,$$

we consider two cases. If $s/t \leq 1/(2C^2)$ we choose $\tau = 0$ and compute

$$\inf_{\xi} \frac{|y-\xi|^2}{C\,s} - C\frac{|x-\xi|^2}{t} \geq -2C\frac{|x-y|^2}{t-s}.$$

If instead $s/t \in]1/(2C^2), 1]$, we set $\tau = s - (t-s)/(2C^2) > 0$ obtaining

$$\inf_{\xi} \frac{1}{C}\frac{|y-\xi|^2}{s-\tau} - C\frac{|x-\xi|^2}{t-\tau} \geq -2C\frac{|x-y|^2}{t-s}.$$

while $(s-\tau)/(t-\tau) = 1/(1+2C^2)$. Therefore the kernel bounds (3.9) imply the following Harnack inequality *at large*, often called *sub-potential lower bound*, for positive solutions u of (3.3) on $\mathbb{R}^N \times]0,T[$: there exists a constant $C = C(N,\Lambda,\lambda) > 1$ such that

$$u(x,t) \geq \frac{1}{C}u(y,s)\left(\frac{s}{t}\right)^{\frac{N}{2}} e^{-C\frac{|x-y|^2}{t-s}} \qquad \text{for all } T > t > s > 0. \tag{3.10}$$

A similar global estimate, with a non-optimal exponent $\alpha = \alpha(N,\Lambda,\lambda) > N/2$ for the ratio s/t, has already been derived through the so-called *Harnack chain* technique by Moser in [71].

3.4 Riemannian Manifolds and Beyond

Following the differential approach of [94], Li and Yau proved in [65] their celebrated parabolic differential Harnack inequality.

Theorem 3.5 *Let M be a complete Riemannian manifold of dimension $N \geq 2$ and* $\mathrm{Ric} \geq 0$*, and let $u > 0$ solve the heat equation on $M \times \mathbb{R}_+$. Then it holds*

$$|D \log u|^2 - \partial_t \log u \leq \frac{N}{2t}. \tag{3.11}$$

In the same paper, many variants of the previous inequality are considered, including one for local solutions in $B_R(x_0)\times\,]t_0 - T, t_0[$ much in the spirit of [16], and several consequences are also derived. Integrating inequality (3.11) along geodesics provides, for any positive solution of the heat equation of $M \times \mathbb{R}_+$

$$u(x,t) \geq u(y,s)\left(\frac{s}{t}\right)^{\frac{N}{2}} e^{-\frac{d^2(x,y)}{4(t-s)}}, \qquad t > s > 0, \tag{3.12}$$

where $d(x,y)$ is the geodesic distance between two points $x, y \in M$. This, in turn, gives the heat kernel estimate (see [80, Ch. 5])

$$\frac{1}{CV(x,\sqrt{t-s})}\, e^{-C\frac{d^2(x,y)}{t-s}} \leq \Gamma(x,t;y,s) \leq \frac{C}{V(x,\sqrt{t-s})}\, e^{-\frac{1}{C}\frac{d^2(x,y)}{t-s}}, \tag{3.13}$$

where $V(x,r)$ is the Riemannian volume of a geodesic ball $B(x,r)$. Notice that, in a general Riemannian manifold of dimension $N \geq 2$,

$$V(x,r) \simeq r^N \qquad \text{for small } r > 0,$$

but, under the sôle assumption $\mathrm{Ric} \geq 0$, the best one can say is

$$\frac{r}{C} \leq V(x,r) \leq C r^N, \qquad \text{for large } r > 0.$$

Therefore, while the Li-Yau estimate on the heat kernel coincides with Aronsson's one locally, it is genuinely different at the global level.

Other parabolic differential Harnack inequalities were then found by Hamilton in [46] for compact Riemannian manifolds with $\mathrm{Ric} \geq 0$, and were later extended in [59, 83] to complete, non-compact manifolds. Actually, far more general differential Harnack inequalities are available under suitable conditions on the Riemannian manifold, see the book [73] for the history and applications of the latters.

Again, the differential Harnack inequality (3.11) requires a good deal of smoothness both on the operator and on the ambient manifold. Yet, the corresponding pointwise inequality (3.12) doesn't depend on the smoothness of the metric g_{ij} but only on its induced distance and the dimension, hence one is lead to believe that a smoothness-free proof exists. Indeed, the papers [44, 78] showed that the parabolic Harnack inequality (and the corresponding heat kernel estimates) can still be obtained through a Moser-type approach based solely on the Doubling and Poincaré condition (2.12). Indeed, [44, 78] independently showed the following *equivalence*.

Theorem 3.6 (Parabolic Harnack Principle) *For any Riemannian manifold the following are equivalent:*

(1) The parabolic Harnack inequality (3.1).
(2) The heat kernel estimate (3.13).
(3) The Doubling and Poincaré condition (2.12).

Since Doubling and Poincaré are stable with respect to quasi-isometries, the previous theorem ensures the stability of the parabolic Harnack inequality with respect to the latters, and thus its validity in a much wider class of Riemannian manifolds than those with $\mathrm{Ric} \geq 0$. Condition *(3)* also ensures that the parabolic Harnack inequality holds for general parabolic equations with elliptic and merely measurable coefficients, see [79]. Actually, under local regularity conditions, it can be proved for *metric spaces* which are roughly isometric to a Riemannian manifold with $\mathrm{Ric} \geq 0$, such as suitable graphs or singular limits of Riemannian manifolds.

3.5 The Nonlinear Setting

An analysis of Moser's proofs reveals that the linearity of the second order operator is immaterial, and that essentially the same arguments can be applied as well to nonnegative weak solutions of a wide family of quasilinear equations. In [2, 87], the Harnack inequality in the form (3.4) was proved to hold for nonnegative solutions of the quasilinear equation

$$u_t = \mathrm{div} A(x, u, Du) \tag{3.14}$$

where the function $A : \Omega \times \mathbb{R} \times \mathbb{R}^N \to \mathbb{R}^N$ is only assumed to be measurable and satisfying

$$\begin{cases} A(x, s, z) \cdot z \geq C_0 |z|^2, \\ |A(x, s, z)| \leq C_1 |z|, \end{cases} \tag{3.15}$$

for some given positive constants $0 < C_0 \leq C_1$. These structural conditions are very weak, as, for example, the validity of the comparison principle holds in general under the so-called *monotonicity condition*

$$\big(A(x,s,z)-A(x,s,w)\big)\cdot(z-w)\geq 0 \tag{3.16}$$

which does not follow from (3.15). To appreciate the generality of (3.15), consider the toy model case $N=1$, $A(x,s,z)=\varphi(z)$, so that a smooth solution of (3.14) fulfills $u_t=\varphi'(u_x)u_{xx}$. Assuming (3.15) alone gives no information on the sign of φ' except at 0 (where $\varphi'(0)\geq C_0>0$), so that (3.15) is actually a backward parabolic equation in the region $\{u_x\in\{\varphi'<0\}\}$.

Trudinger noted in [87] that the Harnack inequality for the case of general p-growth conditions (2.6) with $p\neq 2$ seemed instead a difficult task. He stated the validity of the Harnack inequality (3.4) for positive solutions of the doubly nonlinear equation

$$(u^{p-1})_t=\operatorname{div}A(x,t,u,Du)$$

where A obeys (2.6) with the same p as the one appearing on the left hand side, thus recovering a form of homogeneity in the equation which is lacking in (3.14). The (homogeneous) doubly nonlinear result has later been proved in [40, 41, 56], (see also the survey [55]), but it took around 40 years to obtain the right form of the Harnack inequality for solutions of (3.14) under the general p-growth condition (2.6) on the principal part. The next chapter will be dedicated to this development.

It is worth noting that another widely studied parabolic equation which presented the same kind of difficulties is the *porous medium equation*, namely

$$u_t=\Delta u^m,\qquad m>0.$$

In fact, most of the results in the following sections have analogue statements and proofs for positive solutions of the porous medium equation. The interested reader may consult the monographs [31, 91, 92] for the corresponding results for porous media and related literature. More generally, the doubly nonlinear inhomogeneous equation

$$u_t=\operatorname{div}(u^{m-1}|Du|^{p-2}Du)$$

has found applications in describing polytropic flows of a non-newtonian fluid in porous media [5] and soil science [4, 60, 82], see also the survey article [52]. Regularity results can be found in [50, 76] and Harnack inequalities in [37] for the degenerate and in [39] in the singular case, respectively. To keep things as simple as possible, we chose not to treat these equations, limiting our exposition to (3.14).

4 Singular and Degenerate Parabolic Equations

4.1 The Prototype Equation

Let us consider the parabolic p-Laplace equation

$$u_t = \mathrm{div}(|Du|^{p-2}Du), \quad p > 1, \tag{4.1}$$

which can be seen as a parabolic elliptic equation with $|Du|^{p-2}$ as (intrinsic) isotropic coefficient. The coefficient vanishes near a point where $Du = 0$ when $p > 2$, while it blows up near such a point when $p < 2$. For this reasons we call (4.1) *degenerate* when $p > 2$ and *singular* if $p < 2$.

In the fifties, the seminal paper [3] by Barenblatt was the starting point of the study of the p-Laplacian equation (4.1). The following family of explicit solutions to (4.1) where found, and are since then called *Barenblatt solution* to (4.1).

Theorem 4.1 *For any $p > \frac{2N}{N+1}$ and $M > 0$, there exist constants $a, b > 0$ depending only on N and p such that the function*

$$\mathcal{B}_{p,M}(x,t) := \begin{cases} t^{-\frac{N}{\lambda}}\left[aM^{\frac{p}{\lambda}\frac{p-2}{p-1}} - b\big(|x|\,t^{-\frac{1}{\lambda}}\big)^{\frac{p}{p-1}}\right]_+^{\frac{p-1}{p-2}}, & \text{if } p > 2,\\[2ex] t^{-\frac{N}{\lambda}}\left[aM^{\frac{p}{\lambda}\frac{p-2}{p-1}} + b\big(|x|\,t^{-\frac{1}{\lambda}}\big)^{\frac{p}{p-1}}\right]^{\frac{p-1}{p-2}} & \text{if } 2 > p, \end{cases} \tag{4.2}$$

where $\lambda = N(p-2) + p > 0$, solves the problem

$$\begin{cases} u_t = \mathrm{div}(|Du|^{p-2}Du) & \text{in } \mathbb{R}^N \times\,]0, +\infty[,\\ u(\cdot, t) \rightharpoonup^* M\delta_0 & \text{as } t \downarrow 0. \end{cases}$$

The functions $\mathcal{B}_{p,M}$ are also called *fundamental solutions of mass M*, or simply fundamental solutions when $M = 1$, in which case one briefly writes $\mathcal{B}_{p,1} = \mathcal{B}_p$. Uniqueness of the fundamental solution for the prototype equation was proved by Kamin and Vázquez in [53] (the uniqueness for general monotone operators is still not known).

The Barenblatt solutions show that, when (4.1) is degenerate, the diffusion is very slow and the speed of the propagation of the support is finite, while in the singular case the diffusion is very fast and the solution may become extinct in finite time. These two phenomena are incompatible with a parabolic Harnack inequality of the form (3.1) or (3.4), (suitably modified taking account of the natural scaling) such as

$$C^{-1} \sup_{B_\rho(x_0)} u(\cdot, t_0 - \rho^p) \le u(x_0, t_0) \le C \inf_{B_\rho(x_0)} u(\cdot, t_0 + \rho^p) \tag{4.3}$$

with a constant C depending only on N. Indeed, in the degenerate case the Barenblatt solution has compact support for any positive time, violating the strong minimum principle dictated by (4.3) (the proof of Corollary 3.2 still works). Regarding the singular case, this incompatibility is not immediately apparent from the Barenblatt profile itself and in fact the strong minimum principle still holds for solutions defined in $\mathbb{R}^N \times]0, T[$ when $p > \frac{2N}{N+1}$. However, consider the solution of the Cauchy problem associated to (4.1) in a cylindrical domain $\Omega \times \mathbb{R}_+$ with $u(x, 0) = u_0(x) \in C_c^\infty(\Omega)$ and Dirichlet boundary condition on $\partial\Omega \times \mathbb{R}_+$, with Ω *bounded*. An elementary energetic argument (see [21, Ch VII]) gives a suitable *extinction time* $T^*(\Omega, u_0)$ such that $u(\cdot, t) \equiv 0$ for $t > T^*$, again violating the strong minimum principle which would follow from (4.3).

Let us remark here that for $1 < p \le \frac{2N}{N+1} =: p_*$ the Barenblatt profiles cease to exists. The exponent p_* is called the *critical exponent* for singular parabolic equations and, as it will be widely discussed in the following, the theory is mostly complete in the *supercritical* case $p > p_*$. Solutions of critical and subcritical equations (i.e. with $p \in]1, p_*]$) on the other hand, even in the model case (4.1), exhibit odd and, in some aspects, still unclear (i.e., move the comma at the beginning) features.

4.2 *Regularity*

Let us consider equations of the type

$$u_t = \mathrm{div} A(x, u, Du) \tag{4.4}$$

with general measurable coefficients obeying

$$\begin{cases} A(x, s, z) \cdot z \ge C_0 |z|^p, \\ |A(x, s, z)| \le C_1 |z|^{p-1}. \end{cases} \tag{4.5}$$

We are concerned with weak solutions in $\Omega \times [0, T]$, namely those satisfying

$$\int u\varphi \, dx \Big|_{t_1}^{t_2} + \int_{t_1}^{t_2} \int_\Omega [-u\varphi_t + A(x, u, Du) \cdot \varphi] \, dx \, dt = 0$$

where φ is an arbitrary function such that $\varphi \in W^{1,2}_{\mathrm{loc}}(0, T; L^2(\Omega)) \cap L^p(0, T; W^{1,p}_0(\Omega))$. This readily implies that

$$u \in C_{\mathrm{loc}}(0, T; L^2_{\mathrm{loc}}(\Omega)) \cap L^p_{\mathrm{loc}}(0, T; W^{1,p}_{\mathrm{loc}}(\Omega)).$$

In the case $p = 2$, the local Hölder continuity of solutions to (4.4) has been proved in [62] through a parabolic De Giorgi approach. The case $p \ne 2$ was considered a major open problem in the theory of quasilinear parabolic equations for over two decades. The main obstacle to its solution was that the energy and logarithmic estimates for (4.4) are non-homogeneous when $p \neq 2$. It was solved by DiBenedetto [19] in the degenerate case and Chen and DiBenedetto in [15] in the singular case through an approach nowadays called *method of intrinsic scaling* (see the monograph [90] for a detailed description). Roughly speaking, in order to recover from the lack of homogeneity in the integral estimates, one works in cylinders whose natural scaling is modified by the oscillation of the solution itself. In the original proof, these rescaled cylinders are then sectioned in smaller sub-cylinders and the so-called *alternative* occurs: either there exists a sub-cylinder where u is sizeably (in a measure-theoretic sense) away from its infimum or in each sub-cylinder it is sizeably away from its supremum. In both cases a reduction in oscillation can be proved, giving the claimed Hölder continuity.

Stemming from recent techniques built to deal with the Harnack inequality for (4.4), simpler proofs are nowadays available, avoiding the analysis of said alternative. In the last section we will provide such a simplified proof, chiefly based on [29] and [42].

As it turned out, Hölder continuity of *bounded* solutions to (4.5) (in fact, to much more general equations) always holds. In the degenerate case $p \geq 2$, a-priori boundedness follows from the natural notion of weak solution given above, but in the singular case there is a precise threshold: local boundedness is guaranteed only for $p > p_{**} := \frac{2N}{N+2}$, which is therefore another critical exponent for the singular equation, smaller than p_*. However, when $1 < p < p_{**}$, weak solutions may be unbounded: for example, a suitable multiple of

$$v(x,t) = (T-t)_+^{\frac{1}{2-p}} |x|^{\frac{p}{p-2}}$$

solves the model equation (4.1) in the whole $\mathbb{R}^N \times \mathbb{R}$.

The critical exponents $p_* > p_{**}$ arise from the so-called $L^r - L^\infty$-estimates for sub-solutions, which are parabolic analogues of (2.7). Namely, when $p > p^*$, a $L^1 - L^\infty$ estimate holds true, eventually giving the intrinsic parabolic Harnack inequality. If only $p > p_{**}$ is assumed, one can still obtain a weaker $L^r - L^\infty$ estimate with $r > 1$ being the optimal exponent in the parabolic embedding

$$L^\infty(0,T;L^2(B_R)) \cap L^p(0,T;W^{1,p}(B_R)) \hookrightarrow L^r(0,T;L^r(B_R)), \qquad r = p\frac{N+2}{N}$$

which is ensured by the notion of weak solution.

Finally, we briefly comment on the regularity theory for parabolic *systems*. The general measurable coefficient condition dictated by (4.5) is not enough to ensure continuity, and either some additional structure is required (the so-called *Uhlenbeck structure*, due to the seminal paper [89] in the elliptic setting) or regularity holds everywhere except in a small *singular set*. The parabolic counterpart of [89] has first been proved in [22] and systematized in the monograph [21] for a large class of nonlinear parabolic system with Uhlenbeck structure. For the more recent developments on the partial regularity theory for parabolic system with general structure we refer to the memoirs [7, 32].

4.3 Intrinsic Harnack Inequalities

DiBenedetto and DiBenedetto and Kwong in [20] and [24] found and proved a suitable form of the parabolic Harnack inequality for the prototype equation (4.1), respectively in the degenerate and singular case. The critical value $p_* = 2N/(N+1)$ was shown to be the threshold below which no Harnack inequality, even in intrinsic form, may hold. However, comparison theorems where essential tools for the proof. A similar statement was later found to hold for general parabolic quasilinear equations of p-growth in [26] (degenerate case) and in [27] (singular supercritical case), with no monotonicty assumption. We will now describe these results, starting from the degenerate case.

Theorem 4.2 (Intrinsic Harnack Inequality, Degenerate Case) *Let $p \geq 2$ and u be a non negative weak solution in $B_{2r} \times [-T, T]$ of* (4.4) *under the growth conditions* (4.5). *There exists $C > 0$ and $\theta > 0$, depending only on N, p, C_0, C_1 such that if $0 < \theta\, u(0,0)^{2-p}\, r^p \leq T$, then*

$$C^{-1} \sup_{B_r} u(\cdot, -\theta\, u(0,0)^{2-p}\, r^p) \leq u(0,0) \leq C \inf_{B_r} u(\cdot, \theta\, u(0,0)^{2-p}\, r^p). \tag{4.6}$$

Clearly, for $p = 2$ we recover (3.1). For $p > 2$, the waiting time is larger the smaller $u(0,0)$ is; in other terms $u(0,0)$ bounds from below u on p-paraboloids of opening proportional to $u(0,0)^{p-2}$. It is worth noting here two additional difficulties in the Harnack inequality theory with respect to the linear (or more generally, homogeneous) setting. While it is still true that the forward form in the quasilinear setting implies the backward one, this is no more trivial due to the intrinsic waiting time depending on u_0.

The Harnack inequality in the singular setting turns out to be much more rich and subtle than in the degenerate case. A natural guess would be that (4.6) holds also in the singular case. However, consider the function

$$u(x,t) = (T-t)_+^{\frac{N+2}{2}} \left(a + b|x|^{\frac{2N}{N-2}}\right)^{-\frac{N}{2}}, \tag{4.7}$$

which is a bounded solution in $\mathbb{R}^N \times \mathbb{R}$ of the prototype equation (4.1) for any $p \in]1, p_*[$, $N > 2$ and suitably chosen $a, b > 0$. The latter violates both the forward and backward Harnack inequality in (4.6), as the right hand side vanishes for sufficiently large r, while the left hand side goes to $+\infty$ for $r \to +\infty$.

A similar phenomenon persists at the critical value $p = p_*$, as is shown by the entire solution

$$u(x,t) = \left(e^{ct} + |x|^{\frac{2N}{N-1}}\right)^{-\frac{N-1}{2}} \tag{4.8}$$

for suitable $c > 0$: the left hand side of (4.6) goes to $+\infty$ while the right hand one vanishes as $r \to +\infty$. It turns out that for $p \in]p_*, 2[$, Theorem 4.2 has a corresponding statement.

Theorem 4.3 (Intrinsic Harnack Inequality, Singular Supercritical Case) *Let* $2 > p > \frac{2N}{N+1}$ *and* u *be a non negative weak solution in* $B_{4r} \times [-T, T]$ *of* (4.4) *under the growth conditions* (4.5). *There exists* $C > 0$ *and* $\theta > 0$, *depending only on* N, p, C_0, C_1 *such that if* $u(0, 0) > 0$ *and*

$$r^p \sup_{B_{2r}} u(\cdot, 0)^{2-p} \leq T, \tag{4.9}$$

then

$$C^{-1} \sup_{B_r} u(\cdot, -\theta\, u(0,0)^{2-p}\, r^p) \leq u_0 \leq C \inf_{B_r} u(\cdot, \theta\, u(0,0)^{2-p}\, r^p). \tag{4.10}$$

Assumption (4.9) seems technical, however no proof is known at the moment without it. Following the procedure in [24], it can be removed for solutions of monotone equations fullfilling (3.16) (and thus obeying the comparison principle). The proof of the intrinsic Harnack inequality for supercritical singular equations is considerably more difficult than in the degenerate case and crucially relies on the following L^1-form of the Harnack inequality, first observed in [48] for the porous medium equation, which actually holds in the full singular range.

Theorem 4.4 (L^1-Harnack Inequality for Singular Equations) *Let* $p \in]1, 2[$ *and* u *be a non negative weak solution in* $B_{4r} \times [0, T]$ *of* (4.4) *under the growth conditions* (4.5). *There exists* $C > 0$ *depending only on* N, p, C_0, C_1 *such that*

$$\sup_{t \in [0,T]} \int_{B_r} u(x,t)\, dx \leq C \inf_{t \in [0,T]} \int_{B_{2r}} u(x,t)\, dx + C \left(T/r^{p+N(p-2)}\right)^{\frac{1}{2-p}}.$$

Notice that $p + N(p-2) > 0$ if and only if $p > p_*$. Thanks to this deep result, an elliptic form of the intrinsic Harnack inequality can be proved.

Theorem 4.5 (Elliptic Harnack Inequality for Singular Supercritical Equations) *Let* $p \in]\frac{2N}{N+1}, 2[$ *and* u *be a non negative weak solution in* $B_{4r} \times [-T, T]$ *of* (4.4)–(4.5). *There exists* $C > 0$ *and* $\theta > 0$, *depending on* N, p, C_0, C_1 *such that* *if* $u(0,0) > 0$ *and* (4.9) *holds, then*

$$C^{-1} \sup_{Q_r} u \le u(0,0) \le C \inf_{Q_r} u,$$

$$Q_r = B_r \times [-\theta\, u(0,0)^{2-p}\, r^p, \theta\, u(0,0)^{2-p}\, r^p]. \tag{4.11}$$

Recall that an elliptic form of the Harnack inequality such as (4.11) cannot hold for the classical heat equation. This forces the constants appearing in the previous theorem to blow-up as $p \uparrow 2$, hence, while this last form of the intrinsic Harnack inequality clearly implies (4.10), the constants in (4.10) are instead stable as $p \uparrow 2$. The previous examples also show that both constants must blow-up for $p \downarrow p_*$. The same comments following Theorem 4.3 on the rôle of hypothesis (4.9) can be made.

In the subcritical case, different forms of the Harnack inequality have been considered. Here we mention the one obtained in [38] generalizing to monotone operators a result of Bonforte and Vázquez [11, 12] on the porous medium equation.

Theorem 4.6 (Subcritical Case) *Let* $p \in]1, 2[$, u *be a positive, locally bounded weak solution in* $B_{2r} \times [-T, T]$ *of* (4.4) *under the growth conditions* (4.5) *and the monotonicty assumption* (3.16). *For any* $s \ge 1$ *such that* $\lambda_s := N(p-2) + ps > 0$ *there exists* $C, \delta, \theta > 0$, *depending on* N, p, s, C_0, C_1 *such that letting*

$$\widetilde{Q}_r(u) = B_r \times \left[\theta \left(\fint_{B_r} u(x,0)\, dx\right)^{2-p} r^p, \theta \left(\fint_{B_r} u(x,0)\, dx\right)^{2-p} (2r)^p\right],$$

if $u(0,0) > 0$ *and* $\widetilde{Q}_{2r}(u) \subseteq B_{2r} \times [0, T]$, *then*

$$\sup_{\widetilde{Q}_r(u)} u \le C\, A_u^{\delta} \inf_{\widetilde{Q}_r(u)} u, \qquad A_u = \left[\frac{\fint_{B_r} u(x,0)\, dx}{\left(\fint_{B_r} u^s(x,0)\, dx\right)^{\frac{1}{s}}}\right]^{\frac{ps}{\lambda_s}} \tag{4.12}$$

Notice that (4.12) is an elliptic Harnack inequality for later times, intrinsic in terms of the size of u at the initial time $t = 0$. In the singular supercritical case one can take $r = 1$ and thus $A_u \equiv 1$ in the previous statement to recover partially Theorem 4.5. The main weakness of (4.12) lies in the dependence of the Harnack constant from the solution itself. In general, a constant depending on u won't allow to deduce Hölder continuity but, as noted in [38], the peculiar structure of A_u permits such a deduction.

Other weaker forms not requiring the monotonicity assumption (3.16) are available (see [28]), however the complete picture in the subcritical case is not completely clear up to now.

4.4 Liouville Theorems

As for the classical heat equation, a one sided bound is not sufficient to ensure triviality of the solutions of the prototype equation (4.1). Indeed, a suitable positive multiple of the function

$$u(x,t) = (1 - x + ct)_+^{\frac{p-1}{p-2}} \tag{4.13}$$

solves (4.1) on $\mathbb{R} \times \mathbb{R}$ whenever $c > 0$ and $p > 2$. As is natural with parabolic Liouville theorems, a convenient setting is the one of ancient solutions and it turns out that a two-sided bound at a fixed time is sufficient to conclude triviality. The basic tools to prove the following results are the previously discussed Harnack inequalities and the following results are contained in [30].

Theorem 4.7 *Let $p > 2$ and u be a non-negative solution of*

$$u_t = \operatorname{div} A(x, u, Du) \qquad \text{on } \mathbb{R}^N \times]-\infty, T[\tag{4.14}$$

under the growth condition (4.5). *If for some $t_0 < T$, $u(\cdot, t_0)$ is bounded above, then u is constant.*

Notice that no monotonicity assumption on the principal part of the operator is needed. An optimal Liouville condition such as the one of Theorem 3.4 is unknown and clearly the example in (4.13) shows that it must involve a polynomial growth condition instead of a sub-exponential one. For the prototype parabolic p-Laplacian equation, a polynomial growth condition on both x and t more in the spirit of [83] is considered in [85].

On the complementary side, boundedness for *fixed* x_0 can also be considered, yielding:

Theorem 4.8 *Let $p > 2$ and u be a nonnegative solution in $\mathbb{R}^N \times \mathbb{R}$ of* (4.14), (4.5). *If*

$$\limsup_{t \to +\infty} u(x_0, t) < +\infty \qquad \text{for some } x_0 \in \mathbb{R}^N,$$

u is constant.

In the singular, supercritical case, the elliptic form (4.11) of the Harnack inequality directly ensures that, contrary to what happens for classical heat equation, a one-sided bound suffices to obtain a Liouville theorem. This is no longer true in the critical and subcritical case, as the functions in (4.8) and (4.7) show. However, again a two sided bound suffices.

Theorem 4.9 *Let $1 < p < 2$ and u be a weak solution on $\mathbb{R}^N \times]-\infty, T[$ of* (4.14) *under condition* (4.5). *If u is bounded, it is constant.*

4.5 Harnack Estimates at Large

By Harnack estimates at large, we mean *global* results such as the sub-potential lower bound (3.10) or the two-sided Kernel estimate (3.9). For the quasilinear equation

$$u_t = \operatorname{div} A(x, u, Du) \tag{4.15}$$

with p-growth assumptions (4.5), the natural candidates to state analogous inequalities are the Barenblatt profiles $\mathcal{B}_{p,M}$ given in (4.2). When A satisfies smoothness and monotonicity assumptions such as

$$\begin{cases} (A(x,s,z) - A(x,s,w)) \cdot (z - w) \geq 0 & \forall s \in \mathbb{R},\ x, z, w \in \mathbb{R}^N, \\ |A(x,s,z) - A(x,r,z)| \leq \Lambda(1 + |z|)^{p-1}|s - r| & \forall s, r \in \mathbb{R},\ x, z \in \mathbb{R}^N. \end{cases} \tag{4.16}$$

a comparison principle for weak solutions is available, as well as existence of solutions of the Cauchy problem with L^1 initial datum.

We start by considering the singular supercritical case, since the diffusion is fast and positivity spreads instantly on the whole $\mathbb{R}^N$, giving a behaviour similar to the one of the heat equation. The next result is contained in [13].

Theorem 4.10 (Sub-potential Lower Bound, Singular Case) *Let* $\frac{2N}{N+1} < p < 2$ *and* u *be a nonnegative solution of* (4.15) *in* $\mathbb{R}^N \times]0, +\infty[$ *under assumptions* (4.5), (4.16)*. There are constants* $C, \delta > 0$*, depending on the data, such that if* $u(x_0, t_0) > 0$*, then*

$$u(x,t) \geq \gamma\, u(x_0, t_0) \mathcal{B}_p \left(u(x_0, t_0)^{\frac{p-2}{p}} \frac{x - x_0}{t_0^{1/p}}, \frac{t}{t_0} \right), \tag{4.17}$$

for all $(x,t) \in \mathbb{R}^N \times [t_0(1-\delta), +\infty[$.

As an example, assume $x_0 = 0$, $t_0 = 1$ and $u(0,1) = 1$. Then, the previous sub-potential lower bound becomes

$$u(x,t) \geq \gamma \mathcal{B}_p(x,t)$$

for any $(x,t) \in \mathbb{R}^N \times [1 - \delta, +\infty[$. As a corollary, for any fundamental solution of (4.15), one obtains the two-sided kernel bounds (proved in [77] for the first time)

$$C^{-1} \mathcal{B}_{p,M_1}(x,t) \leq \Gamma(x,t) \leq C \mathcal{B}_{p,M_2}(x,t)$$

for some $C, M_1, M_2 > 0$ depending on the data. Notice how the elliptic nature of (4.15) for $p \in]p_*, 2[$, as expressed by the forward-backward Harnack inequality (4.11), allows to obtain the bound (4.17) also for some $t < t_0$. Previously known sub-potential lower bounds correspond to the case $\delta = 0$ above. As shown in [14], the phenomenon of propagation of positivity for $t < t_0$ not only happens in the near past but, as long as the spatial diffusion has had enough room to happen, it also hold for arbitrarily remote past times. More precisely, in [14] it is proved that (4.17) holds for all

$$(x, t) \in \mathcal{P}^c := \left\{ t > 0, |x - x_0|^p u(x_0, t_0)^{2-p} > 1 - \frac{t}{t_0} \right\},$$

while a weaker, but still optimal, lower bound holds in $\mathcal{P}$.

In the degenerate case $p > 2$, the finite speed of propagation implies that if the initial datum u_0 has compact support, then any solution of (4.15) keeps having compact support for any time $t > 0$. The finite speed of propagation has been quantified in [8], under the sôle p-growth assumption (4.5).

Theorem 4.11 (Speed of Propagation of the Support) *Let $p > 2$ and u be a weak solution of the Cauchy problem*

$$\begin{cases} u_t = \operatorname{div} A(x, u, Du) & \text{in } \mathbb{R}^N \times]0, +\infty[, \\ u(x, 0) = u_0 \end{cases}$$

under assumption (4.5). *If* $R_0 = \operatorname{diam}(\operatorname{supp} u_0) < +\infty$, *then*

$$\operatorname{diam}(\operatorname{supp} u(\cdot, t)) \le 2R_0 + Ct^{1/\lambda} \|u_0\|_{L^1(\mathbb{R}^N)}^{\frac{p-2}{\lambda}},$$

where $\lambda = N(p-2) + p$ *and* C *depend only on* N, p, C_0 *and* C_1.

Such an estimate actually holds for a suitable class of degenerate systems, see [84]. Sub-potential lower bounds are obtained in [8] as well.

Theorem 4.12 (Sub-potential Lower Bound, Degenerate Case) *Let $p > 2$ and u be a nonnegative solution of* (4.15) *in* $\mathbb{R}^N \times]0, +\infty[$ *under assumptions* (4.5), (4.16). *Then there are constants* $C, \varepsilon > 0$ *such that if* $u(x_0, t_0) > 0$, *then* (4.17) *holds in the region*

$$t > t_0, \qquad |x - x_0|^p \le \varepsilon\, u(x_0, t_0)^{p-2} t_0 \min\left\{ \frac{t - t_0}{t_0}, \left(\frac{t - t_0}{t_0}\right)^{p/\lambda} \right\},$$

with $\lambda = N(p-2) + p$.

The last condition on the region of validity of (4.17) is sharp, especially when $t \simeq t_0$ and the minimum is the first one (see [8, Remark 1.3] for details).

Under the additional assumptions (4.5) and (4.16) fundamental solutions exist and, as in the singular case, the sub-potential lower bound implies a two-sided estimate on the kernel in terms of the Barenblatt solution.

5 The Expansion of Positivity Approach

In this section we provide detailed proofs of some of the Harnack inequalities stated until now. Historically, Hölder regularity and Harnack inequalities have always been intertwined, with the former usually proved before the latter. The reason behind this is that Hölder regularity is a statement about a reduction in oscillation of u in B_r as $r \downarrow 0$, i.e. on the *difference* $\sup_{B_r} u - \inf_{B_r} u$. Thus it reduces to prove that *either* $\sup_{B_r} u$ decreases *or* $\inf_{B_r} u$ increases in a quantitative way. On the other hand, a Harnack inequality implies the stronger statement that both $\sup_{B_r} u$ decreases *and* $\inf_{B_r} u$ increases at a certain rate (see the nice discussion in [63, Ch. 1, §10]).

The modern approach thus often shifted the statements, first proving a Harnack inequality and then deducing from it the Hölder continuity of solutions. We instead revert to the historical roadmap, for two main reasons. The first one is pedagogical, as it feels satisfactory to reach an important stepping-stone result such as Hölder regularity, which would anyway follow from the techniques needed to prove the Harnack inequality. The second one is practical, since without continuity assumptions some of the arguments to reach, or even state, the Harnack inequality would be technically involved: for example, one would need to give a precise meaning to $u(0, 0)$ in (4.6).

We start in Sect. 5.1 by considering the elliptic setting. The proof of the Hölder continuity follows closely the original De Giorgi approach, then we introduce the notion of expansion of positivity. A technique due to Landis allows us to construct a largeness point from which to spread the positivity, thus giving the Harnack inequality. These are the common ingredients to all subsequent sections. In Sect. 5.2 we apply this technique to homogeneous parabolic equations with only minor modifications. Then we start discussing degenerate and singular parabolic equation. Section 5.3 is devoted to the proof of common tools to both, Sect. 5.4 to the degenerate case and the last one to singular supercritical equations.

While we won't prove basic propositions such as Energy estimates or Sobolev inequalities, the presentation will be mostly self contained. The only exception will be Theorem 5.32, which is the core tool to treat the singular supercritical Harnack inequality. Its proof is rather technical and since we could not find any simplification we would simply rewrite [31, Appendix A] word-by-word. Incidentally, this will also be the only sup estimate we will use. In striking contrast with the Moser method, in all the other subsections we will only assume *qualitative* boundedness of the solution (which certainly holds, as discussed in the previous section) without ever proving or using a quantitative integral sup-bound.

Since some arguments will be ubiquitous, a detailed discussion will be given at their first appearance, but we will only sketch the relevant modifications on subsequent occurrences. For this reason, the non-expert is advised to follow the path presented here from its very beginning, rather than skipping directly to the desired result.

5.1 Elliptic Equations

We now describe the De Giorgi technique to prove C^α-regularity and Harnack inequality for solutions of elliptic equations of the form

$$\operatorname{div} A(x, u, Du) = 0 \text{ with } \begin{cases} A(x, s, z) \cdot z \geq C_0 |z|^p \\ |A(x, s, z)| \leq C_1 |z|^{p-1} \end{cases} \qquad p \in]1, N[. \tag{5.1}$$

We will not treat boundedness statements (which actually hold true in this setting) and always assume that solutions are locally bounded.

Roughly speaking, the approach of De Giorgi consisted in deriving pointwise estimates on a solution u by analizing the behaviour of $|\{u \leq k\} \cap B_r|$ with respect to the level $k > 0$. First, he proved that the relative size of the sublevel set shrinks as k decreases, at a certain (logarithmic) rate. Then he showed that, when a suitable smallness threshold is reached, it starts decaying exponentially fast, so that it vanishes at a strictly positive level. This procedure produces a pointwise bound from below for u in terms of the size of its sublevel set in a larger ball and is thus called a *measure-to-point estimate* in the literature. This estimate, moreover, expands in space, since the relative size of a sublevel set in a larger ball B_R can also be bounded from below (polynomially in r/R) by its size in $B_r \subseteq B_R$. The quantitative statement arising from this simple observation is called *expansion of positivity* and is the basis for our proof of the Harnack inequality.

With a certain abuse of notation, we will say that u is a (sub-) super-solution of (5.1) if there exists and A obeying the prescribed growth condition for which $-\operatorname{div} A(x, u, Du)(\leq) \geq 0$. Observe that, being (5.1) homogeneous, the class of (sub-/super-) solutions of (5.1) is invariant by scaling, translation and (positive) scalar multiplication. More precisely, performing such transformations to a subsolution of (5.1) for some A results in a subsolution of (5.1) for a possibly different $\tilde{A}$, which nevertheless obeys the same bounds. We will use the following notations: $K_r(x_0)$ will denote a cube of side r and center x_0, $K_r = K_r(0)$ and, respectively,

$$P(K;\ u \lesseqgtr k) = \frac{|K \cap \{u \lesseqgtr k\}|}{|K|},$$

thus, for example, $P(K;\ u \geq 1)$ is the percentage of the cube K where $u \geq 1$. In the following, the dependence from p in the constants will always be omitted, and any constant only depending on N, p, C_0 and C_1 (the "data") will be denoted with a bar over it. Often we will also consider functions $f : \mathbb{R}_+ \to \mathbb{R}_+$ which will also depend on the data, and we will omit such a dependence. We first recall some basic facts.

Proposition 5.1

1) *[31, Lemma II.5.1] Let $X_n \geq 0$ obey for some $\alpha > 0$, $b, C > 0$, the iterative inequality*

$$X_{n+1} \leq C\, b^n\, X_n^{1+\alpha}.$$

Then

$$X_0 \leq C^{-1/\alpha} b^{-1/\alpha^2} \quad \Rightarrow \quad \lim_n X_n = 0. \tag{5.2}$$

2) **De Giorgi-Poincaré inequality***: [31, Lemma II.2.2] For any $u \in W^{1,1}(K_r)$ and $k \leq h$*

$$(h-k)|\{u \leq k\}| \leq \frac{C(N)\, r^{N+1}}{|\{u \geq k\}|} \int_{\{k<u\leq h\}} |Du|\, dx.$$

3) **Energy inequality***: Let u be a supersolution to* (5.1) *in K. Then there exists $\bar{C}$ such that for any $k \in \mathbb{R}$ and $\eta \in C_c^\infty(K)$*

$$\int_K |D(\eta(u-k)_-))|^p \leq \bar{C} \int_K (u-k)_-^p |D\eta|^p\, dx. \tag{5.3}$$

Lemma 5.2 (Shrinking Lemma) *Let $u \geq 0$ be a supersolution in K_R. For any $\mu > 0$ there exists $\beta(\mu) > 0$ such that*

$$P(K_{R/2};\ u \geq 1) \geq \mu \quad \Rightarrow \quad P(K_{R/2};\ u \leq 1/2^n) \leq \beta(\mu)/n^{1-\frac{1}{p}}.$$

Proof Rescale to $R = 2$ and let $k_j = 2^{-j}$. By the De Giorgi-Poincaré inequality

$$\begin{aligned}(k_j - k_{j+1})|K_1 \cap \{u \leq k_{j+1}\}| &\leq \frac{\bar{C}}{|K_1 \cap \{u \geq k_j\}|} \int_{K_1 \cap \{k_{j+1} < u\}} |D(u-k_j)_-| dx \\ &\leq \frac{\bar{C}}{\mu} \int_{K_1 \cap \{k_{j+1} < u\}} |D(u-k_j)_-|\, dx. \end{aligned} \tag{5.4}$$

If $\eta \in C_c^\infty(K_2)$ is such that $0 \le \eta \le 1$, $\eta \equiv 1$ on $K_{1/2}$ and $|\nabla\eta| \le C(N)$, (5.3) gives

$$\int_{K_{1/2}} |D(u-k_j)_-|^p\,dx \le \bar{C}\int_{K_2}(u-k_j)_-^p\,dx,$$

so that the last integral in (5.4) can be bounded through Hölder's inequality as

$$\begin{aligned}&\int_{K_{1/2}\cap\{k_{j+1}<u\}}|D(u-k_j)_-|\,dx\\ &\quad\le\left(\int_{K_{1/2}\cap\{k_{j+1}<u\}}|D(u-k_j)_-|^p\,dx\right)^{\frac{1}{p}}|K_{1/2}\cap\{k_{j+1}<u\le k_j\}|^{1-\frac{1}{p}}\\ &\quad\le\bar{C}\left(\int_{K_2}(u-k_j)_-^p\,dx\right)^{\frac{1}{p}}\big(|K_{1/2}\cap\{u\le k_j\}|-|K_{1/2}\cap\{u\le k_{j+1}\}|\big)^{1-\frac{1}{p}}\end{aligned}$$

Insert the latter into (5.4), use $(u-k_j)_-\le k_j$ and $k_j-k_{j+1}=k_j/2$ to get

$$\frac{k_j}{2}|K_{1/2}\cap\{u\le k_{j+1}\}|\le\frac{\bar{C}}{\mu}k_j\big(|K_{1/2}\cap\{u\le k_j\}|-|K_{1/2}\cap\{u\le k_{j+1}\}|\big)^{1-\frac{1}{p}}.$$

Simplify the k_j's, raise both sides to the power $p/(p-1)$ and sum over $j=0,\ldots,n-1$. Since $|K_{1/2}\cap\{u\le k_j\}|$ is decreasing and $|K_{1/2}\cap\{u\le k_j\}|-|K_{1/2}\cap\{u\le k_{j+1}\}|$ telescopic, we obtain

$$\begin{aligned}n\,|K_{1/2}\cap\{u\le k_n\}|^{\frac{p}{p-1}}&\le\sum_{j=0}^{n-1}|K_{1/2}\cap\{u\le k_{j+1}\}|^{\frac{p}{p-1}}\\ &\le\frac{\bar{C}}{\mu^{\frac{p}{p-1}}}\big(|K_{1/2}\cap\{u\le k_0\}|-|K_{1/2}\cap\{u\le k_n\}|\big)\le\bar{C}\frac{1-\mu}{\mu^{\frac{p}{p-1}}}.\end{aligned}$$

□

Lemma 5.3 (Critical Mass) *Let $u\ge 0$ be a supersolution in K_R. There exists $\bar{\nu}$ such that*

$$P(K_R;\,u\le 1)\le\bar{\nu}\quad\Rightarrow\quad u\ge 1/2\quad\text{in } K_{R/2}. \tag{5.5}$$

Proof Scale back to $R=1$ and define for $n\ge 1$ $k_n=r_n=1/2+1/2^n$, $K_n=K_{r_n}$. Let moreover

$$\eta_n\in C_c^\infty(K_n),\qquad 0\le\eta_n\le 1,\qquad \eta_n|_{K_{n+1}}\equiv 1,\qquad |D\eta_n|\le\bar{C}\,2^n$$

and chain the Sobolev inequality with (5.3) with $k = k_n$, $\eta = \eta_n$, to obtain

$$\int |(u-k_n)_-\eta_n|^{p^*}\,dx \le \bar C\left(\int |D(u-k_n)_-\eta_n|^p\,dx\right)^{\frac{p^*}{p}}$$
$$\le \bar C\left(\int_{K_n} 2^{np}(u-k_n)_-^p\,dx\right)^{\frac{p^*}{p}}. \tag{5.6}$$

On the right we use $(u-k_n)_- \le k_n$ and $|K_n| \le 1$ to bound

$$\int_{K_n} (u-k_n)_-^p\,dx \le k_n^p\,|K_n \cap \{u \le k_n\}| \le 2^{-np}\,P(K_n;\,u \le k_n)$$

while by $\eta_n \equiv 1$ on K_{n+1} and Tchebicev's inequality,

$$\int |(u-k_n)_-\eta_n|^{p^*}\,dx \ge \int_{K_{n+1}} (u-k_n)_-^{p^*}\,dx \ge \int_{K_{n+1}\cap\{u\le k_{n+1}\}} (u-k_n)_-^{p^*}\,dx$$
$$\ge (k_n-k_{n+1})^{p^*}\,|K_{n+1}\cap\{u\le k_{n+1}\}| \ge 2^{-(n+1)p^*}2^{-N}\,P(K_{n+1};\,u\le k_{n+1}).$$

Use the previous two inequalities into (5.6) to get

$$P(K_{n+1};\,u \le k_{n+1}) \le \bar C\,2^{np^*}\,P(K_n;\,u\le k_n)^{\frac{p^*}{p}}.$$

The claim now follows from (5.2) applied to the sequence $X_n = P(K_n;\,u \le k_n)$. □

Lemma 5.4 (Measure-to-Point Estimate) *Let $u \ge 0$ be a supersolution in K_R. For any $\mu > 0$ there exists $m(\mu) > 0$ such that*

$$P(K_{R/2};\,u \ge k) \ge \mu \quad \Rightarrow \quad \inf_{K_{R/4}} u \ge m(\mu)\,k. \tag{5.7}$$

Proof Given $\mu > 0$, choose $n_\mu \ge 1$ in Lemma 5.2 such that $\beta(\mu)/n_\mu^{1-1/p} \le \bar\nu$, so that $P(K_{R/2};\,u \ge k/2^{n_\mu}) \le \bar\nu$. Then apply (5.5) to u/k, obtaining (5.7) with $m(\mu) = 2^{-n_\mu-1}$. □

Theorem 5.5 (Hölder Regularity) *Let u solve* (5.1) *in K_{2R}. There exists $\bar C, \bar\alpha > 0$ such that*

$$\operatorname{osc}(u;\,K_\rho) \le \bar C\,\operatorname{osc}(u;\,K_R)\,(\rho/R)^{\bar\alpha} \quad \textit{for } 0 \le \rho \le R/2. \tag{5.8}$$

Proof Rescaling to $R = 1$ and considering $u/\operatorname{osc}(u;\,K_1)$ we can suppose $\operatorname{osc}(u,K_1) = 1$. Both $u_+ = u - \inf_{K_1} u$ and $u_- = \sup_{K_1} u - u$ are non-negative solutions with $\operatorname{osc}(u_\pm;\,K_1) = 1$. Since

$$P(K_1;\,u_+ \ge 1/2) = P(K_1;\,u_- \le 1/2) = 1 - P(K_1;\,u_- > 1/2),$$

at least one of $P(K_1;\ u_\pm \geq 1/2)$ is at least $1/2$ and we can suppose without loss of generality that it is u_+. Then (5.7) with $R = 2$, $k = 1/2$ provides

$$\inf_{K_{1/2}} u_+ \geq m(1/2)/2 =: \bar{m} \quad \Rightarrow \quad \inf_{K_{1/2}} u \geq \inf_{K_1} u + \bar{m} \quad \Rightarrow \quad \operatorname{osc}(u;\ K_{1/2}) \leq 1 - \bar{m}.$$

Scaling back we obtained $\operatorname{osc}(u;\ K_{R/2}) \leq \operatorname{osc}(u;\ K_R)(1-\bar{m})$ which, iterated for $R_n = R/2^n$ gives

$$\operatorname{osc}(u;\ K_{R_n}) \leq \operatorname{osc}(u;\ K_R)(1-\bar{m})^n.$$

For $\rho \leq R/2$, let $n \geq 1$ obey $R_{n+1} \leq \rho \leq R_n$ and $\bar{\alpha} := -\log_2(1-\bar{m})$. Then, by monotonicity,

$$\begin{aligned}\operatorname{osc}(u;\ K_\rho) \leq \operatorname{osc}(u;\ K_{R_n}) &\leq \operatorname{osc}(u;\ K_R)(1-\bar{m})^n = \operatorname{osc}(u;\ K_R)2^{-n\bar{\alpha}} \\ &= 2^{\bar{\alpha}} \operatorname{osc}(u;\ K_R)(2^{-(n+1)})^{\bar{\alpha}}\end{aligned}$$

giving the claim due to $2^{-(n+1)} \leq \rho/R$. □

Theorem 5.6 (Expansion of Positivity, See Fig. 3) *Let $u \geq 0$ be a supersolution in K_R. There exists $\bar{\lambda} > 1$ and, for any $\mu > 0$, $c(\mu) > 0$ such that*

$$P(K_r;\ u \geq 1) \geq \mu \quad \Rightarrow \quad \inf_{K_\rho} u \geq c(\mu)\,(r/\rho)^{\bar{\lambda}} \quad \text{if } r \leq \rho \leq R/2. \tag{5.9}$$

Proof Using the notations of Lemma 5.4, we let $c = c(\mu) := m(\mu/2^N)$ and iterate (5.7) as follows. From $P(K_r;\ u \geq 1) \geq \mu$ we infer $P(K_{2r};\ u \geq 1) \geq \mu/2^N$ thus (5.7) gives $\inf_{K_r} u \geq c$. If $\bar{\delta} := m(4^{-N})$ and $\rho_n = 2^n r$, we thus have $P(K_{\rho_0};\ u \geq c\,\bar{\delta}^0) = 1$. Moreover

$$P(K_{\rho_n}; u \geq c\bar{\delta}^n) = 1 \;\Rightarrow\; P(K_{\rho_{n+2}}; u \geq c\bar{\delta}^n) \geq 4^{-N} \underset{(5.7)}{\Rightarrow} P(K_{\rho_{n+1}}; u \geq c\bar{\delta}^{n+1}) = 1.$$

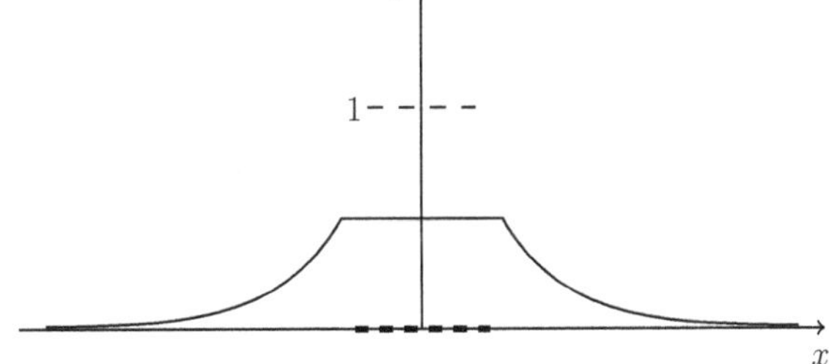

Fig. 3 The expansion of positivity. If $u \geq 1$ on the dashed part of the cube, it is bounded below by a negative power of the distance from the cube

Thus, by induction, $u \ge c\,\bar{\delta}^n$ in K_{ρ_n} for all $n \ge 0$ such that $\rho_{n+2} = 2^{n+2}r \le R$. Given $\rho \in [r, R/2]$ let n be such that $2^{n-1} \le \rho/r \le 2^n$. Then we obtained the claim with $\bar{\lambda} = -\log_2 \bar{\delta}$, since

$$\inf_{K_\rho} u \ge \inf_{K_{\rho_n}} u \ge c\,\bar{\delta}^n \ge c\,\bar{\delta}\,(\rho/r)^{\log_2 \bar{\delta}}.$$

□

We call the exponent $\bar{\lambda}$ the *expansion of positivity rate*.

Theorem 5.7 (Harnack Inequality) *There exists $\bar{C} > 0$ such that for any locally bounded solution $u \ge 0$ to* (5.1) *in K_{8R} it holds*

$$\sup_{K_R} u \le \bar{C}\,\inf_{K_R} u.$$

Proof Rescaling to $R = 1$ and considering $u/\sup_{K_1} u$ we are reduced to prove that

$$\sup_{K_1} u = 1 \quad \Rightarrow \quad \inf_{K_1} u \ge \bar{m} > 0 \tag{5.10}$$

for any solution $u \ge 0$ in K_8. We will find $\bar{m} > 0$, $x_0 \in K_1$ and $r > 0$ such that

$$u(x_0)\,r^{\bar{\lambda}} \ge \bar{m}, \qquad P(K_r(x_0);\; u \ge u(x_0)/2) \ge \bar{\nu} \tag{5.11}$$

for $\bar{\lambda}$ given (5.9) and some universal $\bar{\nu}$. Theorem 5.6 applied to $u/u(x_0)$ will then prove (5.10) for such r, with the choices $R = 8$, $k = u(x_0)/2$, $\mu = \bar{\nu}$ and $\rho = 2$, as $K_1 \subseteq K_2(x_0) \subseteq K_4$.

To choose x_0 and r, observe that Theorem 5.5 implies that the function

$$[0, 1] \ni \rho \mapsto \psi(\rho) = (1-\rho)^{\bar{\lambda}} \sup_{K_\rho} u$$

is continuous and vanishes at $\rho = 1$, thus it attains its maximum at some $\rho_0 < 1$ and we set

$$\max_{[0,1]} \psi = (1-\rho_0)^{\lambda} \sup_{K_{\rho_0}} u = (1-\rho_0)^{\lambda}\,u(x_0)$$

for some $x_0 \in K_{\rho_0}$. Let $\xi \in\,]0, 1[$ to be chosen and define $r = \xi\,(1-\rho_0)$. Then

$$u(x_0)\,r^{\bar{\lambda}} = \xi^{\bar{\lambda}}\,u(x_0)\,(1-\rho_0)^{\bar{\lambda}} = \xi^{\bar{\lambda}}\,\psi(\rho_0) \ge \xi^{\bar{\lambda}}\,\psi(0) = \xi^{\bar{\lambda}}. \tag{5.12}$$

Since $K_r(x_0) \subseteq K_{\rho_0+r}$, we infer from $\psi(\rho_0 + r) \le \psi(\rho_0)$ that

$$\sup_{K_r(x_0)} u \le \sup_{K_{\rho_0+r}} u = \frac{\psi(\rho_0 + r)}{(1 - \rho_0 - r)^{\bar{\lambda}}}$$

$$\le \frac{\psi(\rho_0)}{(1 - \rho_0 - r)^{\bar{\lambda}}} = \frac{(1 - \rho_0)^{\bar{\lambda}}}{(1 - \rho_0 - r)^{\bar{\lambda}}} \sup_{K_{\rho_0}} u = \frac{u(x_0)}{(1 - \xi)^{\bar{\lambda}}}.$$

Choose ξ as per $(1 - \xi)^{-\bar{\lambda}} = 2$, so that $u \le 2\,u(x_0)$ in $K_r(x_0)$, while (5.12) gives the first condition in (5.11) with $\bar{m} = \xi^{\bar{\lambda}}$. Apply (5.8) for $R = r$, $\rho = \bar{\eta} r$ with $\bar{\eta}$ s. t. $4\bar{C}\bar{\eta}^{\bar{\alpha}} \le 1$, so that

$$\operatorname{osc}(u;\, K_{\bar{\eta}r}(x_0)) \le \bar{C} \operatorname{osc}(u;\, K_r(x_0))\, \bar{\eta}^{\bar{\alpha}} \le 2\,\bar{C}\, u(x_0)\, \bar{\eta}^{\bar{\alpha}} \le u(x_0)/2,$$

implying $u \ge u(x_0)/2$ in $K_{\bar{\eta}r}(x_0)$. Thus, the second condition in (5.11) holds for $\bar{\nu} = \bar{\eta}^N$. □

5.2 *Homogeneous Parabolic Equations*

In the forthcoming subsections we will provide the extension of the previous techniques to the parabolic setting. In order to highlight the similarities with the elliptic case, we will proceed step-by-step in increasing generality, gradually introducing the modifications needed to cater with the evolutionary framework.

First we will deal with homogeneous equations, i.e. those for which scalar multiplication still gives a solution of the same (from the structural point of view) type of equation. We chose for simplicity to deal with the quadratic case, i.e., with equations of the form

$$u_t = \operatorname{div} A(x, u, Du), \qquad \begin{cases} A(x, s, z) \cdot z \ge C_0 |z|^2 \\ |A(x, s, z)| \le C_1 |z|. \end{cases} \tag{5.13}$$

As in the previous subsection, we say that u is a (sub-) super-solution if there is some A obeying the growth conditions and such that $u_t (\le) \ge \operatorname{div} A(x, u, Du)$. An important feature of (5.13) is that the class of its solutions is invariant by space/time translations, by the scaling $u_\lambda(x, t) = u(\lambda x, \lambda^2 t)$, $\lambda > 0$ and, more substantially, by scalar multiplication. More generally, homogeneous problems of the form

$$|u_t|^{p-2} u_t = \operatorname{div} A(x, u, Du), \qquad \begin{cases} A(x, s, z) \cdot z \ge C_0 |z|^p \\ |A(x, s, z)| \le C_1 |z|^{p-1} \end{cases}$$

can be dealt in the same way. In fact, as will be apparent from the proofs, in this homogeneous setting the Harnack inequality follows solely from the energy inequality. Indeed, in [40], it has been proved for non-negative functions belonging to the so-called *parabolic De Giorgi classes* i.e., roughly speaking, functions obeying the energy inequality for truncations.

In the following, we set $Q_{R,T} = K_R \times [0,T]$. Given a rectangle $Q = K \times [a,b] \subseteq \mathbb{R}^N \times \mathbb{R}$, $u : Q \to \mathbb{R}$ and $k \in \mathbb{R}$ we define, respectively

$$P\left(Q;\, u \lesseqgtr k\right) = \frac{\left|Q \cap \left\{u \lesseqgtr k\right\}\right|}{|Q|}, \qquad P_t\left(K;\, u \lesseqgtr k\right) = \frac{\left|K \cap \left\{u(\cdot,t) \lesseqgtr k\right\}\right|}{|K|}.$$

The dependence on N, C_0 and C_1 will always be omitted, and a constant c depending only on the latters will be denoted by $\bar{c}$. We also recall the relevant functional analytic tools.

Proposition 5.8

1) **Parabolic Sobolev Embedding:** *[31, Lemma II.4.1] If* $u \in L^2(0,T; W_0^{1,2}(\Omega))$, *then*

$$\int_0^T \int_\Omega |u|^{2\frac{N+2}{N}}\, dx\, dt \le C_N \left(\sup_{t\in[0,T]} \int_\Omega u^2(x,t)\, dx \right)^{\frac{2}{N}} \int_0^T \int_\Omega |Du|^2\, dx\, dt.$$

2) **Energy inequality:** *[31, Prop. III.2.1] Let u be a supersolution to* (5.13) *in* $Q = K \times [0,T]$. *There exists* $\bar{C} > 0$ *s. t. for any* $k \ge 0$ *and* $\eta \in C^\infty(a,b; C_c^\infty(K))$, $0 \le \eta \le 1$ *it holds*

$$\begin{aligned} \sup_{t\in[0,T]} \int_K (u(x,t)-k)_-^2 \eta^2\, dx &+ \frac{1}{\bar{C}} \iint_Q |D(\eta(u-k)_-)|^2\, dx\, dt \\ &\le \int_K (u(x,0)-k)_-^2 \eta^2\, dx + \bar{C} \iint_Q (u-k)_-^2 |\nabla\eta|^2\, dx\, dt \\ &\quad + \bar{C} \iint_Q (u-k)_-^2 |\eta_t|\, dx\, dt. \end{aligned} \tag{5.14}$$

The first lemma shows how initial measure-theoretic positivity propagates at future times.

Lemma 5.9 *Let* $u \ge 0$ *be a supersolution in* Q_{R,R^2}. *For any* $\mu > 0$ *there are* $k, \theta \in]0,1[$ *such that*

$$P_0(K_R; u \ge 1) \ge \mu \;\Rightarrow\; P_t\left(K_R; u \ge k(\mu)\right) > \mu/2 \quad \forall t \in [0, \theta(\mu)R^2]. \tag{5.15}$$

Proof Rescale to $R = 1$ and, for any $\delta, \theta \in]0,1[$, employ the energy inequality (5.14) on $K_1 \times [0,\theta]$ with $\eta \in C_c^\infty(K_1)$ independent of t and such that

$$0 \le \eta \le 1, \qquad \eta|_{K_\delta} \equiv 1, \qquad |D\eta| \le C_N/(1-\delta). \tag{5.16}$$

obtaining for any $t \in [0, \theta]$

$$\int_{K_\delta} (u(x,t)-1)_-^2\,dx \le \int_{K_1} (u(x,0)-1)_-^2\,dx + \frac{\bar{C}}{(1-\delta)^2}\int_0^t\int_{K_1}(u-1)_-^2\,dx\,dt$$
$$\le 1-\mu+\frac{\bar{C}\,\theta}{(1-\delta)^2},$$

where we used the assumption in (5.15) in the last inequality. For $k \in]0,1[$ we have

$$\int_{K_\delta}(u(x,t)-1)_-^2\,dx \ge \int_{K_\delta\cap\{u(\cdot,t)<k\}}(1-k)^2\,dx \ge (1-k)^2|K_\delta\cap\{u(\cdot,t)<k\}|.$$

Insert the latter into the previous one to obtain, for all $t \in [0,\theta]$

$$1-P_t(K_1;\,u\ge k) \le 1-|K_\delta\cap\{u(\cdot,t)\ge k\}| = 1-\delta^N+|K_\delta\cap\{u(\cdot,t)<k\}|$$
$$\le 1-\delta^N+\frac{1}{(1-k)^2}\left(1-\mu+\frac{\bar{C}\,\theta}{(1-\delta)^2}\right). \tag{5.17}$$

Successively choose $\delta, k \in]0,1[$ and, consequently, $\theta \in]0,1[$ so that:

$$1-\delta^N=\frac{\mu}{8},\qquad \frac{1-\mu}{(1-k)^2}=1-\frac{3}{4}\mu,\qquad \frac{1}{(1-k)^2}\frac{\bar{C}\,\theta}{(1-\delta)^2}\le\frac{\mu}{8}$$

to obtain that the right hand side in (5.17) is less than $1-\mu/2$, proving the claim.

□

The next two steps are fully in the spirit of the De Giorgi approach.

Lemma 5.10 (Shrinking Lemma) *Suppose $u \ge 0$ is a supersolution in $Q_{2R,T}$ obeying*

$$P_t(K_R;\,u\ge k)\ge\mu,\qquad \forall t\in[0,T] \tag{5.18}$$

for some $\mu \in]0,1[$, $k > 0$. There exists $\beta = \beta(\mu) > 0$ such that

$$P\left(Q_{R,T};\,u\le k/2^n\right)\le\beta(\mu)\left(1+\frac{R^2}{T}\right)^{1/2}\frac{1}{n^{1/2}},$$

Proof Let $k_j = k/2^j$, $j \ge 0$. The inequality (5.14) with $\eta \in C_c^\infty(K_{2R})$ such that

$$0\le\eta\le 1,\qquad \eta|_{K_R}\equiv 1,\qquad |D\eta|\le C_N/R \tag{5.19}$$

gives

$$\iint_{Q_{R,T}} |D(u-k_j)_-|^2 dxdt \le \bar{C}\int_{K_{2R}} (u(x,0)-k_j)_-^2 dx + \frac{\bar{C}}{R^2}\iint_{Q_{2R,T}} (u-k_j)_-^2 dxdt$$
$$\le \bar{C}\, k_j^2\, R^N (1+T/R^2). \tag{5.20}$$

For any $t \in [0,T]$, apply the De Giorgi-Poincaré inequality and (5.18) to obtain

$$\begin{aligned}(k_j - k_{j+1})|K_R \cap \{u(\cdot,t) \le k_{j+1}\}| &\le \frac{C_N\, R^{N+1}}{|K_R \cap \{u(\cdot,t) < k_j\}|}\int_{K_R\cap\{k_{j+1}\le u(\cdot,t)\}} |D(u(x,t)-k_j)_-|\, dx \\ &\underset{(5.18)}{\le} \frac{C_N\, R}{\mu}\int_{K_R\cap\{k_{j+1}\le u(\cdot,t)\}} |D(u(x,t)-k_j)_-|\, dx\end{aligned}$$

Integrate over $[0,T]$, divide by $|Q_{R,T}|$ and use Hölder's inequality to get

$$\frac{k_j}{2} P(Q_{R,T};\, u \le k_{j+1}) \le \frac{C_N\, R}{\mu\, |Q_{R,T}|}\iint_{Q_{R,T}\cap\{k_{j+1}\le u\}} |D(u-k_j)_-|\, dx\, dt$$

$$\le \frac{R}{\mu}\left(\frac{C_N^2}{|Q_{R,T}|}\iint_{Q_{R,T}\cap\{k_{j+1}\le u\}} |D(u-k_j)_-|^2 dxdt\right)^{\frac{1}{2}} \frac{|Q_{R,T}\cap\{k_{j+1}\le u\le k_j\}|^{\frac{1}{2}}}{|Q_{R,T}|^{\frac{1}{2}}}$$

$$\underset{(5.20)}{\le} \frac{\bar{C}\, R}{\mu} k_j \frac{1}{T^{\frac{1}{2}}}\left(1+\frac{T}{R^2}\right)^{\frac{1}{2}} \left(P(Q_{R,T};\, u\le k_j) - P(Q_{R,T};\, u \le k_{j+1})\right)^{\frac{1}{2}},$$

The latter reads

$$P^2(Q_{R,T}; u \le k_{j+1}) \le C(\mu)\big(1+\frac{R^2}{T}\big)\left(P(Q_{R,T}; u\le k_j) - P(Q_{R,T}; u\le k_{j+1})\right),$$

which, being the right hand side telescopic, can be summed over $j \le n-1$ to get the claim:

$$n\left(P(Q_{R,T};\, u\le k_n)\right)^2 \le \sum_{j=0}^{n-1}\left(P(Q_{R,T};\, u \le k_n)\right)^2 \le C(\mu)\,(1+R^2/T).$$

□

Lemma 5.11 (Critical Mass) *For any $\theta > 0$ there exists $\nu(\theta) > 0$ such that any supersolution $u \ge 0$ on $Q_{R,\theta R^2}$ fulfills*

$$P\left(Q_{R,\theta R^2};\, u \le k\right) \le \nu(\theta) \quad \Rightarrow \quad u \ge k/2 \quad \text{on } K_{R/2}\times[\theta\, R^2/8, \theta\, R^2]. \tag{5.21}$$

Proof Use homogeneity and scaling to reduce to $R = 1, k = 1$. Define for $n \geq 1$

$$r_n = \frac{1}{2} + \frac{1}{2^n}, \qquad k_n = \frac{1}{2} + \frac{1}{2^n}, \qquad \theta_n = \frac{\theta}{8} - \frac{\theta}{2^{n+3}},$$

let $K_n = K_{r_n}$, $Q_n = K_n \times [\theta_n, \theta]$ and choose $\eta_n \in C^\infty([\theta_n, \theta]; C_c^\infty(K_n))$ s. t.

$$\eta_n(\cdot, \theta_n) \equiv 0, \;\; 0 \leq \eta_n \leq 1, \;\; \eta_n|_{Q_{n+1}} \equiv 1, \;\; |D\eta_n| \leq C_N 2^n, \;\; |(\eta_n)_t| \leq C_N \frac{2^n}{\theta}. \tag{5.22}$$

Inserting into the energy inequality (5.31) and noting that $k_n \leq k$, we get

$$\sup_{t\in[\theta_{n+1},\theta]} \int_{K_{n+1}} (u(x,t) - k_n)_-^2 \, dx + \iint_{Q_n} |D(\eta_n(u - k_n)_-)|^2 \, dx \, dt$$
$$\leq \bar{C}\, 2^{2n}(1 + \theta^{-2}) \iint_{Q_n} (u - k_n)_-^2 \, dx \, dt \leq \bar{C}\, 2^{2n}(1 + \theta^{-2}) k_n^2 |Q_n \cap \{u \leq k_n\}|.$$

By the parabolic Sobolev embedding

$$\iint_{Q_{n+2}} (u - k_{n+1})_-^{2\frac{N+2}{N}} \, dx \, dt \leq \iint_{Q_{n+1}} ((u - k_{n+1})_- \eta_{n+1})^{2\frac{N+2}{N}} \, dx \, dt$$

$$\leq C \iint_{Q_{n+1}} |D\, \eta_{n+1}(u - k_{n+1})_-^2|^2 dx dt \left[\sup_{t\in[\theta_{n+1},\theta]} \int_{K_{n+1}} \eta_{n+1}^2(x,t)(u(x,t) - k_{n+1})_-^2 dx \right]^{\frac{2}{N}}$$

$$\leq \bar{C}\, 2^{2n\frac{N+2}{N}} (1 + \theta^{-2})^{\frac{N+2}{N}} h_n^{2\frac{N+2}{N}} \, |Q_n \cap \{u \leq k_n\}|^{\frac{N+2}{N}},$$

while, being $(u - k_{n+1})_- \geq k_{n+1} - k_{n+2} = k_n/4$ when $u \leq k_{n+2}$,

$$\iint_{Q_{n+1}} (u - k_{n+1})_-^{2\frac{N+2}{N}} \, dx \, dt \geq (k_n/4)^{2\frac{N+2}{N}} |Q_{n+2} \cap \{u \leq k_{n+2}\}|.$$

Chaining these latter two estimates and simplifying k_n gives the iterative inequality

$$|Q_{n+2} \cap \{u \leq k_{n+2}\}| \leq \bar{C}\, b_N^n (1 + \theta^{-2})^{1+\frac{2}{N}} |Q_n \cap \{u \leq k_n\}|^{1+\frac{2}{N}}$$

and (5.2) for $X_n := |Q_{2n} \cap \{u \leq k_{2n}\}|$ gives the claim. □

Lemma 5.12 (Measure-to-Point Estimate) *Let* $u \geq 0$ *be a supersolution in* Q_{R,R^2}. *For all* $\mu \in]0, 1[$ *there are* $c(\mu) > 0$ *and* $\theta(\mu) \in]0, 1[$ *such that*

$$P_0(K_\rho; u \geq h) \geq \mu \;\; \Rightarrow \;\; u \geq c(\mu)\, h \;\; \text{in } K_{\rho/2} \times [\theta(\mu)\rho^2/8, \theta(\mu)\rho^2]. \tag{5.23}$$

Proof By homogeneity we can let $h = 1$. Let $\theta(\cdot), k(\cdot)$ be given in Lemma 5.9, so that $P_t(K_\rho;\ u \geq k) \geq \mu/2$ for $t \in [0, \theta\,\rho^2]$, $\theta = \theta(\mu)$ and $k = k(\mu)$. Apply Lemma 5.10, choosing $n = n(\mu)$ such that

$$\beta(\mu/2)\,(1 + \theta(\mu)^{-1})^{1/2}\,n^{-1/2} \leq \nu(\theta),$$

($\nu(\cdot)$ given in (5.21)), to get $P(Q_{\rho,\theta\rho^2};\ u \leq k\,2^{-n}) \leq \nu(\theta)$. Then (5.21) proves (5.23). □

Theorem 5.13 (Hölder Regularity) *Any locally bounded solution of* (5.13) *is locally Hölder continuous, with Hölder exponent depending only on* N, C_0, C_1.

Proof By translation and scaling it suffices to prove an oscillation decay on the cubes $Q_n = K_{2^{-n}} \times [-\bar{\theta}\,2^{-2n}, 0]$ with $\bar{\theta} = \theta(1/2)$ given in (5.23). Suppose $\operatorname{osc}(u, Q_0) = 1$. Then, one of

$$P_{-\bar{\theta}}\big(K_1;\ \sup_{Q_0} u - u \geq 1/2\big) \geq 1/2, \qquad \text{or} \qquad P_{-\bar{\theta}}\big(K_1;\ u - \inf_{Q_0} u \geq 1/2\big) \geq 1/2$$

holds. If it is the first one, apply (5.23) to $\sup_{Q_0} u - u \geq 0$ translated in time to get $\sup_{Q_0} u - u \geq m(1/2)/2 =: \bar{m}$ in Q_1, i.e. $\sup_{Q_1} u \leq \sup_{Q_0} u - \bar{m}$. Therefore

$$\operatorname{osc}(u, Q_1) \leq \sup_{Q_1} u - \inf_{Q_0} u \leq \sup_{Q_0} u - \inf_{Q_0} u = 1 - \bar{m}.$$

The same holds in the other case and by homogeneity we have $\operatorname{osc}(u; Q_1) \leq (1 - \bar{m})\operatorname{osc}(u, Q_0)$. By scaling and induction, $\operatorname{osc}(u, Q_n) \leq \operatorname{osc}(u, Q_0)(1-\bar{m})^n$. Finally, for $(x,t) \in Q_1$ let $n \geq 1$ such that

$$2^{-n-1} \leq \max\{|x|, (|t|/\bar{\theta})^{1/2}\} \leq 2^n,$$

so that we have $(x,t) \in Q_n$ and, for $\bar{\alpha} = -\log_2(1 - \bar{m})$,

$$\begin{aligned} |u(x,t) - u(0,0)| \leq \operatorname{osc}(u, Q_n) &\leq \frac{\operatorname{osc}(u, Q_0)}{1 - \bar{m}}(1 - \bar{m})^{n+1} \\ &\leq \frac{\operatorname{osc}(u, Q_0)}{1 - \bar{m}} \max\left\{|x|, (|t|/\bar{\theta})^{1/2}\right\}^{\bar{\alpha}}. \end{aligned}$$

□

Lemma 5.14 (Expansion of Positivity, See Fig. 4) *Let* $u \geq 0$ *be a supersolution in* Q_{R,R^2}. *There exists* $\bar{\lambda} > 1$, $\bar{\gamma} \in\,]0, 1/4[$ *and, for any* $\mu > 0$ *a constant* $c(\mu) > 0$, *such that*

$$P_0(K_r;\ u \geq 1) \geq \mu \quad \Rightarrow \quad \inf_{K_\rho} u(\cdot, \bar{\gamma}\,\rho^2) \geq c(\mu)\,(r/\rho)^{\bar{\lambda}} \qquad \forall \rho \in [r, R/8].$$

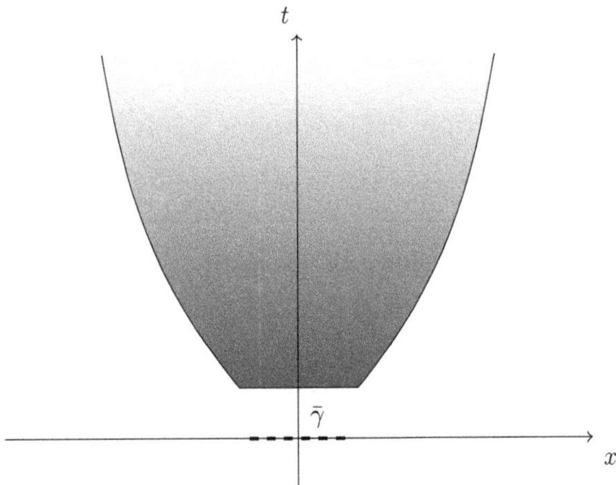

Fig. 4 The parabolic expansion of positivity. If at time $t = 0$ $u \geq 1$ on the dotted part of given measure, after a waiting time $\bar{\gamma}$, u is pointwise bounded from below in the paraboloid by a large negative power of t

Proof First expand (5.23) in space observing that $P_0(K_\rho;\ u \geq h) \geq \mu$ implies $P_0(K_{4\rho};\ u \geq h) \geq \mu\, 4^{-N}$, so that by changing the constants $\theta(\mu)$ and $c(\mu)$, we get

$$P_0(K_\rho; u \geq h) \geq \mu \ \Rightarrow\ u \geq c(\mu)\, h \ \text{ in } K_{2\rho} \times [\theta(\mu)\rho^2/8, \theta(\mu)\rho^2]. \tag{5.24}$$

Let $\bar{c} = c(1), \bar{\theta} = \theta(1)$, $\rho_n - 2^n r$ and define recursively the sequences

$$t_0 = \theta(\mu)r^2/8,\ \ t_{n+1} = t_n + \bar{\theta}\rho_{n+1}^2/8,\ \ s_0 = \theta(\mu)r^2,\ \ s_n = t_{n-1} + \bar{\theta}\rho_n^2,\ \ n \geq 1.$$

Letting furthermore $Q_n = K_{\rho_{n+1}} \times [t_n, s_n]$, apply recursively (5.24) as

$$P_0(K_r; u \geq 1) \geq \mu \ \Rightarrow\ P(Q_0; u \geq c(\mu)) = 1 \ \Rightarrow\ P(Q_1; u \geq c(\mu)\bar{c}) = 1 \ \ldots$$

to get by induction $P(Q_n;\ u \geq c(\mu)\,\bar{c}^n) = 1$ for all $n \geq 1$. It is easily checked that $s_n > t_{n+1}$ for $n \geq 1$, hence

$$\inf_{K_{\rho_{n+1}}} u(\cdot, t) \geq c(\mu)\,\bar{c}^n \qquad t_n \leq t \leq t_{n+1}, \qquad n \geq 1.$$

Notice that we can suppose that $\theta(\mu) \leq \bar{\theta} \leq 1/16$, so that it holds $\bar{\theta}\,\rho_{n-1}^2 \leq t_n \leq \bar{\theta}\,\rho_n^2$ for $n \geq 1$ and a monotonicity argument gives

$$\inf_{K_{\rho_n}} u(\cdot, t) \geq c(\mu)\,\bar{c}^{n+2} \qquad \bar{\theta}\,\rho_n^2 \leq t \leq \bar{\theta}\,\rho_{n+1}^2, \quad n \geq 0.$$

For $\rho \geq r$, let n be such that $\rho_n \leq \rho \leq \rho_{n+1}$, hence $\bar{\theta}\, \rho_{n+1}^2 \leq 4\bar{\theta}\, \rho^2 \leq \bar{\theta}\, \rho_{n+2}^2$. Then the lemma is proved for $\bar{\lambda} = -\log_2 \bar{c}$, and $\bar{\gamma} = 4\bar{\theta}$, since

$$\inf_{K_\rho} u(\cdot, 4\bar{\theta}\, \rho^2) \geq \inf_{K_{\rho_{n+1}}} u(\cdot, 4\bar{\theta}\, \rho^2) \geq c(\mu)\, \bar{c}^{n+3} \geq c(\mu)\, \bar{c}^3\, (r/\rho_n)^{\bar{\lambda}} \geq c(\mu)\, \bar{c}^3\, (r/\rho)^{\bar{\lambda}}.$$

□

Theorem 5.15 (Harnack Inequality) *Let $u \geq 0$ be a locally bounded solution of* (5.13) *in $K_{2R} \times [-(2R)^2, (2R)^2]$. There exists $\bar{C}$ such that $u(0, 0) \leq \bar{C}\, \inf_{K_R} u(\cdot, R^2)$.*

Proof By homogeneity, scaling and a Harnack chain argument it suffices to prove

$$u(0, 0) = 1 \quad \Rightarrow \quad \inf_{K_1} u(\cdot, 1) \geq \bar{c} > 0 \tag{5.25}$$

for any solution u of (5.13), nonnegative in $K_{\bar{L}} \times [-\bar{L}^2, \bar{L}^2]$ for $\bar{L}$ to be chosen. Let

$$\psi(\rho) = (1-\rho)^{\bar{\lambda}} \sup_{Q_\rho^-} u, \qquad Q_\rho^- := K_\rho \times [-\rho^2, 0], \quad \rho \in [0, 1]$$

where $\bar{\lambda}$ is given in Lemma 5.14. By continuity, we can choose $\rho_0 \in [0, 1]$, $(x_0, t_0) \in Q_{\rho_0}^-$ such that

$$\max_{[0,1]} \psi(\rho) = (1-\rho_0)^{\bar{\lambda}} u_0 \qquad u_0 := u(x_0, t_0).$$

For $\xi \in]0, 1[$ to be determined let $r = \xi\, (1-\rho_0)$, so that, being $\psi(0) = u(0, 0) = 1$,

$$u_0\, r^{\bar{\lambda}} = \xi^{\bar{\lambda}}\, u_0 (1-\rho_0)^{\bar{\lambda}} = \xi^{\bar{\lambda}}\, \psi(\rho_0) \geq \xi^{\bar{\lambda}}\, \psi(0) = \xi^{\bar{\lambda}}. \tag{5.26}$$

If $\widetilde{Q}_r = K_r(x_0) \times [t_0 - r^2, t_0]$, it holds $\widetilde{Q}_r \subseteq Q_{\rho_0+r}^-$ and being ρ_0 maximum for ψ,

$$\sup_{\widetilde{Q}_r} u \leq \sup_{Q_{\rho_0+r}^-} u = \frac{\psi(\rho_0 + r)}{(1-\rho_0-r)^{\bar{\lambda}}} \leq \frac{(1-\rho_0)^{\bar{\lambda}}}{(1-\rho_0-r)^{\bar{\lambda}}}\, u_0 = (1-\xi)^{-\bar{\lambda}}\, u_0. \tag{5.27}$$

Choose $\bar{\xi}$ as per $(1-\bar{\xi})^{-\bar{\lambda}} \leq 2$, so that $u \leq 2\, u_0$ in $\widetilde{Q}_r$, and let $\bar{\theta} = \theta(1/2) \in]0, 1[$ be given in (5.23). Since $K_r(x_0) \times [t_0 - \bar{\theta}\, r^2, t_0] \subseteq \widetilde{Q}_r$, the previous proof shows that for all $\rho \leq r/2$

$$\mathrm{osc}(u(\cdot, t_0), K_\rho(x_0)) \leq \bar{C}\, \sup_{\widetilde{Q}_r} u\, (r/\rho)^{\bar{\alpha}} \leq 2\, \bar{C}\, u_0\, (r/\rho)^{\bar{\alpha}},$$

and choosing $\rho = \bar{\eta} r$ with $\bar{C}\,\bar{\eta}^{\bar{\alpha}} \le 1/4$ gives $\mathrm{osc}(u(\cdot, t_0), K_{\bar{\eta} r}(x_0)) \le u_0/2$. The latter ensures $P_{t_0}(K_r(x_0);\, u \ge u_0/2) \ge \bar{\eta}^N$ and the expansion of positivity Lemma 5.14 for $2\,u/u_0$ implies

$$\inf_{K_\rho(x_0)} u(\cdot, t_0 + \bar{\gamma}\rho^2) \ge c(\bar{\eta}^N)\frac{u_0}{2}\frac{r^{\bar{\lambda}}}{\rho^{\bar{\lambda}}} \underset{(5.26)}{\ge} \frac{c(\bar{\eta}^N)\bar{\xi}^{\bar{\lambda}}}{2\rho^{\bar{\lambda}}}, \quad r \le \rho \le \bar{L}/8. \tag{5.28}$$

Solve $t_0 + \bar{\gamma}\,\rho = 1$ in ρ: from $\bar{\gamma} \le 1/4$ and $t_0 \in [-1, 0]$ we infer $2 \le \rho \le 2/\bar{\gamma}$. Therefore $K_\rho(x_0) \supseteq K_1$ and we can let $\bar{L}/8 := 2/\bar{\gamma}$ in (5.28), giving (5.25) and completing the proof. □

5.3 *Inhomogeneous Parabolic Equations*

In the last subsection, we heavily took advantage of the homogeneous structure of the equation. The situation is quite different for inhomogeneous equations whose model is

$$u_t = \mathrm{div} A(x, u, Du), \qquad \begin{cases} A(x, s, z) \cdot z \ge C_0 |z|^p \\ |A(x, s, z)| \le C_1 |z|^{p-1} \end{cases} \tag{5.29}$$

for $p \ne 2$, as it is no longer true that λu is a solution of a similar equation for $\lambda \ne 1$. The translation invariance still holds, and the scale invariance says that if u solves (5.29) then $u_\lambda(x, t) = u(\lambda x, \lambda^p t)$ is a solution (in the usual sense that there exists an A obeying the growth condition such that u solves the corresponding equation). More generally, given $R, T > 0$ and a (sub-) super-solution of (5.29),

$$u_{R,T}(x, t) = R^{\frac{p}{2-p}}\, T^{\frac{1}{p-2}}\, u(R\,x, T\,t) \tag{5.30}$$

is still a (sub-) super-solution (in the structural sense) an equation of the kind (5.29). This shows that statements for λu can be derived from those for u by scaling the space-time variables conveniently (actually, with one degree of freedom).

It is worth noting that, in the inhomogeneous setting, it is not known wether the energy inequality alone suffices to prove the Harnack inequality. In our proof, we will indeed use a clever change of variable introduced in [29], which crucially relies on the equation. Moreover, as extensively discussed in the previous chapter, the degenerate ($p > 2$) and singular ($p < 2$) cases require different treatments. We thus first derive some common tools in this subsection, and discuss in details the two families of equations in the following ones. The notation will be the same as in the previous one, with the additional dependence on p omitted in constants.

Proposition 5.16

1) **Parabolic Sobolev Embedding:** *If $u \in L^p(0, T; W_0^{1,p}(\Omega))$, and $p^* = p(1 + 2/N)$, then*

$$\int_0^T \int_\Omega |u|^{p^*}\, dx\, dt \le C(N) \left(\sup_{t\in[0,T]} \int_\Omega u^2(x,t)\, dx \right)^{\frac{p}{N}} \int_0^T \int_\Omega |Du|^p\, dx\, dt.$$

2) **Energy inequality**: *[31, Prop. III.2.1] Let v be a supersolution to* (5.29) *in Q_T under condition* (4.5). *There exists $C = C(C_0, C_1) > 0$ s. t. for any $k \ge 0$ and $\eta \in C^\infty(0, T; C_c^\infty(K))$, $0 \le \eta \le 1$ it holds*

$$\begin{aligned}
&\sup_{t\in[0,T]} \int_K (v(x,t) - k)_-^2 \eta^p\, dx + \frac{1}{C} \iint_{Q_T} |D(\eta(v-k)_-)|^p\, dx\, dt \\
&\le \int_K (v(x,0) - k)_-^2 \eta^p\, dx + C \iint_{Q_T} (v-k)_-^p |D\eta|^p\, dx\, dt \\
&\quad + C \iint_{Q_T} (v-k)_-^2 |\eta_t|\, dx\, dt.
\end{aligned} \tag{5.31}$$

We start by sketching the proof of the relevant critical mass lemma.

Lemma 5.17 (Critical Mass) *Let $v \ge 0$ be a supersolution of* (5.29) *on $Q_{R,T}$ for $p \ne 2$ and let $h \ge 0$. There exists $\nu > 0$ s.t.*

$$P\left(Q_{R,T}; v \le h\right) \le \nu(h R^{\frac{p}{2-p}} T^{\frac{1}{p-2}}) \;\Rightarrow\; v \ge \frac{h}{2} \text{ on } K_{\frac{R}{2}} \times \left[\frac{T}{2}, T\right]. \tag{5.32}$$

Proof Consider the supersolution $v_{R,T}(x,t) = R^{\frac{p}{2-p}} T^{\frac{1}{p-2}} v(Rx, Tt)$: as (5.32) is invariant by this transformation, it suffices to prove it for $R = T = 1$. Define

$$r_n = \frac{1}{2} + \frac{1}{2^n}, \qquad h_n = \frac{h}{2} + \frac{h}{2^n}, \qquad t_n = \frac{1}{2} - \frac{1}{2^{n+1}}$$

$$K_n = K_{r_n}, \qquad Q_n = K_n \times [t_n, 1], \qquad A_n = Q_n \cap \{v \le h_n\}.$$

Fix η_n as per (5.22) with $\theta = 4$. Inserting into (5.31) and noting that $h_n \le h$, we get

$$\begin{aligned}
&\sup_{t\in[t_{n+1},1]} \int_{K_{n+1}} (v(x,t) - h_n)_-^2\, dx + \iint_{Q_n} |D(\eta_n(v-h_n)_-)|^p\, dx\, dt \\
&\le \bar{C}\, 2^{np} \iint_{Q_n} (v-h_n)_-^p\, dx\, dt + \bar{C}\, 2^n \iint_{Q_n} (v-h_n)_-^2\, dx\, dt \\
&\le \bar{C}(h^p + h^2)\, 2^{np}\, |A_n|.
\end{aligned}$$

Use $h_{n+1} - h_{n+2} = h/2^{n+3}$, Tchebicev and the parabolic Sobolev embedding to get

$$\frac{h^{p^*}|A_{n+2}|}{2^{p^*(n+3)}} \le \iint_{A_{n+2}} (v - h_{n+1})_-^{p^*} \, dxdt \le \iint_{Q_{n+1}} ((v - h_{n+1})_- \eta_{n+1})^{p^*} \, dxdt$$

$$\le \bar{C} \iint_{Q_{n+1}} |D(\eta_{n+1}(v - h_{n+1})_-^2)|^p dxdt \left[\sup_{t \in [t_{n+1},1]} \int_{K_{n+1}} (v(x,t) - h_{n+1})_-^2 dx \right]^{\frac{p}{N}}$$

$$\le \bar{C}\, \bar{b}^n \, (h^p + h^2)^{1+\frac{p}{N}} |A_{n+1}||A_n|^{\frac{p}{N}} \le \bar{C}\, \bar{b}^n \, (h^p + h^2)^{1+\frac{p}{N}} |A_n|^{1+\frac{p}{N}}.$$

This amounts to $|A_{n+2}| \le \bar{b}^n \, \bar{C}(h)|A_n|^{1+\frac{p}{N}}$ and (5.2) for $X_n = |A_{2n}|$ gives the conclusion. □

Lemma 5.18 *Let $v \ge 0$ be a supersolution in $Q_{R,T}$ of* (5.29). *There exists $\bar{\sigma}$ s. t.*

$$\inf_{K_R} v(x,0) \ge h \quad \Rightarrow \quad v \ge h/2 \ \text{ on } K_{R/2} \times \left[0, \min\{\bar{\sigma}\, R^p\, h^{2-p}, T\}\right].$$

Proof Consider the supersolution $\tilde{v}(x,t) = R^{\frac{p}{2-p}} v(Rx,t)$ to reduce to the case $R = 1$, $\tilde{v}(\cdot,0) \ge \tilde{h} = h\, R^{\frac{p}{2-p}}$ on K_1. Proceed as in the previous proof with $t_n \equiv 0$, η_n independent of t and $Q_n = K_{r_n} \times [0,T]$. Since $\tilde{v}(\cdot,0) \ge \tilde{h}_n$ and $(\eta_n)_t \equiv 0$, the first and third term on the right of (5.31) vanish, giving

$$\sup_{t \in [0,T]} \int_{K_{n+1}} (\tilde{v}(x,t) - \tilde{h}_n)_-^2 \, dx + \iint_{Q_n} |D(\eta_n(\tilde{v} - \tilde{h}_n)_-)|^p \, dx\, dt \le \bar{C}\, 2^{np} h^p \, |A_n|.$$

where $A_n = Q_n \cap \{\tilde{v} \le \tilde{h}_n\}$. As before, we get the iterative inequality

$$\tilde{h}^{p^*}/2^{p^*(n+3)}|A_{n+2}| \le \bar{C}\, \bar{b}^n \, h^{p\frac{N+p}{N}} |A_n|^{1+\frac{p}{N}}$$

which, recalling that $p^* = p(N+2)/N$ and enlarging $\bar{b}$, reads

$$|A_{n+2}| \le \bar{C}\, \bar{b}^n \, \tilde{h}^{\frac{p}{N}(p-2)} |A_n|^{1+\frac{p}{N}}.$$

Since $|A_0| \le T$, (5.2) ensures the existence of $\bar{\sigma}$ such that

$$T \le \bar{\sigma}\tilde{h}^{2-p} \quad \Rightarrow \quad \lim_n |A_{2n}| = 0 \quad \Rightarrow \quad \inf_{Q_{1/2,T}} \tilde{v} \ge \tilde{h}/2 \quad \Leftrightarrow \quad \inf_{Q_{R/2,T}} v \ge h/2.$$

□

We conclude this section with a useful tool to prove Hölder continuity of solutions.

Lemma 5.19 *Suppose there exist $\bar{T} > 0$ and $\bar{m}, \bar{\theta} \in \,]0, 1[$, depending only on the data, such that any solution u of* (5.29) *with $p \neq 2$ in $Q_{2,\bar{T}}$ fulfills*

$$P_0(K_1;\ u \geq 1/2) \geq 1/2 \quad \Rightarrow \quad u \geq \bar{m} \text{ on } K_{1/4} \times [(1-\bar{\theta})\bar{T}, \bar{T}]. \tag{5.33}$$

There exists $\bar{C}, \bar{\alpha}$ depending on $\bar{m}$ and $\bar{\theta}$ such that any solution with $\|u\|_{L^\infty(Q_{2,\bar{T}})} \leq 1$ obeys

$$\operatorname{osc}(u, K_r \times [\bar{T}(1-r^p), \bar{T}]) \leq \bar{C}\, r^{\bar{\alpha}}, \qquad 0 \leq r \leq 1. \tag{5.34}$$

Proof Fix $\delta \in \,]0, 1/4]$, $\theta \in \,]0, \bar{\theta}]$ so that $\theta^{\frac{1}{2-p}} \delta^{\frac{p}{p-2}} = \gamma := (1+\bar{m})^{-1} < 1$. We claim by induction

$$\operatorname{osc}(u, Q_n) \leq (1+\bar{m})\gamma^n, \ \ \forall n \geq 0, \text{ where } \ Q_n := K_{\delta^n} \times [\bar{T}(1-\theta^n), \bar{T}]. \tag{5.35}$$

Since $\|u\|_{L^\infty(Q_{2,\bar{T}})} \leq 1$, (5.35) holds true for $n = 0$, so suppose by contradiction that

$$\operatorname{osc}(u, Q_n) \leq (1+\bar{m})\, \gamma^n \quad \& \quad \operatorname{osc}(u, Q_{n+1}) > (1+\bar{m})\, \gamma^{n+1} \tag{5.36}$$

for some $n \geq 1$. Being $\operatorname{osc}(u, Q_n) \geq \operatorname{osc}(u, Q_{n+1})$ we infer

$$\operatorname{osc}(u, Q_n) > (1+\bar{m})\, \gamma^{n+1}. \tag{5.37}$$

By scaling and translation invariance, the function $v(x,t) = \gamma^{-n} u\big(\delta^n x, (t-\bar{T})\theta^n + \bar{T}\big)$ solves in Q_0 an equation of the type (5.29) and, recalling that $\gamma = 1/(\bar{m}+1)$, we have

$$\operatorname{osc}(v, Q_0) = \gamma^{-n} \operatorname{osc}(u, Q_n) \underset{(5.37)}{>} (1+\bar{m})\, \gamma = 1.$$

We infer from the latter that the assumption in (5.33) holds for at least one of the nonnegative supersolutions $v_+ := v - \inf_{Q_0} v$ or $v_- = \sup_{Q_0} v - v$: indeed, for example, $P_0(K_1;\ v_+ \geq 1/2) < 1/2$ is equivalent to $P_0(K_1;\ v_+ < 1/2) \geq 1/2$ and then $\operatorname{osc}(v, Q_0) \geq 1$ ensures

$$P_0(K_1;\ v_- \geq 1/2) \geq P_0(K_1;\ v_- > \operatorname{osc}(v, Q_0) - 1/2) = P_0(K_1;\ v_+ < 1/2) \geq 1/2.$$

Suppose, without loss of generality, that $P_0(K_1;\ v_+ \geq 1/2) \geq 1/2$: then, since $\theta \leq \bar{\theta}$ and $\delta \leq 1/4$, (5.33) implies $\inf_{Q_1} v_+ \geq \bar{m}$ and thus

$$\operatorname{osc}(v, Q_1) = \operatorname{osc}(v_+, Q_1) \leq \operatorname{osc}(v_+, Q_0) - \bar{m} = \operatorname{osc}(v, Q_0) - \bar{m}.$$

Scaling back to u and using the relations in (5.36), we obtained the contradiction

$$(1+\bar{m})\,\gamma < \gamma^{-n}\mathrm{osc}(u, Q_{n+1}) = \mathrm{osc}(v, Q_1) \leq \mathrm{osc}(v, Q_0) - \bar{m}$$
$$= \gamma^{-n}\mathrm{osc}(u, Q_n) - \bar{m} \leq 1 + \bar{m} - \bar{m}.$$

To prove (5.34) let $\eta = \delta\, \min\{1, \gamma^{\frac{p-2}{p}}\}$ and suppose $\eta^{n+1} \leq r \leq \eta^n$ for some n. Then, using $\theta^{\frac{1}{2-p}}\,\delta^{\frac{p}{p-2}} = \gamma$, we infer $K_r \times [\bar{T}(1-r^p)] \subseteq Q_n$. Letting $\bar{\alpha} = \log_\eta \gamma$ and using (5.35) we have

$$\mathrm{osc}(u, K_r \times [\bar{T}(1-r^p), \bar{T}]) \leq \gamma^n = \gamma^{-1}\,(\eta^{n+1})^{\bar{\alpha}} \leq \gamma^{-1}\, r^{\bar{\alpha}}.$$

□

5.4 Degenerate Parabolic Equations

This subsection is devoted to the case $p > 2$ of (5.29). Compared to the homogeneous case $p = 2$, the most delicate part is the proof of the measure-to-point estimate, Lemma 5.22 below.

Lemma 5.20 *Assume that $u \geq 0$ is a supersolution in $Q_{1,T}$ of (5.29) with $p > 2$. For any $\mu > 0$ there exists $k(\mu) \in\,]0, 1[$ such that*

$$P_0(K_1; u \geq 1) \geq \mu \;\Rightarrow\; P_t\left(K_1; u \geq \frac{k(\mu)}{(t+1)^{\frac{1}{p-2}}}\right) > \frac{\mu}{2} \quad \forall t \in [0, T] \tag{5.38}$$

Proof For any $k, \delta \in\,]0, 1[$, we employ (5.31) with η as in (5.16), obtaining for $t \in [0, T]$

$$\int_{K_\delta} (u(x,t) - k)_-^2\, dx \leq \int_{K_1} (u(x,0) - k)_-^2\, dx + \frac{\bar{C}}{(1-\delta)^p} \iint_{Q_{1,t}} (u-k)_-^p\, dx\, dt$$
$$\leq k^2(1-\mu) + \frac{\bar{C}\, k^p\, t}{(1-\delta)^p}.$$

For $\varepsilon \in\,]0, 1[$ we have

$$\int_{K_\delta} (u(x,t) - k)_-^2\, dx \geq \int_{K_\delta \cap \{u(\cdot,t) < \varepsilon k\}} (k - \varepsilon k)^2\, dx \geq k^2(1-\varepsilon)^2 |K_\delta \cap \{u(\cdot,t) < \varepsilon k\}|$$

which, inserted into the previous estimate and dividing by $k^2(1-\varepsilon)^2$ gives

$$|K_\delta \cap \{u(\cdot, t) < \varepsilon k\}| \le \frac{1}{(1-\varepsilon)^2}\left(1 - \mu + \frac{\bar{C}\, k^{p-2}\, t}{(1-\delta)^p}\right). \tag{5.39}$$

Therefore

$$\begin{aligned} 1 - P_t(K_1;\ u \ge \varepsilon k) &\le 1 - |K_\delta \cap \{u(\cdot, t) \ge \varepsilon k\}| = 1 - \delta^N + |K_\delta \cap \{u(\cdot, t) < \varepsilon k\}| \\ &\le 1 - \delta^N + \frac{1}{(1-\varepsilon)^2}\left(1 - \mu + \frac{\bar{C}\, k^{p-2}\,(t+1)}{(1-\delta)^p}\right). \end{aligned}$$

Choose $\delta, \varepsilon \in \,]0, 1[$ and, for each $t \in [0, T]$, $k_t \in \,]0, 1[$ such that

$$1 - \delta^N = \frac{\mu}{8}, \qquad \frac{1-\mu}{(1-\varepsilon)^2} = 1 - \frac{3}{4}\mu, \qquad \frac{\bar{C}\, k_t^{p-2}(t+1)}{(1-\varepsilon)^2(1-\delta)^p} = \frac{\mu}{8}.$$

Clearly it holds $\delta = \delta(\mu)$, $\varepsilon = \varepsilon(\mu)$ and therefore $k_t = k(\mu)/(t+1)^{\frac{1}{p-2}}$. With these choices we have $1 - P_t(K_1;\ u \ge \varepsilon k_t) \le 1 - \mu/2$, proving the claim. □

The previous Lemma suggests to consider the function $(t+1)^{\frac{1}{p-2}} u(x, t)$, which is a supersolution to an equation similar to (5.29), but with structural constants depending on t (and degenerating for large times). In order to keep the structural conditions independent of t, it turns out that the change of time variable $t + 1 = e^\tau$ suffices, so that we consider instead

$$v(x, e^\tau) = e^{\frac{\tau}{p-2}} u(x, e^\tau - 1). \tag{5.40}$$

A straightforward calculation shows that v is a solution on $Q_{1,\log(T+1)}$ of

$$v_t = \operatorname{div} \tilde{A}(x, v, Dv) + v/(p-2)$$

with $\tilde{A}(x, s, z) := e^{\frac{\tau}{p-2}} A\left(x, s e^{\frac{-\tau}{p-2}}, z e^{\frac{-\tau}{p-2}}\right)$ obeying the structural conditions in (5.29). In particular, if $u \ge 0$, v belongs to the class of nonnegative supersolution of (5.29).

Lemma 5.21 (Shrinking Lemma) *Suppose $v \ge 0$ is a supersolution in $Q_{2,S}$ of* (5.29) *for $p \ge 2$ such that*

$$P_t(K_1;\ v \ge k) \ge \mu \qquad \forall t \in [0, S] \tag{5.41}$$

for some $\mu > 0$, $k \geq 0$. *There exists* $\beta = \beta(\mu)$ *such that*

$$P\left(K_1 \times [0, \left(\frac{2^n}{k}\right)^{p-2}]; v \leq \frac{k}{2^n}\right) \leq \frac{\beta(\mu)}{n^{1-\frac{1}{p}}}, \qquad \text{if } \left(\frac{2^n}{k}\right)^{p-2} \leq S \tag{5.42}$$

Proof The proof is very similar to the one of Lemma 5.10 and we only sketch it. Suppose $n \geq 1$ satisfies $2^{n(p-2)} \leq S\,k^{p-2}$ and let $k_j = k/2^j$ for $j = 0, \ldots, n$. Let $Q := Q_{1,S}$ and use (5.31) with η as in (5.19) with $R = 1$, to get

$$\begin{aligned}\iint_Q |D(v - k_j)_-|^p\,dx\,dt &\leq \bar{C}\int_{K_2}(v(x,0)-k_j)_-^2\,dx + \bar{C}\iint_{Q_{2,S}}(v - k_j)_-^p\,dx\,dt \\ &\leq \bar{C}(k_j^2 + Sk_j^p).\end{aligned}$$

For any $t \in [0, S]$ apply the De Giorgi-Poincaré inequality and (5.41) to obtain

$$(k_j - k_{j+1})P_t(K_1;\ v(\cdot,t) \leq k_{j+1}) \leq \frac{\bar{C}}{\mu}\int_{K_1\cap\{k_{j+1}\leq v(\cdot,t)\}} |D(v(x,t) - k_j)_-|\,dx.$$

Integrate over $[0, S]$, use Hölder's inequality and the energy estimate to get for $j = 0, \ldots, n-1$

$$\frac{k_j}{2}P(Q;\ v \leq k_{j+1}) \leq \frac{\bar{C}}{\mu}\left(\frac{k_j^2}{S} + k_j^p\right)^{\frac{1}{p}} \left(P(Q;\ v \leq k_j) - P(Q;\ v \leq k_{j+1})\right)^{1-\frac{1}{p}}. \tag{5.43}$$

As $j \leq n$, it holds $2^{j(p-2)} \leq S\,k^{p-2}$ as well, implying $k_j^2/S \leq k_j^p$. Thus we can simplify all the factors involving k_j above, giving for all $j \leq n-1$

$$\left(P(Q;\ v \leq k_{j+1})\right)^{\frac{p}{p-1}} \leq \bar{C}\,\mu^{\frac{p}{1-p}}\left(P(Q;\ v \leq k_j) - P(Q;\ v \leq k_{j+1})\right).$$

which, summed over $j \leq n-1$ gives (5.42) by the usual telescopic argument. □

Lemma 5.22 (Measure-to-Point Estimate) *For any* $\mu \in]0, 1[$ *there exists* $m(\mu) \in]0, 1[$, $T(\mu) > 1$ *such that any supersolution* $u \geq 0$ *in* $Q_{2,T(\mu)}$ *fulfills*

$$P_0(K_1; u \geq 1) \geq \mu \quad \Rightarrow \quad u \geq m(\mu) \quad \text{in } K_{1/2} \times [T(\mu)/2, T(\mu)]. \tag{5.44}$$

Proof Let T to be determined and suppose $u \geq 0$ is a supersolution in $Q_{1,T}$. By Lemma 5.20,

$$P_t(K_1;\ (t+1)^{\frac{1}{p-2}}\,u \geq k(\mu)) \geq \mu/2, \qquad \forall t \in [0, T].$$

If v is defined as per (5.40), the previous condition reads

$$P_\tau(K_1;\ v \geq k(\mu)) \geq \mu/2, \qquad \forall \tau \in [0, S], \quad S = \log(T+1) > 0$$

so that, being $v \geq 0$ a supersolution, Lemma 5.21 implies

$$P(Q_{1,S_n};\ v \leq S_n^{\frac{1}{2-p}}) \leq \beta(\mu/2)/n^{1-\frac{1}{p}}, \qquad \text{for } S_n := (2^n/k(\mu))^{p-2}.$$

Next choose $n = n(\mu)$, (and thus $S = S(\mu) := S_n$ and $T = T(\mu) := e^S - 1$) so that $\beta(\mu/2)/n^{1-\frac{1}{p}} \leq \nu(1)$, with ν given in (5.32). Lemma 5.17 applied on $Q_{1,S}$ with $h = S^{\frac{1}{2-p}}$ thus gives

$$v \geq S^{\frac{1}{2-p}}/2 \quad \text{on } K_{1/2} \times [S/2, S].$$

Recalling the definition (5.40) of v, in terms of u the latter implies

$$u \geq e^{-\frac{T}{p-2}} \log^{\frac{1}{2-p}}(T+1)/2 \quad \text{on } K_{1/2} \times \left[\sqrt{T+1} - 1, T\right] \supseteq K_{1/2} \times [T/2, T].$$

□

Theorem 5.23 (Hölder Regularity) *Any $L^\infty_{\mathrm{loc}}(\Omega_T)$ solution u of (5.29) in Ω_T for $p > 2$ belongs to $C^{\bar{\alpha}}_{\mathrm{loc}}(\Omega_T)$, with $\bar{\alpha}$ depending only on N, p, C_0 and C_1. Moreover, there exist $\bar{T} \geq 1$ and $\bar{C} > 0$ such that if $Q_R^-(\bar{T}) := K_{2R} \times [-\bar{T} R^p, 0] \subseteq \Omega_T$, for any $r \in [0, R]$ it holds*

$$\operatorname{osc}(u(\cdot, 0), K_r) \leq \bar{C} \max\left\{1, \|u\|_{L^\infty(Q_R^-(\bar{T}))}\right\} \left(\frac{r}{R}\right)^{\bar{\alpha}}. \tag{5.45}$$

Proof Let $\bar{T} = T(1/2)$ be given in the previous Lemma. By space/time translation, it suffices to prove an oscillation decay near $(0, 0)$, with $Q_{r_0}^-(\bar{T}) \subseteq \Omega_T$ for some $r_0 > 0$. By (5.30), $u(x\, r_0, t\, r_0^p)$ (still denoted by u) solves (5.29) on $Q_1^-(\bar{T})$. Let $M = \|u\|_{L^\infty(Q_1^-(\bar{T}))}$: if $M > 1$ consider $v(x, t) = M^{-1}\, u(x, M^{2-p}\, t)$, which, being $p > 2$, solves (5.29) on $Q_1^-(\bar{T})$ and $\|v\|_{L^\infty(Q_1^-(\bar{T}))} \leq 1$. Applying Lemma 5.19 to $v(\cdot, \bar{T} + t)$ (notice that $Q_1^-(\bar{T})$ translates to $Q_{2,\bar{T}}$) proves the Hölder continuity of u, while (5.45) is obtained from (5.34) for v, scaling back to u. □

The next lemma shows that the geometry of the expansion of positivity in the degenerate setting is very similar to the nondegenerate case. Compared to Fig. 4, the only difference is in the shape of the paraboloid which is thinner for larger p.

Lemma 5.24 (Expansion of Positivity) *There exists $\bar{\lambda} > 0$ and, for any $\mu > 0$, $c(\mu) \in]0, 1[$, $\gamma(\mu) \geq 1$, such that if $u \geq 0$ is a supersolution to (5.29) in $Q_{4R,T}$,*

$$P_0(K_r;\ u \geq k) \geq \mu \quad \Rightarrow \quad \inf_{K_\rho} u\Big(\cdot, \gamma(\mu)\Big(\frac{k\, r^{\bar{\lambda}}}{\rho^{\bar{\lambda}}}\Big)^{2-p} \rho^p\Big) \geq c(\mu)\, \frac{k\, r^{\bar{\lambda}}}{\rho^{\bar{\lambda}}}$$

whenever $r \leq \rho \leq R$ and $\gamma(\mu)\big(k\, r^{\bar{\lambda}}/\rho^{\bar{\lambda}}\big)^{2-p} \rho^p \leq T/c(\mu)$.

Proof We first generalize (5.44) as follows: there exists $\theta(\mu) > 0$ such that for any $\eta \geq 1, h > 0$

$$P_0(K_\rho; u \geq h) \geq \mu \ \Rightarrow\ u \geq c(\mu) \frac{h}{\eta^{\frac{1}{p-2}}} \ \text{ in } K_{2\rho} \times \left[\frac{\theta(\mu)}{2} \frac{\rho^p}{h^{p-2}}, \eta\theta(\mu) \frac{\rho^p}{h^{p-2}}\right]. \tag{5.46}$$

By considering $v(x, t) = h^{-1} u(\rho\, x, h^{2-p}\, \rho^p\, t)$ and recalling (5.30), it suffices to prove the claim for $\rho = h = 1$. By Lemma 5.20, (5.38) holds true, implying $P_s(K_4; u \geq k(\mu)/(s+1)^{\frac{1}{p-2}}) \geq \mu\, 4^{-N-1}$ where s is a parameter in $[0, \eta - 1]$. Rescale (5.44) considering

$$v(x, t) = k_s(\mu)^{-1}\, u(4\, x, k_s(\mu)^{2-p}\, 4^p\, t), \qquad k_s(\mu) := k(\mu)/(s+1)^{\frac{1}{p-2}}$$

which fulfills $P_s(K_1;\ v \geq 1) \geq \mu\, 4^{-N-1}$, to obtain, with the notations of (5.44)

$$v \geq m(\mu\, 4^{-N-1}) \quad \text{in} \quad K_{1/2} \times \big[s + T(\mu\, 4^{-N-1})/2, s + T(\mu\, 4^{-N-1})\big],$$

If $c(\mu) := k(\mu)\, m(\mu\, 4^{-N-1})$, using $s \in [0, \eta - 1]$, the latter reads in terms of u

$$\inf_{K_2} u(\cdot, t) \geq k_s(\mu)\, m(\mu\, 4^{-N-1}) \geq c(\mu)\, \eta^{\frac{1}{2-p}} \qquad \text{if } t \in I_s \text{ for some } s \in [0, \eta - 1]$$

$$I_s := \big[4^p\, k_s(\mu)^{2-p}\, (s + T(\mu\, 4^{-N-1})/2), 4^p\, k_s(\mu)^{2-p}\, (s + T(\mu\, 4^{-N-1}))\big].$$

Finally, let $\theta(\mu) = 4^p\, k(\mu)^{2-p} T(\mu\, 4^{-N-1})$ and observe that $\cup_{s\in[0,\eta-1]} I_s \supseteq [\theta(\mu)/2, \eta\, \theta(\mu)]$,[2] proving (5.46). Notice that all the argument goes through as long as it holds $\sup_{s\in[0,\eta-1]} I_s = 4^p k(\mu)^{2-p} \eta(\eta - 1 + T(\mu\, 4^{-N-1})) \leq T$ which, scaling back, is ensured e.g. by $\eta^2\, \theta(\mu)\, \rho^p\, h^{2-p} \leq T$.

---

[2]Both $a(s) = \inf I_s$ and $b(s) = \sup I_s$ are continuous, hence $\cup_{s\in[0,\eta-1]} I_s = \big[\inf_{s\in[0,\eta-1]} a(s), \sup_{s\in[0,\eta-1]} b(s)\big]$. Then observe that $a(0) = \theta(\mu)/2$ while $b(\eta - 1) \geq \eta\, \theta(\mu)$.

To prove the lemma, again we can suppose $k = 1$, (otherwise consider $v(x,t) = k^{-1}u(x, k^{2-p}t)$). Let, as per (5.46), $\bar\theta = \theta(1)$ and $\bar c = c(1)$, $\rho_n = 2^n r$ and define recursively

$$t_0 = \frac{\theta(\mu)}{2} r^p, \qquad t_{n+1} = t_n + \frac{\bar\theta}{2}(c(\mu)\,\bar c^n)^{2-p}\,\rho_n^p. \tag{5.47}$$

Applying (5.46) with $\eta = 1$, we get

$$P_0(K_r; u \ge 1) \ge \mu \;\Rightarrow\; P_{t_0}(K_{\rho_1}; u \ge c(\mu)) = 1 \;\Rightarrow\; P_{t_1}(K_{\rho_2}; u \ge c(\mu)\bar c) = 1$$

and, proceeding by induction, we infer $P_{t_n}(K_{\rho_{n+1}};\ u \ge c(\mu)\,\bar c^n) = 1$, for all $n \ge 0$. In particular $P_{t_n}(K_{\rho_n};\ u \ge c(\mu)\,\bar c^n) = 1$, so we again use (5.46) for η to be determined to obtain

$$u \ge c(\mu)\,\bar c^{n+1}\,\eta^{\frac{1}{2-p}} \quad \text{in } K_{\rho_{n+1}} \times [t_{n+1}, t_n + \eta\,\bar\theta\,(c(\mu)\,\bar c^n)^{2-p}\rho_n^p]. \tag{5.48}$$

Choose η so that

$$t_n + \eta\,\bar\theta\,(c(\mu)\,\bar c^n)^{2-p}\rho_{n+1}^p = t_{n+2} \quad\Leftrightarrow\quad \eta = \bar\eta := (1 + \bar c^{2-p}\,2^p)/2.$$

For this choice (5.48) holds in the time interval $[t_{n+1}, t_{n+2}]$ giving, by monotonicity,

$$\inf_{K_{\rho n}} u(\cdot, t) \ge c(\mu)\,\bar c^{n+m} \qquad \text{for all } t_n \le t \le t_{n+m},\ n, m \ge 0$$

for a smaller $c(\mu)$. Let $s_n := \bar c^{n(2-p)}\rho_n^p$; computing t_n we find $t_n \simeq_\mu s_n$ with constants depending on μ, therefore, for sufficiently large $m = m(\mu) \in \mathbb{N}$, $t_n \le \gamma(\mu)\,s_n \le \gamma(\mu)\,s_{n+1} \le t_{n+m}$ and

$$\inf_{K_{\rho n}} u(\cdot, t) \ge c(\mu)\,\bar c^n \qquad \text{for all } \gamma(\mu)\,s_n \le t \le \gamma(\mu)\,s_{n+1},\ n \ge 0.$$

The same argument as in the end of the proof of Lemma 5.14 gives the thesis. □

Theorem 5.25 (Forward Harnack Inequality) *Let u be a nonnegative solution of (5.29) in $K_{16R} \times [-T, T]$. Then there exists $\bar C > \bar\theta > 0$ such that if $\bar C\,u(0,0)^{2-p}R^p \le T$, then*

$$u(0,0) \le \bar C \inf_{K_R} u(\cdot, \bar\theta\,u(0,0)^{2-p}R^p). \tag{5.49}$$

Proof Thanks to (5.30), the function $v(x,t) = u(0,0)^{-1}u(R^p x, u(0,0)^{2-p}R^p t)$ solves (5.29) in $K_{16} \times [-T\,u(0,0)^{p-2}R^{-p}, T\,u(0,0)^{p-2}R^{-p}]$ and $v(0,0) = 1$. It then suffices to prove the existence of $\bar\theta \ge 1$, $\bar c \in\,]0,1[$ such that any solution $u \ge 0$

of (5.29) in $K_{16} \times [-2, \bar{\theta}/\bar{c}]$ obeys

$$u(0,0) = 1 \quad \Rightarrow \quad \inf_{K_1} u(\cdot, \bar{\theta}) \geq \bar{c}. \tag{5.50}$$

As in Theorem 5.15, let $Q_\rho^- := K_\rho \times [-\rho^p, 0]$ and consider $\psi(\rho) := (1 - \rho)^{\bar{\lambda}} \sup_{Q_\rho^-} u$ for $\rho \in [0, 1]$, where $\bar{\lambda}$ is given in Lemma 5.24. Let by continuity $\rho_0 \in [0, 1]$, $(x_0, t_0) \in Q_{\rho_0}^-$ such that

$$\max_{[0,1]} \psi(\rho) = (1 - \rho_0)^{\bar{\lambda}} u_0 \qquad u_0 := u(x_0, t_0),$$

choose $\bar{\xi} \in]0, 1[$ such that $(1 - \bar{\xi})^{-\bar{\lambda}} = 2$ and let $r = \bar{\xi}\,(1 - \rho_0)$. As in (5.26), it holds $u_0\, r^{\bar{\lambda}} \geq \bar{\xi}^{\bar{\lambda}}$. Let $\bar{T}$ be given in Theorem 5.23 and let $\widetilde{Q}_r := K_{\bar{T}^{-1/p} r}(x_0) \times [t_0 - r^p, t_0]$. Since $\bar{T} \geq 1$, it holds $\widetilde{Q}_r \subseteq Q_{\rho_0 + r}^-$ and we can deduce as in (5.27) that $\sup_{\widetilde{Q}_r} u \leq (1 - \bar{\xi})^{-\bar{\lambda}}\, u_0$. Then (5.45) ensures

$$\operatorname{osc}(u(\cdot, t_0), K_\rho(x_0)) \leq 2\,\bar{C}\,u_0\,(\rho/r)^{\bar{\alpha}} \quad \text{for} \quad \rho \leq \bar{T}^{-1/p} r.$$

Since $u(x_0, t_0) = u_0$, we infer $u(\cdot, t_0) \geq u_0/2$ in $K_{\bar{\eta} r}(x_0)$ for some $\bar{\eta} > 0$. Therefore $P_{t_0}(K_r(x_0);\ u \geq u_0/2) \geq \bar{\eta}^N$ and being $u_0\, r^{\bar{\lambda}} \geq \bar{\xi}^{\bar{\lambda}}$, *a fortiori* it holds $P_{t_0}(K_r(x_0);\ u \geq \bar{\xi}^{\bar{\lambda}}\, r^{-\bar{\lambda}}/2) \geq \bar{\eta}^N$. Since $K_{12}(x_0) \subseteq K_{16}$, Lemma 5.24 with $k = \bar{\xi}^{\bar{\lambda}}\, r^{-\bar{\lambda}}/2$ gives for suitable $\bar{\gamma} \geq 1 > \bar{c} > 0$

$$\inf_{K_\rho(x_0)} u\big(\cdot, t_0 + \bar{\gamma}\bar{\xi}^{\bar{\lambda}(2-p)} \rho^{p+\bar{\lambda}(p-2)}\big) \geq \bar{c}\frac{\xi^{\bar{\lambda}}}{\rho^{\bar{\lambda}}}, \quad r \leq \rho \leq 3, \quad \bar{\gamma}\xi^{\bar{\lambda}(2-p)} \rho^{p+\bar{\lambda}(p-2)} \leq \frac{T}{\bar{c}}.$$

In (5.50) we let $\bar{\theta} := \bar{\gamma}\, 2^{p+\bar{\lambda}(p-2)}$ and choose ρ such that $t_0 + \bar{\gamma}\,\xi^{\bar{\lambda}(2-p)} \rho^{p+\bar{\lambda}(p-2)} = \bar{\theta}$. From $t_0 \leq 0$ we get $\rho \geq 2$ (and thus $K_\rho(x_0) \supseteq K_1$) and from $t_0 \geq -1$ we infer

$$\bar{\gamma}\xi^{\bar{\lambda}} \rho^{p+\bar{\lambda}(p-2)} \leq 1 + \bar{\gamma}\xi^{\bar{\lambda}(2-p)} 2^{p+\bar{\lambda}(p-2)} \leq \bar{\gamma}\xi^{\bar{\lambda}(2-p)}(1 + 2^{p+\bar{\lambda}(p-2)}) \quad \Rightarrow \quad \rho \leq 3.$$

Hence (by eventually lowering $\bar{c}$), such ρ is admissible and its upper bound proves (5.50). □

Theorem 5.26 (Backward Harnack Inequality) *Let u be a nonnegative solution of (5.29) in $K_{16R} \times [-T, T]$. Then there exists $\bar{C}' > \bar{\theta}' > 0$ such that if $\bar{C}'\, u(0, 0)^{2-p}\, R^p \leq T$, then*

$$\sup_{K_R} u(\cdot, -\bar{\theta}'\, u(0, 0)^{2-p}\, R^p) \leq \bar{C}'\, u(0, 0). \tag{5.51}$$

Proof By the same scaling argument as before, we can reduce to the case $R = 1$, $u(0, 0) = 1$. Let, for $t \geq 0$, $w(t) := u(0, -t)$ and apply (5.49) to u with $(0, -t)$

instead of $(0,0)$ to get

$$u(0, -t + \bar{\theta}\, w^{2-p}(t)\, \rho^p) \geq w(t)/\bar{C}, \qquad 0 < \rho \leq 1.$$

If $w(t) \leq 2\,\bar{C}$ for some $t \leq \bar{\theta}/(2\,\bar{C})^{p-2}$, we can choose $\rho(t) > 0$ such that $\rho(t)^p = t\, w^{p-2}(t)/\bar{\theta} \leq 1$, obtaining $u(0,0) = u(0, -t + \bar{\theta}\, w^{2-p}(t)\, \rho^p(t))$. Therefore we proved

$$0 \leq t \leq \bar{\theta}/\bar{(}2C)^{p-2} \quad \& \quad w(t) \leq 2\,\bar{C} \qquad \Rightarrow \qquad w(t) \leq \bar{C}$$

which implies $w(t) \leq 2\,\bar{C}$ for all $0 \leq t \leq \bar{\theta}/\bar{(}2\,\bar{C})^{p-2}$ by a continuity argument. Letting $\bar{\theta}' = \bar{\theta}/\bar{(}2\,\bar{C})^{p-2}$, $\bar{C}' = 2\,\bar{C}$ we prove (5.51) by contradiction: from $u(0, -\bar{\theta}') \leq \bar{C}$ and $\sup_{K_1} u(\cdot, -\bar{\theta}') > 2\,\bar{C}$, by continuity there exists $\bar{x} \in K_1$ such that $u(\bar{x}, -\bar{\theta}') = 2\,\bar{C}$. Since $0 \in K_1(\bar{x})$ and $\bar{\theta}\,(2\,\bar{C})^{2-p} = \bar{\theta}'$, the Harnack inequality (5.49) for u at the point $(\bar{x}, -\theta')$ implies

$$1 = u(0,0) \geq \inf_{K_1(\bar{x})} u(\cdot, -\bar{\theta}' + \bar{\theta}(2\,\bar{C})^{2-p}) \geq u(\bar{x}, -\bar{\theta}')/\bar{C} = 2.$$

□

5.5 *Singular Parabolic Equations*

We conclude with the Harnack inequality for solutions of parabolic singular supercritical equations. The measure-to-point estimate will be treated through a change of variable analogous to the degenerate case, but requires a little bit more care. From this we'll derive a Hölder continuity result for all *bounded* solutions in the full range $p \in \,]1,2[$. As mentioned in the introduction of the section, the proof of the Harnack inequality will rely on Theorem 5.32, which we state without proof.

Lemma 5.27 *Let $u \geq 0$ be a supersolution in $Q_{1,T}$ of* (5.29) *with $p \leq 2$. For any $\mu > 0$ there exists $k(\mu) \in \,]0,1[$, $T(\mu) \in \,]0, \min\{1,T\}]$ such that*

$$P_0(K_1;\, u \geq 1) \geq \mu \quad \Rightarrow \quad P_t\,(K_1;\, u \geq k(\mu)) > \mu/2 \qquad \forall t \in [0, T(\mu)].$$

Proof Proceed as in Lemma 5.20 to get (5.39) for $k = 1$, $\delta, \varepsilon, \in \,]0,1[$ and $t \in [0,T]$. Thus

$$1 - P_t(K_1;\, u \geq \varepsilon) \leq 1 - \delta^N + \frac{1}{(1-\varepsilon)^2}\left(1 - \mu + \frac{\bar{C}\, t}{(1-\delta)^p}\right).$$

Choose $\delta = \delta(\mu)$ and $\varepsilon = \varepsilon(\mu)$ as per $1 - \delta^N = \mu/8$ and $(1-\mu)/(1-\varepsilon)^2 = 1 - 3\mu/4$, so that

$$P_t(K_1;\ u \geq \varepsilon(\mu)) \geq \frac{5}{8}\mu - C(\mu)\,t, \qquad \text{for any } t \in [0, T].$$

Choosing $T(\mu) \leq T$ such that $C(\mu)\,T(\mu) \leq \mu/8$ gives the claim. □

Lemma 5.28 (Shrinking Lemma) *Let $v \geq 0$ be a supersolution in $Q_{2,S}$ of* (5.29) *with $p \in\,]1, 2[$ such that for some $\mu, k \in\,]0, 1[$*

$$P_t\left(K_1;\ v \geq k\right) > \mu \qquad \forall t \in [0, S].$$

Then there exists $\beta = \beta(\mu) > 0$ such that for any $n \geq 1$

$$P\left(Q_{1,S};\ v \leq k/2^n\right) \leq \beta(\mu)\left(1 + k^{2-p}/S\right)^{\frac{1}{p-1}} / n^{1-\frac{1}{p}}.$$

Proof Proceed as in the proof of Lemma 5.21 up to (5.43) with $Q = Q_{1,S}$. As $j \geq 1$ and $p < 2$, it holds $k_j^p + k_j^2/S \leq k_j^p(1 + k^{2-p}/S)$, so that

$$P(Q;\, v \leq k_{j+1})^{\frac{p}{p-1}} \leq \bar{C}\mu^{\frac{p}{p-1}}\left(1 + \frac{k^{2-p}}{S}\right)^{\frac{1}{p-1}}\left(P(Q;\, v \leq k_j) - P(Q;\, v \leq k_{j+1})\right),$$

which yields the conclusion summing over $j \leq n-1$. □

Lemma 5.29 (Measure-to-Point Estimate) *Let $u \geq 0$ be a supersolution of* (5.29) *for $p \in\,]1, 2]$. For any $\mu \in\,]0, 1]$ there exists $m(\mu), T(\mu) \in\,]0, 1[$ such that*

$$P_0(K_1;\ u \geq 1) \geq \mu \quad \Rightarrow \quad u \geq m(\mu) \quad \text{in } K_{1/4} \times [T(\mu)/2, T(\mu)]. \tag{5.52}$$

Moreover, $T(\mu)$ can be chosen arbitrarily small by decreasing $m(\mu)$.

Proof Let $T(\mu)$, $k(\mu)$ be given in Lemma 5.27: clearly $T(\mu)$ can be chosen arbitrarily small. Since $p < 2$, an explicit computation shows that for any fixed $T \in [T(\mu)/2, T(\mu)]$, the function

$$v(x, \tau) = e^{\frac{\tau}{2-p}} u(x, T - e^{-\tau}), \qquad x \in K_1, \tau \geq -\log T$$

is a supersolution to (5.29). The conclusion for u of Lemma 5.27 becomes for v

$$P_\tau(K_1;\ v \geq e^{\frac{\tau}{2-p}} k(\mu)) \geq \mu/2, \qquad \forall \tau \geq -\log T,$$

and for $s \geq -\log T$ to be chosen, the latter implies (thanks to $p < 2$)

$$P_\tau(K_1;\ v \geq e^{\frac{s}{2-p}} k(\mu)) \geq \mu/2, \qquad \forall \tau \geq s \geq -\log T. \tag{5.53}$$

For $\nu(\mu)$ and $\beta(\mu)$ given in Lemmata 5.17 and 5.28, let $n = n(\mu) \geq 1$ be such that

$$\beta(\mu)\, n^{\frac{1}{p}-1} \left(k(\mu)^{2-p} + 1\right)^{\frac{1}{p}} \leq \nu(k(\mu)),$$

and for $s \geq -\log T$ let $I_s = [e^s, 2e^s]$. Due to (5.53), Lemma 5.28 applies to v on $K_1 \times I_s$ for $k = k(\mu)\, e^{\frac{s}{2-p}}$, giving, by the choice of $n = n(\mu)$,

$$P(K_1 \times I_s;\ v \leq k(\mu)\, e^{\frac{s}{2-p}}/2^n) \leq \nu(k(\mu)). \tag{5.54}$$

Subdivide I_s in $[2^{n(2-p)}] + 1$ disjoint intervals, each of length $\lambda \in [e^s\,(2^{-n(2-p)} - 1), e^s\, 2^{-n(2-p)}]$. On at least one of them, say $J = [a, a + \lambda] \subseteq I_s$, (5.54) holds for J instead of I_s, thus *a fortiori*

$$P(K_1 \times J;\ v \leq k(\mu)\, \lambda^{\frac{1}{2-p}}) \leq P(K_1 \times J;\ v \leq k(\mu)\, e^{\frac{s}{2-p}}/2^n) \leq \nu(k(\mu)).$$

Apply (5.32) to v on $K_1 \times J$ to obtain

$$v(x, \tau) \geq k(\mu)\lambda^{\frac{1}{2-p}}/2 \qquad \forall\, \tau \in [a + \lambda/2, a + \lambda] \subseteq I_s,\ x \in K_{1/2}.$$

Since $\lambda \geq e^s\,(2^{-n(2-p)} - 1)$, in terms of u and s, the latter implies that for some $\tau_s \in J \subseteq [e^s, 2e^s]$

$$\inf_{K_{1/2}} u(\cdot, T - e^{-\tau_s}) = e^{-\frac{\tau_s}{2-p}} \inf_{K_{1/2}} v(\cdot, \tau_s) \geq \frac{k(\mu)\, e^{\frac{s-\tau_s}{2-p}}}{2^{2n}} =: c(\mu)\, e^{\frac{s-\tau_s}{2-p}}.$$

Apply Lemma 5.18 to u in $K_{1/2} \times [T - e^{-\tau_s}, T]$ with $h = c(\mu)\, e^{\frac{s-\tau_s}{2-p}}$ to get

$$\inf_{K_{1/4}} u(\cdot, t) \geq c(\mu)\frac{e^{\frac{s-\tau_s}{2-p}}}{2} \quad \forall t \in [T - e^{-\tau_s}, T - e^{-\tau_s} + \min\{e^{-\tau_s}, \frac{\bar{\sigma}}{2^p} c(\mu) e^{s-\tau_s}\}]. \tag{5.55}$$

Finally, let $\tilde{s} = s(\mu) = \max\{-\log(T(\mu)/2), -\log(\bar{\sigma}\, 2^{-p}\, c(\mu)\}$, so that it holds

$$\tilde{s} \geq -\log T \qquad \text{and} \qquad \bar{\sigma}\, 2^{-p}\, c(\mu)\, e^{\tilde{s}-\tau_{\tilde{s}}} \geq e^{-\tau_{\tilde{s}}}.$$

Therefore (5.55) holds for $t = T$ and from $\tau_{\tilde{s}} \leq 2e^{\tilde{s}}$ we deduce a lower bound on $e^{\tilde{s}-\tau_{\tilde{s}}}$ depending only on μ, which proves (5.52) by the arbitrariness of $T \in [T(\mu)/2, T(\mu)]$. □

Theorem 5.30 (Hölder Regularity) *Any $L^\infty_{\mathrm{loc}}(\Omega_T)$ solution u of* (5.29) *in Ω_T for $p \in\,]1, 2[$ belongs to $C^{\bar{\alpha}}_{\mathrm{loc}}(\Omega_T)$, with $\bar{\alpha}$ depending only on the data. Moreover there exists $\bar{S}$, also depending on the data, with the following property: if $S \geq \bar{S}$ there* *exist* $\bar{C}(S) > 0$ *such that*

$$\sup_{K_{2R}\times[-k^{2-p}R^p,0]} u \le Sk \;\Rightarrow\; \operatorname{osc}(u, K_r \times [-k^{2-p}r^p, 0]) \le \bar{C}(S)k\Big(\frac{r}{R}\Big)^{\bar{\alpha}}, \quad r \le R, \tag{5.56}$$

for any $k, R > 0$ *for which* $K_{2R} \times [-k^{2-p} R^p, 0] \subseteq \Omega_T$.

Proof Let $\bar{T} = T(1/2) \in]0, 1]$ given in the previous Lemma. By space-time translations and rescaling we are reduced to prove Hölder continuity near (0, 0) with $\bar{Q} := K_2 \times [-\bar{T}, 0] \subseteq \Omega_T$. If $M := \|u\|_{L^\infty(\bar{Q})} > 1$ consider $M^{-1} u(M^{(p-2)/p}x, t)$ which, being $p \in]1, 2[$, solves (5.29) in $\bar{Q}$ and fulfills $\|v\|_{L^\infty(\bar{Q})} \le 1$. Applying Lemma 5.19 gives the first statement. To prove (5.56), suppose that $S \ge \bar{T}^{\frac{1}{p-2}} =: \bar{S}$, rescale to $R = 1$, then let $\bar{\gamma}(S) := S^{p-2}\,\bar{T}^{-1}$ and consider

$$v(x, t) = (S\,k)^{-1}u(\rho\, x, \tau\, t) \qquad \rho = \bar{\gamma}(S)^{1/p}, \quad \tau = k^{2-p}\,\bar{T}^{-1}.$$

Thanks to (5.30), it is readily verified that v solves (5.29) in $\bar{Q}$ and by the assumption in (5.56) it is bounded by 1. Applying (5.34) (notice that $\bar{T}$ is the same) and rescaling back gives (5.56) for all $r \le \bar{\gamma}(S)^{1/p}R$ and hence for all $r \le R$ with eventually a bigger constant. □

Lemma 5.31 (Expansion of Positivity, See Fig. 5) *There exists* $\bar{\lambda} > p/(2-p)$ *and, for any* $\mu > 0$, $c(\mu), \gamma_1(\mu), \gamma_2(\mu) \in]0, 1[$ *s. t. if* $u \ge 0$ *is a supersolution in* $Q_{8R,T}$

$$P_0(K_r;\, u \ge k) \ge \mu \quad \Rightarrow$$
$$\inf_{K_\rho} u\Big(\cdot, k^{2-p}r^p\big(\gamma_1(\mu) + \gamma_2(\mu)\big(1 - \Big(\frac{r}{\rho}\Big)^{\bar{\lambda}(2-p)-p}\big)\big)\Big) \ge c(\mu)\frac{k\,r^{\bar{\lambda}}}{\rho^{\bar{\lambda}}} \tag{5.57}$$

whenever $r \le \rho \le R$ *and* $k^{2-p}\,r^p\,(\gamma_1(\mu) + \gamma_2(\mu)(1 - (r/\rho)^{\bar{\lambda}(2-p)-p})) \le T$. *Moreover, the* $\gamma_i(\mu)$ *can be chosen arbitrarily small by lowering* $c(\mu)$.

Proof The proof is very similar (and in fact simpler) to the one of Lemma 5.24 and we only sketch it. First expand in space (5.52) through

$$P_0(K_1; u \ge 1) \ge \mu \Rightarrow P_0(K_8; u \ge 1) \ge \frac{\mu}{8^N} \Rightarrow u \ge c(\mu) \text{ in } K_2\times[\frac{\theta(\mu)}{2}, \theta(\mu)],$$

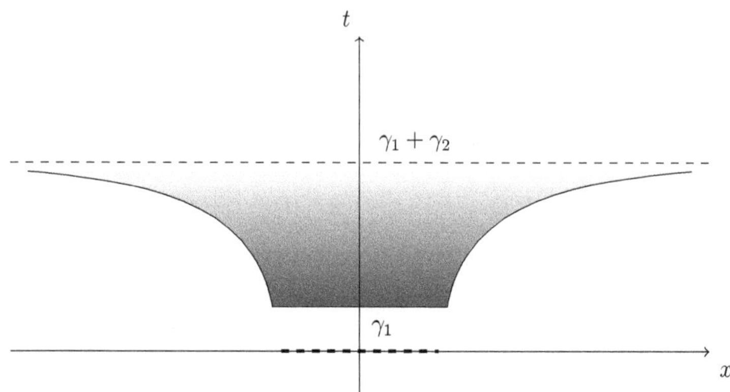

Fig. 5 The expansion of positivity in the singular case. If at time $t = 0$, $u \geq 1$ on the dotted part of given measure, after a waiting time γ_1, u is pointwise bounded from below in the shaded region by a large power of $(\gamma_1 + \gamma_2 - t)$

where we have set, with the notations in (5.52), $\theta(\mu) := T(\mu/8^N)$, $c(\mu) := m(\mu/8^N)$. Notice that, since $p < 2$, we can suppose that $2^p\, c(1)^{2-p} \leq 1$. Through a scaling argument, we infer that for any supersolution $u \geq 0$ in $K_{8\rho} \times [0, \theta(\mu)\, h^{2-p}\, \rho^p]$ it holds

$$P_0(K_\rho; u \geq h) \geq \mu \;\Rightarrow\; u \geq c(\mu)h \text{ in } K_{2\rho} \times \Big[\frac{\theta(\mu)}{2} h^{2-p} \rho^p, \theta(\mu) h^{2-p} \rho^p\Big], \tag{5.58}$$

To prove (5.57), we can suppose that $k = 1$ by scaling and define

$$c(\mu) := c(\mu, 1/2), \quad \bar{\theta} := \theta(1), \quad \bar{c} := c(1) \leq 2^{\frac{p}{p-2}}, \quad \rho_n = 2^n\, r$$

and t_n as per (5.47). Since by assumption $P_0(K_r; u \geq 1) \geq \mu$, a first application of (5.58) implies $P_{t_0}(K_r; u \geq c(\mu)) = 1$. Iterating (5.58) with $\mu = 1$ we thus obtain

$$u \geq c(\mu)\, \bar{c}^n \qquad \text{in} \quad K_{\rho_n} \times \big[t_n, t_n + \frac{\bar{\theta}}{2}\, (c(\mu)\, \bar{c}^{n-1})^{2-p}\, \rho_{n-1}^p\big]$$

for all $n \geq 1$. From $2^p\, \bar{c}^{2-p} \leq 1$ we infer $t_n + 2^{-1}\, \bar{\theta}\, (c(\mu)\, \bar{c}^{n-1})^{2-p}\, \rho_{n-1}^p \geq t_{n+1}$, so that

$$u \geq c(\mu)\, \bar{c}^n \qquad \text{in} \quad K_{\rho_n} \times [t_n, t_{n+1}], \qquad n \geq 1.$$

Finally, an explicit calculation shows that for suitable $\gamma_1(\mu), \gamma_2(\mu) > 0$ it holds

$$t_n = \gamma_1(\mu)\, r^p + \gamma_2(\mu)\big(1 - (2^p\, \bar{c}^{2-p})^n\big) = r^p\, \big(\gamma_1(\mu) + \gamma_2(\mu)\big(1 - (r/\rho_n)^{\bar{\lambda}(2-p)-p}\big)\big)$$

where $\bar{\lambda} = -\log_2 \bar{c} > p/(2-p)$. A monotonicity argument then gives the claim for any $\rho \geq r$. □

Theorem 5.32 (Appendix A of [31]) *Let $u \geq 0$ solve (5.29) in $K_{2R} \times [t-2h, t]$ for some $p \in]p_*, 2[$. Then*

$$\sup_{K_R \times [t-h,t]} u \leq \frac{\bar{c}}{h^{\frac{N}{N(p-2)+p}}} \left(\inf_{s \in [t-2h,t]} \int_{K_{2R}} u(x,s)\, dx \right)^{\frac{p}{N(p-2)+p}} + \bar{c}\Big(\frac{h}{R^p}\Big)^{\frac{1}{2-p}}. \tag{5.59}$$

Theorem 5.33 (Harnack Inequality) *Let $p \in]p_*, 2[$. There exists constants $\bar{C} \geq 1, \bar{\theta} > 0$ such that any solution $u \geq 0$ of (5.29) in $K_{8R} \times [-T, T]$ obeying $u(0,0) > 0$ and*

$$4\, R^p \sup_{K_{2R}} u(\cdot, 0)^{2-p} \leq T \tag{5.60}$$

satisfies the following Harnack inequality

$$\bar{C}^{-1} \sup_{K_R} u(\cdot, s) \leq u(0,0) \leq \bar{C} \inf_{K_R} u(\cdot, t), \quad -\bar{\theta} u(0,0)^{2-p} R^p \leq s, t \leq \bar{\theta} u(0,0)^{2-p} R^p. \tag{5.61}$$

Proof Consider the solution $u(0,0)^{-1}\, u(R\,x, R^p\, u(0,0)^{2-p}\, t)$ in $K_8 \times [-T', T']$ (still denoted by u) with $T' = T\, R^{-p}\, u(0,0)^{p-2}$. This reduces us to $u(0,0) = 1$, $R = 1$, $T' \geq 4$ and (5.60) implies

$$1 \leq M^{2-p} := \sup_{K_1} u(\cdot, 0)^{2-p} \leq T'/4. \tag{5.62}$$

We first prove the inf bound in (5.61). Let $\bar{\lambda} \geq p/(2-p)$ be the expansion of positivity exponent, define $\psi(\rho) = (1-\rho)^{\bar{\lambda}} \sup_{K_\rho} u(\cdot, 0)$ for $\rho \in [0,1]$ and choose $\rho_0, x_0 \in K_{\rho_0}$ such that

$$\max_{[0,1]} \psi = \psi(\rho_0) = (1-\rho_0)^{\bar{\lambda}}\, u_0, \qquad u_0 := u(x_0, 0) \geq 1.$$

As in the proof of Theorem 5.7, we can let $\bar{\xi} \in [0,1]$ obey $(1-\bar{\xi})^{-\bar{\lambda}} = 2$ to find for $r = \bar{\xi}\,(1-\rho_0)$

$$u_0\, r^{\bar{\lambda}} \geq \bar{\xi}^{\bar{\lambda}}, \qquad \sup_{K_r(x_0)} u(\cdot, 0) \leq (1-\bar{\xi})^{-\bar{\lambda}}\, u_0 = 2\, u_0. \tag{5.63}$$

Let $a := u_0^{2-p}\, r^p$. By construction $u_0 \leq M$ and by (5.62), u solves (5.29) in $K_r(x_0) \times [-4\,a, 4\,a]$. Apply (5.59) for $R = r/2$, $t = a$, $s = 0$ and $h = 2\,a$ to get

$$\sup_{K_{\frac{r}{2}}(x_0)\times[-a,a]} u \leq \frac{\bar{c}}{a^{\frac{N}{N(p-2)+p}}} \left(\int_{K_r(x_0)} u(x,0)\, dx \right)^{\frac{p}{N(p-2)+p}} + \bar{c}\, a^{\frac{1}{2-p}}\, r^{\frac{p}{p-2}}$$

$$\leq \bar{c}\, \frac{(2\,u_0\, r^N)^{\frac{p}{N(p-2)+p}}}{(u_0^{2-p}\, r^p)^{\frac{N}{N(p-2)+p}}} + \bar{c}\, u_0 \leq \bar{c}\, u_0, \tag{5.64}$$

where we used the second inequality in (5.63) to bound the integral. Since $a = u_0^{2-p}\, r^p$, we can apply (5.56) with $k = u_0$ in both $K_{r/2}(x_0)\times[-a, 0]$ and $K_{r/2}(x_0)\times [0, a]$ to get

$$\operatorname{osc}(u, K_\rho(x_0) \times [-a, a]) \leq \bar{c}\, u_0\, (\rho/r)^{\bar{\alpha}}, \qquad \rho \leq r/2.$$

As $u(x_0, 0) = u_0$ we infer that $u \geq u_0/2$ in $K_{\bar{\eta} r}(x_0) \times [-\bar{\eta}^p\, a, \bar{\eta}^p\, a]$ for suitable $\bar{\eta} \in]0, 1/2[$, so that $P_t(K_r(x_0);\ u \geq u_0/2) \geq \bar{\eta}^N$ for all $|t| \leq \bar{\eta}\, u_0^{2-p}\, r^p$. Apply the expansion of positivity Lemma 5.31 at an arbitrary time t such that $|t| \leq \bar{\eta}\, u_0^{2-p}\, r^p$, choosing the $\gamma_i(\bar{\eta}^N)$ so small that $\gamma_1(\bar{\eta}^N) + \gamma_2(\bar{\eta}^N) < \bar{\eta}/2$. Its conclusion for $k = u_0/2$, $\rho = 2$ implies, thanks to $K_2(x_0) \supseteq K_1$,

$$\inf_{K_1} u(\cdot, t + \gamma_r u_0^{2-p} r^p) \geq \bar{c} u_0 r^{\bar{\lambda}}, \quad \gamma_r := \gamma_1(\bar{\eta}^N) + \gamma_2(\bar{\eta}^N)\Big(1 - \Big(\frac{r}{2}\Big)^{\bar{\lambda}(2-p)-p}\Big) < \frac{\bar{\eta}}{2}$$

for all $|t| \leq \bar{\eta}\, u_0^{2-p}\, r^p$. The latter readily gives $u(x, t) \geq \bar{c}\, u_0\, r^{\bar{\lambda}}$ for $x \in K_1$ and $|t| \leq \bar{\eta}\, u_0^{2-p}\, r^p/2$. Finally, observe that since $r \leq 1$ and $\bar{\lambda} \geq p/(2-p)$, it holds $u_0^{2-p}\, r^p \geq (u_0\, r^{\bar{\lambda}})^{2-p}$, so that the first inequality in (5.63) yields $u(x, t) \geq \bar{c}\, \bar{\xi}^{\bar{\lambda}} =: 1/\bar{C}$ for $x \in K_1$ and $|t| \leq \bar{\eta}\, \bar{\xi}^{\bar{\lambda}(2-p)}/2 =: \bar{\theta}$.

To prove the sup bound we proceed similarly. Indeed, let $x_* \in K_R$ be such that $u(x_*, 0) = \sup_{K_R} u$. Notice that $K_R(x_*) \subseteq K_{2R}$, hence (5.60) still implies (5.62) for the rescaled (and translated) function. Hence, the same proof as before carries over, giving, after rescaling back, $\inf_{K_R} u(\cdot, 0) \geq c\, u(x_*, 0)$. This implies $\sup_{K_1} u(\cdot, 0) \leq C\, u(0, 0)$ and we can proceed as in (5.64) for $r = 2R$, $x_0 = 0$ and $a = R^p \sup_{K_{2R}} u$ to get the final sup estimate. □

Acknowledgements We would like to thank an anonymous referee for helping us improve the quality of a first version of the paper. S. Mosconi and V. Vespri are members of GNAMPA (Gruppo Nazionale per l'Analisi Matematica, la Probabilità e le loro Applicazioni) of INdAM (Istituto Nazionale di Alta Matematica). F. G. Düzgün is partially funded by Hacettepe University BAP through project FBI-2017-16260; S. Mosconi is partially funded by the grant PdR 2016–2018 - linea di intervento 2: "Metodi Variazionali ed Equazioni Differenziali" of the University of Catania.

References

1. D.G. Aronson, Bounds for the fundamental solution of a parabolic equation. Bull. Am. Math. Soc. **73**, 890–896 (1967)
2. D.G. Aronson, J. Serrin, Local behavior of solutions of quasilinear parabolic equations. Arch. Ration. Mech. Anal. **25**, 81–122 (1967)
3. G.I. Barenblatt, On some unsteady motions of a liquid or a gas in a porous medium. Prikl. Mat. Mech. **16**, 67–78 (1952)
4. G.I. Barenblatt, A.S. Monin, Flying sources and the microstructure of the ocean: a mathematical theory. Uspekhi Mat. Nauk. **37**, 125–126 (1982)
5. G.I. Barenblatt, V.M. Entov, V.M. Rizhnik, *Motion of Fluids and Gases in Natural Strata* (Nedra, Moscow, 1984)
6. M. Barlow, M. Murugan, Stability of the elliptic Harnack inequality. Ann. Math. **187**, 777–823 (2018)
7. V. Bögelein, F. Duzaar, G. Mingione, The regularity of general parabolic systems with degenerate diffusion. Mem. Am. Math. Soc. **221**(1041), x+143 pp. (2013)
8. V. Bögelein, F. Ragnedda, S. Vernier Piro, V. Vespri, Moser-Nash kernel estimates for degenerate parabolic equations. J. Funct. Anal. **272**, 2956–2986 (2017)
9. E. Bombieri, E. Giusti, Harnack's inequality for elliptic differential equations on minimal surfaces. Invent. Math. **15**, 24–46 (1972)
10. E. Bombieri, E. De Giorgi, M. Miranda, Una maggiorazione a priori relativa alle ipersuperfici minimali non parametriche. Arch. Rat. Mech. Anal. **32**, 255–267 (1965)
11. M. Bonforte, J.L. Vazquez, Positivity, local smoothing and Harnack inequalities for very fast diffusion equations. Adv. Math. **223**, 529–578 (2010)
12. M. Bonforte, R.G. Iagar, J.L. Vazquez, Local smoothing effects, positivity, and Harnack inequalities for the fast p-Laplacian equation. Adv. Math. **224**, 2151–2215 (2010)
13. M.V. Calahorrano Recalde, V. Vespri, Harnack estimates at large: sharp pointwise estimates for nonnegative solutions to a class of singular parabolic equations. Nonlinear Anal. **121**, 153–163 (2015)
14. M.V. Calahorrano Recalde, V. Vespri, Backward pointwise estimates for nonnegative solutions to a class of singular parabolic equations. Nonlinear Anal. **144**, 194–203 (2016)
15. Y.Z. Chen, E. DiBenedetto, On the local behaviour of solutions of singular parabolic equations. Arch. Ration. Mech. Anal. **103**, 319–346 (1988)
16. S.Y. Cheng, S.T. Yau, Differential equations on Riemannian manifolds and their geometric applications. Commun. Pure Appl. Math. **28**, 333–354 (1975)
17. T.H. Colding, W.P. Minicozzi II, Harmonic functions on manifolds. Ann. Math. **146**, 725–747 (1997)
18. E. De Giorgi, Sulla differenziabilità e l'analiticità delle estremali degli integrali multipli regolari. Mem. Accad. Sci. Torino. Cl. Sci. Fis. Mat. Nat. **3**, 25–43 (1957)
19. E. DiBenedetto, On the local behaviour of solutions of degenerate parabolic equations with measurable coefficients. Ann. Sc. Norm. Sup. Pisa Cl. Sc. Serie IV **13**, 487–535 (1986)
20. E. DiBenedetto, Intrinsic Harnack type inequalities for solutions of certain degenerate parabolic equations. Arch. Ration. Mech. Anal. **100**, 129–147 (1988)
21. E. DiBenedetto, *Degenerate Parabolic Equations*. Universitext (Springer, New York, 1993)
22. E. DiBenedetto, A. Friedman, Hölder estimates for non-linear degenerate parabolic systems. J. Reine Angew. Math. **357**, 1–22 (1985)
23. E. DiBenedetto, U. Gianazza, Some properties of De Giorgi classes. Rend. Istit. Mat. Univ. Trieste **48**, 245–260 (2016)
24. E. DiBenedetto, Y.C. Kwong, Intrinsic Harnack estimates and extinction profile for certain singular parabolic equations. Trans. Am. Math. Soc. **330**, 783–811 (1992)
25. E. DiBenedetto, N.S. Trudinger, Harnack inequalities for quasi-minima of Variational integrals. Ann. Inst. H. Poincaré Anal. Non Linéaire **1**, 295–308 (1984)

26. E. DiBenedetto, U. Gianazza, V. Vespri, Harnack Estimates for quasi-linear degenerate parabolic differential equation. Acta Math. **200**, 181–209 (2008)
27. E. DiBenedetto, U. Gianazza, V. Vespri, Forward, backward and elliptic Harnack inequalities for non-negative solutions to certain singular parabolic partial differential equations. Ann. Scuola Norm. Sup. Pisa Cl. Sci. (5) **9**, 385–422 (2010)
28. E. DiBenedetto, U. Gianazza, V. Vespri, Harnack estimates and Hölder continuity for solutions to singular parabolic partial differential equations in the sub-critical range. Manuscripta Math. **131**, 231–245 (2010)
29. E. DiBenedetto, U. Gianazza, V. Vespri, A new approach to the expansion of positivity set of non-negative solutions to certain singular parabolic partial differential equations. Proc. Am. Math. Soc. **138**, 3521–3529 (2010)
30. E. DiBenedetto, U. Gianazza, V. Vespri, Liouville-type theorems for certain degenerate and singular parabolic equations. C. R. Acad. Sci. Paris Ser. I **348**, 873–877 (2010)
31. E. DiBenedetto, U. Gianazza, V. Vespri, *Harnack's Inequality for Degenerate and Singular Parabolic Equations*. Springer Monographs in Mathematics (Springer, New York/Heidelberg, 2012)
32. F. Duzaar, G. Mingione, K. Steffen, Parabolic systems with polynomial growth and regularity. Mem. Am. Math. Soc. **214**(1005), x+118 pp. (2011)
33. E.B. Fabes, N. Garofalo, Parabolic B.M.O. and Harnack's inequality. Proc. Am. Math. Soc. **50**, 63–69 (1985)
34. E.B. Fabes, D.W. Stroock, A new proof of Moser's parabolic Harnack inequality via the old ideas of Nash. Arch. Rat. Mech. Anal. **96**, 327–338 (1986)
35. A. Farina, A Bernstein-type result for the minimal surface equation. Ann. Sc. Norm. Super. Pisa Cl. Sci. **14**, 1231–1237 (2015)
36. E. Ferretti, M.V. Safonov, Growth theorems and Harnack inequality for second order parabolic equations, in *Harmonic Analysis and Boundary Value Problems (Fayetteville, AR, 2000)*. Contemporary Mathematics, vol. 277 (American Mathematical Society, Providence, 2001), pp. 87–112
37. S. Fornaro, M. Sosio, Intrinsic Harnack estimates for some doubly nonlinear degenerate parabolic equations. Adv. Differ. Equ. **13**, 139–168 (2008)
38. S. Fornaro, V. Vespri, Harnack estimates for non negative weak solutions of singular parabolic equations satisfying the comparison principle. Manuscrpita Math. **141**, 85–103 (2013)
39. S. Fornaro, M. Sosio, V. Vespri, Harnack type inequalities for some doubly nonlinear singular parabolic equations. Discrete Contin. Dyn. Syst. Ser. A **35**, 5909–5926 (2015)
40. U. Gianazza, V. Vespri, Parabolic De Giorgi classes of order p and the Harnack inequality. Calc. Var. Partial Differ. Equ. **26**, 379–399 (2006)
41. U. Gianazza, V. Vespri, A Harnack inequality for solutions of doubly nonlinear parabolic equations. J. Appl. Funct. Anal. **1**, 271–284 (2006)
42. U. Gianazza, M. Surnachev, V. Vespri, On a new proof of Hölder continuity of solutions of p-Laplace type parabolic equations. Adv. Calc. Var. **3**, 263–278 (2010)
43. E. Giusti, *Direct Methods in the Calculus of Variations* (World Scientific Publishing Co., Inc., River Edge, 2003)
44. A. Grigor'yan, The heat equation on non-compact Riemannian manifolds. Matem. Sbornik **182**, 55–87 (1991). Engl. transl. Math. USSR Sb. **72**, 47–77 (1992)
45. J. Hadamard, Extension à l' équation de la chaleur d' un theoreme de A. Harnack. Rend. Circ. Mat. Palermo **3**, 337–346 (1954)
46. R.S. Hamilton, A matrix Harnack estimate for the heat equation. Commun. Anal. Geom. **1**, 113–126 (1993)
47. C.G.A. von Harnack, *Die Grundlagen der Theorie des logaritmischen Potentiales und der eindeutigen Potentialfunktion in der Ebene* (Teubner, Leipzig, 1887)
48. M.A. Herrero, M. Pierre, The Cauchy problem for $u_t = \Delta(u^m)$ when $0 < m < 1$. Trans. Am. Math. Soc. **291**, 145–158 (1985)

49. C. Imbert, L. Silvestre, An introduction to fully nonlinear parabolic equations, in *An Introduction to the Khler-Ricci Flow*. Lecture Notes in Mathematics vol. 2086 (Springer, Cham, 2013), pp. 7–88
50. A.V. Ivanov, Regularity for doubly nonlinear parabolic equations. J. Math. Sci. **83**, 22 (1997)
51. F. John, L. Nirenberg, On functions of bounded mean oscillation. Commun. Pure Appl. Math. **14**, 415–426 (1961)
52. A.S. Kalashnikov, Some problems of the qualitative theory of nonlinear degenerate second order parabolic equations. Russ. Math. Surv. **42**, 169–222 (1987)
53. S. Kamin, J.L. Vázquez, Fundamental solutions and asymptotic behavior for the p-Laplacian equation. Rev. Mat. Iberoamericana **4**, 339–354 (1988)
54. M. Kassmann, Harnack inequalities: an introduction. Bound. Value Probl. **2007**, 81415 (2007)
55. J. Kinnunen, Regularity for a doubly nonlinear parabolic equation, in *Geometric Aspects of Partial Differential Equations*. RIMS Kôkyûroku, vol. 1842 (Kyoto University, 2013), pp. 40–60
56. J. Kinnunen, T. Kuusi, Local behavior of solutions to doubly nonlinear parabolic equations. Math. Ann. **337**, 705–728 (2007)
57. N.V. Krylov, Fully nonlinear second order elliptic equations: recent development. Dedicated to Ennio De Giorgi. Ann. Scuola Norm. Sup. Pisa Cl. Sci. Ser. 4 **25**, 569–595 (1997)
58. N.V. Krylov, M.V. Safonov, A certain property of solutions of parabolic equations with measurable coefficients. Izv. Akad. Nauk SSSR Ser. Mat. **44**, 161–175 (1980) (in Russian). Translated in Math. of the USSR Izv. **16**, 151–164 (1981)
59. B.L. Kotschwar, Hamilton's gradient estimate for the heat kernel on complete manifolds. Proc. Am. Math. Soc. **135**, 3013–3019 (2007)
60. M. Küntz, P. Lavallée, Experimental evidence and theoretical analysis of anomalous diffusion during water infiltration in porous building materials. J. Phys. D: Appl. Phys. **34**, 2547–2554 (2001)
61. T. Kuusi, G. Mingione, Nonlinear potential theory of elliptic systems. Nonlinear Anal. **138**, 277–299 (2016)
62. O.A. Ladyzenskaya, N.A. Solonnikov, N.N. Uraltzeva, Linear and quasilinear equations of parabolic type. Translations of Mathematical Monographs, vol. 23 (American Mathematical Society, Providence, RI, 1967)
63. E.M. Landis, *Second Order Equations of Elliptic and Parabolic Type*. Translations of Mathematical Monographs, vol. 171 (American Mathematical Society, Providence, RI, 1998; Nauka, Moscow, 1971)
64. P. Li, Harmonic functions on complete Riemannian manifolds, in *Handbook of Geometric Analysis, vol. I*. Advanced Lectures in Mathematics, vol. 7 (Higher Education Press and International Press, Beijing/Boston, 2008), pp. 195–227
65. P. Li, S.T. Yau, On the parabolic kernel of the Schrödinger operator. Acta Math. **156**, 153–201 (1986)
66. F. Lin, Q.S. Zhang, On ancient solutions of the heat equation. Arxiv preprint. arXiv:1712.04091v2
67. D. Maldonado, On the elliptic Harnack inequality. Proc. Am. Math. Soc. **145**, 3981–3987 (2017)
68. J.H. Michael, L.M. Simon, Sobolev and mean-value inequalities on generalized submanifolds of $\mathbb{R}^N$. Commun. Pure Appl. Math. **26**, 361–379 (1973)
69. G. Mingione, Regularity of minima: an invitation to the Dark Side of the Calculus of Variations. Appl. Math. **51**, 355–426 (2006)
70. J. Moser, On Harnack's theorem for elliptic differential equations. Commun. Pure Appl. Math. **14**, 577–591 (1961)
71. J. Moser, A Harnack inequality for parabolic differential equations. Commun. Pure Appl. Math. **17**, 101–134 (1964)
72. J. Moser, On a pointwise estimate for parabolic differential equations. Commun. Pure Appl. Math. **24**, 727–740 (1971)

73. R. Müller, *Differential Harnack Inequalities and the Ricci Flow*. EMS Series of Lectures in Mathematics (European Mathematical Society (EMS), Zürich, 2006)
74. J. Nash, Continuity of solutions of parabolic and elliptic equations. Am. J. Math. **80**, 931–954 (1958)
75. B. Pini, Sulla soluzione generalizzata di Wiener per il primo problema di valori al contorno nel caso parabolico. Rend. Sem. Mat. Univ. Padova **23**, 422–434 (1954)
76. M.M. Porzio, V. Vespri, Hölder estimates for local solutions of some doubly nonlinear degenerate parabolic equations. J. Differ. Equ. **103**, 146–178 (1993)
77. F. Ragnedda, S. Vernier Piro, V. Vespri, Pointwise estimates for the fundamental solutions of a class of singular parabolic problems. J. Anal. Math. **121**, 235–253 (2013)
78. L. Saloff-Coste, A note on Poincare, Sobolev and Harnack inequalities. Duke Math. J. **65**, 27–38 (1992)
79. L. Saloff-Coste, Uniformly elliptic operators on Riemannian manifolds. J. Differ. Geom. **36**, 417–450 (1992)
80. L. Saloff-Coste, *Aspects of Sobolev-Type Inequalities*. London Mathematical Society Lecture Notes Series, vol. 289 (Cambridge University Press, Cambridge, 2001)
81. J. Serrin, Local behavior of solutions of quasi-linear equations. Acta Math. **111**, 247–302 (1964)
82. R.E. Showalter, N.J. Walkington, Diffusion of fluid in a fissured medium with micro-stricture. SIAM J. Mat. Anal. **22**, 1702–1722 (1991)
83. Ph. Souplet, Q.S. Zhang, Sharp gradient estimate and Yau's Liouville theorem for the heat equation on noncompact manifolds. Bull. Lond. Math. Soc. **38**, 1045–1053 (2006)
84. A.F. Tedeev, V. Vespri, Optimal behavior of the support of the solutions to a class of degenerate parabolic systems. Interfaces Free Bound. **17**, 143–156 (2015)
85. E.V. Teixeira, J.M. Urbano, An intrinsic Liouville theorem for degenerate parabolic equations. Arch. Math. **102**, 483–487 (2014)
86. N.S. Trudinger, On Harnack type inequalities and their application to quasilinear elliptic partial differential equations. Commun. Pure Appl. Math. **20**, 721–747 (1967)
87. N.S. Trudinger, Pointwise estimates and quasilinear parabolic equations. Commun. Pure Appl. Math. **21**, 205–226 (1968)
88. N.S. Trudinger, A new proof of the interior gradient bound for the minimal surface equation in N dimensions. Proc. Nat. Acad. Sci. U.S.A. **69**, 821–823 (1972)
89. K. Uhlenbeck, Regularity for a class of nonlinear elliptic systems. Acta Math. **138**, 219–240 (1977)
90. J.M. Urbano, *The Method of Intrinsic Scaling*. Lecture Notes in Mathematics, vol. 1930 (Springer, Berlin, 2008)
91. J.L. Vázquez, *Smoothing and Decay Estimates for Nonlinear Diffusion Equations. Equations of Porous Medium Type*. Oxford Lecture Series in Mathematics and its Applications, vol. 33 (Oxford University Press, Oxford, 2006)
92. J.L. Vázquez, *The Porous Medium Equation: Mathematical Theory*. Oxford Mathematical Monographs (Oxford Science Publications, Clarendon Press, Oxford, 2012)
93. D.V. Widder, The role of the Appell transformation in the theory of heat conduction. Trans. Am. Math. Soc. **109**, 121–134 (1963)
94. S.T. Yau, Harmonic functions on complete Riemannian manifolds. Commun. Pure Appl. Math. **28**, 201–228 (1975)

Lectures on Curvature Flow of Networks

Carlo Mantegazza, Matteo Novaga, and Alessandra Pluda

Abstract We present a collection of results on the evolution by curvature of networks of planar curves. We discuss in particular the existence of a solution and the analysis of singularities.

Keywords Curvature flow · Networks · Singularity formation

1 Introduction

These notes have been prepared for a course given by the second author within the INdAM Intensive Period *Contemporary Research in elliptic PDEs and related topics*, organized by Serena Dipierro at the University of Bari from April to June 2017. We warmly thank the organizer for the invitation, the INdAM for the support, and the Department of Mathematics of the University of Bari for the kind hospitality.

The aim of this work is to provide an overview on the motion by curvature of a network of curves in the plane. This evolution problem attracted the attention of several researchers in recent years, see for instance [9–12, 20, 23, 24, 29, 31, 33, 35, 37, 43]. We refer to the extended survey [32] for a motivation and a detailed analysis of this problem.

This geometric flow can be regarded as the L^2-gradient flow of the length functional, which is the sum of the lengths of all the curves of the network (see [10]). From the energetic point of view it is then natural to expect that configurations with multi-points of order greater than three or 3-points with angles different from

C. Mantegazza
Dipartimento di Matematica e Applicazioni, Università di Napoli Federico II, Napoli, Italy

M. Novaga (✉) · A. Pluda
Dipartimento di Matematica, Università di Pisa, Pisa, Italy
e-mail: matteo.novaga@unipi.it

S. Dipierro (ed.), *Contemporary Research in Elliptic PDEs and Related Topics*, Springer INdAM Series 33, https://doi.org/10.1007/978-3-030-18921-1_9

120°, being unstable for the length functional, should be present only at a discrete set of times, during the flow. Therefore, we shall restrict our analysis to networks whose junctions are composed by exactly three curves, meeting at 120°. This is the so-called **Herring condition**, and we call **regular** the networks satisfying this condition at each junction.

The existence problem for the curvature flow of a regular network with only one triple junctions was first considered by L. Bronsard and F. Reitich in [11], where they proved the local existence of the flow, and by D. Kinderlehrer and C. Liu in [24], who showed the global existence and convergence of a smooth solution if the initial network is sufficiently close to a minimal configuration (Steiner tree).

We point out that the class of regular networks is not preserved by the flow, since two (or more) triple junctions might collide during the evolution, creating a multiple junction composed by more than three curves. It is then natural to ask what is the subsequent evolution of the network. A possibility is restarting the evolution at the collision time with a different set of curves, describing a non-regular network, with multi-points of order higher than three. A suitable short time existence result has been worked out by T. Ilmanen, A. Neves and F. Schulze in [23], where it is shown that there exists a flow of networks which becomes immediately regular for positive times.

These notes are organizes as follows: In Sect. 2 we introduce the notion of regular network and the geometric evolution problem we are interested in. In Sect. 3 we recall the short time existence and uniqueness result by Bronsard and Reitich, and we sketch its proof. We also show that the embeddedness of the network is preserved by the evolution (till the maximal time of smooth existence). In Sect. 4 we describe some special solution which evolve self-similarly. More precisely, we discuss translating, rotating and homothetically shrinking solutions. The latter ones are particularly important for our analysis since they describe the blow-up limit of the flow near a singularity point. In Sect. 5 we derive the evolution equation for the L^2-norm of the curvature and of its derivatives. As a consequence, we show that, at a singular point, either the curvature blows-up or there is a collision of triple junctions. Finally, in Sect. 6 we recall Huisken's Monotonicity Formula for mean curvature flow, which holds also for the evolution of a network, and we introduce the rescaling procedures used to get blow-up limits at the maximal time of smooth existence, in order to describe the singularities of the flow. In particular, we show that the limits of the rescaled networks are self-similar shrinking solutions of the flow, possibly with multiplicity greater than one, and we identify all the possible limits under the assumption that the length of each curve of the network is uniformly bounded from below.

2 Notation and Setting of the Problem

2.1 *Curves and Networks*

Given an interval $I \subset \mathbb{R}$, we consider planar curves $\gamma : I \to \mathbb{R}^2$.

The interval I can be both bounded and unbounded depending whether one wants to parametrize a bounded or an unbounded curve. In the first case we restrict to consider $I = [0, 1]$.

By **curve** we mean both image of the curve in $\mathbb{R}^2$ and parametrization of the curve, we will be more specific only when the meaning cannot be got by the context.

- A curve is of class C^k if it admits a parametrization $\gamma : I \to \mathbb{R}^2$ of class C^k.
- A C^1 curve, is **regular** if it admits a regular parametrization, namely $\gamma_x(x) = \frac{d\gamma}{dx}(x) \neq 0$ for every $x \in I$.
- It is then well defined its **unit tangent vector** $\tau = \gamma_x/|\gamma_x|$.
- We define its **unit normal vector** as $\nu = R\tau = R\gamma_x/|\gamma_x|$, where $R : \mathbb{R}^2 \to \mathbb{R}^2$ is the anticlockwise rotation centred in the origin of $\mathbb{R}^2$ of angle $\pi/2$.
- The **arclength parameter** of a curve γ is given by

$$s := s(x) = \int_0^x |\gamma_x(\xi)|\, d\xi\,.$$

  We use the letter s to indicate the arclength parameter and the letter x for any other parameter. Notice that $\partial_s = |\gamma_x|^{-1}\partial_x$.
- If the curve γ is C^2 and regular, we define the curvature $k := |\tau_s| = |\gamma_{ss}|$ and the **curvature vector** $\boldsymbol{k} := \tau_s = \gamma_{ss}$. We get:

$$\boldsymbol{k} = \frac{1}{|\gamma_x|}\left(\frac{\gamma_x}{|\gamma_x|}\right)_x = \frac{\gamma_{xx}|\gamma_x|^2 - \gamma_x\langle\gamma_{xx}, \gamma_x\rangle}{|\gamma_x|^4}\,.$$

  As we are in $\mathbb{R}^2$ we remind that $\boldsymbol{k} = \tau_s = k\nu$.
- The **length** L of a curve γ is given by

$$L(\gamma) := \int_I |\gamma_x(x)|\, dx = \int_\gamma 1\, ds\,.$$

A curve is injective if for every $x \neq y \in I$ we have $\gamma(x) \neq \gamma(y)$.

A curve $\gamma : [0, 1] \to \mathbb{R}^2$ of class C^k is **closed** if $\gamma(0) = \gamma(1)$ and if γ has a 1-periodic C^k extension to $\mathbb{R}$ (Fig. 1).

In what follows we will consider time-dependent families of curves $(\gamma(t, x))_{t\in[0,T]}$. We let $\tau = \tau(t, x)$ be the unit tangent vector to the curve, $\nu = \nu(t, x)$ the unit normal vector and $\boldsymbol{k} = \boldsymbol{k}(t, x)$ its curvature vector as previously defined.

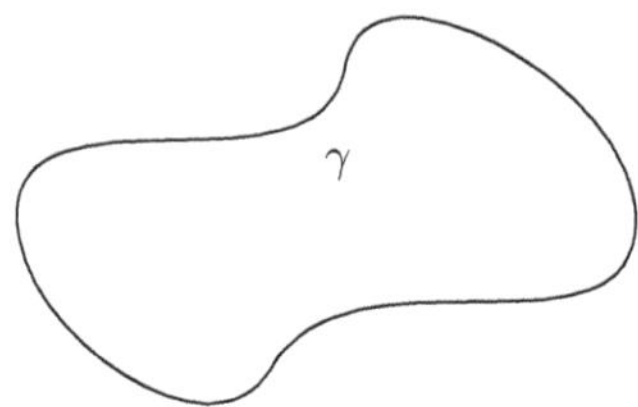

Fig. 1 A simple closed curve

We denote with $\partial_x f$, $\partial_s f$ and $\partial_t f$ the derivatives of a function f along a curve γ with respect to the x variable, the arclength parameter s on such curve and the time, respectively. Moreover $\partial_x^n f$, $\partial_s^n f$, $\partial_t^n f$ are the higher order partial derivatives, possibly denoted also by $f_x, f_{xx} \dots, f_s, f_{ss}, \dots$ and $f_t, f_{tt}, \dots$.

We adopt the following convention for integrals:

$$\int_{\gamma_t} f(t, \gamma, \tau, \nu, k, k_s, \dots, \lambda, \lambda_s \dots)\, ds = \int_0^1 f(t, \gamma^i, \tau^i, \nu^i, k^i, k_s^i, \dots, \lambda^i, \lambda_s^i \dots)\, |\gamma_x^i|\, dx$$

as the arclength measure is given by $ds = |\gamma_x^i|\, dx$ on every curve γ.

Let now Ω be a smooth, convex, open set in $\mathbb{R}^2$.

Definition 2.1 A **network** $\mathcal{N}$ in $\overline{\Omega}$ is a connected set described by a finite family of regular C^1 curves contained in $\overline{\Omega}$ such that

1. the interior of every curve is injective, a curve can self-intersect only at its end-points;
2. two different curves can intersect each other only at their end-points;
3. a curve is allowed to meet $\partial\Omega$ only at its end-points;
4. if an end-point of a curve coincide with $P \in \partial\Omega$, then no other end-point of any curve can coincide with P.

The curves of a network can meet at **multi-points** in Ω, labeled by $O^1, O^2, \dots, O^m$. We call **end-points** of the network, the vertices (of order one) $P^1, P^2, \dots, P^l \in \partial\Omega$.

Condition 4 keeps things simpler implying that multi-points can be only inside Ω, not on the boundary.

We say that a network is of class C^k with $k \in \{1, 2, \dots\}$ if all its curves are of class C^k.

Remark 2.2 With a slightly modification of Definition 2.1 we could also consider networks in the whole $\mathbb{R}^2$ with unbounded curves. In this case we require that every non compact branch of $\mathcal{N}$ is asymptotic to an half line and its curvature is uniformly bounded. We call these unbounded networks **open networks**.

Definition 2.3 We call a network **regular** if all its multi-points are triple and the sum of unit tangent vectors of the concurring curves at each of them is zero.

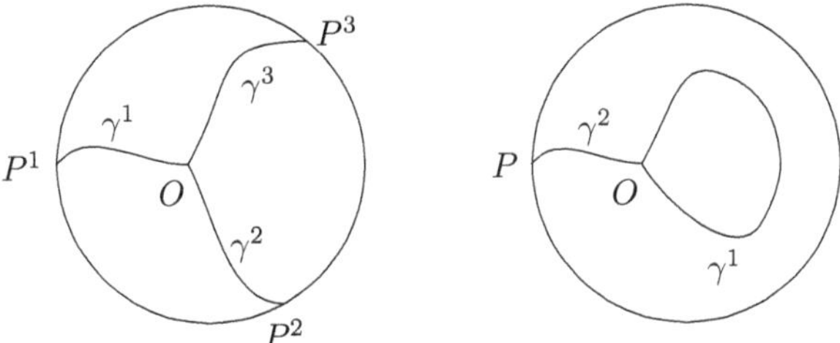

Fig. 2 A triod and a spoon

Example 2.4

- A network could consists of a single closed embedded curve.
- A network could be composed of a single embedded curve with fixed end-points on $\partial\Omega$.
- There are two possible (topological) structures of networks with only one triple junction: the **triod** $\mathbb{T}$ or the **spoon** $\mathbb{S}$. A triod is a tree composed of three curves that intersects each other at a 3-point and have their other end-points on the boundary of Ω. A spoon is the union of two curves: a closed one attached to the other at a triple junction. The "open" curve of the spoon has an end-point on $\partial\Omega$ (Fig. 2).

2.2 *The Evolution Problem*

Given a network composed of n curves we define its **global length** as

$$L = L^1 + \cdots + L^n .$$

The evolution we have in mind is the L^2-gradient flow of the global length L. Therefore, geometrically speaking, this means that the normal velocity of the curves is the curvature. In the case of the curves (curve shortening flow) this condition fully defines the evolution, at least geometrically. In the case of networks another condition at the junctions comes from the variational formulation of the evolution as we will see below.

2.2.1 Formal Derivation of the Gradient Flow

We begin by considering one closed embedded C^2 curve, parametrized by $\gamma : [0, 1] \to \mathbb{R}^2$. Then $\gamma(0) = \gamma(1)$, $\tau(0) = \tau(1)$ and $k(0) = k(1)$. We want to compute the directional derivative of the length. Given $\varepsilon \in \mathbb{R}$ and $\psi : [0, 1] \to \mathbb{R}^2$ a smooth function satisfying $\psi(0) = \psi(1)$, we take $\widetilde{\gamma} = \gamma + \varepsilon\psi$ a variation of γ.

From now on we neglect the dependence on the variable x to maintain the notation simpler. We have

$$\frac{\partial}{\partial \varepsilon} L(\widetilde{\gamma})_{|\varepsilon=0} = \frac{\partial}{\partial \varepsilon} \int_0^1 |\gamma_x + \varepsilon \psi_x| \, dx = \int_0^1 \frac{\langle \psi_x, \gamma_x \rangle}{|\gamma_x|} \, dx = \int_\gamma \langle \psi_s, \tau \rangle \, ds$$

$$= - \int_\gamma \langle \psi, \tau_s \rangle \, ds + \langle \psi(1), \tau(1) \rangle - \langle \psi(0), \tau(0) \rangle \, .$$

As γ is a simple closed embedded curve, then the boundary terms are equal zero. We get

$$\frac{\partial}{\partial \varepsilon} L(\widetilde{\gamma})_{|t=0} = \int_\gamma \langle \psi, -\boldsymbol{k} \rangle \, ds \, .$$

Since we have written the directional derivative of L in the direction ψ as the scalar product of ψ and $-\boldsymbol{k}$, we conclude (at least formally) that $-\boldsymbol{k}$ is the gradient of the length. Hence we can understand the curve shortening flow as the gradient flow of the length.

We considering now a triod $\mathbb{T}$ in a convex, open and regular set $\Omega \subset \mathbb{R}^2$, whose curves are parametrized by $\gamma^i : [0,1] \to \mathbb{R}^2$ of class C^2 with $i \in \{1,2,3\}$. Without loss of generality we can suppose that $\gamma^1(0) = \gamma^2(0) = \gamma^3(0)$ and $\gamma^i(1) = P^i \in \partial\Omega$ with $i \in \{1,2,3\}$. We consider again a variation $\tilde{\gamma}^i = \gamma^i + \varepsilon \psi^i$ of each curve with $\psi^i : [0,1] \to \mathbb{R}^2$ three smooth functions. We require that $\psi^1(0) = \psi^2(0) = \psi^3(0)$ and $\psi^i(1) = 0$ because we want that the set $\tilde{\mathbb{T}}$ parametrized by $\tilde{\gamma} = (\tilde{\gamma}^1, \tilde{\gamma}^2, \tilde{\gamma}^3)$ is a triod with end point on $\partial\Omega$ fixed at P^i. In such a way we are asking two (Dirichlet) boundary conditions. By definition of total length L of a network, we have

$$L(\widetilde{\mathbb{T}}) = \sum_{i=1}^3 L(\gamma^i) = \sum_{i=1}^3 \int_0^1 |\widetilde{\gamma}_x^i| \, dx = \sum_{i=1}^3 \int_0^1 |\gamma_x^i + \varepsilon \psi_x^i| \, dx \, .$$

Repeating the previous computation and using the hypothesis on ψ^i we have

$$\frac{\partial}{\partial \varepsilon} L(\widetilde{\mathbb{T}})_{|\varepsilon=0} = \sum_{i=1}^3 \int_\gamma \left\langle \psi^i, -\boldsymbol{k}^i \right\rangle ds + \sum_{i=1}^3 \left\langle \psi^i(1), \tau^i(1) \right\rangle - \sum_{i=1}^3 \left\langle \psi^i(0), \tau^i(0) \right\rangle$$

$$= \sum_{i=1}^3 \int_\gamma \left\langle \psi^i, -\boldsymbol{k}^i \right\rangle ds + \sum_{i=1}^3 - \left\langle \psi^1(0), \tau^i(0) \right\rangle .$$

Imposing that the boundary term equals zero we get

$$0 = \sum_{i=1}^{3} \left\langle \psi^1(0), \tau^i(0) \right\rangle = \left\langle \psi^1(0), \sum_{i=1}^{3} \tau^i(0) \right\rangle \Longrightarrow \sum_{i=1}^{3} \tau^i(0) = 0 \,.$$

Hence, we have derived a further boundary condition at the junctions.

2.2.2 Geometric Problem

We define the **motion by curvature** of regular networks.

Problem 2.5 Given a regular network we let it evolve by the L^2-gradient flow of the (total) length functional L in a maximal time interval $[0, T)$. That is:

- each curve of the network has a normal velocity equal to its curvature at every point and for all times $t \in [0, T)$—**motion by curvature**;
- the curves that meet at junctions remains attached for all times $t \in [0, T)$—**concurrency**;
- the sum of the unit tangent vectors of the three curves meeting at a junction is zero for all times $t \in [0, T)$—**angle condition**.

Moreover we ask that the end-points $P^r \subset \partial\Omega$ stay fixed during the evolution—**Dirichlet boundary condition**.

As a possible variant one lets the end-points free to move on the boundary of Ω but asking that the curves intersect orthogonally $\partial\Omega$—**Neumann boundary condition**.

Although our problem is **geometric** (as we want to describe the flow of a set moving in $\mathbb{R}^2$), to solve we will turn to a **parametric approach**. As a consequence we will work often at the level of parametrization.

Definition 2.6 (Geometric Admissible Initial Data) A network $\mathcal{N}_0$ is a geometrically admissible initial data for the motion by curvature if it is regular, at each junction the sum of the curvature is zero, the curvature at each end-point on $\partial\Omega$ is zero and each of its curve can be parametrized by a regular curve $\gamma_0^i : [0, 1] \to \mathbb{R}^2$ of class $C^{2+\alpha}$ with $\alpha \in (0, 1)$.

We introduce a way to label the curves: given a network composed by n curves with l end-points $P^1, P^2, \dots, P^l \in \partial\Omega$ (if present) and m triple points $O^1, O^2, \dots O^m \in \Omega$, we denote with γ^{pi} the curves of this network concurring at the multi-point O^p with $p \in \{1, 2, \dots, m\}$ and $i \in \{1, 2, 3\}$.

Definition 2.7 (Solution of the Motion by Curvature of Networks) Consider a geometrically admissible initial network $\mathcal{N}_0$ composed of n curves parametrized by $\gamma_0^i : [0, 1] \to \overline{\Omega}$, with m triple points $O^1, O^2, \dots O^m \in \Omega$ and (if present) l end-points $P^1, P^2, \dots, P^l \in \partial\Omega$. A time dependent family of networks $(\mathcal{N}_t)_{t\in[0,T)}$ is a solution of the motion by curvature in the maximal time interval $[0, T)$ with initial data $\mathcal{N}_0$ if it admits a time dependent family of parametrization $\gamma = (\gamma^1, \dots, \gamma^n)$ such that each curve $\gamma^i \in C^{\frac{2+\alpha}{2}, 2+\alpha}([0, T) \times [0, 1])$ is regular and the following system of conditions is satisfied for every $x \in [0, 1]$, $t \in [0, T)$, $i, j \in \{1, 2, \dots, n\}$

$$\begin{cases} (\gamma^i)_t^\perp(t,x) = \boldsymbol{k}^i(t,x) & & \text{motion by curvature,} \\ \gamma^{pi} = \gamma^{pj} & \text{at every 3-point } O^p & \text{concurrency,} \\ \sum_{i=1}^3 \tau^{pi} = 0 & \text{at every 3-point } O^p & \text{angle condition,} \\ \gamma^r(t,1) = P^r & \text{with } 0 \leq r \leq l & \text{Dirichlet boundary condition,} \end{cases} \tag{2.1}$$

where we assumed conventionally that the end-point P^r of the network is given by $\gamma^r(t, 1)$.

Remark 2.8 The boundary conditions in system (2.1) are consistent with a second order flow of three curves. Indeed we expect three vectorial conditions at the junctions and one for each curve at the other end points.

Remark 2.9 We have defined solutions in $C^{\frac{2+\alpha}{2}, 2+\alpha}$ but the natural class seems to be $C^{1,2}$. It is indeed possible to define a solution to the motion by curvature of networks asking less regularity on the parametrization. Our choice simplify the proof of the short time existence result. We will see in the sequel that it is based on linearization and on a fixed point argument. The classical theory for system of linear parabolic equations developed by Solonnikov [39] is a Hölder functions setting (see [39, Theorem 4.9]).

Remark 2.10 Suppose that $(\mathcal{N}(t))_{t\in[0,T]}$ is a solution to the motion by curvature as defined in Definition 2.7. We will see later that at $t > 0$ the curvature at the end-points and the sum of the three curvatures at every 3-point are automatically zero. Then a necessary condition for $(\mathcal{N}(t))_{t\in[0,T]}$ to be C^2 in space till $t = 0$ is that these properties are satisfied also by the initial regular network. These conditions on the curvatures are **geometric**, independent of the parametrizations of the curves, but intrinsic to the **set** and they are not satisfied by a generic regular, C^2 network.

Remark 2.11 Notice that in the geometric problem we specify only the the **normal component** of the velocity of the curves (their curvature). This does not mean that there is not a tangential component of the velocity, rather a tangential motion is needed to allow the junctions move in any direction.

Example 2.12

- The motion by curvature of a single closed embedded curve was widely studied by many authors [3–5, 16–19, 28]. In particular the curve evolves smoothly,

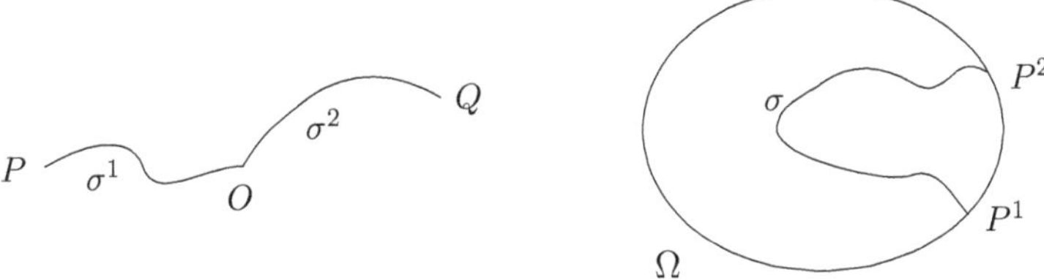

Fig. 3 Two special cases: two curves forming an angle at their junction and a single curve with two end-points on the boundary of Ω

becoming convex and getting rounder and rounder. In finite time it shrinks to a point.

- The case of a curve with either an angle or a cusp can be dealt by the works of Angenent [3–5]. Actually the curve becomes immediately smooth and then for all positive time we come back to the evolution described in the previous example.
- The evolution of a single embedded curve with fixed end-points (Fig. 3) is discussed in [22, 40, 41]. The curve converges to the straight segment connecting the two fixed end-points as the time goes to infinity.
- Two curves that concur at a 2-point forming an angle (or a cusp, if they have the same tangent) can be regarded as a single curve with a singular point, which will vanish immediately under the flow (Fig. 3).

2.2.3 The System of Quasilinear PDEs

In this section we actually work by defining the evolution in terms of differential equations for the parametrization of the curves. For sake of presentation we restrict to the case of the triod. This allows us maintaining the notation simpler.

Let us start focusing on the geometric evolution equation $\gamma_t^{\perp} = \boldsymbol{k}$, that can be equivalently written as

$$\langle \gamma_t(t,x)\,,\ \nu(t,x)\rangle\, \nu(t,x) = \left\langle \frac{\gamma_{xx}(t,x)}{|\gamma_x(t,x)|^2}\,,\ \nu(t,x)\right\rangle \nu(t,x)\,.$$

This equation specify the velocity of each curve only in direction of the normal ν.

Curve shortening flow for closed curve is not affected by tangential velocity. In the evolution by curvature of a smooth closed curve it is well known that any tangential contribution to the velocity actually affects only the "inner motion" of the "single points" (Lagrangian point of view), but it does not affect the motion of the whole curve as a subset of $\mathbb{R}^2$ (Eulerian point of view). Indeed the classical mean curvature flow for hypersurfaces is invariant under tangential perturbations (see for instance [30, Proposition 1.3.4]). In particular in the case of curves it can be shown that a solution of the curve shortening flow satisfying the equation $\gamma_t = k\nu + \lambda\tau$ for some continuous function λ can be globally reparametrized (dynamically in time) in order to satisfy $\gamma_t = k\nu$ and vice versa.

As already anticipated, in the case of networks it is instead **necessary** to consider an extra **tangential term** (as for the case of the single curve that is not closed). It allows the motion of the 3-points. At the junctions the sum of the unit normal vectors is zero. If the velocity would be in normal direction to the three curves concurring at a 3-point, this latter should move in a direction which is normal to all of them, then the only possibility would be that the junction does not move at all.

Saying that a junction cannot move is equivalent to fix it, hence to add a condition in the system (2.1). Thus, from the PDE point of view, the system becomes overdetermined as at the junctions we have already required the concurrency and the angle conditions.

Therefore solving the problem of the motion by curvature of regular networks means that we require the concurrency and the angle condition (regular networks remain regular networks for all the times) and that the main equation for each curve is

$$\gamma^i_t(t,x) = k^i(t,x)\nu^i(t,x) + \lambda^i(t,x)\tau^i(t,x)$$

for some λ continuous function not specified. To the aim of writing a non-degenerate PDE for each curve we consider the tangential velocity

$$\lambda^i = \frac{\langle \gamma^i_{xx} | \tau^i \rangle}{|\gamma^i_x|^2}\,.$$

Then the velocity of the curves is

$$\begin{aligned}\gamma^i_t(t,x) &= \left\langle \frac{\gamma^i_{xx}(t,x)}{|\gamma^i_x(t,x)|^2} \,\Big|\, \nu^i(t,x) \right\rangle \nu^i(t,x) + \left\langle \frac{\gamma^i_{xx}(t,x)}{|\gamma^i_x(t,x)|^2} \,\Big|\, \tau^i(t,x) \right\rangle \tau^i(t,x) \\ &= \frac{\gamma^i_{xx}(t,x)}{|\gamma^i_x(t,x|^2}\,. \end{aligned} \tag{2.2}$$

A family of networks evolving according to (2.2) will be called a **special flow**.

We are finally able to write explicitly the system of PDE we consider.

Without loss of generality any triod $\mathbb{T}$ can be parametrized by $\gamma = (\gamma^1, \gamma^2, \gamma^3)$ in such a way that the triple junction is $\gamma^1(0) = \gamma^2(0) = \gamma^3(0)$ and that the other end-points P^i on $\partial\Omega$ are given by $\gamma^i(1) = P^i$ with $i \in \{1, 2, 3\}$.

Definition 2.13 Given an admissible initial parametrization $\varphi = (\varphi^1, \varphi^2, \varphi^2)$ of a geometrically admissible initial triod $\mathbb{T}_0$ the family of time-dependent parametrizations $\gamma = (\gamma^1, \gamma^2, \gamma^3)$ is a **solution of the special flow** in the time interval $[0, T]$ if the functions γ^i are of class $C^{\frac{2+\alpha}{2}, 2+\alpha}([0, T] \times [0, 1])$ and the following system is satisfied for every $t \in [0, T]$, $x \in [0, 1]$, $i \in \{1, 2, 3\}$

$$\begin{cases} \gamma_t^i(t,x) = \frac{\gamma_{xx}^i(t,x)}{|\gamma_x^i(t,x)|^2} & \text{motion by curvature,} \\ \gamma^1(t,0) = \gamma^2(t,0) = \gamma^3(t,0) & \text{concurrency,} \\ \sum_{i=1}^3 \tau^i(t,0) = 0 & \text{angle condition,} \\ \gamma^i(t,1) = P^i & \text{Dirichlet boundary condition} \\ \gamma^i(0,x) = \varphi^i(x) & \text{initial data} \end{cases} \tag{2.3}$$

Definition 2.14 (Admissible Initial Parametrization of a Triod) We say that a parametrization $\varphi = (\varphi^1, \varphi^2, \varphi^3)$ is admissible for the system (2.3) if:

1. $\cup_{i=1}^3 \varphi^i([0, 1])$ is a triod;
2. each curve φ^i is regular and of class $C^{2+\alpha}([0, 1])$;
3. $\varphi^i(0) = \varphi^j(0)$ for every $i, j \in \{1, 2, 3\}$;
4. $\frac{\varphi_x^1(0)}{|\varphi_x^1(0)|} + \frac{\varphi_x^2(0)}{|\varphi_x^2(0)|} + \frac{\varphi_x^3(0)}{|\varphi_x^3(0)|} = 0$;
5. $\frac{\varphi_{xx}^i(0)}{|\varphi_x^i(0)|^2} = \frac{\varphi_{xx}^j(0)}{|\varphi_x^j(0)|^2}$ for every $i, j \in \{1, 2, 3\}$;
6. $\varphi^i(1) = P^i$ for every $i \in \{1, 2, 3\}$;
7. $\varphi_{xx}^i(1) = 0$ for every $i \in \{1, 2, 3\}$.

Remark 2.15 Notice that in the literature one refers to conditions 3. to 7. in Definition 2.14 as **compatibility conditions** for system (2.3). In particular conditions 5. and 7. are called compatibility conditions of order 2.

We want to stress the fact that choosing the tangential velocity and so passing to consider the special flow allows us to turn the geometric problem into a non degenerate PDE's system. The goodness of our choice will be revealed when one verifies the well posedness of the system (2.3).

Once proved existence and uniqueness of solution for the PDE's system, it is then crucial to come back to the geometric problem and show that we have solved it in a "geometrically" unique way. This can be done in two step: first one shows that for any geometrically admissible initial data there exists an admissible initial parametrization for system (2.3) (and consequently a unique solution related to that parametrization). In the second step one supposes that there exist two different solutions of the geometric problem and then proves that it is possible to pass from one to another by time-dependent reparametrization.

However from the previous discussion we have understood that in our situation of motion of networks the invariance under tangential terms of the curve shortening flow is not trivially true. To prove existence and uniqueness of the motion by curvature of networks starting from existence and uniqueness of the PDE's system solution a key role will be played again by our good choice of the tangential velocity.

3 Short Time Existence and Uniqueness

We now deal with the problem of short time existence and uniqueness of the flow.

3.1 *Existence and Uniqueness for the Special Flow*

We restrict again to a triod in $\Omega \subset \mathbb{R}^2$. We consider first system (2.3). The short time existence result is due to Bronsard and Reitich [11].

We look for classical solutions in the space $C^{\frac{2+\alpha}{2},2+\alpha}([0,T]\times[0,1])$ with $\alpha \in (0,1)$. We recall the definition of this function space and of the norm it is endowed with (see also [39, §11, §13]).

For a function $u : [0,T]\times[0,1] \to \mathbb{R}$ we define the semi-norms

$$[u]_{\alpha,0} := \sup_{(t,x),(\tau,x)} \frac{|u(t,x)-u(\tau,x)|}{|t-\tau|^{\alpha}},$$

and

$$[u]_{0,\alpha} := \sup_{(t,x),(t,y)} \frac{|u(t,x)-u(t,y)|}{|x-y|^{\alpha}}.$$

The classical parabolic Hölder space $C^{\frac{2+\alpha}{2},2+\alpha}([0,T]\times[0,1])$ is the space of all functions $u : [0,T]\times[0,1] \to \mathbb{R}$ that have continuous derivatives $\partial_t^i\partial_x^j u$ (where $i, j \in \mathbb{N}$ are such that $2i + j \leq 2$) for which the norm

$$\|u\|_{C^{\frac{2+\alpha}{2},2+\alpha}} := \sum_{2i+j=0}^{2} \left\|\partial_t^i\partial_x^j u\right\|_\infty + \sum_{2i+j=2} \left[\partial_t^i\partial_x^j u\right]_{0,\alpha} + \sum_{0<2+\alpha-2i-j<2} \left[\partial_t^i\partial_x^j u\right]_{\frac{2+\alpha-2i-j}{2},0}$$

is finite.

The boundary terms are in spaces of the form $C^{\frac{k+\alpha}{2},k+\alpha}([0,T]\times\{0,1\},\mathbb{R}^m)$ with $k \in \{1,2\}$ which we identify with $C^{\frac{k+\alpha}{2}}([0,T],\mathbb{R}^{2m})$ via the isomorphism $f \mapsto (f(t,0), f(t,1))^t$.

Calling B_r the ball of radius r centred at the origin the short time existence result reads as follows:

Theorem 3.1 (Bronsard and Reitich) *For any admissible initial parametrization there exists a positive radius M and a positive time T such that the system* (2.3) *has a unique solution in $C^{\frac{2+\alpha}{2},2+\alpha}([0,T)\times[0,1]) \cap \overline{B}_M$.*

Remark 3.2 Actually in [11] the authors do not consider exactly system (2.3), but the analogous Neumann problem. They require that the end-points of the three curves intersect the boundary of Ω with a prescribed angle (of 90°).

Bronsard and Reitich approach, based on linearising the problem around the initial data, nowadays is considered classical. We explain here their strategy.

Step 1: Linearization
Fix an admissible initial datum $\sigma = (\sigma^1, \sigma^2, \sigma^3)$. We linearise the system (2.3) around σ getting

$$\gamma^i_t - \frac{1}{|\sigma^i_x|^2}\gamma^i_{xx} = \left(\frac{1}{|\gamma^i_x|^2} - \frac{1}{|\sigma^i_x|^2}\right)\gamma^i_{xx} =: f^i(\gamma^i_{xx}, \gamma^i_x)\,. \tag{3.1}$$

The concurrency condition and the Dirichlet boundary condition are already linear. The angle condition instead is not linear, so one has to take into account the linear version of it:

$$-\sum_{i=1}^{3} \frac{\gamma^i_x}{|\sigma^i_x|} - \frac{\sigma^i_x \langle \gamma^i_x, \sigma^i_x\rangle}{|\sigma^i_x|^3} = \sum_{i=1}^{3}\left(\frac{1}{|\gamma^i_x|} - \frac{1}{|\sigma^i_x|}\right)\gamma^i_x + \frac{\sigma^i_x \langle \gamma_x, \sigma^i_x\rangle}{|\sigma^i_x|^3} =: b(\gamma_x)\,.$$

The linearized system associated to (2.3) is the following: for $i \in \{1, 2, 3\}$, $t \in [0, T]$ and $x \in [0, 1]$

$$\begin{cases} \gamma^i_t(t,x) - \frac{\gamma^i_{xx}(t,x)}{|\sigma^i_x|^2} & = f^i(t,x) \quad \text{motion,} \\ \gamma^1(t,0) - \gamma^2(t,0) & = 0 \quad \text{concurrency} \\ \gamma^1(t,0) - \gamma^3(t,0) & = 0 \quad \text{concurrency} \\ -\sum_{i=1}^3 \frac{\gamma^i_x(t,0)}{|\sigma^i_x|} - \frac{\sigma^i_x \langle \gamma(t,x)^i_x, \sigma^i_x\rangle}{|\sigma^i_x|^3} & = b(t,0) \quad \text{angles condition} \\ \gamma^i(t,1) & = P^i \quad \text{Dirichlet boundary condition} \\ \gamma^i(0,x) & = \varphi^i(x) \quad \text{initial data} \end{cases} \tag{3.2}$$

We remind that the initial data for the system has to satisfy some linear compatibility conditions.

Step 2: Existence and Uniqueness of Solution for the Linearized System
We have linearized system (2.3) to obtain system (3.2). We now want to show that this latter admits a unique solution in $C^{\frac{2+\alpha}{2}, 2+\alpha}([0, T] \times [0, 1])$. This is due to general results by Solonnikov [39], provided the so-called **complementary conditions** hold (see [39, p. 11]). The theory of Solonnikov is a generalization to parabolic systems of the elliptic theory by Agmon, Douglis and Nirenberg.

The complementary conditions are algebraic conditions that the matrices that represent the boundary operator and the initial datum have to satisfy (see also [39, p. 97]). Showing this conditions for a particular system can be heavy from the computational point of view. For instance in [15, pp. 11–15] it is proved that the complementary condition follows from the **Lopatinskii–Shapiro condition**. We state here the definition of Lopatinskii–Shapiro condition at the triple junction, it is similar at the end-points on $\partial\Omega$.

Definition 3.3 Let $\lambda \in \mathbb{C}$ with $\Re(\lambda) > 0$ be arbitrary. The Lopatinskii–Shapiro condition for system (3.2) is satisfied at the triple junction if every solution $(\gamma^i)_{i=1,2,3} \in C^2([0,\infty), (\mathbb{C}^2)^3)$ to

$$\begin{cases} \lambda\gamma^i(x) - \frac{1}{|\sigma_x^i(0)|^2}\gamma_{xx}^i(x) & = 0 \quad x \in [0,\infty), i \in \{1,2,3\} \quad \text{motion,} \\ \gamma^1(0) - \gamma^2(0) & = 0 \quad \text{concurrency,} \\ \gamma^2(0) - \gamma^3(0) & = 0 \quad \text{concurrency,} \\ \sum_{i=1}^3 \frac{\gamma_x^i(x)}{|\sigma_x^i(0)|} - \frac{\sigma_x^i(0)\langle\gamma_x^i(x), \sigma_x^i(0)\rangle}{|\sigma_x^i(0)|^3} & = 0 \quad \text{angle condition,} \end{cases}$$

which satisfies $\lim_{x\to\infty}|\gamma^i(x)| = 0$ is the trivial solution.

The angle condition in the previous system can be equivalently written as

$$\sum_{i=1}^{3} \frac{1}{|\sigma_x(0)^i|^3}\left\langle \gamma_x^i(x), \nu_0^i(0)\right\rangle \nu_0^i(0) = 0\,.$$

It can be proved that Lopatinskii–Shapiro condition for system (3.2) is satisfied testing the motion equation by $|\sigma(0)_x^i|\overline{\langle\gamma^i(x), \nu^i(0)\rangle\nu^i(0)}$ and then by $|\sigma(0)_x^i|\overline{\langle\gamma^i(x), \tau^i(0)\rangle\tau^i(0)}$ and using the concurrency and the angle conditions.

Once it is shown that the complementary conditions are fulfilled, then [39, Theorem 4.9] guarantees existence and uniqueness of a solution of system (3.2).

For $T > 0$ we define the map $L_T : X_T \to Y_T$ as

$$L(\gamma) = \begin{pmatrix} \left(\gamma_t^i - \frac{1}{|\sigma_x^i|^2}\gamma_{xx}^i\right)_{i\in\{1,2,3\}} \\ -\sum_{i=1}^3 \frac{\gamma_x^i}{|\sigma_x^i|} - \frac{\sigma_x^i\langle\gamma_x^i, \sigma_x^i\rangle}{|\sigma_x^i|^3}\Big|_{x=0} \\ \gamma^i{}_{|x=1} \\ \gamma_{|t=0} \end{pmatrix}$$

where the linear spaces X_T and Y_T are

$$X_T := \{\gamma \in C^{\frac{2+\alpha}{2}, 2+\alpha}([0,T]\times[0,1]; (\mathbb{R}^2)^3) \text{ such that for } t \in [0,T]\,, i \in \{1,2,3\}$$
$$\text{it holds } \gamma^1(t,1) = \gamma^2(t,2) = \gamma^3(t,3)\}\,,$$

$$Y_T := \{(f, b, \psi) \in C^{\frac{\alpha}{4},\alpha}([0,T]\times[0,1]; (\mathbb{R}^2)^3) \times C^{\frac{1+\alpha}{2}}([0,T]; \mathbb{R}^4)$$
$$\times\, C^{2+\alpha}\left([0,1]; \left(\mathbb{R}^2\right)^3\right)$$
$$\text{such that the linear compatibility conditions hold}\}\,,$$

endowed with the induced norms. Then as a consequence of the existence and uniqueness of a solution of system (3.2) we get that L_T is a continuous isomorphism.

Remark 3.4 The linearized version of $\frac{\gamma_{xx}}{|\gamma_x|^2}$ (linearising around σ) is

$$\frac{1}{|\sigma_x|^2}\gamma_{xx} - 2\frac{\sigma_{xx}\langle\gamma_x,\sigma_x\rangle}{|\sigma_x|^4}. \tag{3.3}$$

As the well posedness of system (3.2) depends only on the highest order term we can restrict to consider (3.1) instead of (3.3).

Step 3: Fixed Point Argument
In the last step of the proof we deduce existence of a solution for system (2.3) from the linear problem by a **contraction argument**.

Let us define the operator N that "contains the information" about the non-linearity of our problem. The two components of this map are the following:

$$N_1 : \begin{cases} X_T^{\varphi,P} & \to C^{\frac{\alpha}{2},\alpha}([0,T]\times[0,1];(\mathbb{R}^2)^3), \\ \gamma & \mapsto f(\gamma), \end{cases}$$

$$N_2 : \begin{cases} X_T^{\varphi,P} & \to C^{\frac{1+\alpha}{2}}([0,T];\mathbb{R}^4), \\ \gamma & \mapsto b(\gamma) \end{cases}$$

where $X_T^{\varphi,P} = \{\gamma \in X_T \text{ such that } \gamma_{|t=0} = \varphi \text{ and } \gamma^i(t,1) = P^i \text{ for } i \in \{1,2,3\}\}$.

Then γ is a solution for system (2.3) if and only if $\gamma \in X_T^{\varphi}$ and

$$L_T(\gamma) = N_T(\gamma) \qquad \Longleftrightarrow \qquad \gamma = L_T^{-1}N_T(\gamma) := K_T(\gamma).$$

Hence there exists a unique solution to system (2.3) if and only if $K_T : X_T^{\varphi,P} \to X_T^{\varphi,P}$ has a unique fixed point. By the contraction mapping principle it is enough to show that K is a contraction. This result conclude the proof of Theorem 3.1. □

The method of Bronsard and Reitich extends to the case of a networks with several 3-points and end-points. Indeed such method relies on the uniform parabolicity of the system (which is the same) and on the fact that the complementary and compatibility conditions are satisfied.

We have only to define what is an admissible initial parametrization of a network.

Definition 3.5 (Admissible Initial Parametrization of a Network) We say that a parametrization $\varphi = (\varphi^1,\dots,\varphi^n)$ of a geometric admissible network $\mathcal{N}_0$ composed by n curves (hence such that $\cup_{i=1}^n \varphi^i([0,1]) = \mathcal{N}_0$) is an admissible initial one if each curve φ^i is regular and of class $C^{2+\alpha}([0,1])$, at the end-points $\varphi^i(1) = P^i$ it holds $\varphi^i_{xx}(1) = 0$ and at any 3-point O^p we have

$$\varphi^{p1}(O^p) = \varphi^{p2}(O^p) = \varphi^{p3}(O^p),$$

$$\frac{\varphi^{p1}_x(O^p)}{|\varphi^{p1}_x(O^p)|} + \frac{\varphi^{p2}_x(O^p)}{|\varphi^{p2}_x(O^p)|} + \frac{\varphi^{p3}_x(O^p)}{|\varphi^{p3}_x(O^p)|} = 0,$$

$$\frac{\varphi^{p1}_{xx}(O^p)}{|\varphi^{p1}_x(O^p)|^2} = \frac{\varphi^{p2}_{xx}(O^p)}{|\varphi^{p2}_x(O^p)|^2} = \frac{\varphi^{p3}_{xx}(O^p)}{|\varphi^{p3}_x(O^p)|^2}$$

where we abused a little the notation as in Definition 2.13.

Theorem 3.6 *Given an admissible initial parametrization $\varphi = (\varphi^1, \dots, \varphi^n)$ of a geometric admissible network $\mathcal{N}_0$, there exists a unique solution $\gamma = (\gamma^1, \dots, \gamma^n)$ in $C^{\frac{2+\alpha}{2}, 2+\alpha}([0,T] \times [0,1])$ of the following system*

$$\begin{cases} \gamma^i_t(t,x) = \frac{\gamma^i_{xx}(t,x)}{|\gamma^i_x(t,x)|^2} & & \textit{motion by curvature} \\ \gamma^{pj}(t,O^p) = \gamma^{pk}(t,O^p) & \textit{at every 3-point } O^p & \textit{concurrency} \\ \sum_{j=1}^3 \frac{\gamma^{pj}_x(t,O^p)}{|\gamma^{pj}_x(t,O^p)|} = 0 & \textit{at every 3-point } O^p & \textit{angles condition} \\ \gamma^r(t,1) = P^r & \textit{with } 0 \le r \le l & \textit{Dirichlet boundary condition} \\ \gamma^i(x,0) = \varphi^i(x) & & \textit{initial data} \end{cases} \tag{3.4}$$

(where we used the notation of Definition 2.13) for every $x \in [0,1]$, $t \in [0,T]$ and $i \in \{1,2,\dots,n\}$, $j \neq k \in \{1,2,3\}$ in a positive time interval $[0,T]$.

3.2 Existence and Uniqueness

In the previous section we have explained how to obtain a unique solution for short time to system (2.3) and more in general to system (3.4), but till now we have not solved our original problem yet. Indeed in Definition 2.7 of solution of the motion by curvature appears a slightly different system. Moreover Theorem 3.1 (and Theorem 3.6) provides a solution given an **admissible initial parametrization** but in Definition 2.7 we speak of **geometrically admissible initial network**. It is then clear that we have to establish a relation between this two notions.

To this aim the following lemma will be useful.

Lemma 3.7 *Consider a triple junction O where the curves γ^1, γ^2 and γ^3 concur forming angles of 120° (that is $\sum_{i=1}^3 \tau^i = \sum_{i=1}^3 \nu^i = 0$). Then*

$$k^1\nu^1 + \lambda^1\tau^1 = k^2\nu^2 + \lambda^2\tau^2 = k^3\nu^3 + \lambda^3\tau^3,$$

is satisfied if and only if

$$k^1 + k^2 + k^3 = 0 \qquad \textit{and} \qquad \lambda^1 + \lambda^2 + \lambda^3 = 0\,.$$

Proof Suppose that for $i \neq j \in \{1, 2, 3\}$ we have

$$k^i \nu^i + \lambda^i \tau^i = k^j \nu^j + \lambda^j \tau^j\,.$$

Multiplying these vectorial equalities by τ^l and ν^l and varying i, j, l, thanks to the conditions $\sum_{i=1}^{3} \tau^{pi} = \sum_{i=1}^{3} \nu^{pi} = 0$, we get the relations

$$\begin{aligned}
\lambda^i &= -\lambda^{i+1}/2 - \sqrt{3}k^{i+1}/2\\
\lambda^i &= -\lambda^{i-1}/2 + \sqrt{3}k^{i-1}/2\\
k^i &= -k^{i+1}/2 + \sqrt{3}\lambda^{i+1}/2\\
k^i &= -k^{i-1}/2 - \sqrt{3}\lambda^{i-1}/2
\end{aligned}$$

with the convention that the second superscripts are to be considered "modulus 3". Solving this system we get

$$\begin{aligned}
\lambda^i &= \frac{k^{i-1} - k^{i+1}}{\sqrt{3}}\\
k^i &= \frac{\lambda^{i+1} - \lambda^{i-1}}{\sqrt{3}}
\end{aligned}$$

which implies

$$\sum_{i=1}^{3} k^i = \sum_{i=1}^{3} \lambda^i = 0\,. \tag{3.5}$$

□

It is also possible to prove that at each triple junction the following properties hold

$$\sum_{i=1}^{3} (k^i)^2 = \sum_{i=1}^{3} (\lambda^i)^2 \qquad \text{and} \qquad \sum_{i=1}^{3} k^i \lambda^{pi} = 0\,,$$

$$\partial_t^l \sum_{i=1}^{3} k^{pi} = \sum_{i=1}^{3} \partial_t^l k^{pi} = \partial_t^l \sum_{i=1}^{3} \lambda^{pi} = \sum_{i=1}^{3} \partial_t^l \lambda^{pi} = \partial_t \sum_{i=1}^{3} k^{pi} \lambda^{pi} = 0\,,\,,$$

$$\sum_{i=1}^{3}(\partial_t^l k^{pi})^2=\sum_{i=1}^{3}(\partial_t^l \lambda^{pi})^2 \quad \text{for every } l\in\mathbb{N},$$

$$\partial_t^m(k_s^{pi}+\lambda^{pi}k^{pi})=\partial_t^m(k_s^{pj}+\lambda^{pj}k^{pj}) \quad \text{for every pair } i,j \text{ and } m\in\mathbb{N}.,$$

$$\sum_{i=1}^{3}\partial_t^l k^{pi}\,\partial_t^m(k_s^{pi}+\lambda^{pi}k^{pi})=\sum_{i=1}^{3}\partial_t^l\lambda^{pi}\,\partial_t^m(k_s^{pi}+\lambda^{pi}k^{pi})=0 \quad \text{for every } l,m\in\mathbb{N}. \tag{3.6}$$

We are ready now to establish the relation between geometrically admissible initial networks and admissible parametrizations.

Lemma 3.8 *Suppose that $\mathbb{T}_0$ is a geometrically admissible initial triod parametrized by $\gamma=(\gamma^1,\gamma^2,\gamma^3)$. Then there exist three smooth functions $\theta^i:[0,1]\to[0,1]$ such that the reparametrization $\varphi:=\left(\gamma^1\circ\theta^1,\gamma^2\circ\theta^2,\gamma^3\circ\theta^3\right)$ is an admissible initial parametrization.*

Proof Consider $\gamma=(\gamma^1,\gamma^2,\gamma^3)$ the parametrization of class $C^{2+\alpha}$ of $\mathbb{T}_0$ (that exists as $\mathbb{T}_0$ is a geometrically admissible initial triod). It is not restrictive to suppose that $\gamma^1(0)=\gamma^2(0)=\gamma^3(0)$ is the triple junction and that $\gamma^i(1)=P^i\in\partial\Omega$ with $i\in\{1,2,3\}$.

We look for smooth maps $\theta^i:[0,1]\to[0,1]$ such that $\theta^i_x(x)\neq 0$ for every $x\in[0,1]$, $\theta^i(0)=0$ and $\theta^i(1)=1$. Then conditions 1. 2. 3. and 6. of Definition 2.14 are satisfied.

Condition 4. at the triple junction is true for any choice of the θ^i as it involves the unit tangent vectors that are invariant under reparametrization.

We pass now to Condition 5. namely we want that

$$\frac{\varphi^1_{xx}}{|\varphi^1_x|^2}=\frac{\varphi^2_{xx}}{|\varphi^2_x|^2}=\frac{\varphi^3_{xx}}{|\varphi^3_x|^2}. \tag{3.7}$$

We indicate with the subscript γ or φ the geometric quantities computed for the parametrization γ or φ, respectively. We define $\lambda^i:=\frac{\langle\varphi^i_{xx}\,|\,\varphi^i_x\rangle}{|\varphi^i_x|^3}$. Then (3.7) can be equivalently written as

$$k^1_\varphi\nu^1_\varphi+\lambda^1\tau^1_\varphi=k^2_\varphi\nu^2_\varphi+\lambda^2\tau^2_\varphi=k^3_\varphi\nu^3_\varphi+\lambda^3\tau^3_\varphi, \tag{3.8}$$

and, as all the geometric quantities involved are invariant under reparametrization, the equality (3.8) is nothing else than

$$k^1_\gamma\nu^1_\gamma+\lambda^1\tau^1_\gamma=k^2_\gamma\nu^2_\gamma+\lambda^2\tau^2_\gamma=k^3_\gamma\nu^3_\gamma+\lambda^3\tau^3_\gamma,$$

that by Lemma 3.7 is satisfied if and only if

$$k^1_\gamma+k^2_\gamma+k^3_\gamma=0 \qquad \text{and} \qquad \lambda^1+\lambda^2+\lambda^3=0. \tag{3.9}$$

To satisfy Condition 7. we need a similar request. Indeed $\varphi^i_{xx} = 0$ at every end-point of the network is equivalent to the condition $k^i_\gamma \nu^i_\gamma + \lambda^i \tau^i_\gamma = 0$, that is satisfied if and only if

$$k^i_\gamma = 0 \qquad \text{and} \qquad \lambda^i = 0 \tag{3.10}$$

at every end-point of the network.

Hence, we only need to find C^∞ reparametrizations θ^i such that at the borders of $[0, 1]$ the values of λ^i are given by the relations in (3.9) and (3.10). This can be easily done since at the borders of the interval $[0, 1]$ we have $\theta^i(0) = 0$ and $\theta^i(1) = 1$, hence

$$\begin{aligned}\lambda^i &= \frac{\langle \varphi^i_{xx} \,|\, \varphi^i_x \rangle}{|\varphi^i_x|^3} = -\partial_x \frac{1}{|\varphi^i_x|} = -\partial_x \frac{1}{|\gamma^i_x \circ \theta^i| \theta^i_x} = \frac{\langle \gamma^i_{xx} \,|\, \gamma^i_x \rangle}{|\gamma^i_x|^3} + \frac{\theta^i_{xx}}{|\sigma^i_x||\theta^i_x|^2} \\ &= \lambda^i_\gamma + \frac{\theta^i_{xx}}{|\sigma^i_x||\theta^i_x|^2}\end{aligned}$$

where $\lambda^i_\gamma = \frac{\langle \gamma^i_{xx} \,|\, \gamma^i_x \rangle}{|\gamma^i_x|^3}$.

Choosing any C^∞ functions θ^i with $\theta^i_x(0) = \theta^i_x(1) = 1$, $\theta(1)^i_{xx} = -\lambda^i_\gamma |\gamma^i_x||\theta^i_x|^2$ and

$$\theta(0)^i_{xx} = \left(\frac{k^{i-1}_\gamma - k^{i+1}_\gamma}{\sqrt{3}} - \lambda^i_\gamma \right) |\gamma^i_x||\theta^i_x|^2$$

(for instance, one can use a polynomial function) the reparametrization $\varphi = (\varphi^1, \varphi^2, \varphi^3)$ satisfies Conditions 1. to 7. of Definition 2.14 and the proof is completed. □

Remark 3.9 Vice versa if φ is an admissible initial parametrization, then the triod $\cup_{i=1}^3 \varphi^i([0, 1])$ is clearly a geometrically admissible initial network. Indeed one uses Lemma 3.7 to get that the sum of the curvature at the junction is zero. The other properties are trivially verified.

We are ready now to discuss existence and uniqueness of solution of the geometric problem. We need to introduce the notion of geometric uniqueness because even if the solution γ of system (2.3) is unique, there are anyway several solutions of Problem 2.5 obtained by reparametrizing γ.

Definition 3.10 We say that Problem 2.5 admits a **geometrically unique** solution if there exists a unique family of time-dependent networks (sets) $(\mathcal{N}_t)_{t \in [0,T]}$ satisfying the definition of solution 2.7.

In particular this means that all the solutions (functions) satisfying system (2.1) can be obtained one from each other by means of time-depending reparametrization.

Theorem 3.11 (Geometric Uniqueness) *Let $\mathbb{T}_0$ be a geometrically admissible initial triod. Then there exists a geometrically unique solution of Problem* (2.3) *in a positive time interval* $[0, T]$.

Proof Let $\mathbb{T}_0$ be a geometrically admissible initial triod parametrized by $\gamma_0 = (\gamma_0^1, \gamma_0^2, \gamma_0^3)$ admissible initial parametrization (that always exists thanks to Lemma 3.8). Then by Theorem 3.1 there exists a unique solution $\gamma = (\gamma^1, \gamma^2, \gamma^3)$ to system (2.3) with initial data $\gamma_0 = (\gamma_0^1, \gamma_0^2, \gamma_0^3)$ in a positive time interval $[0, T]$. In particular $(\mathbb{T}_t)_{t\in[0,T]} = (\cup_{i=1}^3 \gamma^i([0, 1])_t)_{t\in[0,T]}$ is a solution of the motion by curvature.

Suppose by contradiction that there exists another solution $(\widetilde{\mathbb{T}}_t)_{t\in[0,T']}$ to Problem 2.5 with the same initial $\mathbb{T}_0$. Let this solution be parametrized by $\tilde{\gamma} = (\tilde{\gamma}^1, \tilde{\gamma}^2, \tilde{\gamma}^3)$ with

$$\widetilde{\gamma}^i \in C^{\frac{2+\alpha}{2}, 2+\alpha}([0, T'] \times [0, 1]) .$$

We want to show that the sets $\mathbb{T}$ and $\widetilde{\mathbb{T}}$ coincide, namely that $\tilde{\gamma}$ coincides to γ up to a reparametrization of the curves $\widetilde{\gamma}(\cdot, t)$ for every $t \in [0, \min\{T, T'\}]$.

Let $\varphi^i : [0, \min\{T, T'\}] \times [0, 1] \to [0, 1]$ be in $C^{\frac{2+\alpha}{2}, 2+\alpha}([0, \min\{T, T'\}] \times [0, 1])$ and consider the reparametrizations $\overline{\gamma}^i(t, x) = \widetilde{\gamma}^i(t, \varphi^i(t, x))$. We have $\overline{\gamma}^i \in C^{\frac{2+\alpha}{2}, 2+\alpha}([0, \min\{T, T'\}] \times [0, 1])$ and

$$\begin{aligned}
\overline{\gamma}_t^i(t, x) &= \partial_t[\widetilde{\gamma}^i(t, \varphi^i(t, x))] \\
&= \widetilde{\gamma}_t^i(t, \varphi^i(t, x)) + \widetilde{\gamma}_x^i(t, \varphi^i(t, x))\varphi_t^i(t, x) \\
&= \widetilde{k}^i(t, \varphi^i(t, x))\widetilde{\nu}^i(t, \varphi^i(t, x)) + \widetilde{\lambda}^i(t, \varphi^i(t, x))\widetilde{\tau}^i(t, \varphi^i(t, x)) \\
&+ \widetilde{\gamma}_x^i(t, \varphi^i(t, x))\varphi_t^i(t, x) \\
&= \left\langle \frac{\widetilde{\gamma}_{xx}^i\left(t, \varphi^i(t, x)\right)}{\left|\widetilde{\gamma}_x^i\left(t, \varphi^i(t, x)\right)\right|^2} \,\middle|\, \widetilde{\nu}^i(t, \varphi^i(t, x)) \right\rangle \widetilde{\nu}^i(t, \varphi^i(t, x)) \\
&+ \widetilde{\lambda}^i(t, \varphi^i(t, x)) \frac{\widetilde{\gamma}_x^i(t, \varphi^i(t, x))}{\left|\widetilde{\gamma}_x^i\left(t, \varphi^i(t, x)\right)\right|} + \widetilde{\gamma}_x^i(t, \varphi^i(t, x))\varphi_t^i(t, x) .
\end{aligned}$$

We ask now the maps φ^i to be solutions for some positive interval of time $[0, T'']$ of the following quasilinear PDE's

$$\begin{aligned}
\varphi_t^i(t, x) =& \frac{1}{\left|\widetilde{\gamma}_x^i\left(t, \varphi^i(t, x)\right)\right|} \left\langle \frac{\widetilde{\gamma}_{xx}^i\left(t, \varphi^i(t, x)\right)}{\left|\widetilde{\gamma}_x^i\left(t, \varphi^i(t, x)\right)\right|^2} \,\middle|\, \frac{\widetilde{\gamma}_x^i(t, \varphi^i(t, x))}{\left|\widetilde{\gamma}_x^i\left(t, \varphi^i(t, x)\right)\right|} \right\rangle \\
& - \frac{\widetilde{\lambda}^i(t, \varphi^i(t, x))}{\left|\widetilde{\gamma}_x^i\left(t, \varphi^i(t, x)\right)\right|} + \frac{\varphi_{xx}^i(t, x)}{\left|\widetilde{\gamma}_x^i\left(t, \varphi^i(t, x)\right)\right|^2 \left|\varphi_x^i(t, x)\right|^2} ,
\end{aligned}$$

with $\varphi^i(t,0) = 0$, $\varphi^i(t,1) = 1$, $\varphi^i(0,x) = x$ (hence, $\overline{\gamma}^i(0,x) = \gamma^i(0,x) = \sigma^i(x)$) and $\varphi_x(t,x) \neq 0$. The existence of such solutions follows by standard theory of second order quasilinear parabolic equations (see [25, 27]). Then we have

$$
\begin{aligned}
\overline{\gamma}^i_t(t,x) &= \left\langle \frac{\widetilde{\gamma}^i_{xx}\left(t,\varphi^i(t,x)\right)}{\left|\widetilde{\gamma}^i_x\left(t,\varphi^i(t,x)\right)\right|^2} \,\middle|\, \widetilde{\nu}^i(t,\varphi^i(t,x)) \right\rangle \widetilde{\nu}^i(t,\varphi^i(t,x)) \\
&+ \left\langle \frac{\widetilde{\gamma}^i_{xx}\left(t,\varphi^i(t,x)\right)}{\left|\widetilde{\gamma}^i_x\left(t,\varphi^i(t,x)\right)\right|^2} \,\middle|\, \frac{\widetilde{\gamma}^i_x(t,\varphi^i(t,x))}{\left|\widetilde{\gamma}^i_x\left(t,\varphi^i(t,x)\right)\right|} \right\rangle \frac{\widetilde{\gamma}^i_x\left(t,\varphi^i(t,x)\right)}{\left|\widetilde{\gamma}^i_x\left(t,\varphi^i(t,x)\right)\right|} \\
&+ \frac{\varphi^i_{xx}(t,x)\widetilde{\gamma}^i_x\left(t,\varphi^i(t,x)\right)}{\left|\widetilde{\gamma}^i_x\left(t,\varphi^i(t,x)\right)\right|^2 \left|\varphi^i_x(t,x)\right|^2} \\
&= \frac{\widetilde{\gamma}^i_{xx}\left(t,\varphi^i(t,x)\right)}{\left|\widetilde{\gamma}^i_x\left(t,\varphi^i(t,x)\right)\right|^2} + \frac{\varphi^i_{xx}(t,x)\widetilde{\gamma}^i_x\left(t,\varphi^i(t,x)\right)}{\left|\widetilde{\gamma}^i_x\left(t,\varphi^i(t,x)\right)\right|^2 \left|\varphi^i_x(t,x)\right|^2} \\
&= \frac{\overline{\gamma}^i_{xx}(t,x)}{|\overline{\gamma}^i_x(t,x)|^2}\,.
\end{aligned}
$$

By the uniqueness result of Theorem 3.6 we can then conclude that $\overline{\gamma}^i = \gamma^i$ for every $i \in \{1, 2, \dots, n\}$, hence $\gamma^i(t,x) = \widetilde{\gamma}^i(t,\varphi^i(t,x))$ in the time interval $[0,\widetilde{T}]$ where $\widetilde{T} = \min\{T, T', T''\}$. □

3.3 *Geometric Properties of the Flow*

In Definition 2.1 of network we require that the curves are injective and regular. The second assumption is needed to define the flow because $|\gamma_x|$ appears at the denominator. For the short time existence of the flow we did not require that the curves are embedded. We now show that if the initial network is embedded then the evolving networks stay embedded and intersect the boundary of Ω only at the fixed end-points (transversally).

Proposition 3.12 *Let $\mathcal{N}_t$ be the curvature flow of a regular network in a smooth, convex, bounded, open set Ω, with fixed end-points on the boundary of Ω, for $t \in [0,T)$. Then, for every time $t \in [0,T)$, the network $\mathcal{N}_t$ intersects the boundary of Ω only at the end-points and such intersections are transversal for every positive time. Moreover, $\mathcal{N}_t$ remains embedded.*

Proof By continuity, the 3-points cannot hit the boundary of Ω at least for some time $T' > 0$. The convexity of Ω and the strong maximum principle (see [36]) imply that the network cannot intersect the boundary for the first time at an inner regular point. As a consequence, if $t_0 > 0$ is the "first time" when the $\mathcal{N}_t$ intersects the boundary at an inner point, this latter has to be a 3-point. The minimality of t_0 is then easily contradicted by the convexity of Ω, the 120° condition and the nonzero length of the curves of $\mathcal{N}_{t_0}$.

Even if some of the curves of the initial network are tangent to $\partial\Omega$ at the end-points, by the strong maximum principle, as Ω is convex, the intersections become immediately transversal and stay so for every subsequent time.

Finally, if the evolution $\mathcal{N}_t$ loses embeddedness for the first time, this cannot happen neither at a boundary point, by the argument above, nor at a 3-point, by the 120° condition. Hence it must happen at interior regular points, but this contradicts the strong maximum principle. □

Proposition 3.13 *In the same hypotheses of the previous proposition, if the smooth, bounded, open set Ω is strictly convex, for every fixed end-point P^r on the boundary of Ω, for $r \in \{1, 2, \dots, l\}$, there is a time $t_r \in (0, T)$ and an angle α_r smaller than $\pi/2$ such that the curve of the network arriving at P^r form an angle less that α_r with the inner normal to the boundary of Ω, for every time $t \in (t_r, T)$.*

Proof We observe that the evolving network $\mathcal{N}_t$ is contained in the convex set $\Omega_t \subset \Omega$, obtained by letting $\partial\Omega$ (which is a finite set of smooth curves with end-points P^r) move by curvature keeping fixed the end-points P^r (see [22, 40, 41]). By the strict convexity of Ω and strong maximum principle, for every positive $t > 0$, the two curves of the boundary of Ω concurring at P^r form an angle smaller that π which is not increasing in time. Hence, the statement of the proposition follows. □

4 Self-similar Solutions

Once established the existence of solution for a short time, we want to analyse the behavior of the flow in the long time. A good way to understand more about the flow is looking for examples of solutions.

A straight line is perhaps the easiest example. It is also easy to see that an infinite flat triod with the triple junction at the origin (called **standard triod**) is a solution (Fig. 4).

In both these examples the existence is global in time and the set does not change shape during the evolution. From this last observation one could guess that there is

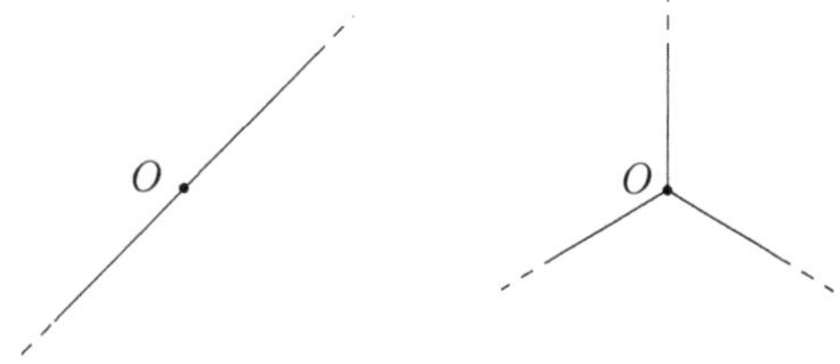

Fig. 4 A straight line and a standard triod are solutions of the motion by curvature

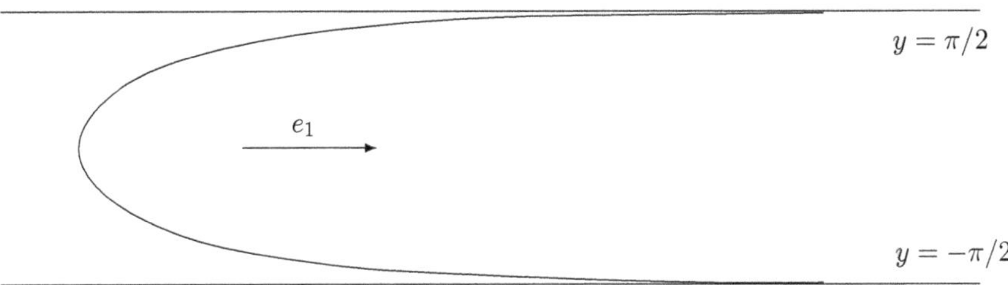

Fig. 5 The *grim reaper* relative to e_1

an entire class of solutions that preserve their shape in time. We try to classify now self-similar solution in a systematic way.

Let us start looking for self-similar **translating** solutions.

Suppose that we have a translating curve γ solving the motion by curvature with initial data σ. We can write $\gamma(t,x) = \eta(x) + w(t)$. The motion by curvature equation $k(t,x) = \langle \gamma_t(t,x), \nu(t,x)\rangle$ in this case reads as $k(x) = \big\langle w'(t), \nu(x)\big\rangle$. As a consequence $w(t)$ is constant, hence we are allowed to write $\gamma(t,x) = \eta(x) + t\boldsymbol{v}$ with $\boldsymbol{v} \in \mathbb{R}^2$, and we obtain

$$k(x) = \langle \boldsymbol{v}, \nu(x)\rangle \ .$$

The reverse is also true: if a curve γ satisfies $k(x) = \langle \boldsymbol{v}, \nu(x)\rangle$, then γ is a translating solution of the curvature flow. By integrating this ODE (with $\boldsymbol{v} = \boldsymbol{e}_1$) one can see that the only translating curve is given by the graph of the function $x = -\log\cos y$ in the interval $(-\frac{\pi}{2}, \frac{\pi}{2})$. Grayson in [19] named this curve the **grim reaper** (Fig. 5).

Passing from a single curve to a regular network, the situation becomes more delicate. Every curve of the translating network has to satisfies $k^i(x) = \big\langle \boldsymbol{v}, \nu^i(x)\big\rangle$. A result for translating triods can be found in [31, Lemma 5.8]: a closed, unbounded and embedded regular triod $\mathbb{R}^2$ self-translating with velocity $\boldsymbol{v} \neq 0$ is composed by halflines parallel to $\boldsymbol{v}$ or translated copies of pieces of the grim reaper relative to $\boldsymbol{v}$, meeting at the 3-point with angles of 120°. Notice that at most one curve is a halfline (Fig. 6).

Among curves there are also **rotating** solutions. Suppose indeed that γ is of the form $\gamma(t,x) = R(t)\eta(x)$ with $R(t)$ a rotation. The motion equation becomes

$$k(x) = \big\langle R'(t)\eta(x), R(t)\nu(x)\big\rangle = \big\langle R^t(t)R'(t)\eta(x), \nu(x)\big\rangle \ .$$

We get that $R^t(t)R'(t)$ is constant. By straightforward computations one also get that $R(t)$ is a anticlockwise rotation by ωt, where ω is a given constant. Then

$$k(x) = \omega\,\langle \eta(x), \tau(x)\rangle \ .$$

A fascinating example can be found in [2]: the Yin–Yang curve.

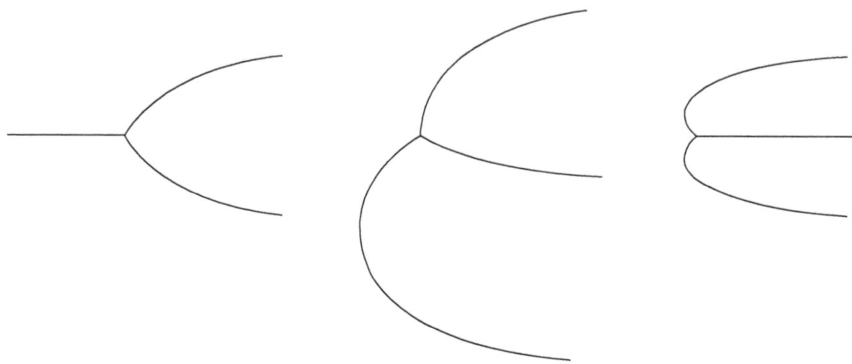

Fig. 6 Some examples of translating triods

We have left at last the more significant case: the **self-similarly shrinking networks**.

Suppose that a solution of the motion by curvature evolves homothetically shrinking in time with center of homothety the origin, namely $\gamma(t,x)=\alpha(t)\eta(x)$ with $\alpha(t)>0$ and $\alpha'(t)<0$. Being a solution of the flow the curve γ satisfies $k(t,x)=\langle\gamma_t(t,x),\nu(t,x)\rangle$. Then

$$k(x)=\alpha(t)\alpha'(t)\left\langle\eta(x),\nu(x)\right\rangle\,.$$

We have that $\alpha(t)\alpha'(t)$ is equal to some constant. Up to rescaling we can suppose $\alpha(t)\alpha'(t)=-1$. Then for every $t\in(-\infty,0]$ we have $\alpha(t)=2\sqrt{t-T}$ and $k(x)=-\left\langle\eta(x),\nu(x)\right\rangle$, or equivalently $\boldsymbol{k}(x)+\eta^{\perp}(x)=0$.

Definition 4.1 A regular C^2 open network $\mathcal{S}$ union of n curves parametrized by η^i is called a **regular shrinker** if for every curve there holds

$$\boldsymbol{k}^i+(\eta^i)^{\perp}=0\,.$$

Remark 4.2 Every curve of a regular shrinker satisfies the equation $\boldsymbol{k}+\eta^{\perp}=0$. As a consequence it must be a piece of a line though the origin or of the so called **Abresch–Langer curves**. Their classification results in [1] imply that any of these non straight pieces is compact. Hence any unbounded curve of a shrinker must be a line or an halfline pointing towards the origin. Moreover, it also follows that if a curve contains the origin, then it is a straight line through the origin or a halfline from the origin.

By the work of Abresch and Langer [1] it follows that the only regular shrinkers without triple junctions (curves) are the lines for the origin and the unit circle. There are two shrinkers with one triple junction [20]: the standard triod and the **Brakke spoon**. The Brakke spoon is a regular shrinker composed by a halfline which intersects a closed curve, forming angles of 120°. It was first mentioned in [10]

Fig. 7 A circle and a Brakke spoon. Together with a straight line and a standard triod, they are all possible regular shrinker with at most one triple junction

Fig. 8 The **standard lens** is a shrinker with two triple junctions symmetric with respect to two perpendicular axes, composed by two halflines pointing the origin, posed on a symmetry axis and opposite with respect to the other. Each halfline intersects two equal curves forming an angle of 120°. The **fish** is a shrinker with the same topology of the standard lens, but symmetric with respect to only one axis. The two halflines, pointing the origin, intersect two different curves, forming angles of 120°

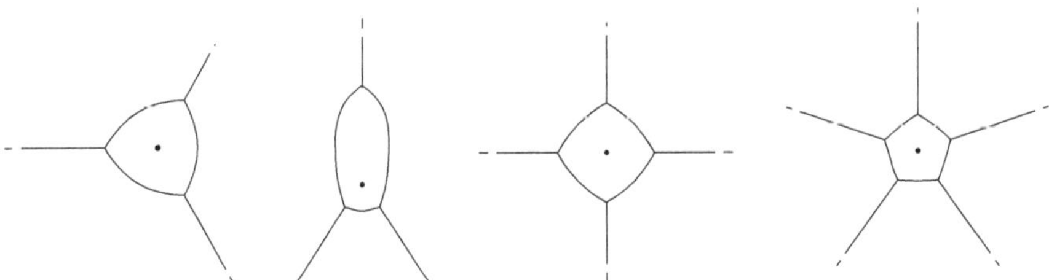

Fig. 9 The regular shrinkers with a single bounded region

as an example of evolving network with a loop shrinking down to a point, leaving a halfline that then, in the framework of Brakke flows, vanishes instantaneously. Up to rotation, this particular spoon-shaped network is unique [12] (Fig. 7).

Also the classification of shrinkers with two triple junctions is complete. It is not difficult to show [7, 8] that there are only two possible topological shapes for a complete embedded, regular shrinker: one is the "lens/fish" shape and the other is the shape of the Greek "Theta" letter (or "double cell"). It is well known that there exist unique (up to a rotation) lens-shaped or fish-shaped, embedded, regular shrinkers which are symmetric with respect to a line through the origin of $\mathbb{R}^2$ [12, 37] (Fig. 8). Instead there are no regular Θ-shaped shrinkers [6].

The classification of (embedded) regular shrinkers is completed for the shrinkers with a single bounded region [6, 12, 13, 37], see Fig. 9.

Several questions (also of independent interest) arise in trying to classify the regular shrinkers. We just mention an open question: does there exist a regular shrinker with more than five unbounded halflines?

Numerical computations, partial results and conjectures can be found in [20].

5 Integral Estimates

A good way to understand what happens during the evolution of a network by curvature is to describe the changing in time of the geometric quantities related to the network. For instance we can write the evolution law of the length of the curves or of area enclosed by the curves. In several situations estimating the evolution of the curvature has revealed a winning strategy to pass from short time to long time existence results.

Differently from the case of the curve shortening flow (and of the mean curvature flow) here to obtain our a priori estimates we cannot use the maximum principle and a comparison principle is not valid because of the presence of junctions. Therefore integral estimates are computed in [31, Section 3] in [32, Section 5] in the case of a triod and a regular network, respectively. An outline for the estimates appeared in [23, Section 7], where the authors consider directly the evolution $\gamma_t = k\nu + \lambda\tau$. We summarise here these calculations focusing on the easier cases.

Form now on we suppose that all the derivatives of the functions that appear exist.

We start showing that if a curve moves by curvature, then its time derivative ∂_t and the arclength derivative ∂_s do not commute.

We have already mentioned that the motion by curvature $\gamma_t^\perp = \boldsymbol{k}$ can be written as

$$\gamma_t = k\nu + \lambda\tau\,,$$

for some continuous function λ.

Lemma 5.1 *If γ is a curve moving by $\gamma_t = k\nu + \lambda\tau$, then we have the following commutation rule:*

$$\partial_t\partial_s = \partial_s\partial_t + (k^2 - \lambda_s)\partial_s\,. \tag{5.1}$$

Proof Let $f : [0, 1] \times [0, T) \to \mathbb{R}$ be a smooth function, then

$$\begin{aligned}\partial_t\partial_s f - \partial_s\partial_t f &= \frac{f_{tx}}{|\gamma_x|} - \frac{\langle \gamma_x \mid \gamma_{xt}\rangle f_x}{|\gamma_x|^3} - \frac{f_{tx}}{|\gamma_x|} = -\langle \tau \mid \partial_s\gamma_t\rangle\partial_s f\\ &= -\,\langle \tau \mid \partial_s(\lambda\tau + k\nu)\rangle\partial_s f = (k^2 - \lambda_s)\partial_s f\end{aligned}$$

and the formula is proved. □

In all this section we will consider a C^∞ solution of the special flow. Hence each curve is moving by

$$\gamma_t^i(t,x) = \frac{\gamma_{xx}^i(t,x)}{\left|\gamma_x^i(t,x)\right|^2},$$

and $\lambda = \frac{\langle \gamma_{xx} \mid \gamma_x \rangle}{|\gamma_x|^3}$.

Using the rule in the previous lemma we can compute

$$\begin{aligned}
\partial_t \tau &= \partial_t \partial_s \gamma = \partial_s \partial_t \gamma + (k^2 - \lambda_s)\partial_s \gamma = \partial_s(\lambda \tau + k\nu) + (k^2 - \lambda_s)\tau = (k_s + k\lambda)\nu\,,\\
\partial_t \nu &= \partial_t(\mathrm{R}\tau) = \mathrm{R}\,\partial_t \tau = -(k_s + k\lambda)\tau\,,\\
\partial_t k &= \partial_t \langle \partial_s \tau \mid \nu \rangle = \langle \partial_t \partial_s \tau \mid \nu \rangle = \langle \partial_s \partial_t \tau \mid \nu \rangle + (k^2 - \lambda_s)\langle \partial_s \tau \mid \nu \rangle\\
&= \partial_s \langle \partial_t \tau \mid \nu \rangle + k^3 - k\lambda_s = \partial_s(k_s + k\lambda) + k^3 - k\lambda_s\\
&= k_{ss} + k_s\lambda + k^3\,,\\
\partial_t \lambda &= -\,\partial_t \partial_x \frac{1}{|\gamma_x|} = \partial_x \frac{\langle \gamma_x \mid \gamma_{tx} \rangle}{|\gamma_x|^3} = \partial_x \frac{\langle \tau \mid \partial_s(\lambda\tau + k\nu) \rangle}{|\gamma_x|} = \partial_x \frac{(\lambda_s - k^2)}{|\gamma_x|}\\
&= \partial_s(\lambda_s - k^2) - \lambda(\lambda_s - k^2) = \lambda_{ss} - \lambda\lambda_s - 2kk_s + \lambda k^2\,.
\end{aligned}$$

5.1 Evolution of Length and Volume

We now compute the evolution in time of the total length.

By the commutation formula (5.1) the time derivative of the measure ds on any curve γ^i of the network is given by the measure $(\lambda_s^i - (k^i)^2)\,ds$. Then the evolution law for the length of one curve is

$$\frac{dL^i(t)}{dt} = \frac{d}{dt}\int_{\gamma^i(\cdot,t)} 1\,ds = \int_{\gamma^i(\cdot,t)} (\lambda_s^i - (k^i)^2)ds = \lambda^i(1,t) - \lambda^i(0,t) - \int_{\gamma^i(\cdot,t)} (k^i)^2 ds.$$

We remind that by relation (3.5) the contributions of λ^{pi} at every 3-point O^p vanish. Suppose that the network has l end-points on the boundary of Ω. With a little abuse of notation we call $\lambda(t, P^r)$ the tangential velocity at the end-point P^r for any $r \in \{1, 2, \ldots, l\}$. Since the total length is the sum of the lengths of all the curves, we get

$$\frac{dL(t)}{dt} = \sum_{r=1}^{l} \lambda(t, P^r) - \int_{\mathcal{N}_t} k^2\,ds\,,$$

In particular if the end-points P^r of the network are fixed during the evolution all the terms $\lambda(t, P^r)$ are zero and we have

$$\frac{dL(t)}{dt} = -\int_{\mathcal{N}_t} k^2\, ds\,.$$

The total length $L(t)$ is decreasing in time and uniformly bounded above by the length of the initial network.

We now discuss the behavior of the area of the regions enclosed by some curves of the evolving regular network. Let us suppose that a region $\mathcal{A}(t)$ is bounded by m curves $\gamma^1, \gamma^2, \dots, \gamma^m$ and let $A(t)$ be its area. We call **loop** ℓ the union of these m curves. The loop ℓ can be regarded as a single piecewise C^2 closed curve parametrized anticlockwise (possibly after reparametrization of the curves that composed it). Hence the curvature of ℓ is positive at the convexity points of the boundary of $\mathcal{A}(t)$. Then we have

$$\begin{aligned} A'(t) &= -\sum_{i=1}^{m}\int_{\gamma^i}\langle x_t \mid \nu\rangle\, ds = -\sum_{i=1}^{m}\int_{\gamma^i}\langle k\nu \mid \nu\rangle\, ds \\ &= -\sum_{i=1}^{m}\int_{\gamma^i} k\, ds = -\sum_{i=1}^{m}\Delta\theta_i\,, \end{aligned} \tag{5.2}$$

where $\Delta\theta_i$ is the difference in the angle between the unit tangent vector τ and the unit coordinate vector $e_1 \in \mathbb{R}^2$ at the final and initial point of the curve γ^i. Indeed supposing the unit tangent vector of the curve γ^i "lives" in the second quadrant of $\mathbb{R}^2$ (the other cases are analogous) there holds

$$\partial_s\theta_i = \partial_s \arccos\langle\tau \mid e_1\rangle = -\frac{\langle\tau_s \mid e_1\rangle}{\sqrt{1-\langle\tau \mid e_1\rangle^2}} = k\,,$$

so

$$A'(t) = -\sum_{i=1}^{m}\int_{\gamma^i}\partial_s\theta_i\, ds = -\sum_{i=1}^{m}\Delta\theta_i\,.$$

Considering that the curves γ^i form angles of $120°$, we have

$$m\pi/3 + \sum_{i=1}^{m}\Delta\theta_i = 2\pi\,.$$

We then obtain the equality (see [44])

$$A'(t) = -(2 - m/3)\pi\,. \tag{5.3}$$

An immediate consequence of (5.3) is that the area of every region bounded by the curves of the network evolves linearly. More precisely it increases if the region has more than six edges, it is constant with six edges and it decreases if its edges are less than six. This implies that if less than six curves of the initial network enclose a region of area A_0, then the maximal time T of existence of a smooth flow is finite and

$$T \leq \frac{A_0}{(2 - m/3)\pi} \leq \frac{3A_0}{\pi}.$$

5.2 *Evolution of the Curvature and Its Derivatives*

We want to estimate the L^2 norm of the curvature and its derivatives, that will result crucial in the analysis of the motion. The main consequence of these computation indeed is that the flow of a regular smooth network with "controlled" end-points exists smooth as long as the curvature stays bounded and none of the lengths of the curves goes to zero (Theorem 5.7).

We consider a regular C^∞ network $\mathcal{N}_t$ in Ω, composed by n curves γ^i with m triple-points $O^1, O^2, \dots, O^m$ and l end-points $P^1, P^2, \dots, P^l$. We suppose that it is a C^∞ solution of the system (3.4). We assume that either the end-points are fixed (the Dirichlet boundary condition in (3.4) is satisfied) or that there exist uniform (in time) constants C_j, for every $j \in \mathbb{N}$, such that

$$|\partial_s^j k(P^r, t)| + |\partial_s^j \lambda(t, P^r)| \leq C_j\,, \tag{5.4}$$

for every $t \in [0, T)$ and $r \in 1, 2, \dots, l$. This second possibility will allow us to localise the estimates if needed.

We are now ready to compute $\frac{d}{dt} \int_{\mathcal{N}_t} |k|^2\, ds$. We get

$$\frac{d}{dt} \int_{\mathcal{N}_t} |k|^2\, ds = 2 \int_{\mathcal{N}_t} k\, \partial_t k\, ds + \int_{\mathcal{N}_t} |k|^2 (\lambda_s - k^2)\, ds\,.$$

Using that $\partial_t k = k_{ss} + k_s \lambda + k^3$ we get

$$\frac{d}{dt} \int_{\mathcal{N}_t} |k|^2\, ds = \int_{\mathcal{N}_t} 2k\, k_{ss} + 2\lambda k\, k_s + k^2 \lambda_s + k^4\, ds = \int_{\mathcal{N}_t} 2k\, k_{ss} + \partial_s(\lambda\, k^2) + k^4\, ds.$$

Integrating by parts and estimating the contributions given by the end-points P^r by means of assumption (5.4) we can write

$$\begin{aligned} \frac{d}{dt} \int_{\mathcal{N}_t} |k|^2\, ds = &- 2 \int_{\mathcal{N}_t} |k_s|^2\, ds + \int_{\mathcal{N}_t} \partial_s(\lambda\, k^2)\, ds + \int_{\mathcal{N}_t} k^4\, ds \\ &- 2 \sum_{p=1}^{m} \sum_{i=1}^{3} k^{pi}\, k_s^{pi} \Big|_{\text{at the 3-point } O^p} + 2 \sum_{r=1}^{l} k^r\, k_s^r \Big|_{\text{at the end-point } P^r} \end{aligned}$$

$$\leq -2\int_{\mathcal{N}_t} |k_s|^2\, ds + \int_{\mathcal{N}_t} k^4\, ds + lC_0C_1$$

$$- \sum_{p=1}^{m}\sum_{i=1}^{3} 2k^{pi}k_s^{pi} + \lambda^{pi}|k^{pi}|^2 \Big|_{\text{at the 3-point } O^p}.$$

Then recalling relation (3.6) at the 3-points we have

$$\sum_{i=1}^{3} k^i k_s^i + \lambda^i |k^i|^2 = 0\,.$$

Substituting it above we lower the maximum order of the space derivatives of the curvature in the 3-point terms

$$\frac{d}{dt}\int_{\mathcal{N}_t} k^2 ds \leq -2\int_{\mathcal{N}_t} |k_s|^2\, ds + \int_{\mathcal{N}_t} k^4\, ds + \sum_{p=1}^{m}\sum_{i=1}^{3}\lambda^{pi}|k^{pi}|^2\Big|_{\text{at the 3-point } O^p} + lC_0C_1.$$

We notice that we can estimate the boundary terms at each 3-point of the form $\sum_{i=1}^{3}\lambda^i|k^i|^2$ by $\sum_{i=1}^{3}\lambda^i|k^i|^2 \leq \|k^3\|_{L^\infty}$ (see [31, Remark 3.9]). Hence

$$\frac{d}{dt}\int_{\mathcal{N}_t} k^2\, ds \leq -2\int_{\mathcal{N}_t} |k_s|^2\, ds + \int_{\mathcal{N}_t} k^4\, ds + \|k\|_{L^\infty}^3 + lC_0C_1\,. \tag{5.5}$$

From now on we do not use any geometric property of our problem. We suppose that the lengths of curves of the networks are equibounded from below by some positive value. We reduce to estimate the L^4 and L^∞ norm of the curvature of any curve γ^i, seen as a Sobolev function defined on the interval $[0, L(\gamma^i)]$.

Lemma 5.2 *Let* $0 < \mathrm{L} < +\infty$ *and* $u \in C^\infty([0,\mathrm{L}],\mathbb{R})$. *Then there exists a uniform constant* C, *depending on* L, *such that*

$$\|u\|_{L^4}^4 + \|u\|_{L^\infty}^3 - 2\|u'\|_{L^2}^2 \leq C\left(\|u\|_{L^2}^2 + 1\right)^3.$$

Proof The key estimates of the proof are Gagliardo–Nirenberg interpolation inequalities [34, Section 3, pp. 257–263] written in the form (see also [31, Proposition 3.11])

$$\|u\|_{L^p} \leq C_p\|u'\|_{L^2}^{\frac{1}{2}-\frac{1}{p}}\|u\|_{L^2}^{\frac{1}{2}+\frac{1}{p}} + \frac{B_p}{\mathrm{L}^{\frac{1}{2}-\frac{1}{p}}}\|u\|_{L^2}\,,$$

$$\|u\|_{L^\infty} \leq C\|u'\|_{L^2}^{\frac{1}{2}}\|u\|_{L^2}^{\frac{1}{2}} + \frac{B}{\mathrm{L}^{\frac{1}{2}}}\|u\|_{L^2}\,.$$

We first focus on the term $\|u\|^4_{L^4}$. We have

$$\|u\|_{L^4} \leq C\left(\|u'\|^{1/4}_{L^2}\|u\|^{3/4}_{L^2} + \frac{\|u\|_{L^2}}{\mathrm{L}^{\frac{1}{2}}}\right),$$

and so

$$\|u\|^4_{L^4} \leq \widetilde{C}\left(\|u'\|_{L^2}\|u\|^3_{L^2} + \|u\|^4_{L^2}\right).$$

Using Young inequality

$$\|u\|^4_{L^4} \leq \widetilde{C}\left(\varepsilon\|u'\|^2_{L^2} + c_\varepsilon\|u\|^6_{L^2} + \|u\|^4_{L^2}\right). \tag{5.6}$$

Similarly we estimate the term $\|u\|^3_{L^\infty}$ by

$$\begin{aligned}\|u\|^3_{L^\infty} &\leq C\left(\|u'\|^{\frac{3}{2}}_{L^2}\|u\|^{\frac{3}{2}}_{L^2} + \|u\|^3_{L^2}\right)\\ &\leq C\left(\varepsilon\|u'\|^2_{L^2} + c_\varepsilon\|u\|^6_{L^2} + \|u\|^3_{L^2}\right).\end{aligned} \tag{5.7}$$

Putting (5.6) and (5.7) together and choosing appropriately ε we obtain

$$\|u\|^4_{L^4} + \|u\|^3_{L^\infty} - 2\|u'\|^2_{L^2} \leq \tilde{c}_\varepsilon\|u\|^6_{L^2} + \widetilde{C}\|u\|^4_{L^2} + C\|u\|^3_{L^2} \leq C\left(\|u\|^2_{L^2} + 1\right)^3.$$

□

Applying Lemma 5.2 to the curvature k^i of each curve γ^i of the network the estimate (5.5) becomes

$$\left|\frac{d}{dt}\int_{\mathcal{N}_t} k^2\,ds\right| \leq C\left(\int_{\mathcal{N}_t} k^2\,ds\right)^3 + C + lC_0C_1. \tag{5.8}$$

The aim now is to repeat the previous computation for $\partial_s^j k$ with $j \in \mathbb{N}$.

Although the calculations are much harder, it is possible to conclude that for every even $j \in \mathbb{N}$ there holds

$$\int_{\mathcal{N}_t} |\partial_s^j k|^2\,ds \leq C\int_0^t\left(\int_{\mathcal{N}_\xi} k^2\,ds\right)^{2j+3} d\xi + C\left(\int_{\mathcal{N}_t} k^2\,ds\right)^{2j+1} + Ct + lC_jC_{j+1}t + C.$$

Passing from integral to L^∞ estimates we have the following proposition.

Proposition 5.3 *If assumption* (5.4) *holds, the lengths of all the curves are uniformly positively bounded from below and the L^2 norm of k is uniformly bounded on $[0, T)$, then the curvature of $\mathcal{N}_t$ and all its space derivatives are uniformly bounded in the same time interval by some constants depending only on the L^2 integrals of the space derivatives of k on the initial network $\mathcal{N}_0$.*

We now derive a second set of estimates where everything is controlled—still under the assumption (5.4)—only by the L^2 norm of the curvature and the inverses of the lengths of the curves at time zero.

As before we consider the C^∞ special curvature flow $\mathcal{N}_t$ of a smooth network $\mathcal{N}_0$ in the time interval $[0, T)$, composed by n curves $\gamma^i(\cdot, t) : [0, 1] \to \overline{\Omega}$ with m triple junctions $O^1, O^2, \ldots, O^m$ and l end-points $P^1, P^2, \ldots, P^l$, satisfying assumption (5.4).

As shown above, the evolution equations for the lengths of the n curves are given by

$$\frac{dL^i(t)}{dt} = \lambda^i(1, t) - \lambda^i(0, t) - \int_{\gamma^i(\cdot,t)} k^2\, ds\,.$$

Then, proceeding as in the computations above, we get

$$\begin{aligned}
&\frac{d}{dt}\left(\int_{\mathcal{N}_t} k^2\, ds + \sum_{i=1}^n \frac{1}{L^i}\right)\\
&\leq -2\int_{\mathcal{N}_t} k_s^2\, ds + \int_{\mathcal{N}_t} k^4\, ds + 6m\|k\|_{L^\infty}^3 + lC_0C_1 - \sum_{i=1}^n \frac{1}{(L^i)^2}\frac{dL^i}{dt}\\
&= -2\int_{\mathcal{N}_t} k_s^2\, ds + \int_{\mathcal{N}_t} k^4\, ds + 6m\|k\|_{L^\infty}^3 + lC_0C_1 - \sum_{i=1}^n \frac{\lambda^i(1,0) - \lambda^i(0,t) + \int_{\gamma^i(\cdot,t)} k^2\, ds}{(L^i)^2}\\
&\leq -2\int_{\mathcal{N}_t} k_s^2\, ds + \int_{\mathcal{N}_t} k^4\, ds + 6m\|k\|_{L^\infty}^3 + lC_0C_1 + 2\sum_{i=1}^n \frac{\|k\|_{L^\infty} + C_0}{(L^i)^2} + \sum_{i=1}^n \frac{\int_{\mathcal{N}_t} k^2\, ds}{(L^i)^2}\\
&\leq -2\int_{\mathcal{N}_t} k_s^2\, ds + \int_{\mathcal{N}_t} k^4\, ds + (6m + 2n/3)\|k\|_{L^\infty}^3 + lC_0C_1 + 2nC_0^3/3\\
&\quad + \frac{n}{3}\left(\int_{\mathcal{N}_t} k^2\, ds\right)^3 + \frac{2}{3}\sum_{i=1}^n \frac{1}{(L^i)^3},
\end{aligned}$$

where we used Young inequality in the last passage. Proceeding as before, but keeping track of the terms where the inverse of the length appear, it is possible to obtain

$$
\begin{aligned}
\frac{d}{dt}\left(\int_{\mathcal{N}_t} k^2\,ds + \sum_{i=1}^{n}\frac{1}{L^i}\right) &\leq -\int_{\mathcal{N}_t} k_s^2\,ds + C\left(\int_{\mathcal{N}_t} k^2\,ds\right)^3 + C\sum_{i=1}^{n}\frac{\left(\int_{\mathcal{N}_t} k^2\,ds\right)^2}{L^i} \\
&\quad + C\sum_{i=1}^{n}\frac{\left(\int_{\mathcal{N}_t} k^2\,ds\right)^{3/2}}{(L^i)^{3/2}} + C\sum_{i=1}^{n}\frac{1}{(L^i)^3} + C \\
&\leq C\left(\int_{\mathcal{N}_t} k^2\,ds\right)^3 + C\sum_{i=1}^{n}\frac{1}{(L^i)^3} + C \\
&\leq C\left(\int_{\mathcal{N}_t} k^2\,ds + \sum_{i=1}^{n}\frac{1}{L^i} + 1\right)^3, \qquad (5.9)
\end{aligned}
$$

with a constant C depending only on the structure of the network and on the constants C_0 and C_1 in assumption (5.4).

5.3 Consequences of the Estimates

Thanks to the just computed estimates on the curvature and on the inverse of the length one can obtain the following result:

Proposition 5.4 *For every $M > 0$ there exists a time $T_M \in (0, T)$, depending only on the* structure *of the network and on the constants C_0 and C_1 in assumption* (5.4), *such that if the square of the L^2 norm of the curvature and the inverses of the lengths of the curves of $\mathcal{N}_0$ are bounded by M, then the square of the L^2 norm of k and the inverses of the lengths of the curves of $\mathcal{N}_t$ are smaller than $2(n+1)M+1$, for every time $t \in [0, T_M]$.*

Proof Consider the positive function $f(t) = \int_{\mathcal{N}_t} k^2\,ds + \sum_{i=1}^{n}\frac{1}{L^i(t)} + 1$. Then by inequality (5.9) f satisfies the differential inequality $f' \leq Cf^3$. After integration it reads as

$$
f^2(t) \leq \frac{f^2(0)}{1 - 2Ctf^2(0)} \leq \frac{f^2(0)}{1 - 2Ct[(n+1)M+1]},
$$

then if $t \leq T_M = \frac{3}{8C[(n+1)M+1]}$ we get $f(t) \leq 2f(0)$. Hence

$$
\int_{\mathcal{N}_t} k^2\,ds + \sum_{i=1}^{n}\frac{1}{L^i(t)} \leq 2\int_{\mathcal{N}_0} k^2\,ds + 2\sum_{i=1}^{n}\frac{1}{L^i(0)} + 1 \leq 2[(n+1)M] + 1\,.
$$

□

The combination of these estimates implies estimates on all the derivatives of the maps γ^i, stated in the next proposition.

Proposition 5.5 *If $\mathcal{N}_t$ is a C^∞ special evolution of the initial network $\mathcal{N}_0 = \bigcup_{i=1}^n \sigma^i$, satisfying assumption* (5.4)*, such that the lengths of the n curves are uniformly bounded away from zero and the L^2 norm of the curvature is uniformly bounded by some constants in the time interval* $[0, T)$*, then*

- *all the derivatives in space and time of k and λ are uniformly bounded in* $[0, 1] \times [0, T)$,
- *all the derivatives in space and time of the curves $\gamma^i(t, x)$ are uniformly bounded in* $[0, 1] \times [0, T)$,
- *the quantities $|\gamma_x^i(t, x)|$ are uniformly bounded from above and away from zero in* $[0, 1] \times [0, T)$.

All the bounds depend only on the uniform controls on the L^2 norm of k, on the lengths of the curves of the network from below, on the constants C_j in assumption (5.4)*, on the L^∞ norms of the derivatives of the curves σ^i and on the bound from above and below on $|\sigma_x^i(t, x)|$, for the curves describing the initial network $\mathcal{N}_0$.*

By means of Proposition 5.4 we can strengthen the conclusion of Proposition 5.5.

Corollary 5.6 *In the hypothesis of the previous proposition, in the time interval $[0, T_M]$ all the bounds in Proposition 5.5 depend only on the L^2 norm of k on $\mathcal{N}_0$, on the constants C_j in assumption* (5.4)*, on the L^∞ norms of the derivatives of the curves σ^i, on the bound from above and below on $|\sigma_x^i(t, x)|$ and on the lengths of the curves of the initial network $\mathcal{N}_0$.*

By means of the a priori estimates we can work out some results about the smooth flow of an initial regular geometrically smooth network $\mathcal{N}_0$.

Theorem 5.7 *If $[0, T)$, with $T < +\infty$, is the maximal time interval of existence of a C^∞ curvature flow of an initial geometrically smooth network $\mathcal{N}_0$, then*

1. *either the inferior limit of the length of at least one curve of $\mathcal{N}_t$ is zero, as $t \to T$,*
2. *or $\overline{\lim}_{t\to T} \int_{\mathcal{N}_t} k^2\, ds = +\infty$.*

Proof We can C^∞ reparametrize the flow $\mathcal{N}_t$ in order that it becomes a special smooth flow $\widetilde{\mathcal{N}}_t$ in $[0, T)$. If the lengths of the curves of $\mathcal{N}_t$ are uniformly bounded away from zero and the L^2 norm of k is bounded, the same holds for the networks $\widetilde{\mathcal{N}}_t$. Then, by Proposition 5.5 and Ascoli–Arzelà Theorem, the network $\widetilde{\mathcal{N}}_t$ converges in C^∞ to a smooth network $\widetilde{\mathcal{N}}_T$ as $t \to T$. We could hence restart the flow obtaining a C^∞ special curvature flow in a longer time interval. Reparametrizing back this last flow, we get a C^∞ "extension" in time of the flow $\mathcal{N}_t$, hence contradicting the maximality of the interval $[0, T)$. □

Proposition 5.8 *If $[0, T)$, with $T < +\infty$, is the maximal time interval of existence of a C^∞ curvature flow of an initial geometrically smooth network $\mathcal{N}_0$. If the lengths of the n curves are uniformly positively bounded from below, then this superior limit* *is actually a limit and there exists a positive constant C such that*

$$\int_{\mathcal{N}_t} k^2\, ds \geq \frac{C}{\sqrt{T-t}}\,,$$

for every $t \in [0, T)$.

Proof Considering the flow $\widetilde{\mathcal{N}}_t$ introduced in the previous theorem. By means of differential inequality (5.8), we have

$$\frac{d}{dt}\int_{\widetilde{\mathcal{N}}_t} \widetilde{k}^2\, ds \leq C\left(\int_{\widetilde{\mathcal{N}}_t} \widetilde{k}^2\, ds\right)^3 + C \leq C\left(1+\int_{\widetilde{\mathcal{N}}_t} \widetilde{k}^2\, ds\right)^3,$$

which, after integration between $t, r \in [0, T)$ with $t < r$, gives

$$\frac{1}{\left(1+\int_{\widetilde{\mathcal{N}}_t} \widetilde{k}^2\, ds\right)^2} - \frac{1}{\left(1+\int_{\widetilde{\mathcal{N}}_r} \widetilde{k}^2\, ds\right)^2} \leq C(r-t)\,.$$

Then, if case (1) does not hold, we can choose a sequence of times $r_j \to T$ such that $\int_{\widetilde{\mathcal{N}}_{r_j}} \widetilde{k}^2\, ds \to +\infty$. Putting $r = r_j$ in the inequality above and passing to the limit, as $j \to \infty$, we get

$$\frac{1}{\left(1+\int_{\widetilde{\mathcal{N}}_t} \widetilde{k}^2\, ds\right)^2} \leq C(T-t)\,,$$

hence, for every $t \in [0, T)$,

$$\int_{\widetilde{\mathcal{N}}_t} \widetilde{k}^2\, ds \geq \frac{C}{\sqrt{T-t}} - 1 \geq \frac{C}{\sqrt{T-t}}\,,$$

for some positive constant C and $\lim_{t\to T} \int_{\widetilde{\mathcal{N}}_t} k^2\, ds = +\infty$.

By the invariance of the curvature by reparametrization, this last estimate implies the same estimate for the flow $\mathcal{N}_t$. □

This theorem obviously implies the following corollary.

Corollary 5.9 *If* $[0, T)$, *with* $T < +\infty$, *is the maximal time interval of existence of a* C^∞ *curvature flow of an initial geometrically smooth network* $\mathcal{N}_0$ *and the lengths of the curves are uniformly bounded away from zero, then*

$$\max_{\mathcal{N}_t} k^2 \geq \frac{C}{\sqrt{T-t}} \to +\infty\,, \tag{5.10}$$

as $t \to T$.

In the case of the evolution γ_t of a single closed curve in the plane there exists a constant $C > 0$ such that if at time $T > 0$ a singularity develops, then

$$\max_{\gamma_t} k^2 \geq \frac{C}{T-t}$$

for every $t \in [0, T)$ (see [21]). It is unknown if this lower bound on the rate of blow-up of the curvature holds also in the case of the evolution of a network.

Remark 5.10 Using more refine estimates it is possible to weaken the assumption of Theorem 5.7: one can suppose to have a $C^{\frac{2+\alpha}{2},2+\alpha}$ curvature flow (see [32, p. 33]).

We conclude this section with the following estimate from below on the maximal time of smooth existence.

Proposition 5.11 *For every $M > 0$ there exists a positive time T_M such that if the L^2 norm of the curvature and the inverses of the lengths of the geometrically smooth network $\mathcal{N}_0$ are bounded by M, then the maximal time of existence $T > 0$ of a C^∞ curvature flow of $\mathcal{N}_0$ is larger than T_M.*

Proof As before, considering again the reparametrized special curvature flow $\widetilde{\mathcal{N}}_t$, by Proposition 5.4 in the interval $[0, \min\{T_M, T\})$ the L^2 norm of $\widetilde{k}$ and the inverses of the lengths of the curves of $\widetilde{\mathcal{N}}_t$ are bounded by $2M^2 + 6M$.

Then, by Theorem 5.7, the value $\min\{T_M, T\}$ cannot coincide with the maximal time of existence of $\widetilde{\mathcal{N}}_t$ (hence of $\mathcal{N}_t$), so it must be $T > T_M$. □

6 Analysis of Singularities

6.1 Huisken's Monotonicity Formula

We shall use the following notation for the evolution of a network in $\Omega \subset \mathbb{R}^2$: let $\mathcal{N} \subset \mathbb{R}^2$ be a network homeomorphic to the all $\mathcal{N}_t$, we consider a map

$$F : (0, T) \times \mathcal{N} \to \mathbb{R}^2$$

given by the union of the maps $\gamma^i : (0, T) \times I_i \to \overline{\Omega}$ (with I_i the intervals $[0, 1]$, $(0, 1]$, $[1, 0)$ or $(0, 1)$) describing the curvature flow of the network in the time interval $(0, T)$, that is $\mathcal{N}_t = F(t, \mathcal{N})$.

Let us start from the easiest case in which the network is composed by a unique closed simple smooth curve. Let $t_0 \in (0, +\infty)$, $x_0 \in \mathbb{R}^2$ and $\rho_{t_0,x_0} : [0, t_0) \times \mathbb{R}^2$ be the one-dimensional **backward heat kernel** in $\mathbb{R}^2$ relative to (t_0, x_0), that is

$$\rho_{t_0,x_0}(t, x) = \frac{e^{-\frac{|x-x_0|^2}{4(t_0-t)}}}{\sqrt{4\pi(t_0-t)}}.$$

Theorem 6.1 (Monotonicity Formula) *Assume* $t_0 > 0$. *For every* $t \in [0, \min\{t_0, T\})$ *and* $x_0 \in \mathbb{R}^2$ *we have*

$$\frac{d}{dt}\int_{\mathcal{N}_t} \rho_{t_0,x_0}(t,x)\,ds = -\int_{\mathcal{N}_t} \left| \boldsymbol{k} + \frac{(x-x_0)^\perp}{2(t_0-t)} \right|^2 \rho_{t_0,x_0}(t,x)\,ds$$

Proof See [21]. □

Then one can wonder if a modified version of this formula holds for networks. Clearly one needs a way to deal with the boundary points (the triple junctions). In [31] the authors gave a positive answer to this question in the case of a triod. With a slight modification of the computation in [31, Lemma 6.3] one can extend the result to any regular network. As before, with a little abuse of notation, we will write $\tau(t, P^r)$ and $\lambda(t, P^r)$ respectively for the unit tangent vector and the tangential velocity at the end-point P^r of the curve of the network getting at such point, for any $r \in \{1, 2, \dots, l\}$.

Proposition 6.2 (Monotonicity Formula) *Assume* $t_0 > 0$. *For every* $t \in [0, \min\{t_0, T\})$ *and* $x_0 \in \mathbb{R}^2$ *the following identity holds*

$$\begin{aligned}\frac{d}{dt}\int_{\mathcal{N}_t} \rho_{t_0,x_0}(t,x)\,ds = &-\int_{\mathcal{N}_t} \left| \boldsymbol{k} + \frac{(x-x_0)^\perp}{2(t_0 \quad t)} \right|^2 \rho_{t_0,x_0}(t,x)\,ds \\ &+ \sum_{r=1}^{l} \left[\left\langle \frac{P^r - x_0}{2(t_0-t)} \,\middle|\, \tau(t,P^r) \right\rangle - \lambda(t,P^r) \right] \rho_{t_0,x_0}(t,P^r)\,.\end{aligned}$$

Integrating between t_1 and t_2 with $0 \le t_1 \le t_2 < \min\{t_0, T\}$ we get

$$\begin{aligned}\int_{t_1}^{t_2}\int_{\mathcal{N}_t} \left| \boldsymbol{k} + \frac{(x-x_0)^\perp}{2(t_0-t)} \right|^2 \rho_{t_0,x_0}(t,x)\,ds\,dt = &\int_{\mathcal{N}_{t_1}} \rho_{t_0,x_0}(x,t_1)\,ds - \int_{\mathcal{N}_{t_2}} \rho_{t_0,x_0}(x,t_2)\,ds \\ &+ \sum_{r=1}^{l} \int_{t_1}^{t_2} \left[\left\langle \frac{P^r - x_0}{2(t_0-t)} \,\middle|\, \tau(t,P^r) \right\rangle - \lambda(t,P^r) \right] \rho_{t_0,x_0}(t,P^r)\,dt\,.\end{aligned}$$

We need the following lemma in order to estimate the end-points contribution (see [31, Lemma 6.5]).

Lemma 6.3 *For every* $r \in \{1, 2, \dots, l\}$ *and* $x_0 \in \mathbb{R}^2$, *the following estimate holds*

$$\left| \int_t^{t_0} \left[\left\langle \frac{P^r - x_0}{2(t_0-\xi)} \,\middle|\, \tau(\xi,P^r) \right\rangle - \lambda(\xi,P^r) \right] \rho_{t_0,x_0}(\xi,P^r)\,d\xi \right| \le C\,,$$

where C *is a constant depending only on the constants* C_l *in assumption* (5.4).

Then for every point $x_0 \in \mathbb{R}^2$, we have

$$\lim_{t\to t_0} \sum_{r=1}^{l} \int_t^{t_0} \left[\left\langle \frac{P^r - x_0}{2(t_0-\xi)} \,\middle|\, \tau(\xi, P^r) \right\rangle - \lambda(\xi, P^r) \right] \rho_{t_0,x_0}(\xi, P^r)\, d\xi = 0\,.$$

As a consequence, the following definition is well posed.

Definition 6.4 (Gaussian Densities) For every $t_0 \in (0, +\infty)$, $x_0 \in \mathbb{R}^2$ we define the **Gaussian density** function $\Theta_{t_0,x_0} : [0, \min\{t_0, T\}) \to \mathbb{R}$ as

$$\Theta_{t_0,x_0}(t) = \int_{\mathcal{N}_t} \rho_{t_0,x_0}(t, \cdot)\, ds$$

and provided $t_0 \leq T$ the **limit Gaussian density** function $\widehat{\Theta} : (0, +\infty) \times \mathbb{R}^2 \to \mathbb{R}$ as

$$\widehat{\Theta}(t_0, x_0) = \lim_{t\to t_0} \Theta_{t_0,x_0}(t)\,.$$

For every $(t_0, x_0) \in (0, T] \times \mathbb{R}^2$, the limit $\widehat{\Theta}(t_0, x_0)$ exists (by the monotonicity of Θ_{t_0,x_0}) it is finite and non negative. Moreover the map $\widehat{\Theta} : \mathbb{R}^2 \to \mathbb{R}$ is upper semicontinuous [29, Proposition 2.12].

6.2 Dynamical Rescaling

We introduce the rescaling procedure of Huisken in [21] at the maximal time T.

Fixed $x_0 \in \mathbb{R}^2$, let $\widetilde{F}_{x_0} : [-1/2 \log T, +\infty) \times \mathcal{N} \to \mathbb{R}^2$ be the map

$$\widetilde{F}_{x_0}(\mathfrak{t}, p) = \frac{F(t, p) - x_0}{\sqrt{2(T-t)}} \qquad \mathfrak{t}(t) = -\frac{1}{2}\log(T-t)$$

then, the rescaled networks are given by

$$\widetilde{\mathcal{N}}_{\mathfrak{t},x_0} = \frac{\mathcal{N}_t - x_0}{\sqrt{2(T-t)}} \tag{6.1}$$

and they evolve according to the equation

$$\frac{\partial}{\partial \mathfrak{t}} \widetilde{F}_{x_0}(\mathfrak{t}, p) = \widetilde{\boldsymbol{v}}(\mathfrak{t}, p) + \widetilde{F}_{x_0}(\mathfrak{t}, p)$$

where

$$\widetilde{\boldsymbol{v}}(\mathfrak{t}, p) = \sqrt{2(T - t(\mathfrak{t}))} \cdot \boldsymbol{v}(t(\mathfrak{t}), p) = \widetilde{k}\nu + \widetilde{\lambda}\tau \qquad \text{and} \qquad t(\mathfrak{t}) = T - e^{-2\mathfrak{t}}\,.$$

Notice that we did not put the sign "$\sim$" over the unit tangent and normal, since they remain the same after the rescaling.

When there is no ambiguity on the point x_0, we will write $\widetilde{P}^r(\mathfrak{t}) = \widetilde{F}_{x_0}(\mathfrak{t}, P^r)$ for the end-points of the rescaled network $\widetilde{\mathcal{N}}_{\mathfrak{t},x_0}$.

The rescaled curvature evolves according to the following equation,

$$\partial_{\mathfrak{t}}\widetilde{k} = \widetilde{k}_{\mathfrak{s}\mathfrak{s}} + \widetilde{k}_{\mathfrak{s}}\widetilde{\lambda} + \widetilde{k}^3 - \widetilde{k}$$

which can be obtained by means of the commutation law

$$\partial_{\mathfrak{t}}\partial_{\mathfrak{s}} = \partial_{\mathfrak{s}}\partial_{\mathfrak{t}} + (\widetilde{k}^2 - \widetilde{\lambda}_{\mathfrak{s}} - 1)\partial_{\mathfrak{s}} ,$$

where we denoted with $\mathfrak{s}$ the arclength parameter for $\widetilde{\mathcal{N}}_{\mathfrak{t},x_0}$.

By straightforward computations (see [21]) we have the following rescaled version of the Monotonicity Formula.

Proposition 6.5 (Rescaled Monotonicity Formula) *Let $x_0 \in \mathbb{R}^2$ and set*

$$\widetilde{\rho}(x) = e^{-\frac{|x|^2}{2}}$$

For every $\mathfrak{t} \in [-1/2 \log T, +\infty)$ the following identity holds

$$\frac{d}{d\mathfrak{t}} \int_{\widetilde{\mathcal{N}}_{\mathfrak{t},x_0}} \widetilde{\rho}(x)\, d\mathfrak{s} = -\int_{\widetilde{\mathcal{N}}_{\mathfrak{t},x_0}} |\widetilde{\boldsymbol{k}} + x^{\perp}|^2 \widetilde{\rho}(x)\, d\mathfrak{s} + \sum_{r=1}^{l} \Big[\Big\langle \widetilde{P}^r(\mathfrak{t}) \,\Big|\, \tau(t(\mathfrak{t}), P^r)\Big\rangle$$
$$-\widetilde{\lambda}(\mathfrak{t}, P^r)\Big]\, \widetilde{\rho}(\widetilde{P}^r(\mathfrak{t}))$$

where $\widetilde{P}^r(\mathfrak{t}) = \frac{P^r - x_0}{\sqrt{2(T - t(\mathfrak{t}))}}$.

Integrating between $\mathfrak{t}_1$ and $\mathfrak{t}_2$ with $-1/2 \log T \leq \mathfrak{t}_1 \leq \mathfrak{t}_2 < +\infty$ we get

$$\int_{\mathfrak{t}_1}^{\mathfrak{t}_2} \int_{\widetilde{\mathcal{N}}_{\mathfrak{t},x_0}} |\widetilde{\boldsymbol{k}} + x^{\perp}|^2 \widetilde{\rho}(x)\, d\mathfrak{s}\, d\mathfrak{t} = \int_{\widetilde{\mathcal{N}}_{\mathfrak{t}_1,x_0}} \widetilde{\rho}(x)\, d\mathfrak{s} - \int_{\widetilde{\mathcal{N}}_{\mathfrak{t}_2,x_0}} \widetilde{\rho}(x)\, d\mathfrak{s} \tag{6.2}$$
$$+ \sum_{r=1}^{l} \int_{\mathfrak{t}_1}^{\mathfrak{t}_2} \Big[\Big\langle \widetilde{P}^r(\mathfrak{t}) \,\Big|\, \tau(t(\mathfrak{t}), P^r)\Big\rangle - \widetilde{\lambda}(\mathfrak{t}, P^r)\Big]\, \widetilde{\rho}(\widetilde{P}^r(\mathfrak{t}))\, d\mathfrak{t}.$$

We have also the analog of Lemma 6.3 (see [31, Lemma 6.7]).

Lemma 6.6 *For every $r \in \{1, 2, \ldots, l\}$ and $x_0 \in \mathbb{R}^2$, the following estimate holds for all $\mathfrak{t} \in \big[-\frac{1}{2}\log T, +\infty\big)$,*

$$\left| \int_{\mathfrak{t}}^{+\infty} \Big[\Big\langle \widetilde{P}^r(\xi) \,\Big|\, \tau(t(\xi), P^r)\Big\rangle - \widetilde{\lambda}(\xi, P^r)\Big]\, d\xi \right| \leq C ,$$

where C is a constant depending only on the constants C_l in assumption (5.4).

As a consequence, for every point $x_0 \in \mathbb{R}^2$, we have

$$\lim_{\mathrm{t}\to+\infty} \sum_{r=1}^{l} \int_{\mathrm{t}}^{+\infty} \Big[\Big\langle \widetilde{P}^r(\xi) \,\Big|\, \tau(t(\xi), P^r)\Big\rangle - \widetilde{\lambda}(\xi, P^r)\Big]\, d\xi = 0\,.$$

6.3 Blow-Up Limits

We now discuss the possible blow-up limits of an evolving network at the maximal time of existence. This analysis can be seen as a tool to exclude the possible arising of singularity in the evolution and to obtain (if possible) global existence of the flow.

Thanks to Theorem 5.7 we know what happens when the evolution approaches the singular time T: either the length of at least one curve of the network goes to zero, or the L^2-norm of the curvature blows-up. When the curvature does not remain bounded, we look at the possible limit networks after (Huisken's dynamical) rescaling procedure. The rescaled Monotonicity Formula 6.5 will play a crucial role. We first suppose that the length of all the curves of the network remains strictly positive during the evolution. In this case the classification of the limits is complete (Proposition 6.8). Without a bound from below on the length of the curves the situation is more involved, and we will see that in general the limit sets are no longer regular networks. For this purpose, we shall introduce the notion of degenerate regular network.

We now describe the blow-up limit of networks under the assumption that the length of each curve is bounded from below by a positive constant independent of time. We start with a lemma due to A. Stone [42].

Lemma 6.7 *Let $\widetilde{\mathcal{N}}_{\mathrm{t},x_0}$ be the family of rescaled networks obtained via Huisken's dynamical procedure around some $x_0 \in \mathbb{R}^2$ as defined in formula* (6.1)*.*

1. *There exists a constant $C = C(\mathcal{N}_0)$ such that, for every $\overline{x} \in \mathbb{R}^2$, $\mathrm{t} \in \big[-\frac{1}{2}\log T, +\infty\big)$ and $R > 0$ there holds*

$$\mathcal{H}^1(\widetilde{\mathcal{N}}_{\mathrm{t},x_0} \cap B_R(\overline{x})) \le CR\,.$$

2. *For any $\varepsilon > 0$ there is a uniform radius $R = R(\varepsilon)$ such that*

$$\int_{\widetilde{\mathcal{N}}_{\mathrm{t},x_0} \setminus B_R(\overline{x})} e^{-|x|^2/2}\, ds \le \varepsilon\,,$$

that is, the family of measures $e^{-|x|^2/2}\mathcal{H}^1 \llcorner \widetilde{\mathcal{N}}_{\mathrm{t},x_0}$ is tight *(see [14]).*

Proposition 6.8 *Let $\mathcal{N}_t = \bigcup_{i=1}^n \gamma^i(t, [0,1])$ be a $C^{1,2}$ curvature flow of regular networks with fixed end-points in a smooth, strictly convex, bounded open set $\Omega \subset \mathbb{R}^2$ in the time interval $[0, T)$. Assume that the lengths $L^i(t)$ of the curves of the* *networks are uniformly in time bounded away from zero for every* $i \in \{1, 2, \dots, n\}$. *Then for every* $x_0 \in \mathbb{R}^2$ *and for every subset* $\mathcal{I}$ *of* $[-1/2 \log T, +\infty)$ *with infinite Lebesgue measure, there exists a sequence of rescaled times* $\mathfrak{t}_j \to +\infty$, *with* $\mathfrak{t}_j \in \mathcal{I}$, *such that the sequence of rescaled networks* $\widetilde{\mathcal{N}}_{\mathfrak{t}_j, x_0}$ *(obtained via Huisken's dynamical procedure) converges in* $C^{1,\alpha}_{\mathrm{loc}} \cap W^{2,2}_{\mathrm{loc}}$, *for any* $\alpha \in (0, 1/2)$, *to a (possibly empty) limit, which is (if non-empty)*

- *a straight line through the origin with multiplicity* $m \in \mathbb{N}$ *(in this case* $\widehat{\Theta}(x_0) = m$*);*
- *a standard triod centered at the origin with multiplicity* 1 *(in this case* $\widehat{\Theta}(x_0) = 3/2$*).*
- *a halfline from the origin with multiplicity* 1 *(in this case* $\widehat{\Theta}(x_0) = 1/2$*).*

Moreover the L^2*-norm of the curvature of* $\widetilde{\mathcal{N}}_{\mathfrak{t}_j, x_0}$ *goes to zero in every ball* $B_R \subset \mathbb{R}^2$, *as* $j \to \infty$.

Proof We divide the proof into three steps. We take for simplicity $x_0 = 0$.

Step 1: Convergence to $\widetilde{\mathcal{N}}_\infty$

Consider the rescaled Monotonicity Formula (6.2) and let $\mathfrak{t}_1 = -1/2 \log T$ and $\mathfrak{t}_2 \to +\infty$. Then thanks to Lemma 6.6 we get

$$\int_{-1/2 \log T}^{+\infty} \int_{\widetilde{\mathcal{N}}_{\mathfrak{t}, x_0}} |\widetilde{\boldsymbol{k}} + x^\perp|^2 \widetilde{\rho} \, d\sigma \, d\mathfrak{t} < +\infty \,,$$

which implies

$$\int_{\mathcal{I}} \int_{\widetilde{\mathcal{N}}_{\mathfrak{t}, x_0}} |\widetilde{\boldsymbol{k}} + x^\perp|^2 \widetilde{\rho} \, d\sigma \, d\mathfrak{t} < +\infty \,.$$

Being the last integral finite and being the integrand a non negative function on a set of infinite Lebesgue measure, we can extract within $\mathcal{I}$ a sequence of times $\mathfrak{t}_j \to +\infty$, such that

$$\lim_{j \to +\infty} \int_{\widetilde{\mathcal{N}}_{\mathfrak{t}_j, x_0}} |\widetilde{\boldsymbol{k}} + x^\perp|^2 \widetilde{\rho} \, d\sigma = 0 \,. \tag{6.3}$$

It follows that for every ball B_R of radius R in $\mathbb{R}^2$ the networks $\widetilde{\mathcal{N}}_{\mathfrak{t}_j, x_0}$ have curvature uniformly bounded in $L^2(B_R)$. Moreover, by Lemma 6.7, for every ball B_R centered at the origin of $\mathbb{R}^2$ we have the uniform bound $\mathcal{H}^1(\widetilde{\mathcal{N}}_{\mathfrak{t}_j, x_0} \cap B_R) \leq CR$, for some constant C independent of $j \in \mathbb{N}$. Then reparametrizing the rescaled networks by arclength, we obtain curves with uniformly bounded first derivatives and with second derivatives uniformly bounded in L^2_{loc}.

By a standard compactness argument (see [21, 26]), the sequence $\widetilde{\mathcal{N}}_{\mathfrak{t}_j,x_0}$ of reparametrized networks admits a subsequence $\widetilde{\mathcal{N}}_{\mathfrak{t}_{j_l},x_0}$ which converges, weakly in $W^{2,2}_{\mathrm{loc}}$ and strongly in $C^{1,\alpha}_{\mathrm{loc}}$, to a (possibly empty) limit $\widetilde{\mathcal{N}}_\infty$ (possibly with multiplicity). The strong convergence in $W^{2,2}_{\mathrm{loc}}$ is implied by the weak convergence in $W^{2,2}_{\mathrm{loc}}$ and Eq. (6.3).

Step 2: The Limit $\widetilde{\mathcal{N}}_\infty$ is a Regular Shrinker
We first notice that the bound from below on the lengths prevents any "collapsing" along the rescaled sequence. Since the integral functional

$$\widetilde{\mathcal{N}} \mapsto \int_{\widetilde{\mathcal{N}}} |\widetilde{\boldsymbol{k}} + x^\perp|^2 \widetilde{\rho}\, d\sigma$$

is lower semicontinuous with respect to this convergence (see [38], for instance), the limit $\widetilde{\mathcal{N}}_\infty$ satisfies $\widetilde{\boldsymbol{k}}_\infty + x^\perp = 0$ in the sense of distributions.

A priori, the limit network is composed by curves in $W^{2,2}_{\mathrm{loc}}$, but from the relation $\widetilde{\boldsymbol{k}}_\infty + x^\perp = 0$, it follows that the curvature $\widetilde{\boldsymbol{k}}_\infty$ is continuous. By a bootstrap argument, it is then easy to see that $\widetilde{\mathcal{N}}_\infty$ is actually composed by C^∞ curves.

Step 3: Classification of the Possible Limits
If the point $x_0 \in \mathbb{R}^2$ is distinct from all the end-points P^r, then $\widetilde{\mathcal{N}}_\infty$ has no end-points, since they go to infinity along the rescaled sequence. If $x_0 = P^r$ for some r, the set $\widetilde{\mathcal{N}}_\infty$ has a single end-point at the origin of $\mathbb{R}^2$.

Moreover, from the lower bound on the length of the original curves it follows that all the curves of $\widetilde{\mathcal{N}}_\infty$ have infinite length, hence, by Remark 4.2, they must be pieces of straight lines from the origin.

This implies that every connected component of the graph underlying $\widetilde{\mathcal{N}}_\infty$ can contain at most one 3-point and in such case such component must be a standard triod (the 120° condition must be satisfied) with multiplicity one since the converging networks are all embedded (to get in the C^1_{loc}-limit a triod with multiplicity higher than one it is necessary that the approximating networks have self-intersections). Moreover, since the converging networks are embedded, if both a triod and a straight line or another triod are present, they would intersect transversally. Hence if a standard toroid is present, a straight line cannot be present and conversely if a straight line is present, a triod cannot be present.

If no end-point is present, that is, we are rescaling around a point in Ω (not on its boundary), and no 3-point is present, the only possibility is a straight line (possibly with multiplicity) through the origin.

If an end-point is present, we are rescaling around an end-point of the evolving network, hence, by the convexity of Ω (which contains all the networks) the limit $\widetilde{\mathcal{N}}_\infty$ must be contained in a halfplane with boundary a straight line H for the origin. This exclude the presence of a standard triod since it cannot be contained in any halfplane. Another halfline is obviously excluded, since they "come" only from end-points and they are all distinct. In order to exclude the presence of a straight line, we observe that the argument of Proposition 3.13 implies that, if $\Omega_t \subset \Omega$ is the evolution by curvature of $\partial\Omega$ keeping fixed the end-points P^r, the blow-up of Ω_t at an end-point must be a cone spanning angle strictly less then π (here we use the fact that three end-points are not aligned) and $\widetilde{\mathcal{N}}_\infty$ is contained in such a cone. It follows that $\widetilde{\mathcal{N}}_\infty$ cannot contain a straight line.

In every case the curvature of $\widetilde{\mathcal{N}}_\infty$ is zero everywhere and the last statement follows by the $W^{2,2}_{\mathrm{loc}}$-convergence.

□

Remark 6.9 In the previous proposition the hypothesis on the length of the curve can be replace by the weaker assumption that the lengths $L^i(t)$ of the curves satisfy

$$\lim_{t\to T} \frac{L^i(t)}{\sqrt{T-t}} = +\infty\,,$$

for every $i \in \{1, 2, \ldots, n\}$.

Lemma 6.10 *Under the assumptions of Proposition 6.8, there holds*

$$\lim_{j\to\infty} \frac{1}{\sqrt{2\pi}} \int_{\widetilde{\mathcal{N}}_{\mathfrak{t}_j,x_0}} \widetilde{\rho}\, d\sigma = \frac{1}{\sqrt{2\pi}} \int_{\widetilde{\mathcal{N}}_\infty} \widetilde{\rho}\, d\overline{\sigma} = \Theta_{\widetilde{\mathcal{N}}_\infty} = \widehat{\Theta}(T, x_0)\,, \tag{6.4}$$

where $d\overline{\sigma}$ denotes the integration with respect to the canonical measure on $\widetilde{\mathcal{N}}_\infty$, counting multiplicities.

Proof By means of the second point of Lemma 6.7, we can pass to the limit in the Gaussian integral and we get

$$\lim_{j\to\infty} \frac{1}{\sqrt{2\pi}} \int_{\widetilde{\mathcal{N}}_{\mathfrak{t}_j,x_0}} \widetilde{\rho}\, d\sigma = \frac{1}{\sqrt{2\pi}} \int_{\widetilde{\mathcal{N}}_\infty} \widetilde{\rho}\, d\overline{\sigma} = \Theta_{\widetilde{\mathcal{N}}_\infty}\,.$$

Recalling that

$$\frac{1}{\sqrt{2\pi}} \int_{\widetilde{\mathcal{N}}_{\mathfrak{t}_j,x_0}} \widetilde{\rho}\, d\sigma = \int_{\mathcal{N}_{t(\mathfrak{t}_j)}} \rho_{T,x_0}(\tau(\mathfrak{t}_j), \cdot)\, ds = \Theta_{x_0}(t(\mathfrak{t}_j)) \to \widehat{\Theta}(T, x_0)$$

as $j \to \infty$, equality (6.4) follows. □

Remark 6.11 If the three end-points P^{r-1}, P^r, P^{r+1} are aligned the argument of Proposition 3.13 does not work and we cannot conclude that the only blow-up at P^r is a halfline with multiplicity 1. It could also be possible that a straight line (possibly with higher multiplicity) is present.

We describe now how Proposition 6.8 allows us to obtain a (conditional) global existence result when the lengths of all the curves of the networks are strictly positive.

Suppose that $T < +\infty$. Since we have assumed that the lengths of all the curves of the network are uniformly positively bounded from below, the curvature blows-up as $t \to T$ (Theorem 5.7 and Proposition 5.8). Performing a Huisken's rescaling at an interior point x_0 of Ω, we obtain as blow-up limit (if not empty) a standard triod or a straight line with multiplicity $m \in \mathbb{N}$. One can argue as in [29] to show that when such limit is a regular triod, the curvature is locally bounded around such point x_0. For the case of a straight line, if we suppose that the multiplicity m is equal to 1, by White's local regularity theorem [45] we conclude that the curvature is bounded uniformly in time, in a neighborhood of the point x_0. If we instead rescale at an end-point P^r we get a halfline. This case can be treated as above by means of a reflection argument. Indeed, for the flow obtain by the union of the original network and the reflection of this latter, the point P^r is no more an end-point. A blow-up at P^r give a straight line, implying that the curvature is locally bounded also around P^r as before by White's theorem.

Supposing that the lengths of the curves of the network are strictly positive and supposing also that any blow-up limit has multiplicity one, it follows that the original network $\mathcal{N}_t$ has bounded curvature as $t \to T$. Hence T cannot be a singular time, and we have therefore global existence of the flow.

In the previous reasoning a key point is the hypothesis that the blow-ups have multiplicity one. Unfortunately, for a general regular network, this is still conjectural and possibly the major open problem in the subject.

Multiplicity–One Conjecture (**M1**) Every possible C^1_{loc}-limit of rescalings of networks of the flow is an embedded network with multiplicity one.

However, in some special situations one can actually prove **M1**.

Proposition 6.12 *If Ω is strictly convex and the evolving network $\mathcal{N}_t$ has at most two triple junctions, every C^1_{loc}-limit of rescalings of networks of the flow is embedded and has multiplicity one.*

Proof See [33, Section 4,Corollary 4.7]. □

Proposition 6.13 *If during the curvature flow of a tree $\mathcal{N}_t$ the triple junctions stay uniformly far from each other and from the end-points, then every C^1_{loc}-limit of rescalings of networks of the flow is embedded and has multiplicity one.*

Proof See [32, Proposition 14.14]. □

We now remove the hypothesis on the lengths of the curves of the network. In this case, nothing prevents a length to go to zero in the limit.

In order to describe the possible limits, we introduce the notion of degenerate regular networks. First of all we define the **underlying graph**, which is an oriented graph G with n edges E^i, that can be bounded and unbounded. Every vertex of G can either have order one (and in this case it is called end-points of G) or order three.

For every edge E^i we introduce an orientation preserving homeomorphisms $\varphi^i : E^i \to I^i$ where I^i is the interval $(0, 1)$, $[0, 1)$, $(0, 1]$ or $[0, 1]$. If E^i is a segment, then $I^i = [0, 1]$. If it is an halfline, we choose $I^i = [0, 1)$ or $I^i = (0, 1]$. Notice that the interval $(0, 1)$ can only appear if it is associated to an unbounded edge E^i without vertices, which is clearly a single connected component of G.

We then consider a family of C^1 parametrizations $\sigma^i : I^i \to \mathbb{R}^2$. In the case I^i is $(0, 1)$, $[0, 1)$ or $(0, 1]$, the map σ^i is a regular C^1 curve with unit tangent vector τ^i. If instead $I^i = [0, 1]$ the map σ^i can be either a regular C^1 curve with unit tangent vector τ^i, or a constant map (**degenerate curves**). In this last case we assign a constant unit vector $\tau^i : I^i \to \mathbb{R}^2$ to the curve σ^i. At the points 0 and 1 of I^i the **assigned exterior unit tangents** are $-\tau^i$ and τ^i, respectively. The exterior unit tangent vectors (real or assigned) at the relative borders of the intervals I^i, I^j, I^k of the concurring curves σ^i, σ^j σ^k have zero sum (**degenerate** 120° **condition**). We require that the map $\Gamma : G \to \mathbb{R}^2$ given by the union $\Gamma = \bigcup_{i=1}^n (\sigma^i \circ \varphi^i)$ is well defined and continuous.

We define a **degenerate regular network** $\mathcal{N}$ as the union of the sets $\sigma^i(I^i)$. If one or several edges E^i of G are mapped under the map $\Gamma : G \to \mathbb{R}^2$ to a single point $p \in \mathbb{R}^2$, we call this sub-network given by the union G' of such edges E^i the **core** of $\mathcal{N}$ at p.

We call multi-points of the degenerate regular network $\mathcal{N}$ the images of the vertices of multiplicity three of the graph G, by the map Γ and end-point of $\mathcal{N}$ the images of the vertices of multiplicity one of the graph G by the map Γ.

A degenerate regular network $\mathcal{N}$ with underlying graph G, seen as a subset in $\mathbb{R}^2$, is a C^1 network, not necessarily regular, that can have end-points and/or unbounded curves. Moreover, self-intersections and curves with integer multiplicities can be present. Anyway, at every image of a multi-point of G the sum (possibly with multiplicities) of the exterior unit tangents is zero.

Definition 6.14 We say that a sequence of regular networks $\mathcal{N}_k = \bigcup_{i=1}^n \sigma_k^i(I_k^i)$ converges in C^1_{loc} to a degenerate regular network $\mathcal{N} = \bigcup_{j=1}^l \sigma_\infty^j(I_\infty^j)$ with underlying graph $G = \bigcup_{j=1}^l E^j$ if:

- letting $O^1, O^2, \dots, O^m$ the multi-points of $\mathcal{N}$, for every open set $\Omega \subset \mathbb{R}^2$ with compact closure in $\mathbb{R}^2 \setminus \{O^1, O^2, \dots, O^m\}$, the networks $\mathcal{N}_k$ restricted to Ω, for k large enough, are described by families of regular curves which, after possibly reparametrizing them, converge to the family of regular curves given by the restriction of $\mathcal{N}$ to Ω;
- for every multi-point O^p of $\mathcal{N}$, image of one or more vertices of the graph G (if a core is present), there is a sufficiently small $R > 0$ and a graph $\widetilde{G} = \bigcup_{r=1}^s F^r$, with edges F^r associated to intervals J^r, such that:
  - the restriction of $\mathcal{N}$ to $B_R(O^p)$ is a regular degenerate network described by a family of curves $\widetilde{\sigma}_\infty^r : J^r \to \mathbb{R}^2$ with (possibly "assigned", if the curve is degenerate) unit tangent $\widetilde{\tau}_\infty^r$,

- for k sufficiently large, the restriction of $\mathcal{N}_k$ to $B_R(O^p)$ is a regular network with underlying graph $\widetilde{G}$, described by the family of regular curves $\widetilde{\sigma}_k^r : J^r \to \mathbb{R}^2$,
- for every j, possibly after reparametrization of the curves, the sequence of maps $J^r \ni x \mapsto \big(\widetilde{\sigma}_k^r(x), \widetilde{\tau}_k^r(x)\big)$ converge in C^0_{loc} to the maps $J^r \ni x \mapsto \big(\widetilde{\sigma}_\infty^r(x), \widetilde{\tau}_\infty^r(x)\big)$, for every $r \in \{1, 2, \dots, s\}$.

We will say that $\mathcal{N}_k$ converges to $\mathcal{N}$ in $C^1_{\mathrm{loc}} \cap E$, where E is some function space, if the above curves also converge in the topology of E.

Removing the hypothesis on the lengths of the curves, we get that the limit networks are degenerate regular networks which are homothetically shrinking under the flow.

Proposition 6.15 *Let $\mathcal{N}_t = \bigcup_{i=1}^n \gamma^i(t, [0, 1])$ be a $C^{1,2}$ curvature flow of regular networks in the time interval $[0, T)$, then, for every $x_0 \in \mathbb{R}^2$ and for every subset $\mathcal{I}$ of $[-1/2 \log T, +\infty)$ with infinite Lebesgue measure, there exists a sequence of rescaled times $\mathfrak{t}_j \to +\infty$, with $\mathfrak{t}_j \in \mathcal{I}$, such that the sequence of rescaled networks $\widetilde{\mathcal{N}}_{\mathfrak{t}_j, x_0}$ (obtained via Huisken's dynamical procedure) converges in $C^{1,\alpha}_{\mathrm{loc}} \cap W^{2,2}_{\mathrm{loc}}$, for any $\alpha \in (0, 1/2)$, to a (possibly empty) limit network, which is a degenerate regular shrinker $\widetilde{\mathcal{N}}_\infty$ (possibly with multiplicity greater than one).*

Moreover, we have

$$\lim_{j\to\infty} \frac{1}{\sqrt{2\pi}} \int_{\widetilde{\mathcal{N}}_{\mathfrak{t}_j, x_0}} \widetilde{\rho}\, d\sigma = \frac{1}{\sqrt{2\pi}} \int_{\widetilde{\mathcal{N}}_\infty} \widetilde{\rho}\, d\overline{\sigma} = \Theta_{\widetilde{\mathcal{N}}_\infty} = \widehat{\Theta}(T, x_0)\,.$$

where $d\overline{\sigma}$ denotes the integration with respect to the canonical measure on $\widetilde{\mathcal{N}}_\infty$, counting multiplicities.

Remark 6.16 Notice that the blow-up limit degenerate shrinker obtained by this proposition a priori depends on the chosen sequence of rescaled times $\mathfrak{t}_j \to +\infty$.

Remark 6.17 Thanks to Proposition 6.12, if the network $\mathcal{N}$ has at most two triple junctions, the degenerate regular shrinker $\widetilde{\mathcal{N}}_\infty$ has multiplicity one.

Assuming that the length of at least one curve of $\mathcal{N}_t$ goes to zero, as $t \to T$, there are two possible situations:

- the curvature stays bounded;
- the curvature is unbounded as $t \to T$.

Suppose that the curvature remains bounded in the maximal time interval $[0, T)$. As $t \to T$ the networks $\mathcal{N}_t$ converge in C^1 (up to reparametrization) to a unique limit degenerate regular network $\widehat{\mathcal{N}}_T$. This network can be **non-regular** seen as a subset of $\mathbb{R}^2$: multi-points can appear, but anyway the sum of the exterior unit tangent vectors of the concurring curves at every multi-point must be zero. Every triple junction satisfies the angle condition. The non-degenerate curves of $\widehat{\mathcal{N}}_T$ belong to $C^1 \cap W^{2,\infty}$ and they are smooth outside the multi-points (for the proof see [32, Proposition 10.11]).

We have seen in Sect. 5.1 that if a region is bounded by less than six curves then its area decreases linearly in time going to zero at $\overline{T}$. Not only the area goes to zero in a finite time, but also the lengths of all the curves that bound the region. Moreover when the lengths of all the curves of the loop go to zero, then the curvature blows up. Let us call the loop ℓ. Combing (5.2) with (5.3) there is a positive constant c such that $\int_\ell |k|ds \geq c$. By Hölder inequality

$$c = \int_\ell |k|ds \leq \left(\int_\ell k^2\, ds\right)^{1/2} L(\ell)^{1/2},$$

where $L(\ell)$ is the total length of the loop. Hence

$$\|k\|_{L^2} \geq \frac{c^2}{L(\ell)} \to \infty \quad \text{as} \quad L(\ell) \to 0\,.$$

Then at time $\overline{T}$ we have a singularity where both the length goes to zero and the curvature explodes.

Developing careful a priori estimates of the curvature one can show that if two triple junctions collapse into a 4-point, then the curvature remains bounded (see [32]). The interest of this result relies on the fact that it describes the formation of a "type zero" singularity: a singularity due to the change of topology, not to the blow up of the curvature. This is a new phenomenon with respect to the classical curve shortening flow and the mean curvature flow more in general. Thanks to this result it is possible to show that given an initial network without loops (a tree), if Multiplicity-One Conjecture **M1** is valid, then the curvature is uniformly bounded during the flow. The only possible "singularities" are given by the collapse of a curve with two triple junctions going to collide. Moreover in the case of a tree we are able to show the uniqueness of the blow up limit (see Remark 6.16).

Although one can find example of global existence of the flow (consider for instance an initial triod contained in the triangle with vertices its three end-points and with all angles less than 120°) our analysis underlines the generic presence of singularities. Then a natural question is if it is possible to go beyond the singularity.

There are results on the short time existence of the flow for non-regular networks, that is, networks with multi-points (not only 3-points), or networks that do not satisfy the 120° condition at the 3-points. Till now the most general result of this kind is the one by Ilmanen, Neves and Schulze [23], which provides short time existence of the flow starting from a non-regular network with **bounded curvature**. Notice that the network arising after the collapse of (exactly) two triple junctions has bounded curvature, and therefore fits with the hypotheses this result.

An ambitious project should be constructing a bridge between the analysis of the long time behavior of networks moving by curvature and short time existence results for non-regular initial data: one can interpret the short time existence results for non-regular data as a âĂIJrestartingâĂİ theorem for the flow after the onset of the first singularity.

References

1. U. Abresch, J. Langer, The normalized curve shortening flow and homothetic solutions. J. Differ. Geom. **23**(2), 175–196 (1986)
2. S.J. Altschuler, Singularities of the curve shrinking flow for space curves. J. Differ. Geom. **34**(2), 491–514 (1991).
3. S. Angenent, Parabolic equations for curves on surfaces. I. Curves with p-integrable curvature. Ann. Math. (2) **132**(3), 451–483 (1990)
4. S. Angenent, On the formation of singularities in the curve shortening flow. J. Differ. Geom. **33**, 601–633 (1991)
5. S. Angenent, Parabolic equations for curves on surfaces. II. Intersections, blow-up and generalized solutions. Ann. Math. (2) **133**(1), 171–215 (1991)
6. P. Baldi, E. Haus, C. Mantegazza, Non-existence of theta-shaped self-similarly shrinking networks moving by curvature. Commun. Partial Differ. Equ. **43**(3), 403–427 (2018)
7. P. Baldi, E. Haus, C. Mantegazza, Networks self-similarly moving by curvature with two triple junctions. Atti Accad. Naz. Lincei Rend. Lincei Mat. Appl. **28**(2), 323–338 (2017)
8. P. Baldi, E. Haus, C. Mantegazza, On the classification of networks self-similarly moving by curvature. Geom. Flows **2**, 125–137 (2017)
9. G. Bellettini, M. Novaga, Curvature evolution of nonconvex lens-shaped domains. J. Reine Angew. Math. **656**, 17–46 (2011)
10. K.A. Brakke, *The Motion of a Surface by Its Mean Curvature* (Princeton University Press, Princeton, 1978)
11. L. Bronsard, F. Reitich, On three-phase boundary motion and the singular limit of a vector-valued Ginzburg–Landau equation. Arch. Ration. Mech. Anal. **124**(4), 355–379 (1993)
12. X. Chen, J.-S. Guo, Self-similar solutions of a 2-D multiple-phase curvature flow. Phys. D **229**(1), 22–34 (2007)
13. X. Chen, J.-S. Guo, Motion by curvature of planar curves with end points moving freely on a line. Math. Ann. **350**(2), 277–311 (2011)
14. C. Dellacherie, P.-A. Meyer, *Probabilities and Potential*. North-Holland Mathematics Studies, vol. 29 (North-Holland Publishing Co., Amsterdam, 1978)
15. S.D. Eidelman, N.V. Zhitarashu, *Parabolic Boundary Value Problems*. Operator Theory: Advances and Applications, vol. 101 (Birkhäuser Verlag, Basel, 1998)
16. M. Gage, An isoperimetric inequality with applications to curve shortening. Duke Math. J. **50**(4), 1225–1229 (1983)
17. M. Gage, Curve shortening makes convex curves circular. Invent. Math. **76**, 357–364 (1984)
18. M. Gage, R.S. Hamilton, The heat equation shrinking convex plane curves. J. Differ. Geom. **23**, 69–95 (1986)
19. M.A. Grayson, The heat equation shrinks embedded plane curves to round points. J. Differ. Geom. **26**, 285–314 (1987)
20. J. Hättenschweiler, Mean curvature flow of networks with triple junctions in the plane. Master's Thesis, ETH Zürich, 2007
21. G. Huisken, Asymptotic behavior for singularities of the mean curvature flow. J. Differ. Geom. **31**, 285–299 (1990)
22. G. Huisken, A distance comparison principle for evolving curves. Asian J. Math. **2**, 127–133 (1998)

23. T. Ilmanen, A. Neves, F. Schulze, On short time existence for the planar network flow. J. Differ. Geom. **111**(1), 39–89 (2019)
24. D. Kinderlehrer, C. Liu, Evolution of grain boundaries. Math. Models Methods Appl. Sci. **11**(4), 713–729 (2001)
25. O.A. Ladyzhenskaya, V.A. Solonnikov, N.N. Ural'tseva, *Linear and Quasilinear Equations of Parabolic Type* (American Mathematical Society, Providence, 1975)
26. J. Langer, A compactness theorem for surfaces with L_p-bounded second fundamental form. Math. Ann. **270**, 223–234 (1985)
27. A. Lunardi, *Analytic Semigroups and Optimal Regularity in Parabolic Problems* (Birkhäuser, Basel, 1995)
28. A. Magni, C. Mantegazza, A note on Grayson's theorem. Rend. Semin. Mat. Univ. Padova **131**, 263–279 (2014)
29. A. Magni, C. Mantegazza, M. Novaga, Motion by curvature of planar networks II. Ann. Sc. Norm. Sup. Pisa **15**, 117–144 (2016)
30. C. Mantegazza, *Lecture Notes on Mean Curvature Flow*. Progress in Mathematics, vol. 290 (Birkhäuser/Springer Basel AG, Basel, 2011)
31. C. Mantegazza, M. Novaga, V.M. Tortorelli, Motion by curvature of planar networks. Ann. Sc. Norm. Sup. Pisa **3**(5), 235–324 (2004)
32. C. Mantegazza, M. Novaga, A. Pluda, F. Schulze, Evolution of networks with multiple junctions. arXiv Preprint Server – http://arxiv.org (2016)
33. C. Mantegazza, M. Novaga, A. Pluda, Motion by curvature of networks with two triple junctions. Geom. flows **2**, 18–48 (2017)
34. L. Nirenberg, On elliptic partial differential equations. Ann. Sc. Norm. Sup. Pisa **13**, 116–162 (1959)
35. A. Pluda, Evolution of spoon–shaped networks. Netw. Heterog. Media **11**(3), 509–526 (2016)
36. M.H. Protter, H.F. Weinberger, *Maximum Principles in Differential Equations* (Springer, New York, 1984)
37. O.C. Schnürer, A. Azouani, M. Georgi, J. Hell, J. Nihar, A. Koeller, T. Marxen, S. Ritthaler, M. Sáez, F. Schulze, B. Smith, Evolution of convex lens–shaped networks under the curve shortening flow. Trans. Am. Math. Soc. **363**(5), 2265–2294 (2011)
38. L. Simon, *Lectures on Geometric Measure Theory*. Proc. Center Math. Anal., vol. 3 (Australian National University, Canberra, 1983)
39. V.A. Solonnikov, The Green's matrices for parabolic boundary value problems. Zap. Naučn. Sem. Leningrad. Otdel. Mat. Inst. Steklov. (LOMI) **14**, 256–287 (1969). Translated in Semin. Math. Steklova Math. Inst. Leningrad 109–121 (1972)
40. A. Stahl, Convergence of solutions to the mean curvature flow with a Neumann boundary condition. Calc. Var. Partial Differ. Equ. **4**(5), 421–441 (1996)
41. A. Stahl, Regularity estimates for solutions to the mean curvature flow with a Neumann boundary condition. Calc. Var. Partial Differ. Equ. **4**(4), 385–407 (1996)
42. A. Stone, A density function and the structure of singularities of the mean curvature flow. Calc. Var. Partial Differ. Equ. **2**, 443–480 (1994)
43. Y. Tonegawa, N. Wickramasekera, The blow up method for Brakke flows: networks near triple junctions. Arch. Ration. Mech. Anal. **221**(3), 1161–1222 (2016)
44. J. von Neumann, Discussion and remarks concerning the paper of C. S. Smith "Grain shapes and other metallurgical applications of topology", in *Metal Interfaces* (American Society for Metals, Materials Park, 1952)
45. B. White, A local regularity theorem for mean curvature flow. Ann. Math. (2) **161**(3), 1487–1519 (2005)

Maximum Principles at Infinity and the Ahlfors-Khas'minskii Duality: An Overview

Luciano Mari and Leandro F. Pessoa

Abstract This note is meant to introduce the reader to a duality principle for nonlinear equations recently discovered in Valtorta (Reverse Khas'minskii condition. Math Z 270(1):65–177, 2011), Mari and Valtorta (Trans Am Math Soc 365(9):4699–4727, 2013), and Mari and Pessoa (Commun Anal Geom, to appear). Motivations come from the desire to give a unifying potential-theoretic framework for various maximum principles at infinity appearing in the literature (Ekeland, Omori-Yau, Pigola-Rigoli-Setti), as well as to describe their interplay with properties coming from stochastic analysis on manifolds. The duality involves an appropriate version of these principles formulated for viscosity subsolutions of fully nonlinear inequalities, called the Ahlfors property, and the existence of suitable exhaustion functions called Khas'minskii potentials. Applications, also involving the geometry of submanifolds, will be discussed in the last sections. We conclude by investigating the stability of these maximum principles when we remove polar sets.

Keywords Potential theory · Liouville theorem · Omori-Yau · Maximum principles · Stochastic completeness · Martingale · Completeness · Ekeland · Brownian motion

2010 Mathematics Subject Classification Primary 31C12, 35B50; Secondary 35B53, 58J65, 58J05, 53C42

L. Mari (✉)
Dipartimento di Matematica, Scuola Normale Superiore, Pisa, Italy
e-mail: luciano.mari@sns.it; mari@mat.ufc.br

L. F. Pessoa
Departamento de Matemática, Universidade Federal do Piauí-UFPI, Teresina, Brazil
e-mail: leandropessoa@ufpi.edu.br

S. Dipierro (ed.), *Contemporary Research in Elliptic PDEs and Related Topics*, Springer INdAM Series 33, https://doi.org/10.1007/978-3-030-18921-1_10

1 Prelude: Maximum Principles at Infinity

Maximum principles at infinity are a powerful tool to investigate problems in Geometry. They arose from the desire to generalize the statement that, on a compact Riemannian manifold $(X, \langle\, , \rangle)$ of dimension $m \geqslant 2$, every function $u \in C^2(X)$ attains a maximum point x_0 and

$$(i)\,:\; u(x_0) = \sup_X u, \qquad (ii)\,:\; |\nabla u(x_0)| = 0, \qquad (iii)\,:\; \nabla^2 u(x_0) \leqslant 0, \tag{1}$$

where $\nabla^2 u$ is the Riemannian Hessian and the last relation is meant in the sense of quadratic forms. If M is noncompact and given $u \in C^2(X)$ bounded from above, although one cannot ensure the existence of a maximum point, there could still exist a sequence $\{x_k\} \subset X$ such that some of the relations in (1) hold in a limit sense as $k \to \infty$. Informally speaking, when this happens we could think that X is "not too far from being compact". The first example of maximum principle at infinity is the famous Ekeland's principle, [23], that can be stated as follows:

Definition 1.1 A metric space (X, d) satisfies the *Ekeland maximum principle* if, for each u upper semicontinuous (USC) on X and bounded from above, there exists a sequence $\{x_k\} \subset X$ with the following properties:

$$u(x_k) > \sup_X u - \frac{1}{k}, \qquad u(y) \leqslant u(x_k) + \frac{1}{k} d(x_k, y) \ \text{ for each } \ y \in X.$$

The full statement of Ekeland's principle contains, indeed, a further property that is crucial in applications, that is, the possibility to create one such $\{x_k\}$ suitably close to a given maximizing sequence $\{\bar{x}_k\}$. We will briefly touch on it later. By works of Ekeland, Weston and Sullivan, cf. [23, 67, 72], the validity of Ekeland's principle in the form given above is in fact *equivalent* to X being a complete metric space. Therefore, in the smooth setting, the (geodesic) completeness of a manifold X enables to find $\{x_k\}$ approximating both (i) and (ii) in (1.1). However, condition (iii) requires further restrictions on the geometry of X, first investigated by Omori [55] and Yau [14, 74] (the second with Cheng). They introduced the following two principles, respectively, in the Hessian case [55] and in the Laplacian case [14, 74]:

Definition 1.2 Let $(X, \langle\, , \rangle)$ be a Riemannian manifold. We say that X satisfies the *strong Hessian (respectively, Laplacian) maximum principle* if, for each $u \in C^2(X)$ bounded from above, there exists a sequence $\{x_k\} \subset X$ with the following properties:

$$\begin{array}{llll} \text{(Hessian)} & u(x_k) > \sup_X u - k^{-1}, & |\nabla u(x_k)| < k^{-1}, & \nabla du(x_k) \leqslant k^{-1}\langle\, , \rangle; \\ \text{(Laplacian)} & u(x_k) > \sup_X u - k^{-1}, & |\nabla u(x_k)| < k^{-1}, & \Delta u(x_k) \leqslant k^{-1}. \end{array} \tag{2}$$

Here, the word "strong" refers to the presence of the condition on the gradient. Historically, these principles are called the *Omori-Yau maximum principles*, and proved to be remarkably effective in a wealth of different geometric problems. Among them, we stress the striking proofs of the generalized Schwarz Lemma for maps between Kahler manifolds in [75], and of the Bernstein theorem for maximal hypersurfaces in Minkovski space in [15].

Geometric conditions to guarantee the Omori-Yau principles are often expressed in terms of growths of the curvatures of X with respect to the distance $\varrho(x)$ from a fixed origin $o \in X$, but are not necessarily depending on them. The most general known condition guaranteeing the Omori-Yau principles is given by Pigola et al. [58] (cf. also improvements in [9, 11]): the principle holds whenever X supports a function w with the following properties[1]:

$$\begin{gathered} 0 < w \in C^2(X \backslash K) \ \text{ for some compact } K, \qquad w \to +\infty \ \text{ as } x \text{ diverges}, \\ |\nabla w| \leqslant G(w), \qquad \text{and} \qquad \begin{cases} \nabla^2 w \leqslant G(w)\langle\, , \rangle & \text{for Omori's principle}, \\ \Delta w \leqslant G(w) & \text{for Yau's principle}, \end{cases} \end{gathered} \tag{3}$$

for some G satisfying

$$0 < G \in C^1(\mathbb{R}^+), \qquad G' \geqslant 0, \qquad \int^{+\infty} \frac{ds}{G(s)} = +\infty. \tag{4}$$

For instance, $G(t) = (1 + t)$ gives the sharp polynomial threshold. The criterion is effective, since the function in (3) can be explicitly found in a number of geometrically relevant applications: for instance, if the radial sectional curvature[2] (respectively, Ricci curvature) is bounded from below as follows:

$$\begin{array}{lll} \mathrm{Sect}_{\mathrm{rad}} \geqslant -G^2(\varrho) & \text{on } X, & \text{for Omori's principle}, \\ \mathrm{Ric}(\nabla\varrho, \nabla\varrho) \geqslant -G^2(\varrho) & \text{on } X \backslash \{\{o\} \cup \mathrm{cut}(o)\} & \text{for Yau's principle}, \end{array} \tag{5}$$

then one can choose $w(x) = \log(1 + \varrho(x))$ in (3) to deduce the validity of the strong Hessian, respectively, strong Laplacian principle (technically, ϱ is not C^2, but one can overcome the problem by using Calabi's trick, see [58]). Note that (5) includes

[1]Here, as usual, if we write "$w(x) \to +\infty$ as x diverges" we mean that the sublevels of w have compact closure in X, that is, that w is an exhaustion.

[2]The radial sectional curvature is the sectional curvature restricted to 2-planes containing $\nabla\varrho$. Inequality $\mathrm{Sect}_{\mathrm{rad}} \geqslant -G^2(\varrho)$ means that $\mathrm{Sect}(\pi_x) \geqslant -G^2\big(\varrho(x)\big)$ for each $x \notin \{o\} \cup \mathrm{cut}(o)$ and $\pi_x \leqslant T_xX$ 2-plane containing $\nabla\rho$.

the case when G is constant, considered in [55, 74]. However, the existence of w could be granted even without bounds like (5): for instance, in the strong Laplacian case, this happens if X is properly immersed with bounded mean curvature in $\mathbb{R}^m$ (or in a Cartan-Hadamard ambient space with bounded sectional curvature), or if X is a Ricci soliton, see [4]. The function w in (3) is an example of what we will call a *Khas'minskii potential*. The reason for the name will be apparent in a moment.

1.1 An Example: Immersions into Cones

There is, by now, a wealth of applications of the Omori-Yau principles in geometry, see for instance [4]. Here, we illustrate how the principles can be effectively used in geometry by means of the following example in [49], that is related to the pioneering paper by Omori [55]. Hereafter, a non-degenerate cone $\mathcal{C}_{o,v,\varepsilon}$ of center $o \in \mathbb{R}^n$, axis $v \in \mathbb{S}^{n-1}$ and width $\varepsilon \in (0, \pi/2)$ is the set of points $x \in \mathbb{R}^n$ such that

$$\langle \frac{x - o}{|x - o|}, v \rangle \geqslant \cos \varepsilon.$$

Theorem 1.3 ([49, Cor. 1.18]) *Let $\varphi : X^m \to \mathbb{R}^{2m-1}$ be an isometric immersion. Denoting with ρ the distance from a fixed origin o, assume that the sectional curvature of X satisfies*

$$-C(1 + \rho^2) \leqslant \mathrm{Sect} \leqslant 0, \qquad \text{on } M, \tag{6}$$

for some constant $C > 0$. Then, $\varphi(X)$ cannot be contained into any non-degenerate cone of $\mathbb{R}^{2m-1}$.

Sketch of the Proof Suppose, by contradiction, that $\varphi(X) \subset \mathcal{C}_{o,v,\varepsilon}$ for some o, v, ε. Without loss of generality, we can assume that o is the origin of $\mathbb{R}^{2m-1}$. The first step is to construct a function that encodes the geometry of the problem at hand: following [55], we fix $x_0 \in X\backslash\{o\}$ and $a \in (0, \cos \varepsilon)$, we set $T = \langle \varphi(x_0), v \rangle$ and we define

$$u(x) = \sqrt{T^2 + a^2|\varphi(x)|^2} - \langle \varphi(x), v \rangle.$$

By construction, it is easy to show that $u < T$ on X and that the non-empty upper level set $\{u > 0\}$ (that contains x_0) has bounded image $\varphi\big(\{u > 0\}\big)$. In view of (6) and the discussion above, the strong Hessian principle holds and can be applied to u.

However, for $x \in \{u > 0\}$ and unit vector $W \in T_xX$, a computation shows that

$$\begin{aligned}\nabla^2 u(W,W) &= \frac{a^2(1+\langle II(W,W),\varphi\rangle}{\sqrt{T^2+a^2|\varphi|^2}} - \langle II(W,W),\nu\rangle - \frac{a^4\langle W,\varphi\rangle^2}{(T^2+a^2|\varphi|^2)^{3/2}} \\ &\geqslant \frac{a^2(1-|II(W,W)||\varphi|)}{\sqrt{T^2+a^2|\varphi|^2}} - |II(W,W)| - \frac{a^4|\varphi|^2}{(T^2+a^2|\varphi|^2)^{3/2}} \\ &\geqslant \frac{a^2T^2}{(T^2+a^2|\varphi|^2)^{3/2}} - |II(W,W)|g(x),\end{aligned}$$

for some continuous function g on $\{u > 0\}$. Since the codimension of φ is strictly less than m and the sectional curvature is non-positive, a useful algebraic lemma due to Otsuki [20, 58] guarantees the existence of $W \in T_xX$ such that $II(W,W) = 0$. Having fixed such W, and taking into account that $\varphi(\{u > 0\})$ is bounded, there exists a uniform constant $c > 0$ such that

$$\sup_{Z\in T_xX,\ |Z|=1} \nabla^2 u(Z,Z) \geqslant \nabla^2 u(W,W) \geqslant \frac{a^2T^2}{(T^2+a^2|\varphi|^2)^{3/2}} \geqslant c > 0 \qquad \text{on } \{u > 0\}.$$

However, evaluating the above inequality on a sequence $\{x_k\}$ realizing the strong Hessian principle we obtain a contradiction. □

Remark 1.4 In its full strength, Ekeland's principle also guarantees that the sequence $\{x_k\}$ satisfying (1.1) can be chosen to be close to a given maximizing sequence $\{\bar{x}_k\}$ with explicit bounds, a fact that is very useful in applications to functional analysis and PDEs. On the other hand, to present no systematic investigation of an analogous property was performed for the Omori-Yau principles. A notable exception, motivated by a geometrical problem involving convex hulls of isometric immersions, appeared in [25]: there, the authors proved that if X is complete and Sect $\geqslant -c$ for some $c \in \mathbb{R}^+$ (resp. Ric $\geqslant -c$), any maximizing sequence $\{\bar{x}_k\}$ has a *good shadow*, that is, a sequence $\{x_k\}$ satisfying Omori (resp. Yau) principle and also $d(x_k, \bar{x}_k) \to 0$ as $k \to \infty$.

The properties in (2) can be rephrased as follows, say in the Hessian case: denoting with $\lambda_1(A) \leqslant \lambda_2(A) \leqslant \ldots \leqslant \lambda_m(A)$ the eigenvalues of a symmetric matrix A, X has the strong Hessian principle if either one of the following properties holds:

(i) it is not possible to find a function $u \in C^2(X)$ bounded from above and such that, on some non-empty upper level-set $\{u > \gamma\}$,

$$\max\Big\{|\nabla u| - 1, \lambda_m(\nabla^2 u) - 1\Big\} \geqslant 0. \tag{7}$$

(ii) for every open set $U \subset M$ and every $u \in C^2(U) \cap C(\overline{U})$ bounded from above and solving the differential inequality

$$\max\left\{|\nabla u| - 1, \lambda_m(\nabla^2 u) - 1\right\} \geqslant 0,$$

it holds $\sup_U u = \sup_{\partial U} u$.

By rescaling, the constant 1 can be replaced by any fixed positive number. It is evident that the strong Hessian principle is equivalent to (i), while (i)⇔(2) can easily be proved by contradiction: if we assume that (i) fails for some u, consider as U the upper level set where (7) holds and contradict (ii), while if (ii) fails for some u, then take any upper level set at height $\gamma \in (\sup_{\partial U} u, \sup_U u)$ to contradict (i).

Conditions (i) and (ii) are invariant by translations $u \mapsto u + \text{const}$. For future use it is important to describe another characterization, not translation invariant, where the constant 1 in (ii) is replaced by a pair of functions

$$\begin{cases} f \in C(\mathbb{R}), \quad f(0) = 0, \quad f > 0 \text{ on } \mathbb{R}^+, \quad f \text{ is odd and strictly increasing;} \\ \xi \in C(\mathbb{R}), \quad \xi(0) = 0, \quad \xi < 0, \text{ on } \mathbb{R}^+, \quad \xi \text{ is odd and strictly decreasing.} \end{cases} \tag{fξ}$$

Namely, (i) and (ii) are also equivalent to

($\mathscr{A}$) for some (*equivalently*, any) pair (f, ξ) satisfying (fξ), the following holds: for every open set $U \subset M$ and every $u \in C^2(U) \cap C(\overline{U})$ bounded from above and solving the differential inequality

$$\max\left\{|\nabla u| - \xi(-u), \ \lambda_m(\nabla^2 u) - f(u)\right\} \geqslant 0 \qquad \text{on } \left\{x \, : \, u(x) > 0\right\} \neq \emptyset, \tag{8}$$

it holds $\sup_U u = \sup_{\partial U} u$.

The choice of the upper level-set $\{u > 0\}$ is related to the vanishing of f, ξ in (fξ) at zero and is, of course, just a matter of convenience. The proof of (ii)⇔($\mathscr{A}$) proceeds by translation and rescaling arguments, and localizing on suitable upper level sets of u, and is given in detail in [48, Prop. 5.1]. From the technical point of view, the dependence of ($\mathscr{A}$) on f, ξ just in terms of the mild properties in (fξ), and especially the possibility to check ($\mathscr{A}$) in terms of a *single* pair f, ξ satisfying (fξ), is useful in applications.

Characterization ($\mathscr{A}$) is in the form of a maximum principle on sets with boundary, for subsolutions of the fully nonlinear inequality

$$\mathscr{F}(x, u(x), \nabla u(x), \nabla^2 u(x)) \doteq \max\left\{|\nabla u| - \xi(-u), \ \lambda_m(\nabla^2 u) - f(u)\right\} \geqslant 0.$$

This point of view relates the principles to another property that can be seen as a replacement of the compactness of X, the *parabolicity* of X. We recall that

Definition 1.5 A manifold X is said to be *parabolic* if each solution of $\Delta u \geqslant 0$ on X that is bounded from above is constant.

Indeed, Ahlfors (see [2, Thm. 6C]) observed that X is parabolic if and only if property ($\mathscr{A}$) holds with (8) replaced by $\Delta u \geqslant 0$. The problem of deciding whether a manifold is parabolic or not is classical, and there is by now a well established theory, see [28] for a thorough account. For surfaces, the theory arose in connection to the *type problem* for Riemann surfaces, cf. [2], and arguments involving parabolicity are still crucial in establishing a number of powerful, recent results in modern minimal surface theory (see for instance [16, 51, 52] for beautiful examples).

The tight relation between parabolicity and potential theory suggests that it might be possible to treat both Ekeland and Omori-Yau principles as well in terms of a fully nonlinear potential theory. In recent years, there has been an increasing interest in the theory of fully-nonlinear PDEs on manifolds, and especially Harvey and Lawson dedicated a series of papers [29–32, 34, 35] to develop a robust geometric approach for fully nonlinear PDEs well suited to do potential theory for those equations. Their work fits perfectly to the kind of problems considered in the present paper, and will be introduced later. Our major concern in the recent [47, 48, 50] is to put the above principles, as well as other properties to be discussed in a moment, into a unified framework where new relations, in particular an underlying duality, could emerge between them.

1.2 *Parabolicity, Capacity and Evans Potentials*

There are a number of equivalent conditions characterizing the parabolicity of X, see [28, Thm. 5.1] and [56], and we now focus on two of them.

The first one describes parabolic manifolds as those for which the 2-capacity of every compact K vanishes. We recall that, for fixed $q \in (1, \infty)$, the q-capacity of a condenser (K, Ω) with $K \subset \Omega \subset X$, K compact, Ω open, is the following quantity:

$$\mathrm{cap}_q(K, \Omega) = \inf \left\{ \int_\Omega |\nabla \phi|^q, \ : \ \phi \in \mathrm{Lip}_c(\Omega), \ \phi \geqslant 1 \text{ on } K \right\}. \tag{9}$$

If K and Ω have Lipschitz boundary, the infimum is realized by the unique solution u of the q-Laplace equation

$$\begin{cases} \Delta_q u \doteq \mathrm{div}\left(|\nabla u|^{q-2} \nabla u\right) = 0 & \text{on } \Omega \backslash K, \\ u = 1 \quad \text{on } K, \qquad u = 0 \quad \text{on } \partial\Omega, \end{cases} \tag{10}$$

called the q-capacitor of (K, Ω). A manifold is called q-parabolic if $\mathrm{cap}_q(K, X) = 0$ for some (equivalently, every) compact set K, see [36, 69]. By extending Ahlfors result for the Laplace-Beltrami operator [59, 62],

$$X \text{ is q-parabolic} \qquad \Longleftrightarrow \qquad \left\{ \begin{array}{c} \forall U \subset X \text{ open}, \ \forall u \in C^2(U) \cap C(\overline{U}), \\ \text{bounded above and solving } \Delta_q u \geqslant 0, \\ \text{it holds} \quad \sup_U u = \sup_{\partial U} u, \end{array} \right\} \tag{11}$$

and both are equivalent to the constancy of solutions of $\Delta_q u \geqslant 0$ that are bounded from above.

There is a further, quite useful characterization of parabolicity (in the linear setting $q = 2$), expressed in terms of suitable exhaustion functions and studied by Kuramochi and Nakai [44, 53, 54], with previous contribution by Khas'minskii [42]. In [44, 53, 54], the authors proved that X is parabolic if and only if, for each compact set K with smooth boundary, there exists a function w solving

$$\begin{array}{lll} w \in C^\infty(X \backslash K), & w > 0 \ \text{ on } X \backslash K, & w = 0 \ \text{ on } \partial K, \\ w(x) \to +\infty \ \text{ as } x \text{ diverges}, & \Delta w = 0 \ \text{ on } X \backslash K. & \end{array} \tag{12}$$

Such a w is named an *Evans potential* on $X \backslash K$. Evans potentials proved to be useful in investigating the topology of X by means of the beautiful Li-Tam-Wang's theory of harmonic functions, cf. [46, 68] (cf. also [45, 61] for comprehensive accounts), and it is therefore of interest to see whether other Liouville type properties could be characterized in terms of Evans potentials. One quickly realizes that, for this to hold, the function $\mathscr{F}$ replacing Δ in (12) must have a very specific form, and in fact, to present, the Laplace-Beltrami is the only operator for which solutions w in (12) *with the equality sign* have been constructed. The reason is that the proof in [44, 53, 54], see also [70], strongly uses the characterization of parabolicity in terms of the 2-capacity and the linearity of the Laplace-Beltrami operator. Even for $q \neq 2$, the equivalence of q-parabolicity with the existence of q-harmonic Evans potentials is still an open problem, although results for more general operators on rotationally symmetric manifolds (cf. the last section in [50]) indicate that it is likely to hold.

Quite differently, if we just require that w be a *supersolution*, that is, $\Delta_q w \leqslant 0$, things are much more flexible and are still worth interest, as we shall see later. In [71] the author proved that the k-parabolicity is equivalent to the existence of w satisfying (12) with the last condition weakened to $\Delta_q w \leqslant 0$. Although part of the proof uses q-capacities and is therefore very specific to the q-Laplacian, the underlying principle is general: the construction of w proceeds by "stacking" solutions of suitable obstacle problems, an idea that will be described later in a more general framework.

It is natural to ask what is the picture for $p = +\infty$, that is, setting,

$$\mathrm{cap}_\infty(K, \Omega) = \inf\left\{ \|\nabla\phi\|_\infty, \ : \ \phi \in \mathrm{Lip}_c(\Omega), \ \ \phi \geqslant 1 \text{ on } K \right\},$$

to study ∞-parabolic manifolds, defined as those for which $\mathrm{cap}_\infty(K, X) = 0$ for every compact K. The problem has recently been addressed in [56], where the authors proved that

$$X \text{ is } \infty\text{-parabolic} \qquad \Longleftrightarrow \qquad X \text{ is (geodesically) complete}$$

and thus, a-posteriori, ∞-parabolicity is equivalent to Ekeland's principle. Below, we shall complement these characterizations as applications of our main duality principle. To do so, we exploit the existence of ∞-capacitors for (K, Ω), that is, *suitable* minimizers (absolutely minimizing Lipschitz extensions, cf. [17, 39]) for $\mathrm{cap}_\infty(K, \Omega)$. Their existence was first considered by Aronsson [6] and proved by Jensen [38] when $X = \mathbb{R}^m$: the ∞-capacitor turns out to be the (unique) solution of

$$\begin{cases} \Delta_\infty u \doteq \nabla^2 u(\nabla u, \nabla u) = 0 & \text{on } \Omega\backslash K, \\ u = 1 \quad \text{on } K, \qquad u = 0 & \text{on } \partial\Omega, \end{cases} \tag{13}$$

where the equation is meant in the viscosity sense (see below). The operator Δ_∞, called the infinity Laplacian, has recently attracted a lot of attention because of its appearance in Analysis, Game Theory and Physics, and its investigation turns out to be challenging because of its high degeneracy (see [17, 19, 38]).

1.3 Link with Stochastic Processes

Before introducing the duality, we mention some other important function-theoretic properties of X, coming from stochastic analysis, that fit well with our setting and give further geometric motivation. We start recalling that parabolicity can be further characterized in terms of the Brownian motion on X. Briefly, on each Riemannian manifold X one can construct the heat kernel $p(x, y, t)$ (cf. [21]), and consequently a stochastic process $\mathscr{B}_t$ whose infinitesimal generator is Δ, called the Brownian motion, characterized by the identity

$$\mathbb{P}\left(\mathscr{B}_t \in \Omega \ : \ \mathscr{B}_0 = x\right) = \int_\Omega p(x, y, t)dy, \tag{14}$$

for every open subset $\Omega \subset X$ (see [7] for a beautiful, self-contained introduction).

Definition 1.6 Let $\mathscr{B}_t$ be the Brownian motion on X.

- $\mathscr{B}_t$ on X is called *recurrent* if, almost surely, its trajectories visit every fixed compact set $K \subset X$ infinitely many times.
- $\mathscr{B}_t$ on X is called *non-explosive* if, almost surely, its trajectories do not escape to infinity in finite time, that is, (14) with $\Omega = X$ is identically 1 for some (equivalently, every) $(x, t) \in X \times \mathbb{R}^+$.

A manifold whose Brownian motion is non-explosive is called *stochastically complete*. The recurrency of $\mathscr{B}_t$ is equivalent to the parabolicity of X, cf. [28], and can therefore be characterized in terms of a property of type $(\mathscr{A})$ above. Similarly, by [28, Thm. 6.2], X is stochastically complete if and only if there exists no open subset $U \subset X$ supporting a solution u of

$$\begin{cases} \Delta u \geqslant \lambda u \qquad \text{on } U, \quad \text{for some } \lambda \in \mathbb{R}^+, \\ u > 0 \quad \text{on } U, \qquad u = 0 \quad \text{on } \partial U. \end{cases} \tag{15}$$

for some (equivalently, any) fixed $\lambda \in \mathbb{R}^+$. The last property is $(\mathscr{A})$ provided that (7) is replaced by $\Delta u - \lambda u \geqslant 0$, that is, choosing $f(r) = \lambda r$ and removing the gradient condition. Again by translation, rescaling and localizing arguments, the function λr can be replaced by any f satisfying $(f\xi)$.

Around 15 years ago, new interest arose around the notion of stochastic completeness, after the observation in [57, 58] that the property is *equivalent* to a relaxed form of the strong Laplacian principle, called the *weak (Laplacian) maximum principle*. Namely, X has the weak Laplacian principle if, for each $u \in C^2(X)$ bounded above, there exists a sequence $\{x_k\}$ such that

$$u(x_k) > \sup_X u - k^{-1}, \qquad \Delta u(x_k) \leqslant k^{-1},$$

that is, (2) holds with no gradient condition. The weak Hessian principle can be defined accordingly. It turns out that, in most geometric applications, the gradient condition in (2) is unnecessary, making thus interesting to study both the possible difference between weak and strong principles, and the geometric conditions guaranteeing the weak principles.

Remark 1.7 (Strong Laplacian $\neq$ Weak Laplacian) It is easy to construct *incomplete* manifolds satisfying the weak Laplacian principle but not the strong one, for instance $X = \mathbb{R}^m \backslash \{0\}$ (cf. [48, Ex. 1.21]). A nice example of a *complete*, radially symmetric surface satisfying the weak Laplacian principle but not the strong one has recently been found in [12]. Therefore, the two principles are really different. Also, the weak Laplacian principle is unrelated to the (geodesic) completeness of X, and in fact, if one removes a compact subset K that is polar for the Brownian motion on X, $X \backslash K$ is still stochastically complete (see Theorem 6.3 below).

Remark 1.8 (Geometric Conditions for Weak Laplacian Principle) Conditions involving just the volume growth of balls in X (that, by Bishop-Gromov volume comparison, are weaker than those in (5), cf. [61]) were first considered in [40, 45], later improved in [28, Thm. 9.1]: X is stochastically complete provided that

$$\int^{+\infty} \frac{r}{\log \mathrm{vol} B_r} dr = +\infty. \tag{16}$$

As a consequence of work of Khas'minskii [42, 58], X is stochastically complete provided that it supports an exhaustion w outside a compact set K that satisfies

$$0 < w \in C^2(X \backslash K), \qquad w(x) \to +\infty \text{ as } x \text{ diverges}, \qquad \Delta w \leqslant \lambda w \text{ on } X \backslash K, \tag{17}$$

for some $\lambda > 0$. The analogy with (3) and (12) is evident, and it is the reason why we call w in (3) a Khas'minskii type potential. It was first observed in [50] that, in fact, the existence of w satisfying (17) is *equivalent* to the stochastic completeness of X. It is therefore natural to ask whether this is specific to the operators Δu and $\Delta u - \lambda u$ or if it is a more general fact, and, in the latter case, how one can take advantage from such an equivalence. This is the starting point of the papers [48, 50].

A further motivation to study Khas'minskii type potentials comes from the desire to understand the link between the Hessian maximum principles and the theory of stochastic processes. It has been suggested in [60, 62] that a good candidate to be a probabilistic counterpart of a Hessian principle is the *martingale completeness of* X. In fact, one can study the non-explosure property for a natural class of stochastic processes that includes the Brownian motion: the class of martingales (cf. [24, 66]).

- X is called *martingale complete* if and only if each martingale on X has infinite lifetime almost surely.

Differently from the case of stochastic completeness, there is not much literature on the interplay between martingale completeness and geometry, with the notable exception of [24]. The picture is still fragmentary and seems to be quite different from the Laplacian case: for instance, a martingale complete manifold must be (geodesically) complete [24, Prop. 5.36]. In [24, Prop. 5.37], by using probabilistic tools Emery proved that X is martingale complete provided that there exists $w \in C^2(X)$ satisfying

$$\begin{aligned} &0 < w \in C^2(X), \qquad w(x) \to +\infty \text{ as } x \text{ diverges}, \\ &|\nabla w| \leqslant C, \qquad \nabla dw \leqslant C\langle\ ,\ \rangle \text{ on } X, \end{aligned} \tag{18}$$

for some $C > 0$. Evidently, this is again a Khas'minskii type property. Although the gradient condition in (18) might suggest that the martingale completeness of X be related to the strong Hessian principle, in [60, 62] the authors give some results to support a tight link to the *weak* Hessian principle. Which Hessian principle relates to martingale completeness, and why? Is (18) equivalent to the martingale completeness of X?

The picture described above for Omori-Yau and Ekeland principles, and for parabolicity, stochastic and martingale completeness, suggest that there might be a general "duality principle" relating an appropriate maximum principle in the form of $(\mathscr{A})$ on open sets, to the existence of suitable Khas'minskii type potentials. This is in fact the case, and the rest of this note aims to settle the problem in the appropriate framework, to explain our main result (the AK-duality) and describe its geometric consequences. An important starting point is to reduce the regularity of solutions of the relevant differential inequalities.

1.4 On Weak Formulations: The Case of Quasilinear Operators

Formulations of maximum principles at infinity for functions with less than C^2 regularity have already been studied in depth in recent years, see [4, 10, 58, 59, 62], by using distributional solutions. Due to the appearance of quasilinear operators in Geometric Analysis, a natural class of inequalities to investigate is the following quasilinear one:

$$\Delta_a u \doteq \mathrm{div}\Big(a(|\nabla u|)\nabla u\Big) \geqslant b(x)f(u)l(|\nabla u|), \tag{19}$$

for $a \in C(\mathbb{R}^+)$, $0 < b \in C(X)$, $f \in C(\mathbb{R})$, $l \in C(\mathbb{R}^+_0)$, considered in [10] in full generality. For instance, the study of graphs with prescribed mean curvature and of mean curvature solitons in warped product ambient space leads to inequalities like

$$\mathrm{div}\left(\frac{\nabla u}{\sqrt{1+|\nabla u|^2}}\right) \geqslant \frac{b(x)f(u)}{\sqrt{1+|\nabla u|^2}},$$

see Section 1 in [10]. The weak and strong maximum principles are stated in terms of functions solving (19) on some upper level set, much in the spirit of (i) at page 423, and are summarized in the next

Definition 1.9 We say that

- $(bl)^{-1}\Delta_a$ satisfies the *weak maximum principle at infinity* if for each non-constant $u \in \mathrm{Lip}_{loc}(X)$ bounded above, and for each $\eta < \sup_X u$,

$$\inf_{\{u>\eta\}} \left\{ \Big(b(x)l(|\nabla u|)\Big)^{-1} \Delta_a u \right\} \leqslant 0,$$

and the inequality has to be intended in the following sense: if u solves

$$\Delta_a u \geqslant Kb(x)l(|\nabla u|) \qquad \text{weakly on } \{u > \eta\}, \tag{20}$$

for some $K \in \mathbb{R}$, then necessarily $K \leqslant 0$.

- $(bl)^{-1}\Delta_a$ satisfies the *strong maximum principle at infinity* if for each non-constant $u \in C^1(X)$ bounded above, and for each $\eta < \sup_X u$, $\varepsilon > 0$,

$$\Omega_{\eta,\varepsilon} = \{x \in X \ : \ u(x) > \eta, \ |\nabla u(x)| < \varepsilon\} \qquad \text{is non-empty,} \tag{21}$$

and

$$\inf_{\Omega_{\eta,\varepsilon}} \left\{ \Big(b(x)l(|\nabla u|) \Big)^{-1} \Delta_a u \right\} \leqslant 0,$$

where, again, the inequality has to be intended in the way explained above.

To present, there exist sharp sufficient conditions both to guarantee the weak and the strong principles for $(bl)^{-1}\Delta_a$. These are explicit, and expressed in terms of the growth of the Ricci curvature (of the type in (5)) or of the volume of geodesic balls in X (resembling (16)). The conditions enable to deduce, among others, sharp Liouville theorems for entire graphs with controlled mean curvature. The interested reader is referred to Theorems 1.2 and 1.7 in [10] for the most up-to-date results, and to [4, 10, 58] for applications. While working with distributional solutions is quite effective for the weak principle, it seems not an optimal choice in the presence of a gradient condition because $\Omega_{\eta,\varepsilon}$ in (21) needs to be open and thus forces to restrict to C^1 functions u. For our purposes, we found more advantageous to work with upper semicontinuous (USC) viscosity solutions.

2 The General Framework

We summarize the picture both for the weak and the strong principles. By property ($\mathscr{A}$) above, they can be rephrased in terms of solutions of a fully nonlinear PDE of the type

$$\mathscr{F}\big(x, u(x), \nabla u(x), \nabla^2 u(x)\big) \geqslant 0 \tag{22}$$

on an open subset $U \subset X$, where $\mathscr{F}$ is continuous in its arguments, elliptic and proper, in the following sense:

(degenerate ellipticity)	$\mathscr{F}(x, r, p, A) \geqslant \mathscr{F}(x, r, p, B)$	if $A \geqslant B$ as a quadratic form,
(properness)	$\mathscr{F}(x, r, p, A) \leqslant \mathscr{F}(x, s, p, A)$	if $r \geqslant s$.

For instance,

$$\mathscr{F} = \mathrm{Tr}(A) - f(r) \qquad \text{for the weak Laplacian case, or}$$

$$\mathscr{F} = \max\left\{\lambda_m(A) - f(r),\ |p| - \xi(-r)\right\} \qquad \text{for the strong Hessian case,}$$

for some (any) f, ξ satisfying (fξ). Note that the 4-ple (x, r, p, A) lies in the set

$$J^2(X) = \left\{(x, r, p, A)\ :\ x \in X,\ r \in \mathbb{R},\ p \in T_xX,\ A \in \mathrm{Sym}^2(T_xX)\right\},$$

called the 2-jet bundle of X, and in what follows, with J^2_{xu} we denote the 2-jet of u at x, i.e. the 4-ple $(x, u(x), \nabla u(x), \nabla^2 u(x))$.

Remark 2.1 (Regularity of Solutions) Although in the above discussion we dealt with C^2 solutions, it will be crucial for us to relax the regularity requirements and consider viscosity solutions: an upper semicontinuous (USC) function $u : X \to [-\infty, +\infty)$ solves (22) in the viscosity sense provided that, for every x and for every test function ϕ of class C^2 in a neighbourhood of x and touching u from above at x, that is, satisfying

$$\begin{cases} \phi \geqslant u & \text{around } x, \\ \phi(x) = u(x), \end{cases} \qquad \text{it holds} \qquad \mathscr{F}\big(x, \phi(x), \nabla\phi(x), \nabla^2\phi(x)\big) \geqslant 0,$$

To state the property that encompasses the maximum principles discussed above, it is more convenient for us to exploit the geometric approach to fully-nonlinear PDEs pioneered by Krylov [43] and systematically developed by Harvey and Lawson Jr. in recent years [29, 30, 32]. To the differential inequality (22), we associate the closed subset

$$F = \left\{(x, r, p, A)\ :\ \mathscr{F}(x, r, p, A) \geqslant 0\right\} \subset J^2(X). \tag{23}$$

The ellipticity and properness of $\mathscr{F}$ imply a positivity and negativity property for F (properties (P) and (N) in [30]). To avoid some pathological behaviour in the existence-uniqueness theory for the Dirichlet problem for $\mathscr{F} = 0$, one also needs a mild topological requirement on F (assumption (T) in [30]). A subset $F \subset J^2(X)$ satisfying (P), (N), (T) is called a *subequation*: it might be given in terms of a function $\mathscr{F} : J^2(X) \to \mathbb{R}$, as in (23), but not necessarily.

A function $u \in C^2(X)$ is said to be F-subharmonic if $J^2_x u \in F$ for each $x \in X$. If $u \in USC(X)$, as in Remark 2.1 we say that u is F-subharmonic if, for every test function $\phi \in C^2$ at any point x, $J^2_x\phi \in F$. Given an open subset $\Omega \subset M$, we define

$$F(\Omega) = \left\{u \in USC(\Omega)\ :\ u \text{ is F-subharmonic on } \Omega\right\},$$

while, for closed K, we set F(K) to denote the functions $u \in USC(K)$ that are F-subharmonics on Int K.

Examples that are relevant for us include the following subsets for $f \in C(\mathbb{R})$ non-decreasing and denoting with $\lambda_1(A) \leqslant \ldots \leqslant \lambda_m(A)$ the eigenvalues of A.

($\mathscr{E}1$) (**Eikonal type**). The eikonal $E = \{|p| \leqslant 1\}$, and its modified version $E_\xi = \{|p| \leqslant \xi(r)\}$ for ξ satisfying ($f\xi$). In view of the properties of ξ, note that E_ξ-subharmonics must be non-positive.

($\mathscr{E}2$) (k-**subharmonics**). $F = \{\lambda_1(A) + \ldots + \lambda_k(A) \geqslant f(r)\}$, $k \leqslant m$. When $f = 0$, F-subharmonic functions are called k-plurisubharmonic; these subequations, which naturally appear in the theory of submanifolds, have been investigated for instance in [31, 32, 65, 73]. The class includes the subequation $\{\mathrm{Tr}(A) \geqslant f(r)\}$, related both to the stochastic completeness of X (if f satisfies ($f\xi$)) and to the parabolicity of X (if $f = 0$).

($\mathscr{E}3$) (**Prescribing eigenvalues**). $F = \{\lambda_k(A) \geqslant f(r)\}$, for $k \in \{1, \ldots, \dim X\}$. For $k = m$ (and, as we shall see by duality, $k = 1$) this is related to the Hessian principles. In particular, if $f \equiv 0$, the subequations

$$F_j = \{\lambda_j(A) \geqslant 0\}$$

describe the m-branches associated to the Monge-Ampère equation $\det(\nabla^2 u) = 0$.

($\mathscr{E}4$) (**Branches of** k-**Hessian subequation**). For $\lambda \doteq (\lambda_1, \ldots, \lambda_m) \in \mathbb{R}^m$ and $k \in \{1, \ldots, m\}$, consider the elementary symmetric function

$$\sigma_k(\lambda) = \sum_{1 \leqslant i_1 < \ldots < i_k \leqslant m} \lambda_{i_1} \lambda_{i_2} \cdots \lambda_{i_k}.$$

Since σ_k is invariant by permutation of coordinates of λ, we can define $\sigma_k(A)$ as σ_k being applied to the ordered eigenvalues $\{\lambda_j(A)\}$. According to Gärding's theory in [26], $\sigma_k(\lambda)$ is a hyperbolic polynomial with respect to the vector $v = (1, \ldots, 1) \in \mathbb{R}^m$. Denote with

$$\mu_1^{(k)}(\lambda) \leqslant \ldots \leqslant \mu_k^{(k)}(\lambda)$$

the ordered eigenvalues[3] of σ_k. Clearly, $\mu_j^{(k)}$ is permutation invariant, thus the expression $\mu_j^{(k)}(A)$ is meaningful. It is nontrivial to prove that

$$F_j = \left\{\mu_j^{(k)}(A) \geqslant f(r)\right\}, \qquad 1 \leqslant j \leqslant k$$

is a subequation, see [33]. In particular, if $f \equiv 0$, $F_1, \ldots, F_k$ are called the branches of the k-Hessian equation $\sigma_k(\nabla^2 u) = 0$. The smallest branch F_1 can

[3]That is, the opposite of the roots of $\mathscr{P}(t) \doteq \sigma_k(\lambda + tv) = 0$.

be equivalently described as

$$F_l = \Big\{\sigma_1(A) \geqslant 0, \ldots, \sigma_k(A) \geqslant 0\Big\}.$$

Many more examples of this kind arise from hyperbolic polynomials $q(\lambda)$, cf. [33].

($\mathscr{E}$5) (**Subequations on complex, quaternionic and Cayley manifolds**). If X is an almost complex, Hermitian manifold, the complexified Hessian matrix A splits into pieces of type $(2,0)$, $(1,1)$ and $(0,2)$, and it makes sense to consider the last three examples in terms of the eigenvalues of the Hermitian symmetric matrix $A^{(1,1)}$. In particular this includes plurisubharmonic functions, that is, solutions of

$$\big\{\lambda_1(A^{(1,1)}) \geqslant 0\big\}.$$

Analogous examples can be given on quaternionic and octonionic manifolds.

($\mathscr{E}$6) (**Pucci operators**). For $0 < \lambda \leqslant \Lambda$, the Pucci operators (cf. [13]) are classically defined as

$$\mathcal{P}^+_{\lambda,\Lambda}(\nabla^2 u) = \sup\Big\{ \mathrm{Tr}(X \cdot \nabla^2 u) \ : \ X \in \mathrm{Sym}^2(TX) \text{ with } \lambda I \leqslant X \leqslant \Lambda I\Big\},$$
$$\mathcal{P}^-_{\lambda,\Lambda}(\nabla^2 u) = \inf\Big\{ \mathrm{Tr}(X \cdot \nabla^2 u) \ : \ X \in \mathrm{Sym}^2(TX) \text{ with } \lambda I \leqslant X \leqslant \Lambda I\Big\}.$$

The subequations describing solutions of $\mathcal{P}^\pm_{\lambda,\Lambda}(\nabla^2 u) \geqslant f(u)$ can be defined as follows: denoting with $A^+ \geqslant 0$ and $A^- \leqslant 0$ the positive and negative part of a symmetric matrix $A = A^+ + A^-$, we can set

$$F^+_{\lambda,\Lambda} = \Big\{\lambda\,\mathrm{Tr}(A^-) + \Lambda\,\mathrm{Tr}(A^+) \geqslant f(r)\Big\},$$
$$F^-_{\lambda,\Lambda} = \Big\{\Lambda\,\mathrm{Tr}(A^-) + \lambda\,\mathrm{Tr}(A^+) \geqslant f(r)\Big\}.$$

($\mathscr{E}$7) (**Quasilinear**). We can also consider viscosity solutions of

$$\Delta_a u \doteq \mathrm{div}\big(a(|\nabla u|)\nabla u\big) \geqslant f(u), \tag{24}$$

for $a \in C^1(\mathbb{R}^+)$ satisfying

$$\theta_1(t) \doteq a(t) + ta'(t) \geqslant 0, \qquad \theta_2(t) \doteq a(t) > 0. \tag{25}$$

Examples include

– the mean curvature operator, describing the mean curvature of the graph hypersurface $\{(x, v(x)) : x \in M\}$ into the Riemannian product $M \times \mathbb{R}$. In this case, $a(t) = (1+t^2)^{-1/2}$;

- the q-Laplacian Δ_q, $q > 1$, where $a(t) = t^{q-2}$;
- the operator of exponentially harmonic functions, where $a(t) = \exp(t^2)$, considered for instance in [22];

Indeed, expanding the divergence we can set

$$F = \overline{\{p \neq 0,\ \mathrm{Tr}(T(p)A) > f(r)\}},$$

where

$$T(p) \doteq a(|p|)\langle\, ,\, \rangle + \frac{a'(|p|)}{|p|} p \otimes p = \theta_1(|p|)\Pi_p + \theta_2(|p|)\Pi_{p^\perp},$$

and $\Pi_p, \Pi_{p^\perp}$ are, respectively, the $(2, 0)$-versions of the orthogonal projections onto the spaces $\langle p \rangle$ and $p^\perp$. Similarly, we can consider the non-variational, *normalized* quasilinear operator given by

$$F = \overline{\left\{p \neq 0,\ \frac{\mathrm{Tr}(T(p)A)}{\max\{\theta_1(|p|), \theta_2(|p|)\}} > f(r)\right\}}.$$

In the case of the mean curvature operator, the last subequation represents viscosity solutions of

$$\mathrm{div}\left(\frac{\nabla u}{\sqrt{1 + |\nabla u|^2}}\right) \geqslant \frac{f(u)}{\sqrt{1 + |\nabla u|^2}},$$

that are related to prescribed mean curvature graphs and mean curvature solitons in warped product spaces, see [10, Chapter 1].

($\mathscr{E}8$) (**∞-Laplacian**). The normalized ∞-Laplacian $F = \overline{\{p \neq 0,\ |p|^{-2}A(p, p) > f(r)\}}$.

Remark 2.2 In ($\mathscr{E}7$) and ($\mathscr{E}8$), the necessity to take as F the closure of its interior is made necessary to match property (T), due to the possible singularity of the operator at $p = 0$. Indeed, (T) also appears, implicitly, in adjusting the classical definition of subsolutions for operators $\mathscr{F}$ such that $\{\mathscr{F} \geqslant 0\}$ is not the closure of $\{\mathscr{F} > 0\}$. A typical example is the unnormalized ∞-Laplacian, cf. [17].

The above examples can be defined on each Riemannian (complex, quaternionic, octonionic) manifold, since there is no explicit dependence of F from the point x, and are therefore called *universal* subequations. To include large classes of subequations with coefficients depending on the point x, that can be seen as "deformations" of universal ones, Harvey and Lawson in [30] introduced the concept of *local jet-equivalence* between subequations. Without going into the details here, we limit to say that, for instance, any semilinear inequality of the type

$$a^{ij}(x)u_{ij} + b^i(x)u_i \geqslant c(x)f(u)$$

for smooth a^{ij}, b^i, c with $c > 0$ on X and $[a^{ij}]$ positive definite at every point, is locally jet-equivalent to the universal example describing solutions of

$$\Delta u \geqslant f(u).$$

The key fact is that local jet-equivalence allows to transfer properties holding for the universal example to the subequations locally jet equivalent to it.

Supersolutions for $\mathscr{F}$, that is, solutions of

$$\mathscr{F}\big(x, u(x), \nabla u(x), \nabla^2 u(x)\big) \leqslant 0, \tag{26}$$

are taken into account starting from the observation that $w = -u$ solves

$$\widetilde{\mathscr{F}}\big(x, w(x), \nabla w(x), \nabla^2 w(x)\big) \geqslant 0, \tag{27}$$

with

$$\widetilde{\mathscr{F}}(x, r, p, A) = -\mathscr{F}(x, -r, -p, -A).$$

This suggests to define the *dual subequation*

$$\widetilde{F} = - \sim \mathrm{Int}(F).$$

In particular, in $(\mathscr{E}1), \ldots (\mathscr{E}8)$,

$$\text{if} \quad F = \overline{\big\{\mathscr{F}(x, r, p, A) > 0\big\}}, \qquad \text{then} \qquad \widetilde{F} = \overline{\big\{\widetilde{\mathscr{F}}(x, r, p, A) > 0\big\}}.$$

Therefore, u is $\widetilde{F}$-subharmonic if $-u$ is a supersolution in the standard, viscosity sense, and u is F-harmonic on Ω if $u \in F(\Omega)$ and $-u \in \widetilde{F}(\Omega)$. The above operator is in fact a duality, in particular

$$\widetilde{F \cap G} = \widetilde{F} \cup \widetilde{G}, \qquad \widetilde{\widetilde{F}} = F \tag{28}$$

for each subequations F, G, and $\widetilde{F}$ is a subequation if F is so. Concerning examples $(\mathscr{E}1)$ to $(\mathscr{E}8)$,

- In $(\mathscr{E}1)$, the dual of the eikonal equation is $\big\{|p| \geqslant 1\big\}$, that of E_ξ is $\widetilde{E_\xi} = \big\{|p| \geqslant \xi(-r)\big\}$.
- In $(\mathscr{E}2)$,

$$\widetilde{F} = \left\{ \sum_{j=m-k+1}^{m} \lambda_j(A) \geqslant f(r) \right\};$$

in particular, $\{\mathrm{Tr}(A) \geqslant f(r)\}$ is self-dual: $\widetilde{F} = F$.

- In ($\mathscr{E}3$), $\widetilde{F} = \{\lambda_{m-k+1}(A) \geqslant f(r)\}$.
- In ($\mathscr{E}4$), if $F = \{\mu_j^{(k)}(A) \geqslant f(r)\}$ then $\widetilde{F} = \{\mu_{k-j+1}^{(k)}(A) \geqslant f(r)\}$.
- In ($\mathscr{E}6$), the dual of $F^{\pm}_{\lambda,\Lambda}$ is $F^{\mp}_{\lambda,\Lambda}$.
- Examples ($\mathscr{E}7$) and ($\mathscr{E}8$) are self-dual: $\widetilde{F} = F$.

Given a subequation F and for $g \in C(X)$ we shall introduce the *obstacle* subequation

$$F^g = F \cap \{r \leqslant g(x)\},$$

that describes F-subharmonic functions lying below the obstacle g. Note that, since the dual of $\{r \leqslant g(x)\}$ is $\{r \leqslant -g(x)\}$, $\widetilde{F^g}$ describes functions u that are $\widetilde{F}$-subharmonic on the upper set $\{u(x) > -g(x)\}$. As we shall see in a moment, functions in $\widetilde{F^0}$ will be used to describe the property that unifies the maximum principles at infinity described above, and it is therefore expectable, by duality, that obstacles subequations play an important role in our main result.

Definition 2.3 Let $F \subset J^2(X)$ be a subequation. Given $\Omega \Subset X$ open, $g \in C(\overline{\Omega})$ and $\phi \in C(\partial\Omega)$ with $\phi \leqslant g$ on $\partial\Omega$, a function $u \in C(\overline{\Omega})$ is said to solve the *obstacle problem* with obstacle g and boundary value ϕ if

$$\begin{cases} u \quad \text{is } F^g\text{-harmonic on } \Omega \\ u = \phi \quad \text{on } \partial\Omega. \end{cases}$$

3 Ahlfors, Khas'minskii Properties and the AK-Duality

The definition of the next property is inspired by the original work of Ahlfors [2], as well as by the recent improvements in [4, 5, 37].

Definition 3.1 A subequation $H \subset J^2(X)$ is said to satisfy the *Ahlfors property* if, having set $H_0 = H \cup \{r \leqslant 0\}$, for each $U \subset X$ open with non-empty boundary and for each $u \in H_0(\overline{U})$ bounded from above and positive somewhere, it holds

$$\sup_{\partial U} u^+ \equiv \sup_{\overline{U}} u.$$

Roughly speaking, when u is H-subharmonic on the set $\{u > 0\}$, the Ahlfors property means that its supremum is attained on the boundary of U.

Example 3.2 We consider the following subequations:

1) If $F = \{\mathrm{Tr}(A) \geqslant 0\}$, in view of Ahlfors' characterization [2], the Ahlfors property for $\widetilde{F}$ (= F) is a version, for viscosity solutions, of the property

characterizing the parabolicity of X in [2]. Similarly, if $F = \{\mathrm{Tr}(A) \geqslant f(r)\}$ for f satisfying (fξ), the Ahlfors property for $\widetilde{F}$ can be seen as a viscosity version of the weak Laplacian principle, that is, of the stochastic completeness of X. As a matter of fact (cf. [48]), in both of the cases the property still *characterizes* the parabolicity, respectively the stochastic completeness, of X.

2) The Ahlfors property for the dual eikonal $\widetilde{E} = \{|p| \geqslant 1\}$ can be viewed as a viscosity version of Ekeland's principle. Its equivalence to the original Ekeland's principle, hence to geodesic completeness, is one of the applications of our main result below.
3) Consider the subequations $F = \{\lambda_1(A) \geqslant f(r)\}$ and $E_\xi = \{|p| \leqslant \xi(r)\}$, for (f, ξ) satisfying (fξ). Then, the Ahlfors property for the dual

$$\widetilde{F \cap E_\xi} = \widetilde{F} \cup \widetilde{E_\xi} = \{\lambda_m(A) \geqslant f(r)\} \cup \{|p| \geqslant \xi(-r)\}$$
$$= \Big\{ \max\{|p| - \xi(-r),\ \lambda_m(A) - f(r)\} \geqslant 0 \Big\}$$

can be seen as a viscosity analogue of the strong Hessian principle. Analogously, the Ahlfors property for $\widetilde{F} \cup \widetilde{E_\xi}$ with $F = \{\mathrm{Tr}(A) \geqslant f(r)\}$ is a natural, viscosity version of Yau's strong Laplacian principle. Differently from the examples in 1), it is *not known* whether these Ahlfors properties are, in fact, equivalent to the classical strong Hessian and Laplacian principles for C^2 solutions.

A comment is in order: although the above viscosity versions might be strictly stronger than the corresponding classical ones, all of the known geometric conditions that guarantee the weak and strong Hessian or Laplacian principles in the C^2 case *also ensure* their viscosity counterparts. Therefore, passing to the viscosity realm does not prevent from geometric applications, and indeed is able to uncover new relations. To see them, we shall introduce the Khas'minskii properties, that generalize (3) and (17). Hereafter, a pair (K, h) consists of

- a smooth, relatively compact open set $K \subset X$;
- a function $h \in C(X\backslash K)$ satisfying $h < 0$ on $X\backslash K$ and $h(x) \to -\infty$ as x diverges.

Definition 3.3 A subequation $F \subset J^2(X)$ satisfies the *Khas'minskii property* if, for each pair (K, h), there exists a function w satisfying:

$$w \in F(X\backslash K), \quad h \leqslant w \leqslant 0 \quad \text{on } X\backslash K, \quad \text{and} \quad w(x) \to -\infty \quad \text{as x diverges.} \tag{29}$$

Such a function w is called a *Khas'minskii potential* for (K, h).

Loosely speaking, F has the Khas'minskii property if it is possible to construct F-subharmonic exhaustions that decay to $-\infty$ as slow as we wish. In practice, checking the Khas'minskii property might be a hard task, and often, from the geometric problem under investigation, one is just able to extract *some* of the Khas'minskii potentials. This motivates the following definition (cf. the recent [47]).

Definition 3.4 A subequation $F \subset J^2(X)$ satisfies the *weak Khas'minskii property* if there exist a relatively compact, smooth open set K and a constant $C \in \mathbb{R} \cup \{+\infty\}$ such that, for each $x_0 \notin \overline{K}$ and each $\varepsilon > 0$, there exists w satisfying

$$w \in F(X \backslash K), \qquad w \leqslant 0 \quad \text{on } X \backslash K, \qquad w(x_0) \geqslant -\varepsilon, \qquad \limsup_{x \to \infty} w(x) \leqslant -C. \tag{30}$$

We call such a w a *weak Khas'minskii potential* for the triple $(\varepsilon, K, \{x_0\})$.

Remark 3.5 When $C = +\infty$ and F is scale invariant (that is, it is fiber-wise a cone), the condition over ε in (30) can be avoided by rescaling w.

Example 3.6 The existence of w in (3), for G satisfying (4), implies that a weak Khas'minskii property holds for the subequation

$$\big\{ \mathrm{Tr}(A) \geqslant G(-r) \big\} \cap \big\{ |p| \leqslant G(-r) \big\}.$$

Indeed, the weak Khas'minskii potentials can be constructed by suitably rescaling and modifying the function $-w$. Up to playing with w and G, the above is equivalent to the weak Khas'minskii property for

$$\big\{ \mathrm{Tr}(A) \geqslant f(r) \big\} \cap \big\{ |p| \leqslant \xi(r) \big\},$$

for some (any) (f, ξ) satisfying $(f\xi)$. Similarly, the existence of w in (17) implies the weak Khas'minskii property for $F = \big\{ \mathrm{Tr}(A) \geqslant f(r) \big\}$.

We are ready to state our main result, the Ahlfors-Khas'minskii duality (shortly, AK-duality), Theorems 4.3 and 4.10 in [48]. It applies to subequations F on X that are locally jet-equivalent to a universal one and satisfy a few further assumptions. Some of them are merely technical and will not be described here. Their validity characterizes the set of *admissible* subequations, that is still quite general. For instance, each of the examples in $(\mathscr{E}2), \ldots, (\mathscr{E}6)$, and the subequations locally jet-equivalent to them, are admissible provided that f satisfies $(f\xi)$.

Theorem 3.7 *Let* $F \subset J^2(X)$ *be an admissible subequation, locally jet-equivalent to a universal one. Assume that*

$(\mathscr{H}1)$ *negative constants are strictly* F*-subharmonic;*
$(\mathscr{H}2)$ F *satisfies the comparison theorem: whenever* $\Omega \Subset X$ *is open,* $u \in F(\Omega)$, $v \in \widetilde{F}(\Omega)$,

$$u + v \leqslant 0 \quad \text{on } \partial\Omega \quad \Longrightarrow \quad u + v \leqslant 0 \quad \text{on } \Omega.$$

Then, AK-duality holds for F *and for* $F \cap E_\xi$ *for some (any)* ξ *satisfying* $(f\xi)$*, i.e.,*

$$\begin{matrix} F \textit{ satisfies } (K) \\ (\textit{Khas'minskii prop.}) \end{matrix} \iff \begin{matrix} F \textit{ satisfies } (K_w) \\ (\textit{weak Khas'minskii prop.}) \end{matrix} \iff \begin{matrix} \widetilde{F} \textit{ satisfies } (A) \\ (\textit{Ahlfors prop.}) \end{matrix},$$

and

$$F \cap E_\xi \text{ satisfies (K)} \iff F \cap E_\xi \text{ satisfies } (K_w) \iff \widetilde{F} \cup \widetilde{E}_\xi \text{ satisfies (A)}.$$

Seeking to clarify the role of each assumption in the AK-duality, we briefly examine the importance of each one.

Remark 3.8 (On Assumption ($\mathscr{H}1$)) This property holds for each of ($\mathscr{E}2$), ..., ($\mathscr{E}6$) provided that f satisfies (fξ), and it is important to ensure the validity of the finite maximum principle: functions $u \in \widetilde{F^0}(Y)$ cannot achieve a local positive maximum. The latter is crucial for our proof to work.

Remark 3.9 (On Assumption ($\mathscr{H}2$)) This is delicate to check, and curiously enough it plays a role just in the proof of $(K_w) \Rightarrow$ (A). Comparison holds for uniformly continuous subequations which are strictly increasing in the r variable, a case that covers examples ($\mathscr{E}2$), ..., ($\mathscr{E}5$) as well as ($\mathscr{E}6$), see [48, Thm. 2.25] for details.[4] The uniform continuity resembles the classical condition 3.14 in [18]. Regarding examples ($\mathscr{E}7$) and ($\mathscr{E}8$), the worse dependence on the gradient term makes comparison much subtler. One can check the comparison theorem for the universal subequation in ($\mathscr{E}8$), even with $f \equiv 0$, by means of other interesting methods, cf. [48, Thm. 2.27]. As for ($\mathscr{E}7$), on Euclidean space the validity of comparison for strictly increasing f is a direct application of the classical theorem on sums (i.e. Ishii's Lemma, [18] and [30, Thm. C.1]). In a Riemannian setting, Ishii's Lemma uses the infimal convolutions with the squared distance function $r^2(x, y)$ on $X \times X$, and in neighbourhoods where the sectional curvature is negative the Hessian of r^2 is *positive* to second order on pairs of parallel vectors. The error term produced by such positivity can be easily controlled for normalized quasilinear operators, but in the unnormalized case one has to require the boundedness of the eigenvalues θ_1, θ_2 in order to avoid further a-priori bounds on the subsolutions and supersolutions (like Lipschitz continuity of either one of them). Nevertheless, we note that the boundedness of θ_1, θ_2 notably includes the mean curvature operator.

Summarizing, we have

[4] In [48], the uniform continuity of the Pucci operators in ($\mathscr{E}6$) is not explicitly stated but can be easily checked. For instance, in the case of $\mathcal{P}^+_{\lambda,\Lambda}$, referring to Definition 2.23 in [48] and using the min-max definition,

$$\mathcal{P}^+_{\lambda,\Lambda}(B) \geqslant \mathcal{P}^+_{\lambda,\Lambda}(A) - \mathcal{P}^+_{\lambda,\Lambda}(A - B) \geqslant \mathcal{P}^+_{\lambda,\Lambda}(A) - \Lambda \operatorname{Tr}\big((A - B)_+\big).$$

If $\|(A - B)_+\| < \delta$, then $\mathcal{P}^+_{\lambda,\Lambda}(B) \geqslant \mathcal{P}^+_{\lambda,\Lambda}(A) - m\Lambda\delta$, that proves the uniform continuity of $F^+_{\lambda,\Lambda}$. The case of $F^-_{\lambda,\Lambda}$ is analogous.

Corollary 3.10 *The AK-duality holds both for* F *and for* $F \cap E_\xi$*, with* ξ *satisfying* (fξ)*, in each of the following cases:*

- F *is locally jet-equivalent to any of* ($\mathscr{E}$2), ..., ($\mathscr{E}$6)*;*
- F *is the normalized quasilinear example in* ($\mathscr{E}$7)*, or* F *is the unnormalized example with* $\theta_1, \theta_2 \in L^\infty(\mathbb{R}^+)$*;*
- F *is the universal example in* ($\mathscr{E}$8)*,*

and, in each case, f *satisfies* (fξ)*. Furthermore, the AK-duality holds for the eikonal subequations in* ($\mathscr{E}$1)*.*

Sketch of the Proof of the AK-Duality Since (K) $\Rightarrow$ (K_w) is obvious, we shall prove (K_w) $\Rightarrow$ (A) and (A) $\Rightarrow$ (K). The proof of the first implication is inspired by a classical approach that dates back to Phrágmen-Lindeloff type theorems in classical complex analysis, and we therefore concentrate on (A) $\Rightarrow$ (K).

Fix a pair (K, h), and a smooth exhaustion $\{D_j\}$ of X with $K \subset D_1$. Our desired Khas'minskii potential w will be constructed as a locally uniform limit of a decreasing sequence of USC functions $\{w_i\}$, such that $w_0 = 0$ and for each $i \geqslant 1$ we have:

$$
\begin{array}{lll}
\text{(a)} & w_i \in F(X\backslash K), \quad w_i = (w_i)_* = 0 \quad \text{on } \partial K; & \\
\text{(b)} & w_i \geqslant -i \quad \text{on } X\backslash K, \quad w_i = -i \quad \text{outside a compact set } C_i \text{ containing } D_i; & \\
\text{(c)} & \left(1 - 2^{-i-2}\right) h < w_{i+1} \leqslant w_i \leqslant 0 \quad \text{on } X\backslash K, \quad \|w_{i+1} - w_i\|_{L^\infty(D_i\backslash K)} \leqslant \frac{\varepsilon}{2^i}. &
\end{array}
\tag{31}
$$

With the above properties, the sequence $\{w_i\}$ is locally uniformly convergent on $X\backslash K$ to some function $w \in F(X\backslash K)$ with $h \leqslant w \leqslant 0$ on $X\backslash K$ and satisfying $w(x) \to -\infty$ as x diverges, that is, to the desired Khas'minskii potential.

Fix $w = w_i$. We build w_{i+1} inductively via a sequence of obstacle problems, an idea inspired by [50, 70]: we fix obstacles $g_j = w + \lambda_j$, for some sequence $\{\lambda_j\} \subset C(X\backslash K)$ such that

$$
\begin{cases}
0 \geqslant \lambda_j \geqslant -1, \quad \lambda_j = 0 \text{ on } K, \quad \lambda_j = -1 \text{ on } X\backslash D_{j-1}, \\
\{\lambda_j\} \text{ is an increasing sequence, and } \lambda_j \uparrow 0 \text{ locally uniformly,}
\end{cases}
\tag{32}
$$

and search for solutions of the obstacle problem

$$
\begin{cases}
u_j \quad \text{is } F^{g_j}\text{-harmonic on } D_j\backslash K, \\
u_j = 0 \quad \text{on } \partial K, \qquad u_j = -i - 1 \quad \text{on } \partial D_j.
\end{cases}
\tag{33}
$$

However, in some relevant cases we cannot fully solve (33). The first problem we shall consider is the absence of barriers, needed to prove the existence of u_j via Perron's method. No problem arise on ∂D_j, since the constant function $-i-1$ is F^{g_j}-subharmonic by ($\mathscr{H}$1). However, since we are working in the complement of a compact set K (think of K being a small geodesic ball, for instance), ∂K might be concave in the outward direction, that in general prevents from having barriers there. To overcome this problem, we modify X inside of K by gluing a compact manifold Y that is Euclidean in a sufficiently small ball $\mathbb{B}$. The gluing only involves small annuli inside of $\mathbb{B}$ and K, with the new metric coinciding with those of X and Y outside of the gluing region. In particular, the new manifold is Euclidean in a neighbourhood of $\partial\mathbb{B}$. In this way, replacing K by $K' = Y\backslash\mathbb{B}$, $X\backslash K$ embeds isometrically into $X\backslash K'$ and the latter has a *convex* boundary isometric to $\partial\mathbb{B}$. Because of a technical assumption included in those defining the admissibility of F, this is enough to produce barriers on $\partial K'$. Once we perform this change, we suitably modify the subequation F preserving it outside the gluing region, and making it, on K', the universal Riemannian subequation to which F is locally jet equivalent. Although these modifications change in several ways the manifold and the subequation, they are stable to preserve the Ahlfors property for $\widetilde{F}$ as well as the assumption ($\mathscr{H}1$). The price to pay is that we may lose the comparison property ($\mathscr{H}2$), since comparison is very sensitive to the geometry of X, at least for some relevant operators like those in ($\mathscr{E}2$), ($\mathscr{E}3$), ($\mathscr{E}4$). This is the main reason why, generally, we cannot fully solve (33). However, with barriers finally available, Perron's method yields an "almost solution" u_j of the obstacle problem on $X\backslash K'$ (see [30], [48, Thm. 3.3]), that is, u_j solves

$$\begin{cases} u_j \in F^{g_j}(\overline{D_j\backslash K'}), & (-u_j)^* \in \widetilde{F^{g_j}}(\overline{D_j\backslash K'}), \\ u_j = (u_j)_* = 0 & \text{on } \partial K', \\ u_j = (u_j)_* = -i-1 & \text{on } \partial D_j. \end{cases} \tag{34}$$

We extend u_j outside D_j by setting $u_j \doteq -i-1$, and define $v_j \doteq (-u_j)^* - i$. By the definition of Perron's solution, the sequence $\{v_j\}$ is decreasing on $X\backslash K'$. Thus, passing to the limit using that $g_j \to w \geqslant -i$ as $j \to \infty$, $w = -i$ outside of C_i, we get

$$v_j \downarrow v \in \widetilde{F^0}(X\backslash K'), \quad \text{with} \quad \begin{cases} -i \leqslant v \leqslant 1 & \text{on } X\backslash K', \\ v = -i < 0 & \text{on } \partial K', \\ v \geqslant 0 & \text{on } X\backslash C_i. \end{cases}$$

Here is the crucial point where the Ahlfors property enters: in fact, using Ahlfors on $X\backslash K'$ we infer that $v \equiv 0$ outside of C_i, and by the USC-version of Dini's theorem,

$$v_j \downarrow 0 \qquad \text{locally uniformly on } X\backslash C_i.$$

Then, the definition of v_j yields

$$u_j \uparrow -i \quad \text{locally uniformly on } X\backslash C_i. \tag{35}$$

It remains to investigate the convergence of u_j on the bounded set $\overline{C_i}\backslash K'$. Although comparison might fail on this set, what guarantees the convergence $u_j \uparrow w$ is that each u_j, being a Perron's solution, is maximal in the set of F^{g_j}-subharmonic functions whose boundary values do not exceed 0 (on $\partial K'$) and $-i-1$ (on ∂D_j). Concluding, $u_j \uparrow w$ locally uniformly on $X\backslash K'$, hence on $X\backslash K$. For j large enough, if we set $w_{i+1} = u_j$ it is therefore possible to meet all of (a)–(c) in (31), as desired.

To treat the case when F is coupled to the eikonal equation E_ξ, the issue is again the absence of barriers on $\partial K'$ to solve the obstacle problem for $F^{g_j} \cap E_\xi$. Indeed, even if, after the gluing, $\partial K'$ is convex in the direction pointing towards $X\backslash K'$, barriers must be E_ξ-subharmonic and the gradient control may prevent to build barriers up to height $-i$ at step i. To overcome this problem, the idea is to modify the subequation E_ξ in the gluing region in a different way at each step i, weakening the bound $\xi(r)$ by means of a cut-off function ϕ_i supported in a neighbourhood of $\partial K'$. The size of ϕ_i depends on the L^∞ norm of the gradient of the barriers on $\partial K'$ joining zero to $-i$, and therefore it diverges as $i \to \infty$. In this way, we clearly lose the gradient control in the limit in a neighbourhood of K', but since $K' \Subset K$, for suitable ϕ_i no property of w_i on $X\backslash K$ get lost. □

It is worth to remark that an important case was left uncover by Theorem 3.7. For instance, when F is independent on r (examples $(\mathscr{E}2), \dots, (\mathscr{E}6)$ with $f \equiv 0$), assumption $(\mathscr{H}1)$ does not hold. However $(\mathscr{H}1)$ is just used to ensure the strong maximum principle for functions in $\widetilde{F^0}$ on any manifold. Therefore, we can state the following alternative version of our main theorem.

Theorem 3.11 *Let* $F \subset J^2(X)$ *be a universal subequation satisfying* $(\mathscr{H}2)$ *and*

$(\mathscr{H}1')$ $\widetilde{F}$ *has the strong maximum principle on each manifold* Y *where it is defined:* $\widetilde{F^0}$*-subharmonic functions on* Y *are constant if they attain a local maximum.*

Then, AK-duality holds for F.[5]

The strong maximum principle for viscosity subsolutions is a classical subject that has been investigated by many authors, in particular we quote [8, 35, 41] (cf. also [10, 63, 64] for the quasilinear case). Particularizing Theorems 3.7 and 3.11 to the mean curvature operator and its normalized version, for which the strong maximum principle is proved in [41], we have the following:

Theorem 3.12 *The AK-duality holds for the subequation in* $(\mathscr{E}7)$ *describing solutions of*

$$\operatorname{div}\left(\frac{\nabla u}{\sqrt{1+|\nabla u|^2}}\right) \geqslant f(u) \qquad \textit{and} \qquad \operatorname{div}\left(\frac{\nabla u}{\sqrt{1+|\nabla u|^2}}\right) \geqslant \frac{f(u)}{\sqrt{1+|\nabla u|^2}}, \tag{36}$$

for every non-decreasing, odd function $f \in C(\mathbb{R})$.

[5] Theorem 3.11 can be stated for F locally jet-equivalent to a universal example, provided that the strong maximum principle in $(\mathscr{H}1')$ holds for each manifold Y and each $\widetilde{F} \subset J^2(Y)$ constructed by gluing as in the theorem.

Other Quasilinear Operators

As said, the lack of a strong enough comparison theorem forces us to require, in the unnormalized version of ($\mathscr{E}7$) of Corollary 3.10, the boundedness of the eigenvalues θ_1, θ_2 on $\mathbb{R}_0^+$. We believe that the AK-duality holds for each subequation locally jet-equivalent to ($\mathscr{E}7$), both normalized and unnormalized, independently of θ_1, θ_2. More information can be found in Section 2.5 and Appendix A of [48], where the authors investigate classes of quasilinear operators where comparison holds. For inequalities of the type

$$\mathrm{div}\mathscr{A}(x, \nabla u) \geqslant \mathscr{B}(x, u),$$

with $\mathscr{A}$ a Caratheódory map that locally behaves like a q-Laplacian, and $\mathscr{B}$ non-decreasing in u with $u\mathscr{B}(x, u) \geqslant 0$, the AK-duality in a slightly less general version was first established in [50]. The use of weak instead of viscosity solutions allows to work with very general $\mathscr{A}$, $\mathscr{B}$, since a comparison theorem is easy to show and the obstacle problem is solvable by classical results. Nevertheless, the method does not allow to include a gradient dependence and thus investigate the "strong" versions of the corresponding Ahlfors property.

Liouville Property

As the cases of parabolicity and stochastic completeness show, the Ahlfors property is also related to the next Liouville one:

Definition 3.13 A subequation $F \subset J^2(X)$ has the *Liouville property* if any $u \in F(X)$ bounded from above and non-negative is constant.

Indeed, in [50] the main result itself is expressed as a duality between Khas'minskii and Liouville properties. It is not difficult to show that the Ahlfors property implies the Liouville one, and that the two are equivalent provided that

$$u \equiv 0 \qquad \text{is F-harmonic,} \tag{37}$$

cf. [48, Prop. 4.2] and previous work in [2, 4, 5, 28]. While (37) holds in many instances, there are notable exceptions, for example the eikonal subequation. For such subequations, it is the Ahlfors property the one that actually realizes duality.

4 Applications

4.1 *Completeness, Viscosity Ekeland Principle and ∞-Parabolicity*

Let $u \in C^1(X)$ be a function bounded from above and assume that there exists a classical C^1-Khas'minskii potential w, that is, satisfying only the exhaustion and the gradient properties in (18). Up to a rescaling, w is a Khas'minskii potential for the eikonal subequation $E = \{|p| \leqslant 1\}$. Following the original argument that goes back to Ahlfors [1], we consider a sequence of functions $u + \frac{1}{k}w$ each of which attains a maximum at some point $x_k \in X$. Up to choosing a subsequence, it is easy to see that

$$u(x_k) > \sup_X u - \frac{1}{k}, \quad \text{and} \quad u(y) \leqslant u(x_k) + \frac{1}{k}d(x_k, y) \quad \text{for } y \text{ nearby } x_k.$$

Thus, recalling the AK-duality, one can see the Ahlfors property for the dual eikonal subequation $\widetilde{E} = \{|p| \geqslant 1\}$ as a sort of *viscosity version of Ekeland principle.* Clearly, the above argument does not give a formal proof of the equivalence between the Ahlfors property for $\widetilde{E}$ and the Ekeland principle stated in Definition 1.1. In fact, it follows from the next application of Theorem 3.7:

Theorem 4.1 *Let X be a Riemannian manifold. Then, the following statements are equivalent:*

(1) X *is complete.*
(2) *the dual eikonal* $\widetilde{E} = \{|p| \geqslant 1\}$ *has the Ahlfors property (viscosity Ekeland principle).*
(3) *the infinity Laplacian* $F_\infty \doteq \overline{\{A(p, p) > 0\}}$ *has the Ahlfors property.*
(4) F_∞ *has the next strengthened Liouville property:*

> *Any* F_∞*-subharmonic function* $u \geqslant 0$ *such that* $|u(x)| = o(\varrho(x))$ *as* x *diverges (*$\varrho(x)$ *the distance from a fixed origin) is constant.*

Sketch of the Proof The key implications are (2) $\Rightarrow$ (1) and (3) $\Rightarrow$ (1). Both proceed by contradiction, so assume the existence of a unit speed geodesic γ defined on a maximal finite interval $[0, T)$, and pick a small compact set K not intersecting $\gamma([0, T))$ (this is possible since γ is diverging).

(2) $\Rightarrow$ (1): Apply the AK-duality to produce a Khas'minskii potential $w \in E(X\backslash K)$. By restriction, the function $u \doteq w \circ \gamma$ is E-subharmonic on $[0, T)$, that is, any C^2 test ϕ touching u from above shall satisfy $|\phi'| \leqslant 1$ at touching points. However, since $u \leqslant 0$ and $T < +\infty$, we can choose a line with derivative strictly less than -1 lying above the graph of u: translating the line downwards up to the first touching point we get a contradiction. Thus, $T = +\infty$ and X is complete.

(3) $\Rightarrow$ (1): Pick an exhaustion of X by smooth, relatively compact domains Ω_j with $K \Subset \Omega_1$. As we said before, we exploit the existence of a (unique) ∞-capacitor u_j for (K, Ω_j) (see [17, 39]), that satisfies

$$\begin{cases} u_j \text{ is } F_\infty\text{-harmonic on } \Omega_j\backslash K, \\ u_j = 1 \quad \text{on } \partial K, \quad u_j = 0 \quad \text{on } \partial\Omega_j. \end{cases} \tag{38}$$

By comparison (Theorem 2.27 in [48]), and since $\{u_j\}$ is equi-Lipschitz because of the minimization properties of u_j, the sequence $v_j = 1 - u_j$ subconverges locally uniformly to a F_∞-harmonic, Lipschitz function $v_\infty \geqslant 0$. Applying the Ahlfors property on $X \backslash K$ we get that $v_\infty = 0$ on $X \backslash K$. Now, setting $w_j = v_j \circ \gamma$, we have $w_j(0) = 0$, $w_j = 1$ after some $T_j < T$, and by integration, $1/T \leqslant \|w_j'\|_\infty \leqslant C$ on $[0, T)$, for each j. This is impossible, since $w_j \to 0$ locally uniformly. □

4.2 The Hessian Principle and Martingale Completeness

According to 3) in Example 3.2, we formally define the viscosity, weak and strong Hessian principles in terms of Ahlfors properties. Let us consider the subequations $F = \{\lambda_1(A) \geqslant -1\}$ and $E = \{|p| \leqslant 1\}$, whose duals are $\widetilde{F} = \{\lambda_m(A) \geqslant 1\}$ and $\widetilde{E} = \{|p| \geqslant 1\}$. Then, X satisfies:

- the viscosity, weak Hessian principle if the Ahlfors property holds for $\widetilde{F}$;
- the viscosity, strong Hessian principle if the Ahlfors property holds for $\widetilde{F} \cup \widetilde{E}$.

The r independence on F and E in the above definition are just for convenience. The properties could be stated as in Example 3.2 by making use of a pair of functions (f, ξ) satisfying $(f\xi)$. As discussed in the introduction, there are evidences that an Hessian principle, either weak or strong, be related with the martingale completeness. Perhaps surprisingly, exploiting the low regularity and the AK-duality, we found that the two Hessian principles are equivalent, and that the martingale completeness is necessary for the validity of them. Apart from a regularity issue, this answers a question (Question 70) raised in [60] (see also [62]).

Theorem 4.2 *Let* X *be a Riemannian manifold. Then, the following properties are equivalent:*

(1) X *satisfies the viscosity, weak Hessian principle;*
(2) X *satisfies the viscosity, strong Hessian principle;*
(3) $F \cap E$ *has the Khas'minskii property with* C^∞ *potentials.*

In particular, all the above assertions imply that X *is martingale (and so, geodesically) complete.*

Idea of the Proof The key implication is (1) ⇒ (3). As a consequence of AK-duality and the flexibility in the choice of (f), (1) is equivalent to the Khas'minskii property for $F = \{\lambda_1(A) \geqslant f(r)\}$. The sought depends on the following facts very specific to this F.

- By exploiting Greene-Wu's techniques in [27], we can approximate a Khas'minskii potential for F with a smooth Khas'miskii potential, call it w. Up to playing with f and extending w on the entire X, we can assume that w satisfies

$$w < 0 \quad \text{on } X, \qquad w(x) \to -\infty \quad \text{if } x \text{ diverges}, \qquad \nabla^2 w \geqslant -w\langle\, ,\, \rangle \quad \text{on } X. \tag{39}$$

- Integrating along the flow lines of ∇w and applying ODE comparison, it is possible to prove that $|\nabla w| \leqslant w$ on X. Starting from w, it is therefore easy to construct a weak Khas'minskii potential for $F \cap E$ that is smooth. AK-duality again and Greene-Wu approximation yield the full Khas'minskii property.

Concluding, by work of Emery [24] property (3) is known to imply the martingale completeness of X. □

Remark 4.3 In order to check the viscosity Hessian principle, we can only consider semiconcave[6] functions, which are locally Lipschitz and 2-times differentiable a.e. Thus, regarding to regularity, the viscosity Hessian principle is very close to the classical C^2 Hessian principle.

4.3 Laplacian Principles

Differently from the Hessian principle, in view of elliptic estimates for semilinear equations, the viscosity weak Laplacian principle is equivalent to its corresponding classical C^2 principle, that is, to the stochastic completeness of X. In this case, the AK-duality improves on the original results in [42, 50]. Regarding the viscosity, strong Laplacian principle, that is, the Ahlfors property for the subequation $\{\mathrm{Tr}(A) \geqslant 1\} \cup \{|p| \geqslant 1\} = \widetilde{F} \cup \widetilde{E}$, its equivalence with the classical, C^2 one (that is, Yau's principle) seems quite delicate and is currently unknown. However, the AK-duality in Corollary 3.10 guarantees the following:

Theorem 4.4 *Let* X *be a Riemannian manifold. Then, the following statements are equivalent:*

(1) X *satisfies the viscosity, strong Laplacian principle;*
(2) $F \cap E$ *has the (weak) Khas'minskii property.*

In particular, any manifold satisfying the viscosity, strong Laplacian principle must be (geodesically) complete.

5 Partial Trace (Grassmannian) Operators

In the context of submanifolds it is interesting to consider extrinsic conditions instead of constrain directly the geometry of the submanifold. For instance, many applications of the Omori-Yau maximum principles (cf. [3, 4, 9]) have been investigated in that spirit. Specifically, when $\sigma : X^m \to Y^n$ is an isometric immersion, and $F \subset J^2(Y)$ is a subequation, the pull-back $\sigma^* F$ induces a subset

[6]By definition, a function u is semiconcave if and only if, locally, there exists $v \in C^2$, such that $u + v$ is concave when restricted to geodesics.

$H \doteq \overline{\sigma^* F}$, maybe only satisfying the conditions (P) and (N). In some relevant examples, like those in ($\mathscr{E}$2), ($\mathscr{E}$3) and their complex analogues in ($\mathscr{E}$5), the induced H is nontrivial and the following question is therefore natural:

can we transplant the Ahlfors property from $\widetilde{F}$ on Y to $\widetilde{H}$ on X?

Trying to address the problem by contradiction, that is, assuming that the Ahlfors property does not hold for $\widetilde{H}$, one would need to extend a nontrivial $\widetilde{H}$-subharmonic function on X to the entire Y. This seems a bit challenging to achieve, especially if X is merely immersed. On the contrary, the use of AK-duality makes the problem feasible. In particular, we can obtain the following result:

Theorem 5.1 *Let* $\sigma : X^m \to Y^n$ *be a proper isometric immersion. Assume that either*

i) F_f *is the universal subequation in* ($\mathscr{E}$2), ($\mathscr{E}$5) *with* $k \leqslant m$, *and* σ *has bounded second fundamental form* II*;*
ii) F_f *is the universal subequation in* ($\mathscr{E}$3) *with* $k \leqslant m$, *and*

$$\sup\Big\{ \big|\operatorname{Tr}_{\mathcal{V}} \mathrm{II}(x)\big| \ : \ x \in X, \ \mathcal{V} \leqslant T_x X \ k\text{-dimensional}\Big\} < +\infty.$$

Then,

$$\widetilde{F}_f \cup \widetilde{E}_\xi \textit{ has the Ahlfors property on } Y \implies \widetilde{F}_f \cup \widetilde{E}_\xi \textit{ has the Ahlfors property on } X,$$

for some (any) pair (f, ξ) *satisfying* $(f\xi)$.

Idea of the Proof It is conceptually quite simple: in our assumptions, AK-duality holds for each of ($\mathscr{E}$2), ($\mathscr{E}$3) and ($\mathscr{E}$5), and therefore, $F_f \cap E_\xi$ has the Khas'minskii property for some (any) such (f, ξ). Given an arbitrary potential $\bar{w}$ for $F_f \cap E_\xi$ on Y, by the flexibility in the choice of (f, ξ) and the properness of σ, the composition $w \doteq \bar{w} \circ \sigma$ should correspond to a weak Khas'minskii potential for $F_g \cap E_\xi$ on X, where g just depends on (f, ξ). To check this claim, one uses the standard chain rule formula

$$\nabla^2 w(X, Y) = \bar{\nabla}^2 w(\sigma_* X, \sigma_* Y) + \langle \bar{\nabla} w, \mathrm{II}(X, Y)\rangle, \tag{40}$$

where $\nabla, \bar{\nabla}$ are the connections on X and Y, respectively. The adaptation to viscosity solutions, however, makes the proof of the claim subtler from the technical point of view. The arbitrariness of $\bar{w}$ and of these choices guarantees the validity of the weak Khas'minskii property on X. Then, AK-duality again implies the desired conclusion. □

Remark 5.2 The presence of $\bar{\nabla} w$ in (40) forces to include the eikonal in the Ahlfors properties, otherwise more restrictive assumptions have to be imposed on X. In fact, without a gradient bound, it is possible to control the last term in (40) if and only if the second fundamental form II is trace-free on suitable subspaces $\mathcal{V}$. For instance, if $\sigma : X^m \to Y^n$ is a proper minimal immersion and F_f is the subequation described in ($\mathscr{E}$2) with $k = m$, that is, $\widetilde{F}_f = \{\lambda_{n-m+1} + \ldots + \lambda_n(A) \geqslant f(r)\}$, the validity of the Ahlfors property for $\widetilde{F}_f$ on Y implies that X is stochastically complete (i.e. X has the viscosity, weak Laplacian principle).

Remark 5.3 A result similar to Theorem 5.1 can be stated for Riemannian submersions, cf. [48, Thm. 7.8].

The class of partial trace operators, example ($\mathscr{E}$2), helps to understand the geometry of submanifolds. Thus, having in mind Theorem 5.1, it is important to investigate sufficient geometric conditions that imply the validity of the Ahlfors property for this kind of operators. Inspired by the seminal papers of Omori [55] and Yau [74] (that correspond to cases $k = 1$ and $k = m$ in ($\mathscr{E}$2), respectively), we will focus on conditions involving the k-th Ricci curvature

Definition 5.4 Let Y^n be an n-dimensional manifold, and let $k \in \{1, \ldots, n-1\}$. The k-th Ricci curvature is the function

$$\begin{aligned} \mathrm{Ric}^{(k)} : TY &\longrightarrow \mathbb{R} \\ v &\longmapsto \inf_{\substack{\mathcal{W}_k \leqslant v^\perp \\ \dim \mathcal{W}_k = k}} \left(\frac{1}{k} \sum_{j=1}^{k} \mathrm{Sect}(v \wedge e_j) \right), \end{aligned}$$

where $\{e_j\}$ is an orthonormal basis of $\mathcal{W}_k$.

We recall that bounding from below the k-th Ricci curvature is an intermediate condition between the corresponding bounding for the sectional and Ricci curvature. In the next result,

$$F_f = \{\lambda_1(A) + \ldots + \lambda_{k+1}(A) \geqslant f(r)\}$$

and the functions f and ξ satisfying (f, ξ). Having fixed an origin o, we denote with $\rho(x)$ the distance from o and with cut(o) the cut-locus of o, cf. [20].

Theorem 5.5 *Let* Y^n *be complete, and assume that*

$$\mathrm{Ric}^{(k)}_x(\nabla \rho) \geqslant -G^2(\rho(x)) \qquad \forall\, x \notin \mathrm{cut}(o), \tag{41}$$

for some $k \in \{1, \ldots, n-1\}$ *and some* G *satisfying*

$$0 < G \in C^1(\mathbb{R}^+_0), \qquad G' \geqslant 0, \qquad G^{-1} \notin L^1(+\infty).$$

Then, Y *has the Ahlfors property for* $\widetilde{F}_f \cap \widetilde{E}_\xi$. *Moreover, if* $\sigma : X^m \to Y^n$ *is a proper isometric immersion,* $k + 1 \leqslant m \leqslant n - 1$ *and the eigenvalues* $\mu_1 \leqslant \ldots \leqslant \mu_m$ *of the second fundamental form* II *satisfy*

$$\max\Big\{\big|\mu_1+\ldots+\mu_{k+1}\big|,\ \big|\mu_{m-k}+\ldots+\mu_m\big|\Big\}\leqslant CG(\rho\circ\sigma)\qquad on\ X,\tag{42}$$

for some constant $C>0$, *then the Ahlfors property for* $\widetilde{F}_f\cap\widetilde{E}_\xi$ *holds on* X. *In particular, if* $m=k+1$ *and the mean curvature satisfies*

$$|H|\leqslant CG(\rho\circ\sigma),$$

then X *has the viscosity, strong Laplacian principle.*

6 AK-Duality and Polar Sets

With the aid of AK-duality, we can characterize polar (hence, removable) sets for subequations in terms of preservation of the Ahlfors property. The study of removable sets for linear and nonlinear equations is a classical subject with a long history, and the interested reader can consult the recent [34] and the references therein for further insight. There, the problem is set in the language of subequations, and to introduce our application we first need to recall some terminology. We say that a subequation $F\subset J^2(X)$ is a

- truncated cone subequation if each fiber F_x over a point $x\in X$ is a truncated cone, that is, it satisfies the following property:

$$\text{if}\quad J\in F_x,\quad\text{then}\quad tJ\in F_x\quad\forall t\in[0,1].$$

- convex cone subequation if each fiber F_x over a point $x\in X$ is a convex cone.

A subequation M is called a *monotonicity cone* for F if M is a convex cone subequation and $F+M\subset F$, that is, $J_1+J_2\in F$ whenever $J_1\in F$ and $J_2\in M$. In this case, we say that F is M-monotone. By duality, also $\widetilde{F}$ is M-monotone, that is,

$$F+M\subset F\qquad\Longrightarrow\qquad\widetilde{F}+M\subset\widetilde{F}.$$

In particular, since M is a convex cone subequation, $M+M\subset M$ and thus $\widetilde{M}$ is M-monotone and $M\subset\widetilde{M}$. In general, $\widetilde{M}$ is a cone subequation much larger than M and it is non-convex. Moreover, it is maximal among M-monotone cone subequations: indeed, if F is a cone subequation that is M-monotone, then $0\in F$ and thus $M=0+M\subset F$. Duality gives $\widetilde{F}\subset\widetilde{M}$.

Example 6.1

1) On an almost complex, Hermitian manifold X, consider the subequations

$$F_j = \{\lambda_j(A^{(1,1)}) \geqslant 0\}, \qquad 1 \leqslant j \leqslant m,$$

that are the branches of the complex Monge-Ampère equation $\det(\nabla^2 u)^{(1,1)} = 0$. Then, F_1 (the only branch that is convex) is a monotonicity cone for each F_j.
2) Let $F_1, \ldots, F_k$ be the branches of the k-Hessian equation $\sigma_k(\nabla^2 u) = 0$. Then, the smallest branch F_1 is a monotonicity cone for F_j for each j.
3) Let F be a universal subequation that is pure second order (i.e. it just depends on the Hessian of a function), and let M_k be the subequation in ($\mathscr{E}2$) describing k-subharmonic functions:

$$M_k = \{\lambda_1(A) + \ldots + \lambda_k(A) \geqslant 0\}.$$

It is proved in [34] that M_k is a monotonicity cone for F if and only if the *Riesz characteristic* of F, p_F, satisfies $p_F \geqslant k$. The Riesz characteristic of a universal subequation F is an explicitably computable quantity defined as follows:

$$p_F = \sup\left\{t > 0 \ : \ I - t\Pi_v \in F \ \ \forall x \in X, \ v \in T_xX \text{ with } |v| = 1\right\},$$

where Π_v is the orthogonal projection onto the span of v. The interested reader can consult Section 11 of [34] for further information.

Definition 6.2 Let F be a subequation. We say that a function $\psi \in USC(X)$ is *polar for a set* Σ if $\Sigma \equiv \{x : \psi(x) = -\infty\}$. A closed subset $\Sigma \subset X$ is called F-*polar* if there exists an open neighbourhood $\Omega \supset K$ and $\psi \in F(\Omega)$ polar for Σ. The set Σ is called C^2 F-polar if, moreover, $\psi \in C^2(\Omega\backslash\Sigma)$.

Sets that are C^2 M-polar are removable for subequations having M as a monotonicity cone, cf. [34, Thm. 6.1]. The result is particularly effective when $F = \widetilde{M}$. Concerning Example 2) above, M_k-polar sets are very well understood in the Euclidean space (and, with some technical modifications, also on manifolds). In particular, if Σ has locally finite $(p-2)$-dimensional Hausdorff measure for some $p < p_F$, then Σ is M_k-polar (cf. [34], Theorems 11.4, 11.5 and 11.13). Moreover, if $\Sigma \subset \mathbb{R}^m$, having locally finite $(p_F - 2)$-dimensional Hausdorff measure is enough to guarantee the M_k-polarity, cf. [34, Thm. A].

In our setting, we consider subequations for which the AK-duality holds and thus we restrict to assume at least ($\mathscr{H}1$). Consequently, since F is a closed subset, the constant function 0 is F-subharmonic. Any monotonicity cone M for F satisfies

$$M = \{0\} + M \subset F,$$

thus any M-polar subset is automatically F-polar.

Theorem 6.3 *Let* F *be an admissible subequation that is locally jet-equivalent to a universal subequation and satisfies* $(\mathscr{H}1)$, $(\mathscr{H}2)$. *Suppose that* $\widetilde{\mathrm{F}}$ *has the Ahlfors property on* X, *and let* $\Sigma \subset \mathrm{X}$ *be a compact subset. Then, the following holds:*

(i) *if* F *is a truncated cone subequation, then*

$$\begin{array}{c}\widetilde{\mathrm{F}} \text{ has the Ahlfors} \\ \text{property on } \mathrm{X}\backslash\Sigma\end{array} \qquad \Longleftrightarrow \qquad \Sigma \text{ is } \mathrm{F}\text{-polar};$$

(ii) *if* M *is a monotonicity cone for* F *and* Σ *is* M*-polar, then* $\widetilde{\mathrm{F}}$ *has the Ahlfors property on* $\mathrm{X}\backslash\Sigma$.

Proof We first prove (i).
($\Rightarrow$) By AK-duality, any fixed pair (K, h) with $\mathrm{K} \Subset \mathrm{X}\backslash\Sigma$ admits a Khas'minskii potential ψ. Since $\psi(\mathrm{x}) \to -\infty$ as $\mathrm{x} \to \Sigma$, extending ψ on $\mathrm{X}\backslash\mathrm{K}$ by setting $\psi = -\infty$ on Σ gives a USC and F-subharmonic function on $\mathrm{X}\backslash\mathrm{K}$. Hence, Σ is F-polar.
($\Leftarrow$) By F-polarity, fix $\Omega \supset \Sigma$ open and $\psi \in \mathrm{F}(\Omega)$ satisfying $\Sigma = \{\psi = -\infty\}$. The upper semicontinuity of ψ implies that $\psi(\mathrm{x}) \to -\infty$ as $\mathrm{x} \to \Sigma$. By AK-duality, we can consider a Khas'minskii potential z for a pair (Ω', h) with $\Omega \Subset \Omega' \Subset \mathrm{X}$. Then, for $\delta \in (0, 1]$, the family of functions $\{\delta \mathrm{w}\}$ with

$$\mathrm{w}(\mathrm{x}) = \begin{cases} \psi(\mathrm{x}) & \text{if } \mathrm{x} \in \Omega, \\ \mathrm{z}(\mathrm{x}) & \text{if } \mathrm{x} \in \mathrm{X}\backslash\Omega', \end{cases} \tag{43}$$

realizes the weak Khas'minskii property on $\mathrm{X}\backslash\Sigma$. By AK-duality, $\widetilde{\mathrm{F}}$ has the Ahlfors property on $\mathrm{X}\backslash\Sigma$, as claimed.
To show (ii), fix $\Omega \supset \Sigma$ open and $\psi \in M(\Omega)$ that satisfies $\Sigma = \{\psi = -\infty\}$. By $(\mathscr{H}1)$ and since F is a closed subset, the constant $0 \in \mathrm{F}(\mathrm{X})$. Therefore, being M a monotonicity cone, $\delta\psi = 0 + \delta\psi \in \mathrm{F}(\Omega)$ for each $\delta \in (0, 1]$. Defining w as in (43), the family $\{\delta \mathrm{w}\}$ give again the desired weak Khas'minskii potentials. □

Acknowledgements This work was completed when the second author was visiting the Abdus Salam International Center for Theoretical Physics (ICTP), Italy. He is grateful for the warm hospitality and for financial support. The authors would also like to thank the organizing and local committees of the INdAM workshop "Contemporary Research in elliptic PDEs and related topics" (Bari, May 30/31, 2017) for the friendly and pleasant environment.

The first author "Luciano Mari" is supported by the grants SNS17_B_MARI and SNS_RB_MARI of the Scuola Normale Superiore.

The second author "Leandro F. Pessoa" was partially supported by CNPq-Brazil.

References

1. L.V. Ahlfors, An extension of Schwarz's lemma. Trans. Am. Math. Soc. **43**, 359–364 (1938)
2. L.V. Ahlfors, L. Sario, *Riemann Surfaces*. Princeton Mathematical Series, vol. 26 (Princeton University Press, Princeton, 1960)
3. L.J. Alías, G.P. Bessa, M. Dajczer, The mean curvature of cylindrically bounded submanifolds. Math. Ann. **345**(2), 367–376 (2009)
4. L.J. Alías, P. Mastrolia, M. Rigoli, *Maximum Principles and Geometric Applications*. Springer Monographs in Mathematics (Springer, Cham, 2016), xvii+570 pp.
5. L.J. Alías, J. Miranda, M. Rigoli, A new open form of the weak maximum principle and geometric applications. Commun. Anal. Geom. **24**(1), 1–43 (2016)
6. G. Aronsson, Extension of functions satisfying Lipschitz conditions. Ark. Mat. **6**, 551–561 (1967)
7. C. Bär, F. Pfäffle, Wiener measures on Riemannian manifolds and the Feynman-Kac formula. Mat. Contemp. **40**, 37–90 (2011)
8. M. Bardi, F. Da Lio, On the strong maximum principle for fully nonlinear degenerate elliptic equations. Arch. Math. (Basel) **73**, 276–285 (1999)
9. G.P. Bessa, B.P. Lima, L.F. Pessoa, Curvature estimates for properly immersed ϕ_h-bounded submanifolds. Ann. Mat. Pura Appl. **194**(1), 109–130 (2015)
10. B. Bianchini, L. Mari, P. Pucci, M. Rigoli, On the interplay among maximum principles, compact support principles and Keller-Osserman conditions on manifolds (2018). Available at arXiv:1801.02102
11. A. Borbely, A remark on the Omori-Yau maximum principle. Kuwait J. Sci. Eng. **39**(2A), 45–56 (2012)
12. A. Borbely, Stochastic completeness and the Omori-Yau maximum principle. J. Geom. Anal. (2017). https://doi.org/10.1007/s12220-017-9802-7
13. L. Caffarelli, X. Cabre, *Fully Nonlinear Elliptic Equations*. Colloquium Publications, vol. 43 (American Mathematical Society, Providence, 1995)
14. S.Y. Cheng, S.T. Yau, Differential equations on Riemannian manifolds and their geometric applications. Commun. Pure Appl. Math. **28**(3), 333–354 (1975)
15. S.Y. Cheng, S.T. Yau, Maximal space-like hypersurfaces in the Lorentz-Minkowski spaces. Ann. Math. (2) **104**(3), 407–419 (1976)
16. P. Collin, R. Kusner, W.H. Meeks III, H. Rosenberg, The topology, geometry and conformal structure of properly embedded minimal surfaces. J. Differ. Geom. **67**(2), 377–393 (2004)
17. M.G. Crandall, A visit with the ∞-Laplace equation, in *Calculus of Variations and Nonlinear Partial Differential Equations*. Lecture Notes in Mathematics, vol. 1927 (Springer, Berlin, 2008), pp. 75–122
18. M.G. Crandall, H. Ishii, P.L. Lions, User's guide to viscosity solutions of second-order partial differential equations. Bull. Am. Math. Soc. **27**, 1–67 (1992)
19. M.G. Crandall, L.C. Evans, R.F. Gariepy, Optimal Lipschitz extensions and the infinity Laplacian. Calc. Var. P.D.E. **13**(2), 123–139 (2001)
20. M.P. Do Carmo, *Riemannian Geometry*. Mathematics: Theory and Applications (Birkäuser Boston Inc., Boston, 1992)
21. J. Dodziuk, Maximum principle for parabolic inequalities and the heat flow on open manifolds. Indiana Univ. Math. J. **32**(5), 703–716 (1983)
22. D.M. Duc, J. Eells, Regularity of exponentially harmonic functions. Int. J. Math. **2**, 395–408 (1991)
23. I. Ekeland, On the variational principle. J. Math. Anal. Appl. **47**, 324–353 (1974)
24. M. Emery, *Stochastic Calculus in Manifolds*. Universitext (Springer, Berlin, 1989)
25. F. Fontenele, F. Xavier, Good shadows, dynamics and convex hulls of complete submanifolds. Asian J. Math. **15**(1), 9–31 (2011)
26. L. Gärding, An inequality for hyperbolic polynomials. J. Math. Mech. **8**, 957–965 (1959)

27. R.E. Greene, H. Wu, C^∞ approximation of convex, subharmonic, and plurisubharmonic functions. Ann. Sci. Ec. Norm. Sup. 4[e] serie t.12, 47–84 (1979)
28. A. Grigor'yan, Analytic and geometric background of recurrence and non-explosion of the Brownian motion on Riemannian manifolds. Bull. Am. Math. Soc. **36**, 135–249 (1999)
29. F.R. Harvey, H.B. Lawson Jr., Dirichlet duality and the non-linear Dirichlet problem. Commun. Pure Appl. Math. **62**, 396–443 (2009)
30. F.R. Harvey, H.B. Lawson Jr., Dirichlet duality and the nonlinear Dirichlet problem on Riemannian manifolds. J. Differ. Geom. **88**, 395–482 (2011)
31. F.R. Harvey, H.B. Lawson Jr., Geometric plurisubharmonicity and convexity: an introduction. Adv. Math. **230**(4–6), 2428–2456 (2012)
32. F.R. Harvey, H.B. Lawson Jr., Existence, uniqueness and removable singularities for nonlinear partial differential equations in geometry, in *Surveys in Differential Geometry 2013*, vol. 18, ed. by H.D. Cao, S.T. Yau (International Press, Somerville, 2013), pp. 102–156
33. F.R. Harvey, H.B. Lawson Jr., Gärding's theory of hyperbolic polynomials. Commun. Pure Appl. Math. **66**(7), 1102–1128 (2013)
34. F.R. Harvey, H.B. Lawson Jr., Removable singularities for nonlinear subequations. Indiana Univ. Math. J. **63**(5), 1525–1552 (2014)
35. F.R. Harvey, H.B. Lawson Jr., Characterizing the strong maximum principle for constant coefficient subequations. Rend. Mat. Appl. (7) **37**(1–2), 63–104 (2016)
36. I. Holopainen, Nonlinear potential theory and quasiregular mappings on Riemannian manifolds. Ann. Acad. Sci. Fenn. Ser. A I Math. Dissertationes **74**, 45 pp. (1990)
37. D. Impera, S. Pigola, A.G. Setti, Potential theory for manifolds with boundary and applications to controlled mean curvature graphs. J. Reine Angew. Math. **733**, 121–159 (2017)
38. R. Jensen, Uniqueness of Lipschitz extensions: minimizing the sup norm of the gradient. Arch. Ration. Mech. Anal. **123**(1), 51–74 (1993)
39. P. Juutinen, Absolutely minimizing Lipschitz extensions on a metric space. Ann. Acad. Sci. Fenn. Math. **27**(1), 57–67 (2002)
40. L. Karp, Differential inequalities on complete Riemannian manifolds and applications. Math. Ann. **272**(4), 449–459 (1985)
41. B. Kawohl, N. Kutev, Strong maximum principle for semicontinuous viscosity solutions of nonlinear partial differential equations. Arch. Math. (Basel) **70**(6), 470–478 (1998)
42. R.Z. Khas'minskii, Ergodic properties of recurrent diffusion processes and stabilization of the solution of the Cauchy problem for parabolic equations. Teor. Verojatnost. i Primenen., Akademija Nauk SSSR. Teorija Verojatnosteĭ i ee Primenenija **5**, 196–214 (1960)
43. N.V. Krylov, On the general notion of fully nonlinear second-order elliptic equations. Trans. Am. Math. Soc. **347**(3), 857–895 (1995)
44. Z. Kuramochi, Mass distribution on the ideal boundaries of abstract Riemann surfaces I. Osaka Math. J. **8**, 119–137 (1956)
45. P. Li, Harmonic functions and applications to complete manifolds. XIV Escola de Geometria Diferencial: Em homenagem a Shiing-Shen Chern (2006)
46. P. Li, L.-F. Tam, Harmonic functions and the structure of complete manifolds. J. Differ. Geom. **35**, 359–383 (1992)
47. M. Magliaro, L. Mari, M. Rigoli, On a paper of Berestycki-Hamel-Rossi and its relations to the weak maximum principle at infinity, with applications. Rev. Mat. Iberoam. **34**, 915–936 (2018)
48. L. Mari, L.F. Pessoa, Duality between Ahlfors-Liouville and Khas'minskii properties for non-linear equations. Commun. Anal. Geom. (to appear)
49. L. Mari, M. Rigoli, Maps from Riemannian manifolds into non-degenerate Euclidean cones. Rev. Mat. Iberoam. **26**(3), 1057–1074 (2010)
50. L. Mari, D. Valtorta, On the equivalence of stochastic completeness, Liouville and Khas'minskii condition in linear and nonlinear setting. Trans. Am. Math. Soc. **365**(9), 4699–4727 (2013)
51. L. Mazet, A general halfspace theorem for constant mean curvature surfaces. Am. J. Math. **135**(3), 801–834 (2013)

52. W.H. Meeks III, H. Rosenberg, Maximum principles at infinity. J. Differ. Geom. **79**(1), 141–165 (2008)
53. M. Nakai, On Evans potential. Proc. Jpn. Acad. **38**, 624–629 (1962)
54. M. Nakai, L. Sario, *Classification Theory of Riemann Surfaces* (Springer, Berlin, 1970)
55. H. Omori, Isometric immersions of Riemannian manifolds. J. Math. Soc. Jpn. **19**, 205–214 (1967)
56. S. Pigola, A.G. Setti, *Global Divergence Theorems in Nonlinear PDEs and Geometry.* Ensaios Matemáticos [Mathematical Surveys], vol. 26 (Sociedade Brasileira de Matemática, Rio de Janeiro, 2014), ii+77 pp.
57. S. Pigola, M. Rigoli, A.G. Setti, A remark on the maximum principle and stochastic completeness. Proc. Am. Math. Soc. **131**(4), 1283–1288 (2003)
58. S. Pigola, M. Rigoli, A.G. Setti, Maximum principles on Riemannian manifolds and applications. Mem. Am. Math. Soc. **174**(822) (2005)
59. S. Pigola, M. Rigoli, A.G. Setti, Some non-linear function theoretic properties of Riemannian manifolds. Rev. Mat. Iberoam. **22**(3), 801–831 (2006)
60. S. Pigola, M. Rigoli, A.G. Setti, Maximum principles at infinity on Riemannian manifolds: an overview. Mat. Contemp. **31**, 81–128 (2006). Workshop on Differential Geometry (Portuguese)
61. S. Pigola, M. Rigoli, A.G. Setti, *Vanishing and Finiteness Results in Geometric Analysis. A Generalization of the Böchner Technique.* Progress in Mathematics, vol. 266 (Birkäuser, Boston, 2008)
62. S. Pigola, M. Rigoli, A.G. Setti, Aspects of potential theory on manifolds, linear and non-linear. Milan J. Math. **76**, 229–256 (2008)
63. P. Pucci, J. Serrin, *The Maximum Principle.* Progress in Nonlinear Differential Equations and Their Applications, vol. 73 (Birkhäuser Verlag, Basel, 2007), p. x+235
64. P. Pucci, M. Rigoli, J. Serrin, Qualitative properties for solutions of singular elliptic inequalities on complete manifolds. J. Differ. Equ. **234**(2), 507–543 (2007)
65. J.P. Sha, p-Convex riemannian manifolds. Invent. Math. **83**, 437–447 (1986)
66. D.W. Stroock, S.R.S. Varadhan, *Multidimensional Diffusion Processes*. Reprint of the 1997 edition. Classics in Mathematics (Springer, Berlin, 2006), xii+338 pp.
67. F. Sullivan, A characterization of complete metric spaces. Proc. Am. Math. Soc. **83**(2), 345–346 (1981)
68. C.-J. Sung, L.-F. Tam, J. Wang, Spaces of harmonic functions. J. Lond. Math. Soc. **61**(3), 789–806 (2000)
69. M. Troyanov, Parabolicity of manifolds. Sib. Adv. Math. **9**(4), 125–150 (1999)
70. D. Valtorta, Potenziali di Evans su varietá paraboliche (2009). Available at arXiv:1101.2618
71. D. Valtorta, Reverse Khas'minskii condition. Math. Z. **270**(1), 65–177 (2011)
72. J.D. Weston, A characterization of metric completeness. Proc. Am. Math. Soc. **64**(1), 186–188 (1977)
73. H. Wu, Manifolds of partially positive curvature. Indiana Univ. Math. J. **36**(3), 525–548 (1987)
74. S.T. Yau, Harmonic functions on complete Riemannian manifolds. Commun. Pure Appl. Math. **28**, 201–228 (1975)
75. S.T. Yau, A general Schwarz lemma for Kähler manifolds. Am. J. Math. **100**(1), 197–203 (1978)

Singularities in the Calculus of Variations

Connor Mooney

Abstract In these notes we discuss the regularity of minimizers of convex functionals in the calculus of variations, with a focus on the vectorial case. We first treat the theory of linear elliptic systems and give some consequences. Then we discuss important singular solutions of De Giorgi, Giusti-Miranda, and Maz'ya to linear elliptic systems, and of Sverak-Yan in the nonlinear case. At the end we discuss the parabolic theory.

Keywords Elliptic and parabolic systems · Singular minimizers · Blowup

1 Introduction

In these notes we discuss regularity results for minimizers in the calculus of variations, with a focus on the vectorial case. We then discuss some important singular examples.

The notes follow a mini-course given by the author for the INdAM intensive period "Contemporary research in elliptic PDEs and related topics" in April 2017. I am very grateful for the generous hospitality of Serena Dipierro and her family, Enrico Valdinoci, and l'Università degli Studi di Bari during this time. I am also grateful to the anonymous referee for helpful suggestions that improved the exposition. This work was partially supported by National Science Foundation fellowship DMS-1501152 and by the ERC grant "Regularity and Stability in Partial Differential Equations (RSPDE)."

C. Mooney (✉)
UC Irvine, Irvine, CA, USA
e-mail: mooneycr@math.uci.edu

S. Dipierro (ed.), *Contemporary Research in Elliptic PDEs and Related Topics*,
Springer INdAM Series 33, https://doi.org/10.1007/978-3-030-18921-1_11

2 Preliminaries

In this section we introduce the main question of the course.

Let $F : M^{m\times n} \to \mathbb{R}$ be a smooth, convex function satisfying

$$\lambda I < D^2 F < \lambda^{-1} I$$

for some positive constant $\lambda \leq 1$. Let $\mathbf{u} = (u^1, \ldots, u^m) \in H^1(B_1 \subset \mathbb{R}^n; \mathbb{R}^m)$ be the unique minimizer of the functional

$$E(\mathbf{u}) = \int_{B_1} F(D\mathbf{u})\, dx, \tag{1}$$

subject to its own boundary data. A classical example is $F(p) = |p|^2$ (the Dirichlet energy), whose minimizers are harmonic maps.

Exercise Show the existence and uniqueness of minimizers in $H^1(B_1)$ of (1), subject to the boundary condition $\mathbf{u}|_{\partial B_1} = \psi \in H^1(B_1)$. Use the direct method (take a minimizing sequence).

Hints: Use the bounds on $D^2 F$ to find a subsequence that converges weakly in H^1. Use the convexity of F to show that the limit is a minimizer, and the strict convexity to show it is unique.

For classical examples like $F(p) = |p|^2$, minimizers are smooth. The main question of the course is:

Are Minimizers Always Smooth?

Our approach to the regularity problem is to study the PDE that minimizers and their derivatives solve. By minimality we have

$$0 \leq \int_{B_1} (F(D\mathbf{u} + \epsilon D\varphi) - F(D\mathbf{u}))\, dx = \epsilon \int_{B_1} \nabla F(D\mathbf{u}) \cdot D\varphi\, dx + O(\epsilon^2)$$

for all ϵ and all $\varphi \in C_0^\infty(B_1; \mathbb{R}^m)$. In particular, $\mathbf{u}$ solves the Euler-Lagrange system

$$\mathrm{div}(\nabla F(D\mathbf{u})) = \partial_i (F_{p_i^\alpha}(D\mathbf{u})) = 0 \tag{2}$$

in the distributional sense.

Exercise Show that if $\mathbf{u} \in H^1(B_1)$ solves the Euler-Lagrange system (2), then it is the unique minimizer of (1).

Remark 1 An interesting question is the uniqueness for (2) in weaker Sobolev spaces. Examples of Šverák-Yan [13] show non-uniqueness in $W^{1,p}$ for $p < 2$. We discuss these examples in Sect. 4.

Equation (2) is invariant under translations, and under the Lipschitz rescalings

$$\mathbf{u} \to \mathbf{u}_r = \frac{1}{r}\mathbf{u}(rx).$$

This scaling invariance plays an important role in regularity results. The classical approach to regularity is to differentiate the Euler-Lagrange system. Formally, we have

$$\operatorname{div}(D^2F(D\mathbf{u})D^2\mathbf{u}) = \partial_i(F_{p_i^\alpha p_j^\beta}(D\mathbf{u})u_{kj}^\beta) = 0. \tag{3}$$

We then treat the problem as a linear system for $D\mathbf{u}$ with coefficients $D^2F(D\mathbf{u})$.

Remark 2 For justification that $D\mathbf{u} \in H^1_{loc}(B_1)$ and solves (3), see the exercises in the next section.

If $D\mathbf{u}$ is continuous, then the coefficients $D^2F(D\mathbf{u})$ are continuous. By perturbation theory from the constant coefficient case (see e.g. [3]), we obtain that $\mathbf{u}$ is smooth. However, we have no a priori regularity for $D\mathbf{u}$, so we can only assume the coefficients are bounded and measurable. As a result, below we will consider the linear system

$$\operatorname{div}(AD\mathbf{v}) = \partial_i(A^{ij}_{\alpha\beta}(x)v^\beta_j) = 0 \tag{4}$$

in B_1, where $A^{ij}_{\alpha\beta}|^{\alpha,\beta=1,\dots,m}_{i,j=1,\dots,n}$ are bounded measurable coefficients satisfying the ellipticity condition

$$\lambda|p|^2 \le A(x)(p,\, p) < \lambda^{-1}|p|^2$$

for all $x \in B_1$ and $p \in M^{m\times n}$, and $\mathbf{v} = (v^1, \dots, v^m) \in H^1(B_1; \mathbb{R}^m)$ solves the system in the distribution sense.

Exercise Show that if $\mathbf{v} \in H^1(B_1)$ solves (4), then $\mathbf{v}$ is a minimizer of the functional

$$J(\mathbf{v}) = \int_{B_1} A(x)(D\mathbf{v},\, D\mathbf{v})\, dx. \tag{5}$$

If one can show that solutions to (4) are continuous, then minimizers of (1) are smooth.

This course consists of two main parts. In the first part (Sect. 3) we discuss estimates for the linear system (4), and consequences for minimizers of (1). In the second part (Sect. 4) we discuss some examples that show the optimality of the linear results, and also the optimality of their consequences for minimizers.

In Sect. 5 we discuss the parabolic case (which was not covered in the lectures). We emphasize some striking differences with the elliptic case.

3 Linear Estimates and Consequences

In this section we discuss the key estimates for solutions to the linear system (4), and their consequences for minimizers of (1).

3.1 Energy Estimate

Recall that solutions to the linear system (4) minimize the energy (5). Thus, the natural quantity controlled by the linear system (4) is the H^1 norm of $\mathbf{v}$. By using minimality or the equation, we can get more precise information.

Exercise Let φ be a cutoff function that is 1 in $B_{1/2}$ and 0 outside B_1. Use $\mathbf{v}\varphi^2$ as a test function in (4) to derive the Caccioppoli inequality

$$\int_{B_{1/2}} |D\mathbf{v}|^2 \, dx < C(\lambda) \int_{B_1} |\mathbf{v}|^2 |\nabla \varphi|^2. \tag{6}$$

Exercise Derive the Caccioppoli inequality by using $\mathbf{v}(1 - \epsilon\varphi^2)$ as a competitor for $\mathbf{v}$ in the energy (5). This gives a perhaps more illuminating way to understand the inequality: the energy density of $\mathbf{v}$ cannot concentrate near the center of B_1, since then the energy lost by dilating $\mathbf{v}$ by a factor less than 1 in $B_{1/2}$ is more than the energy paid to reconnect to the same boundary data.

One consequence of the Caccioppoli inequality is the following energy loss estimate

$$\int_{B_{r/2}} |D\mathbf{v}|^2 \, dx < \gamma(n, \lambda) \int_{B_r} |D\mathbf{v}|^2 \, dx, \tag{7}$$

for some $\gamma < 1$ and all $r < 1$. This inequality says that the energy density must "spread evenly at all scales."

Exercise Prove Inequality (7).

Hints: Reduce to the case $r = 1$ by scaling. Since the system (4) is invariant under adding constant vectors, we can replace $\mathbf{v}$ by $\mathbf{v} - \text{avg.}_{\{B_1 \backslash B_{1/2}\}}\mathbf{v}$ in Inequality (6). (By $\text{avg.}_\Omega \mathbf{v}$ we mean the average of $\mathbf{v}$ in Ω). Finally, note that $\nabla\varphi$ is supported in $B_1 \backslash B_{1/2}$. The result follows by applying the Poincarè inequality in the annulus $B_1 \backslash B_{1/2}$.

As a consequence of the energy loss estimate, we have that the mass of the energy in B_r decays like a power of r:

$$\int_{B_r} |D\mathbf{v}|^2 \, dx < C(n, \lambda) \left(\int_{B_1} |\mathbf{v}|^2 \, dx \right) r^{2\alpha}, \tag{8}$$

for all $r < 1/2$ and some $\alpha > 0$. Inequality (8) is our main result for the linear system (4).

Exercise Prove Inequality (8) by iterating Inequality (7) on dyadic scales.

Remark 3 The energy decay estimate (8) says that $D\mathbf{v}$ behaves as if it were in L^q for q slightly larger than 2. It is in fact true that $\mathbf{v} \in W^{1,2+\delta}$ for some $\delta > 0$. This result is part of the "reverse-Hölder theory" (see e.g. [3]). This stronger result will not be required for our purposes.

The energy decay estimate is particularly powerful in the case $n = 2$, due to the invariance of the H^1 norm under the rescaling $\mathbf{v} \to \mathbf{v}(rx)$. More specifically, by standard embeddings for Morrey-Campanato spaces, if

$$\frac{r^2}{|B_r|}\int_{B_r(x)} |D\mathbf{v}|^2\,dx < Cr^{2\alpha}$$

for all $r < 1/4$ and all $x \in B_{1/2}$, then $\mathbf{v} \in C^\alpha(B_{1/2})$. In particular, in the case $n = 2$, we conclude from the energy decay (8) that $\mathbf{v} \in C^\alpha$.

We conclude by noting that (8) also holds for inhomogeneous systems when the right side is sufficiently integrable.

Exercise Consider the inhomogeneous system

$$\text{div}(A(x)D\mathbf{v}) = \text{div}(\mathbf{g}),$$

and assume that

$$\int_{B_r} |\mathbf{g}|^2\,dx < r^{2\beta}$$

for some $\beta > 0$ and all $r < 1$. Repeat the above line of reasoning to show that

$$\int_{B_r} |D\mathbf{v}|^2\,dx < Cr^{2\gamma}$$

for some $\gamma(n, \lambda, \beta) > 0$ and $C\left(\int_{B_1} |\mathbf{v}|^2\,dx,\, n,\, \lambda\right)$.

Hint: Note that the system solved by the rescaling $\mathbf{v}(rx)$ has right side $\text{div}(r\mathbf{g}(rx))$.

Remark 4 The required condition for $\mathbf{g}$ is satisfied e.g. when $\mathbf{g} \in L^q$ for some $q > 2$. We will use this result when we discuss the parabolic case in Sect. 5.

3.2 Consequences for Minimizers

Now we investigate the consequences (8) for minimizers of (1). Below we assume that $\mathbf{u}$ is a minimizer of the regular functional (1).

Exercise Let $\mathbf{u}$, $\mathbf{w} \in H^1(B_1)$ solve the Euler-Lagrange equation (2). Show using the fundamental theorem of calculus that

$$\partial_i \left(\left(\int_0^1 F_{p_i^\alpha p_j^\beta}(D\mathbf{u} + s(D\mathbf{w} - D\mathbf{u}))\, dx \right) (w_j^\beta - u_j^\beta) \right) = 0,$$

i.e. that the difference $\mathbf{w} - \mathbf{u}$ solves a linear system of the type (4).

Exercise Using the previous exercise for difference quotients $h^{-1}(\mathbf{u}(x+he) - \mathbf{u}(x))$ and the Caccioppoli inequality, justify that $\mathbf{u} \in W^{2,2}_{loc}(B_1)$ and that $D\mathbf{u}$ solves the differentiated Euler-Lagrange equation (3).

As a consequence of the estimate (8) for linear systems, we have

$$\int_{B_r} |D^2\mathbf{u}|^2\, dx < Cr^{2\alpha} \tag{9}$$

for some $\alpha > 0$ and all $r < 1/2$.

Exercise Using embedding theorems from Sobolev and Campanato-Morrey spaces, conclude from Inequality (9) the following results:

- In the case $n = 2$, $D\mathbf{u} \in C^\alpha$, hence $\mathbf{u}$ is smooth.
- In the cases $n = 3$ and $n = 4$, $\mathbf{u} \in C^\beta$ for some $\beta > 0$.
- In the case $n \geq 5$, unbounded minimizers are not ruled out.

Hint: In the case $n = 4$, $W^{2,2}$ embeds into $W^{1,4}$, which nearly embeds to continuous. Using the decay estimate one can improve. Apply the Sobolev-Poincaré inequality to obtain $\int_{B_r} |D\mathbf{u} - (D\mathbf{u})_{B_r}|^4\, dx < Cr^{4\alpha}$. (Here $(D\mathbf{u})_{B_r}$ is the average in B_r). Then use the Cauchy-Schwarz inequality to reduce to a Morrey-Campanato embedding.

We will show in the next section that when $m > 1$, both the decay estimate (8) and the above consequences for minimizers are optimal. We discuss examples of De Giorgi [2], Giusti-Miranda [4], and Šverák-Yan [12, 13].

Remark 5 The energy estimate (8) and its consequences for minimizers are due to Morrey, in the 1930s (see e.g. [7]).

3.3 Scalar Case

The energy decay estimate (8) came from comparison with a simple competitor obtained by slightly deforming $\mathbf{v}$. It is natural to ask whether one can improve upon this result.

As the examples in the next section show, the answer is in general no. However, in the scalar case $m = 1$, one can improve to $\mathbf{v} \in C^{\alpha}$. The key property of solutions to (4) in the scalar case is the maximum principle: $\mathbf{v}$ never goes beyond its maximum or minimum values on the boundary. Indeed, we get competitors with smaller energy by truncating $\mathbf{v}$ where it goes beyond its boundary data (e.g. if $v \geq 0$ on ∂B_1, then consider $\max\{v, 0\}$). In the vectorial case, making truncations of certain components doesn't send the full gradient to 0, so truncations are not always energetically favorable.

Remark 6 It is instructive to consider a simple example. In dimensions $n = m = 2$ let F be the quadratic $|p|^2 - 2\epsilon(p_1^1 p_2^2 + p_2^1 p_1^2)$. It is clear that F is uniformly convex for ϵ small. Direct computation shows that $\mathbf{v} = \left(x_1 x_2, \frac{\epsilon}{2}(|x|^2 - 1)\right)$ is a minimizer of $\int_{B_1} F(D\mathbf{u})\, dx$. However, the second component of $\mathbf{v}$ vanishes on ∂B_1; in particular, the "truncation" $(x_1 x_2, 0)$ has larger energy. One also checks that $|\mathbf{v}|$ has a local maximum at 0.

As a consequence of the maximum principle, solutions exhibit oscillation decay in L^{∞} when we decrease scale. By quantifying the maximum principle, one can obtain C^{α} regularity. This breakthrough result is due to De Giorgi [1], and at the same time Nash [8], in the late 1950s.

To illustrate the role of the maximum principle, it is instructive to consider the two dimensional case. Assume that $v \in H^1(B_1; \mathbb{R})$, with $B_1 \subset \mathbb{R}^2$ and

$$\int_{B_1} |\nabla v|^2\, dx \leq 1.$$

Assume further that the maximum and minimum of v on B_r occur on ∂B_r, for all $r < 1$. (For convenience, assume that v is continuous so that we can make sense of these values, and derive a priori estimates). Such v share the key properties of solutions to (4) in the scalar case. We indicate how to use the maximum principle to find a modulus of continuity for v at 0. Let

$$osc_{B_r} v = \max_{B_r} v - \min_{B_r} v = \max_{\partial B_r} v - \min_{\partial B_r} v.$$

Exercise Show using the fundamental theorem of calculus that

$$\frac{1}{2\pi r}(osc_{B_r} v)^2 \leq \int_{\partial B_r} |\nabla v|^2\, ds.$$

(This is the only place where we use that $n = 2$). Using the maximum principle, show that $osc_{B_r} v$ is increasing with r. Combine with the above inequality to obtain

$$(osc_{B_\delta} v)^2 \frac{|\log(\delta)|}{2\pi} \leq \int_{B_1 \setminus B_\delta} |\nabla v|^2 \, dx.$$

Conclude that

$$osc_{B_\delta} v < \left(\frac{2\pi}{|\log \delta|} \right)^{1/2},$$

for all $\delta < 1/2$.

It is instructive to investigate why this argument doesn't work in higher dimensions. Scaling provides a useful explanation. Roughly, if a function v oscillates order 1 on S^{n-1}, then we expect that $\int_{S^{n-1}} |\nabla v|^2 \, ds$ is order 1 (see the remark below). If v oscillates order 1 on ∂B_r for all $r > 0$, then applying the unit-scale estimate to $v_r = v(rx)$ we obtain that the Dirichlet energy on ∂B_r is order r^{n-3}. In the case $n \geq 3$ this is not enough to contradict H^1 boundedness. De Giorgi's argument overcomes this difficulty by using the Caccioppoli inequality for a sequence of truncations of v.

Remark 7 Even the "expectation" that if v oscillates order 1 on S^{n-1} then $\int_{S^{n-1}} |\nabla v|^2 \, dx$ has order at least 1 is not quite true when $n \geq 3$ (unlike the case $n = 2$). Consider for example the functions on $B_1 \subset \mathbb{R}^2$ (rather than S^2, for simplicity) equal to $-\log r / \log R$ on $B_1 \setminus B_{1/R}$ and equal to 1 in $B_{1/R}$. These have small Dirichlet energy going like $(\log R)^{-1}$.

To conclude the section, we remark that for systems with special structure, we can sometimes find a quantity that solves a scalar equation or inequality. In these cases we have stronger regularity results. Here is an important example due to Uhlenbeck [14].

Assume (like above) that F is a smooth, uniformly convex function on $M^{m \times n}$ with bounded second derivatives. Assume further that F has radial symmetry, i.e. $F(p) = f(|p|)$, with $0 < \lambda \leq f'' \leq \lambda^{-1}$. Let $\mathbf{u}$ be a minimizer to the corresponding functional.

Exercise Show that $\nabla F(p) = \frac{f'}{|p|} p$. Conclude that the Euler-Lagrange equation is

$$\partial_i \left(\frac{f'}{|D\mathbf{u}|} u_i^\alpha \right) = 0,$$

i.e. that the components of $\mathbf{u}$ solve elliptic equations. Give a variational explanation that each component satisfies the maximum principle.

Hint: If we truncate a component, then $|D\mathbf{u}|$ (hence F) decreases.

As a consequence, minimizers of rotationally symmetric functionals are continuous. We can in fact show that $|D\mathbf{u}|^2$ is a subsolution to a scalar equation (it takes its maxima on the boundary):

Exercise Show that

$$D^2F(p) = \frac{f'}{|p|}I + \left(f'' - \frac{f'}{|p|}\right)\frac{p \otimes p}{|p|^2}.$$

Conclude that

$$\partial_i\left(\frac{f'}{|D\mathbf{u}|}u^\alpha_{ik} + \left(f'' - \frac{f'}{|D\mathbf{u}|}\right)\frac{u^\alpha_i u^\beta_j}{|D\mathbf{u}|^2}u^\beta_{jk}\right) = 0.$$

Multiply this equation by u^α_k and sum over α and k to conclude that

$$\text{div}(A(x)\nabla|D\mathbf{u}|^2) \geq \lambda|D^2\mathbf{u}|^2, \tag{10}$$

where $A(x)$ are uniformly elliptic coefficients.

By using De Giorgi's results for the inequality (10), one can show that $\mathbf{u}$ is smooth (see e.g. [3, Chapter 7]). Radial symmetry for F is one of the few structure conditions known to ensure full regularity of minimizers.

4 Singular Examples

We discuss some examples of singular minimizers. The examples show optimality of the linear estimates, and of their consequences for minimizers of (1).

4.1 Linear Elliptic Examples

Here we describe examples of discontinuous homogeneous solutions to (4), that show the optimality of the energy decay estimate (8) in the vectorial case. The examples are due to De Giorgi [2] and Giusti-Miranda [4] in 1968, about 10 years after the De Giorgi proved continuity of solutions in the scalar case.

We first establish some notation. For $\mathbf{a} \in \mathbb{R}^m$ and $\mathbf{b} \in \mathbb{R}^n$ we let $\mathbf{a} \otimes \mathbf{b} \in M^{m\times n}$ act on $\mathbb{R}^n$ by $(\mathbf{a} \otimes \mathbf{b})(x) = (\mathbf{b} \cdot x)\mathbf{a}$. In particular, $(\mathbf{a} \otimes \mathbf{b})^\alpha_i = a^\alpha b_i$. Likewise, if $A,\ B \in M^{m\times n}$ we let $A \otimes B$ be the linear map on $M^{m\times n}$ defined by $(A \otimes B)(p) = (B \cdot p)A$, where the dot product on matrices is defined by $B \cdot p = \text{tr}(B^T p) = B^\alpha_i p^\alpha_i$. In particular, $A \otimes B$ is a four-index tensor with components $(A \otimes B)^{ij}_{\alpha\beta} = A^\alpha_i B^\beta_j$.

It is natural to start the search for singular examples by considering 0-homogeneous maps, which have a bounded discontinuity at the origin. Let $|x| = r$ and let $\mathbf{v} = \nu := r^{-1}x$ be the radial unit vector. The De Giorgi construction is based on the observation that

$$D\nu = r^{-1}(I - \nu \otimes \nu),$$

the matrix that projects tangent to the sphere, is non-vanishing and is perpendicular to $B := \nu \otimes \nu$ in $M^{n\times n}$. In particular, ν clearly minimizes the functional $\int_{B_1} A(x)(D\mathbf{v}, D\mathbf{v})\,dx$ for $A = B \otimes B$. (Note that A is zero-homogeneous, with a discontinuity at the origin). Since this functional is degenerate convex (indeed, $A^{ij}_{\alpha\beta}(x)p^{\alpha}_i p^{\beta}_j = 0$ when p is perpendicular to $\nu \otimes \nu$), we need to make a small perturbation.

We first do some simple calculations. We compute

$$\Delta\nu = \nabla(\Delta r) = -\frac{n-1}{r^2}\nu, \quad \operatorname{div}\left(\nu \otimes \frac{\nu}{r}\right) = \nu\Delta(\log(r)) = \frac{n-2}{r^2}\nu. \tag{11}$$

Remark 8 Note that the last expression vanishes in the plane.

Now we take coefficients

$$A = \delta I_{n^2} + (B + \gamma(I_n - B)) \otimes (B + \gamma(I_n - B)).$$

It is useful to think that δ and γ are small, so that A is a perturbation of $B \otimes B$. We compute

$$AD\nu = \delta D\nu + \gamma(n-1)(\gamma D\nu + \nu \otimes \nu/r).$$

Taking the divergence and using (11), we obtain

$$\operatorname{div}(AD\nu) = [-\delta(n-1) + \gamma(n-1)(n-2-\gamma(n-1))]r^{-2}\nu.$$

The example follows provided

$$\delta = (n-2)\gamma - (n-1)\gamma^2 > 0,$$

which is true when $n \geq 3$ and $\gamma < \frac{n-2}{n-1}$.

Thus, ν solves a system of the type (4) in $\mathbb{R}^n \backslash \{0\}$, for $n \geq 3$ and zero-homogeneous coefficients that are analytic away from the origin. It remains to verify that ν solves the equation globally.

Exercise Show that $\nu \in H^1_{loc}(\mathbb{R}^n)$ when $n \geq 3$, but not $n = 2$. Show that ν solves (4) in B_1 in the sense of distributions when $n \geq 3$.

Hint: use that

$$\int_{B_1} A(D\mathbf{v},\, D\varphi)\, dx = \lim_{r\to 0} \int_{B_1\backslash B_r} A(D\mathbf{v},\, D\varphi)\, dx,$$

and integrate by parts in the expression on the right.

Remark 9 There are no discontinuous H^1_{loc} solutions to (4) in $\mathbb{R}^2$, by the energy decay estimate (8). It is interesting that the above approach doesn't even give a nontrivial zero-homogeneous solution to a uniformly elliptic system in $\mathbb{R}^2\backslash\{0\}$. It is natural to ask whether such solutions exist. In the next section we prove a rigidity result showing that zero-homogeneous solutions to (4) in $\mathbb{R}^2\backslash\{0\}$ are constant. We also prove a higher-dimensional analogue.

The above example shows that De Giorgi's results for the scalar case don't extend to the vectorial case. However, observe that in the differentiated Euler-Lagrange equation (3), the coefficients $D^2F(D\mathbf{u})$ depend smoothly on the solution $D\mathbf{u}$. It is natural to ask whether this structure improves regularity.

The above example answers this question in the negative. If we choose in particular $\gamma = \frac{n-2}{n}$ and divide the coefficients by γ^2 we obtain

$$A = I_{n^2} + \left(I_n + \frac{2}{n-2}B\right) \otimes \left(I_n + \frac{2}{n-2}B\right) = I_{n^2} + C(\mathbf{v}) \otimes C(\mathbf{v}),$$

where

$$C = I_n + \frac{4}{n-2}\frac{\mathbf{v}\otimes\mathbf{v}}{1+|\mathbf{v}|^2}$$

is bounded and depends analytically on $\mathbf{v}$. This example shows that $\mathbf{v} = \nu$ solves a uniformly elliptic system of the form

$$\operatorname{div}(A(\mathbf{v})D\mathbf{v}) = 0,$$

where A depend analytically on $\mathbf{v} \in \mathbb{R}^n$. The example is due to Giusti and Miranda [4].

We now modify the construction to get unbounded examples. (De Giorgi's original example was actually of this type). Let

$$\mathbf{v} = r^{-\epsilon}\nu.$$

Exercise Take A of the same form as above, and compute in a similar way that the equation $\mathrm{div}(AD\mathbf{v}) = 0$ in $\mathbb{R}^n \backslash \{0\}$ gives

$$-\delta(n-1-\epsilon)(1+\epsilon) + ((n-1)\gamma - \epsilon)(n-2-(n-1)\gamma - \epsilon) = 0. \tag{12}$$

Hint: For the first term note that $\Delta \mathbf{v} = \nabla(\Delta(r^{1-\epsilon}))/(1-\epsilon)$ (at least when $\epsilon \neq 1$). For the second term, divide $r^{-\epsilon-1}$ into the pieces $r^{-\epsilon}$ and r^{-1}, and use the computation in the example above.

Exercise Using condition (12), show that $r^{-\epsilon}\nu$ solves a uniformly elliptic system in $\mathbb{R}^n \backslash \{0\}$ for any $\epsilon \neq \frac{n-2}{2}$, and any $n \geq 2$.

Exercise Show that $r^{-\epsilon}\nu \in H^1$ for $\epsilon < \frac{n-2}{2}$, and in this case $\mathbf{v}$ solves $\mathrm{div}(AD\mathbf{v}) = 0$ in B_1 (in particular, across the origin). Finally, show by taking ϵ arbitrarily close to $\frac{n-2}{2}$ that the decay estimate (8) is optimal in the vectorial case, in any dimension n.

Remark 10 Observe in particular that for $n = 3, 4$ the examples are the gradients of bounded non-Lipschitz functions, and for $n \geq 5$ the examples are the gradients of unbounded functions.

Finally, to appreciate fully the vectorial nature of the above examples, it is instructive to make similar constructions in the scalar case.

Exercise Let $v(x) = r^{-\epsilon}g(\nu) : \mathbb{R}^n \to \mathbb{R}$ be a $-\epsilon$-homogeneous function. Let

$$A = a\nu \otimes \nu + (I - \nu \otimes \nu)$$

for some constant $a > 0$. Show that

$$A\nabla v = r^{-\epsilon-1}(-a\epsilon g \nu + \nabla_{S^{n-1}} g),$$

$$\mathrm{div}(A\nabla v) = r^{-\epsilon-2}(\Delta_{S^{n-1}} g - a\epsilon(n-2-\epsilon)g).$$

(Here $\nabla_{S^{n-1}}$ and $\Delta_{S^{n-1}}$ denote the gradient and Laplace-Beltrami operators on S^{n-1}).

Thus, the equation $\mathrm{div}(A\nabla v) = 0$ becomes the eigenvalue problem

$$\Delta_{S^{n-1}} g = a\epsilon(n-2-\epsilon)\, g$$

on the sphere. The maximum principle enters the picture when we consider the solvability of this problem: we need $a\epsilon(n-2-\epsilon) \leq 0$ to find nonzero solutions.

Provided that either $\epsilon < 0$ or $\epsilon > n-2$, we can find many solutions g on the sphere by choosing $a > 0$ appropriately. In the borderline cases $\epsilon = 0$ or $\epsilon = n-2$ we see that v is radial, hence v is constant resp. the fundamental solution to Δ.

Finally, observe that when $\epsilon < 0$ we have $v \in H^1(B_1)$ and the equation $\mathrm{div}(A\nabla u) = 0$ holds across the origin. These examples show the optimality of the De Giorgi result in the scalar case.

4.2 *Rigidity Result for Homogeneity* $-\frac{n-2}{2}$

Above we constructed solutions to (4) in B_1 in the case $n = m \geq 2$ that are homogeneous of degree $-\epsilon$, for any $\epsilon < \frac{n-2}{2}$. This showed the optimality of the energy decay estimate (8). We also found $-\epsilon$-homogeneous maps that solve the system in $\mathbb{R}^n \backslash \{0\}$ for all $\epsilon \neq \frac{n-2}{2}$. It is natural to ask whether there is some rigidity result for this special homogeneity. In this section we verify that this is the case. The main result is:

Theorem 1 *Assume that* $\mathbf{v} : \mathbb{R}^n \to \mathbb{R}^m$ *is* $-\frac{n-2}{2}$*-homogeneous and that* A *are bounded, uniformly elliptic coefficients. If*

$$div(AD\mathbf{v}) = 0$$

in $B_1 \backslash \{0\}$*, then* $\mathbf{v}$ *is constant. (In particular,* $\mathbf{v} = 0$ *when* $n \geq 3$*.)*

Proof Take the dot product of the equation with $\mathbf{v}$ and integrate by parts in $B_1 \backslash B_\epsilon$ to obtain

$$\int_{B_1 \backslash B_\epsilon} A(D\mathbf{v},\, D\mathbf{v})\, dx = \int_{\partial(B_1 \backslash B_\epsilon)} A(D\mathbf{v}, \mathbf{v} \otimes \nu)\, ds.$$

Since $D\mathbf{v} \cdot \mathbf{v}$ is homogeneous of degree $-(n-1)$, the flux of the vector field $AD\mathbf{v} \cdot \mathbf{v}$ through ∂B_r is bounded independently of r. Thus, the right side of the above identity is bounded independently of ϵ. Using the ellipticity of the coefficients, we conclude that

$$\int_{B_1 \backslash B_\epsilon} |D\mathbf{v}|^2\, dx \leq C.$$

However, by the homogeneity of $\mathbf{v}$ we have

$$\int_{B_1 \backslash B_\epsilon} |D\mathbf{v}|^2\, dx \geq |\log \epsilon| \int_{\partial B_1} |D\mathbf{v}|^2\, ds.$$

Taking $\epsilon \to 0$ we conclude from the previous inequalities that $D\mathbf{v} \equiv 0$. □

Exercise Prove Theorem (1) assuming that A are zero-homogeneous, by working only on the sphere, as follows. Write

$$\mathbf{G} = AD\mathbf{v} \cdot \mathbf{v} = r^{-(n-1)}(f(\nu)\nu + \tau(\nu)),$$

where f is a zero-homogeneous function and τ is tangential to S^{n-1}. Show that the first term is divergence-free, and that the divergence of the second term is $r^{-n}\text{div}_{S^{n-1}}\tau$. Integrate the inequality $\lambda|D\mathbf{v}|^2 \leq \text{div}_{S^{n-1}}\tau$ on the sphere to complete the proof.

Observe that in the case $n = 2$, if $\mathbf{v}$ is zero-homogeneous then $D\mathbf{v}$ has rank one.

Exercise Show in the case $n = 2$ that Theorem (1) holds when we replace uniform ellipticity with the condition that $A(x)(p, p) > \lambda|p|^2$ for rank-one matrices p.

4.3 Null Lagrangian Approach of Šverák-Yan

In this section we discuss an approach to constructing singular minimizers due to Šverák-Yan [12, 13]. This approach is based on the concept of null Lagrangian. We will discuss the idea in a simple situation.

A null Lagrangian L is a function on $M^{m\times n}$ such that

$$\int_\Omega L(D\mathbf{u})\,dx = \int_\Omega L(D\mathbf{u} + D\boldsymbol{\varphi})\,dx$$

for all domains Ω and smooth deformations $\boldsymbol{\varphi}$ supported in Ω. In particular, every map solves the Euler-Lagrange system

$$\text{div}(\nabla L(D\mathbf{u})) = 0.$$

Any linear function is a null Lagrangian. The most important nontrivial example is the determinant.

Exercise Let $\mathbf{u} = (u^1, u^2)$ be a map from $\mathbb{R}^2$ to $\mathbb{R}^2$ and let Ω be a smooth bounded domain. Verify using integration by parts that

$$\int_\Omega \det D\mathbf{u}\,dx = \int_{\partial\Omega} u^1\nabla_T u^2\,ds,$$

where ∇_T denote derivative tangential to $\partial\Omega$. Conclude that det is a null Lagrangian. Then compute directly that

$$\text{div}((\nabla\det)(D\mathbf{u})) = \partial_j(\det D\mathbf{u}\,(D\mathbf{u})^{-1}_{ji}) = 0.$$

More generally, sub-determinants are null Lagrangians. Some of the simplest non-trivial null Lagrangians are the quadratic ones. There is a useful characterization of quadratic null Lagrangians:

Exercise Show that a quadratic form A on $M^{m\times n}$ is a null Lagrangian if and only if $A(p, p) = 0$ for all rank-one matrices p. (Recall that p is rank-one if and only if $p = \mathbf{a} \otimes \mathbf{b}$ for some $\mathbf{a} \in \mathbb{R}^m$ and $\mathbf{b} \in \mathbb{R}^n$.)

Hints: To show the "if" direction, use the Fourier transform, and use that $A(b \otimes a, c \otimes a) = 0$ for $a \in \mathbb{R}^n$ and $b, c \in \mathbb{R}^m$. To show the "only if" direction, use Lipschitz rescalings of a simple periodic test function whose gradients lie on a rank-one convex line. More explicitly, take $\Omega = B_1$, take $\mathbf{u} = 0$, and take $\varphi_\lambda(x) = (bf(\lambda a \cdot x)/\lambda)\eta(x)$, where $a \in \mathbb{R}^n$, $b \in \mathbb{R}^m$, f is periodic, and η is a compactly supported function in B_1 equal to 1 on $B_{1-\epsilon}$ with $|\nabla \eta| < 2/\epsilon$. In the definition of null Lagrangian, take $\lambda \to \infty$, then ϵ to 0. (The idea is to build a map whose gradient has size of order 1 and lies on a rank-one convex line, such that the map is very small in L^∞. We accomplish this by making many oscillations.Then we can cut off without changing the integral much.)

Remark 11 The quadratic forms A on $M^{m\times n}$ that are null Lagrangians are in fact linear combinations of 2×2 sub-determinants. Indeed, by the previous exercise, they vanish on rank-one matrices. In particular, they vanish on the subspaces $\mathbb{R}^m \otimes e_i$ and $f^\alpha \otimes \mathbb{R}^n$ (where $\{e_i\}$ are the coordinate directions in $\mathbb{R}^n$ and $\{f^\alpha\}$ are the coordinate directions in $\mathbb{R}^m$), giving $A^{ii}_{\alpha\beta} = A^{ij}_{\alpha\alpha} = 0$. They also vanish on $(f^\alpha + f^\beta)\otimes(e_i + e_j)$, where $\alpha \neq \beta$ and $i \neq j$, giving $A^{ij}_{\alpha\beta} + A^{ji}_{\alpha\beta} = A^{ij}_{\alpha\beta} + A^{ij}_{\beta\alpha} = 0$. It follows that $A(p, p)$ is a sum of terms of the form $c(p^\alpha_i p^\beta_j - p^\beta_i p^\alpha_j)$.

The idea of Šverák-Yan is to find a homogeneous map $\mathbf{u}$ such that the image $K := D\mathbf{u}(B_1)$ lies on or close to a subspace of $M^{m\times n}$ on which a null Lagrangian L is convex. Then one can hope to construct a smooth, convex F with the same first-order Taylor expansion as L on $D\mathbf{u}(B_1)$. The Euler-Lagrange equation $\text{div}(\nabla F(D\mathbf{u})) = \text{div}(\nabla L(D\mathbf{u})) = 0$ is then automatically satisfied.

A trivial example illustrating the idea is $\mathbf{u} = r^{-1}\nu : \mathbb{R}^2 \to \mathbb{R}^2$. Then $D\mathbf{u}(B_1)$ lies in the space of traceless symmetric matrices, and the null Lagrangian $L = -\det$ is uniformly convex when restricted to this subspace. An extension of L with the same values and gradients on the symmetric traceless matrices is $F(p) = |p|^2$. Thus, $\mathbf{u}$ is harmonic away from the origin. (Of course, this was clear from the outset since $\mathbf{u} = \nabla \log$). However, $\text{div}(\mathbf{u})$ has a Dirac mass at the origin.

Motivated by this example, consider singular candidates of the De Giorgi type

$$\mathbf{u} = r^{1-\alpha}\nu$$

from $\mathbb{R}^2 \to \mathbb{R}^2$ with $1 < \alpha < 2$. We investigate the geometry of $D\mathbf{u}(B_1)$ in matrix space.

Exercise Show that the identification

$$(x, y, z) = \begin{pmatrix} z + x & y \\ y & z - x \end{pmatrix}$$

of the symmetric 2×2 matrices with $\mathbb{R}^3$ is an isometry. Show that the surfaces of constant trace are the horizontal planes, and the surfaces of constant determinant are hyperboloid sheets asymptotic to the cone

$$z^2 = x^2 + y^2.$$

Then show that $\mathbf{Du}(S^1)$ is a circle of constant positive trace and constant negative determinant matrices centered on the z-axis, and $\mathbf{Du}(B_1)$ is the cone centered at the origin going through this circle, minus the ball centered at the origin whose boundary passes through the circle.

The above visualization of $\mathbf{Du}(B_1)$ shows that for $\alpha > 1$, the conical surface $\mathbf{Du}(B_1)$ lies "close to" the subspace of traceless symmetric matrices, where $L = -\det$ is convex. We can in fact construct a uniformly convex, smooth function F on $Sym_{2\times 2}$ such that $F = L$ and $\nabla F = \nabla L$ on $\mathbf{Du}(B_1)$.

By homogeneity, it suffices to construct a 2-homogeneous, uniformly convex F such that $F = L$ and $\nabla F = \nabla L$ on the circle $K := \mathbf{Du}(S^1)$. Thus, it suffices to find a smooth, bounded, uniformly convex set $\Sigma \subset Sym_{2\times 2}$ containing the origin, such that $\partial \Sigma$ contains K, and the outer normal to $\partial \Sigma$ is in the direction of ∇L on K. Indeed, then we can let $F = L = \alpha - 1$ on $\partial \Sigma$, and then take the 2-homogeneous extension.

As a first step to constructing Σ, consider the surface $\Gamma = \{L = \alpha - 1\}$. In the coordinates introduced above, Γ is a hyperboloid of revolution around the z axis, asymptotic to $z^2 = x^2 + y^2$. Thus, there is a circular cone centered on the negative z axis that is tangent to Γ on K. This cone divides $Sym_{2\times 2}$ into two components; let Σ_0 be the component containing the origin. Then Σ_0 is convex, and the outer normal to $\partial \Sigma_0$ on K is in the direction of ∇L. From here, it is easy to find a smooth, uniformly convex surface of revolution $\partial \Sigma$ bounding a region $\Sigma \subset \Sigma_0$, such that $\partial \Sigma$ touches $\partial \Sigma_0$ on K and $0 \in \Sigma$. This completes the construction.

Remark 12 To make F smooth, we need to modify it near the origin. This doesn't affect the equation since $\mathbf{Du}(B_1)$ stays outside a ball around the origin. To make a uniformly convex extension G of F to all of $M^{2\times 2}$, decompose $p \in M^{2\times 2}$ into its symmetric and anti-symmetric parts S and A and let $G(p) = F(S) + |A|^2$.

Exercise

- Show that for $1 < \alpha < 2$, the map $\mathbf{u} \in W^{1,p}$ for $p < \frac{2}{\alpha}$. Show that $\mathbf{u}$ solves the Euler-Lagrange equation $\mathrm{div}(\nabla F(\mathbf{Du})) = 0$ in the distribution sense.
- Show that linear maps are minimizers of (1) subject to their own boundary data. Conclude that $\mathbf{w} = x$ is the unique minimizer in $H^1(B_1)$ for the functional corresponding to F.
- Conclude that there is non-uniqueness for the Euler-Lagrange equation (2) in the spaces $W^{1,p}$ for $p < 2$.

Remark 13 The quadratic analogue of $-\det$ in higher dimensions is $L = -\sigma_2$, where $\sigma_2(M) = \frac{1}{2}(tr(M)^2 - |M|^2)$. Thus, L is a uniformly convex, radial quadratic on the constant-trace symmetric matrices.

We can repeat the above procedure for $\mathbf{u} = r^{1-\alpha}\nu$ in higher dimensions provided $L > 0$ on $D\mathbf{u}(B_1)$. This gives the condition $\alpha > \frac{n}{2}$. Again, these maps provide counterexamples to uniqueness for (2) in $W^{1,p}$ when $p < 2$.

The above examples are due to Šverák-Yan [13]. To find examples of singular minimizers to smooth, uniformly convex functionals, more complicated maps are required. In [12], Šverák and Yan use the null Lagrangian technique to show that the one-homogeneous map

$$\mathbf{u}(x) = r\left(\nu \otimes \nu - \frac{1}{3}I\right),$$

viewed as a map from $\mathbb{R}^3$ to the space of symmetric traceless matrices (isomorphic to $\mathbb{R}^5$), is a Lipschitz but not C^1 singular minimizer in the case $n = 3,\ m = 5$. We describe the construction here.

To understand the geometry of $K := D\mathbf{u}(B_1)$, it is useful to use the symmetries of $\mathbf{u}$. Letting $a_{ijk} = \partial_k u^{ij}$ and $R \in SO(3)$, we have

$$\mathbf{u}(Rx) = R\mathbf{u}(x)R^T,$$

$$a_{ijk}(Rx) = R_{il}R_{jm}R_{kn}a_{lmn}(x).$$

It is not hard to check that two invariant subspaces of $\{a_{iik} = 0,\ a_{ijk} = a_{jik}\} \cong M^{5\times 3}$ are the space of traceless tensors $T^0 = \{a_{ikk} = 0\}$, and its orthogonal complement T_3. Among the traceless matrices, two invariant subspaces are the permutation-invariant subspace $T_7 = \{a_{ijk} = a_{jki}\}$, and its orthogonal complement T_5. (The subscripts represent the dimension of the subspace). The quadratic forms invariant under the above action take the simple form $\alpha|X|^2 + \beta|Y|^2 + \gamma|Z|^2$, where $X,\ Y,\ Z$ are the projections to T_3, T_7 and T_5 respectively.

Recall that the quadratic null Lagrangians correspond to quadratic forms that vanish in the rank-one directions. By imposing the condition $L(C_{ij}\eta_k) = 0$ for all symmetric traceless C_{ij} and $\eta \in \mathbb{R}^3$ (and using explicit formulae for the projections of a_{ijk} to T_3, T_7 and T_5), Šverák-Yan compute

$$L = 3|X|^2 - 2|Y|^2 + |Z|^2,$$

up to multiplication by constants.

Ideally, we would be able to say that the Y projection of $D\mathbf{u}$ vanishes, so that K lies in a subspace where L is convex. This is not quite the case. However, a computation (see [12]) shows that

$$|D\mathbf{u}_{T_3}|^2 = \frac{64}{15}, \quad |D\mathbf{u}_{T_7}|^2 = \frac{2}{5}$$

and the remaining projection vanishes. Heuristically, the example works because $D\mathbf{u}$ is closer to the T_3 subspace where L is convex.

One can compute explicitly that $L = 12$ on K, and that L separates from its tangent planes quadratically on K:

$$L(Y) - L(X) - \nabla L(X) \cdot (Y - X) = -L(X - Y) > c|Y - X|^2$$

for $X, Y \in K$ and some $c > 0$. This is enough to construct a uniformly convex, smooth function F on $M^{5\times 3}$ with the same first-order expansion as L on K.

Remark 14 This beautiful example was the first singular minimizer to a smooth, uniformly convex functional in dimension $n = 3$. The first singular minimizer, constructed by Nečas in 1977 [9], was also one-homogeneous and worked in high dimensions.

It is natural to ask whether the regularity results for minimizers obtained from linear theory (Hölder continuity in dimensions 3 and 4, and possible unboundedness in dimension 5) are optimal. Šverák-Yan accomplish this in [13] using modifications of the above example. More precisely, they consider $\mathbf{u} = r^{1-\alpha}(\nu \otimes \nu - (1/n)I) : \mathbb{R}^n \to M^{n(n+1)/2-1}$. In higher dimensions one can perform the same decomposition of $M^{(n(n+1)/2-1)\times n}$. An important observation is that the coefficient of L in the higher-dimensional version of T_3 (the trace part of a_{ijk}) is n, and the other coefficients remain the same. As a result, the higher the dimension, the better the convexity in the trace subspace. Furthermore, the component of $D\mathbf{u}$ in the direction of this subspace grows roughly linearly with n, while the other component remains bounded.

This allows the construction of increasingly singular examples in higher dimensions. They show quadratic separation of L from its tangent planes on $D\mathbf{u}(B_1)$ when $0 \leq \alpha < C(n)$, where $C(n) > 0$ for $n \geq 3$ and increases with n. In the cases $n = 3$ and $n = 4$ this gives non-Lipschitz minimizers. Furthermore, a careful computation shows that $C(5) > 1$, providing examples of unbounded singular minimizers in the optimal dimension.

Remark 15 The most recent approach to constructing singular examples is based on constructing a singular minimizer to a degenerate convex functional in the scalar case, and coupling two such minimizers together in a way that removes the degeneracy. Using this approach, Savin and the author constructed a one-homogeneous singular minimizer in the minimal dimensions $n = 3$, $m = 2$ (see [6]).

5 Parabolic Case

In the final section we discuss the regularity problem for the parabolic case. To emphasize ideas, we assume that solutions are smooth and obtain a priori estimates.

The gradient flow $\mathbf{u} : Q_1 = B_1 \times [-1, 0) \subset \mathbb{R}^n \times \mathbb{R} \to \mathbb{R}^m$ of the regular functional (1) solves

$$\partial_t \mathbf{u} - \operatorname{div}(\nabla F(D\mathbf{u})) = 0. \tag{13}$$

Differentiating (13), we see that the space and time derivatives of $\mathbf{u}$ solve a linear uniformly parabolic system of the form

$$\mathbf{v}_t - \operatorname{div}(A(x, t) D\mathbf{v}) = 0, \tag{14}$$

where $A(x, t)$ are bounded measurable, uniformly elliptic coefficients.

Remark 16 As in the elliptic case, the coefficients of the system obtained by differentiating (13) depend smoothly on $D\mathbf{u}$. By perturbation theory, a continuity result for (14) implies smoothness for gradient flows. Around the same time that De Giorgi proved continuity of solutions to the scalar elliptic problem, Nash showed continuity of solutions to (14) in the scalar case. Again, the maximum principle plays an important role. We will focus on the vectorial case.

We will discuss the key estimate for (14), its consequences for gradient flows, and some singular examples. While much of the theory is motivated by the elliptic case, some of the parabolic results required significant new ideas, and some have no elliptic analogue. We emphasize these differences in the discussion.

5.1 Linear Estimates

The classical energy estimate for (14) says that the L^2 norm of $\mathbf{v}$ is controlled uniformly in time, and the H^1 norm is controlled on average in time:

$$\sup_{\{t>-1/2\}} \int_{B_{1/2}} |\mathbf{v}|^2 \, dx + \int\int_{Q_{1/2}} |D\mathbf{v}|^2 \, dx \, dt < C(n, \lambda) \int\int_{Q_1} |\mathbf{v}|^2 \, dx \, dt. \tag{15}$$

(Here Q_r is the parabolic cylinder $B_r \times [-r^2, 0)$.)

Recall that in the elliptic case, linear estimates give control on the H^2 norms of minimizers, and in the case $n = 2$ the H^2 norm is invariant under the natural rescalings that preserve the Euler-Lagrange equation. In the parabolic case, the natural scaling is

$$\mathbf{u} \to \mathbf{u}_r = r^{-1} \mathbf{u}(rx, r^2 t).$$

The quantities controlled in (15) obtained by taking $\mathbf{v} = D\mathbf{u}$ are not scaling-invariant in the case $n = 2$; under the above scaling, they increase by a factor of r^{-2}. However, for $\mathbf{v} = \partial_t \mathbf{u}$, the quantities controlled are (15) are scaling-invariant in the case $n = 2$. Roughly, one time derivative plays the role of two spatial derivatives.

This observation suggests the following approach: obtain a version of "energy decay" for (14), and apply it to the "second-order" quantity $\mathbf{v} = \mathbf{u}_t$. Nečas and Šverák accomplished this in 1991 (see [10]). Precisely, they show for solutions $\mathbf{v}$ to (14) that

$$\sup_{t>-1/2} \int_{B_{1/2}} |\mathbf{v}|^\gamma \, dx < \infty, \tag{16}$$

for some $\gamma > 2$. We can then treat the parabolic system (13) as an elliptic system for each fixed time.

Remark 17 If we could apply (16) directly to $D^2\mathbf{u}$, we recover the elliptic result uniformly in time. We cannot do this, since the second derivatives don't solve (14). The key observation is that (16) does apply to $\mathbf{u}_t$, and estimates for $\mathbf{u}_t$ are as good as estimates for $D^2\mathbf{u}$ by elliptic theory.

We sketch the argument here.

Exercise

- Derive the energy estimate (15) by taking the time derivative of $\int_{B_1} |\mathbf{v}|^2 \varphi^2 \, dx$, where φ is a spacetime cutoff function that is 1 in $Q_{1/2}$ and vanishes outside Q_1.
- Apply the Sobolev inequality to the second term in (15) and use the interpolation

$$\int_{B_1} w^{2+2/q} \, dx < \left(\int_{B_1} w^{2^*} \, dx \right)^{2/2^*} \left(\int_{B_1} w^2 \, dx \right)^{1/q}$$

  to conclude that $\mathbf{v} \in L^\gamma_{loc}(Q_1)$ for some $\gamma > 2$. (Here q is the Hölder conjugate of $2^*/2$).
- Apply the same procedure as in the first exercise to the integral of $|\mathbf{v}|^\gamma$ to obtain

$$\sup_{t>-1/2} \int_{B_{1/2}} |\mathbf{v}|^\gamma \, dx + \int \int_{Q_{1/2}} |\mathbf{v}|^{\gamma-2} \left(\lambda |D\mathbf{v}|^2 - (\gamma - 2)\gamma A \left(\frac{\mathbf{v} \otimes \mathbf{v}}{|\mathbf{v}|^2} (D\mathbf{v}), D\mathbf{v} \right) \right) dx \, dt$$
$$< C \int \int_{Q_{3/4}} |\mathbf{v}|^\gamma \, dx \, dt.$$

  Conclude that if $\gamma - 2 = \delta > 0$ is small, then $\mathbf{v}$ is bounded in $L^{2+\delta}(B_{1/2})$ uniformly in $t > -1/4$.

Remark 18 Observe that the improved parabolic energy estimate does not imply continuity of solutions to (14) in the case $n = 2$, unlike in the elliptic case.

As a consequence of the Nečas-Šverák result, the regularity results for gradient flows coming from linear theory mirror those of the elliptic case. Let $\mathbf{u}$ solve (13).

Exercise

- Using (15) for $\mathbf{v} = D\mathbf{u}$, show that $D^2\mathbf{u} \in L^2_{loc}(Q_1)$. Conclude from the Eq. (13) that $\mathbf{u}_t \in L^2_{loc}(Q_1)$.
- Using the previous exercise, show that $\mathbf{u}_t \in L^{\gamma}_{loc}(Q_1)$ for some $\gamma > 2$.
- Using the previous exercise, conclude that $\mathbf{u}_t$ is bounded in $L^{2+\delta}(B_{1/2})$, independently of $t > -1/4$.
- Conclude that $D\mathbf{u}$ solves at each time an inhomogeneous linear elliptic system with right side $D(\mathbf{g})$, where $\mathbf{g}$ is bounded in $L^{2+\delta}(B_{1/2})$ uniformly in $t > -1/4$. (Here $\mathbf{g} = \mathbf{u}_t$). Conclude from the elliptic theory that $\int_{B_r} |D^2\mathbf{u}|^2\,dx < Cr^{2\alpha}$ for some $\alpha > 0$ and all $r < 1/2$, with C independent of $t > -1/4$.

As a consequence, Nečas-Šverák show that $\mathbf{u}$ is smooth in the case $n = 2$, and continuous in the case $n \leq 4$, as in the elliptic case.

5.2 *Singularities from Smooth Data*

The elliptic examples of De Giorgi, Giusti-Miranda, and Šverák-Yan are of course parabolic examples, with singularities on the cylindrical set $\{x = 0\}$. It is natural to ask for examples that develop a singularity from smooth data. In addition, a difference between the elliptic and parabolic theory is that the energy estimate (15) does not imply continuity of solutions to (14) in the case $n = 2$. (However, a version of it implies that there are no examples of a singularity that persists in time in the case $n = 2$, unlike in higher dimensions.)

In this last section we discuss examples of finite time singularity in the case $n = m \geq 3$ due to Stará-John-Malý [11]. We then describe a more recent example in the case $n = m = 2$ [5].

To find examples of discontinuity from smooth data, it is natural to seek examples that are invariant under parabolic rescalings that preserve zero-homogeneous maps:

$$\mathbf{v}(x,\, t) = \mathbf{V}\left(\frac{x}{\sqrt{-t}}\right), \quad A(x,\, t) = A\left(\frac{x}{\sqrt{-t}}\right). \tag{17}$$

Exercise Show that imposing the self-similarity (17) reduces (14) to the elliptic system

$$\mathrm{div}(AD\mathbf{V}) = \frac{1}{2}D\mathbf{V}\cdot x. \tag{18}$$

This approach reduces the problem to constructing a global, bounded solution to the elliptic system (18). In [11] Stará-John-Malý use a perturbation of the De Giorgi example of the form $\varphi(r)\nu$, with φ asymptotic to 1 near ∞. The resulting solution $\mathbf{v}$ becomes the De Giorgi example at time $t = 0$.

To simplify computations they make the useful observation that

$$\left(\delta I_{n^2} + \frac{B - \delta D\mathbf{V}}{[(B - \delta D\mathbf{V}) \cdot D\mathbf{V}]^{1/2}} \otimes \frac{B - \delta D\mathbf{V}}{[(B - \delta D\mathbf{V}) \cdot D\mathbf{V}]^{1/2}}\right) \cdot D\mathbf{V} = B.$$

This reduces the problem further to constructing a matrix field $B(x)$ whose divergence is the right side of (18), such that $B \cdot D\mathbf{V} \sim |D\mathbf{V}|^2$ and $|B| \sim |D\mathbf{V}|$. (Here $\sim$ denotes equivalent up to multiplication by positive constants.)

Exercise

- Use the observation that $B := r^{-1}((n-2)I_{n\times n} + \nu \otimes \nu)$ is divergence-free and $B \cdot D\nu \sim |D\boldsymbol{\nu}|^2$ when $n \geq 3$, to re-derive the De Giorgi example.
- Now take $\mathbf{V} = \varphi(r)\nu$ and take $B = r^{-1} f(r)\nu \otimes \nu + h(r)D\nu$. With the choice of coefficients above, show that the system (18) becomes

$$\frac{f'}{r} + (n-2)\frac{f}{r^2} - (n-1)\frac{h}{r^2} = \frac{1}{2}r\varphi'.$$

  (Hint: The left side is just the divergence of B.)
- If φ is asymptotically homogeneous of degree zero, then the left side of the system (18) scales like r^{-2}. It is thus natural to take $r\varphi' \sim r^{-2}$. Show that for the the choice $\varphi = \frac{r}{(1+r^2)^{1/2}}$ and $f = \varphi$, we have in dimension $n \geq 3$ that

$$B \cdot D\mathbf{V} \sim |D\mathbf{V}|^2.$$

- Show that f and h depend analytically on φ. Show similarly that the coefficients depend analytically on $\mathbf{V}$ in a neighborhood of the image of $\mathbf{V}$ $(= B_1)$. Conclude that $\mathbf{v} = \mathbf{V}(x/\sqrt{-t})$ solves an equation of the form $\partial_t \mathbf{v} - \text{div}(A(\mathbf{v})D\mathbf{v}) = 0$, with coefficients that depend smoothly on $\mathbf{v}$.
- Taking $\mathbf{V}$ and $\mathbf{B}$ of the above form, show that there are smooth solutions to the uniformly elliptic system $\text{div}(AD\mathbf{V}) = 0$ that approximate $r^{-\epsilon}\nu$ for all $\epsilon < \frac{n-2}{2}$. (Hint: Take φ linear near the origin, and smoothly connect to $r^{-\epsilon}$ near $r = 1$. Then rescale.)

This gives a parabolic analogue of the De Giorgi example in dimension $n = m \geq 3$.

It is natural to ask whether a map of the form $\varphi(r)\nu$ can work in two dimensions. The following exercise reveals an important restriction on the "shape" of possible φ:

Exercise Observe that in the above example, $|\mathbf{V}|$ is radially increasing. For any such map solving the system (18), show that

$$\mathbf{V} \cdot \text{div}(AD\mathbf{V}) \geq 0.$$

Multiply by a cutoff ψ that agrees with 1 in B_1 and integrate by parts to conclude that if $\mathbf{V}$ is bounded, then

$$\int_{B_1} |D\mathbf{V}|^2\, dx < C \inf_{\psi|_{\partial B_1}=1,\ \psi|_{\partial B_R}=0} \int |\nabla\psi|^2\, dx,$$

for each $R > 2$. Show that the quantity on the right approaches 0 as $R \to \infty$ in the case $n = 2$, by taking ψ to be the harmonic function with the given boundary data.

Thus, bounded solutions to (18) with radially increasing modulus are constant in two dimensions.

In [5] we construct a solution to (18) in the case $n = m = 2$, using a different perspective. The idea is to show that for the correct choice of $\varphi(r)$, each component of $\varphi\nu$ solves the scalar version of (18) away from an annulus where the error in each equation is small. By introducing off-diagonal coefficients in this region, we cancel the errors without breaking the uniform ellipticity of the coefficients.

In view of the previous exercise, the function $\varphi(r)$ is not increasing. In fact, the equations fail to hold exactly where φ has a local maximum. The philosophy of the example is to capture in an explicit way how coupling can cancel the regularizing effect of the maximum principle.

Remark 19 In this example, the coupling coefficients are changing near the maximum of φ. Thus, the coefficients can not be written as functions of the solution. However, by considering a pair of similar maps $(\varphi\nu,\ \tilde{\varphi}\nu)$ we obtain a solution to (18) in the case $n = 2$, $m = 4$ that is injective into $\mathbb{R}^4$ (see [5]). In this way we get an example with coefficients that depend smoothly on the solution.

Remark 20 To obtain examples of L^∞ blowup from smooth data, look for self-similar solutions that are invariant under parabolic rescalings that preserve $-\epsilon$-homogeneous maps:

$$\mathbf{v}(x,\, t) = \frac{1}{(-t)^{\epsilon/2}}\mathbf{V}\left(\frac{x}{\sqrt{-t}}\right).$$

This reduces the problem to finding asymptotically $-\epsilon$-homogeneous solutions to a certain elliptic system. The methods described above adapt to this case.

References

1. E. De Giorgi, Sulla differenziabilità e l'analicità delle estremali degli integrali multipli regolari. Mem. Accad. Sci. Torino cl. Sci. Fis. Fat. Natur. **3**, 25–43 (1957)
2. I. De Giorgi, Un esempio di estremali discontinue per un problema variazionale di tipo ellittico. Boll. UMI **4**, 135–137 (1968)
3. M. Giaquinta, *Multiple Integrals in the Calculus of Variations and Nonlinear Elliptic Systems* (Princeton University Press, Princeton, 1983)

4. E. Giusti, M. Miranda, Un esempio di soluzione discontinua per un problem di minimo relativo ad un integrale regolare del calcolo delle variazioni. Boll. Un. Mat. Ital. **2**, 1–8 (1968)
5. C. Mooney, Finite time blowup for parabolic systems in two dimensions. Arch. Ration. Mech. Anal. **223**, 1039–1055 (2017)
6. C. Mooney, O. Savin, Some singular minimizers in low dimensions in the calculus of variations. Arch. Ration. Mech. Anal. **221**, 1–22 (2016)
7. C.B. Morrey, *Multiple Integrals in the Calculus of Variations* (Springer, Heidelberg, 1966)
8. J. Nash, Continuity of solutions of parabolic and elliptic equations. Am. J. Math. **80**, 931–954 (1958)
9. J. Nečas, Example of an irregular solution to a nonlinear elliptic system with analytic coefficients and conditions of regularity. *Theory of Non Linear Operators*, Abhandlungen Akad. der Wissen. der DDR (1997), Proceedings of a Summer School held in Berlin (1975)
10. J. Nečas, V. Šverák, On regularity of solutions of nonlinear parabolic systems. Ann. Sc. Norm. Super. Pisa Cl. Sci. (4) **18**, 1–11 (1991)
11. J. Stará, O. John, J. Malý, Counterexample to the regularity of weak solution of the quasilinear parabolic system. Comment. Math. Univ. Carol. **27**, 123–136 (1986)
12. V. Šverák, X. Yan, A singular minimizer of a smooth strongly convex functional in three dimensions. Cal. Var. PDE **10**(3), 213–221 (2000)
13. V. Šverák, X. Yan, Non-Lipschitz minimizers of smooth uniformly convex functionals. Proc. Natl. Acad. Sci. USA **99**(24), 15,269–15,276 (2002)
14. K. Uhlenbeck, Regularity for a class of nonlinear elliptic systems. Acta Math. **138**, 219–240 (1977)

Comparison Among Several Planar Fisher-KPP Road-Field Systems

Andrea Tellini

Abstract In this chapter we consider several reaction-diffusion systems—known as road-field systems—which describe the effect that one (or two) line(s) with heterogeneous diffusion has (have) on the speed of propagation in a planar domain, where the classical Fisher-KPP equation is considered. We recall the results by Berestycki et al. (J. Math. Biol. 66:743–766, 2013) for the case of a line in a half-plane, and those obtained in collaboration with Rossi et al. (SIAM J. Math. Anal. 49, 4595–4624, 2017) for two lines bounding a strip. The main goal is to compare the speed of propagation in the direction of the line(s) of these situations with the cases of a plane with one and two lines on which the diffusion is different with respect to the rest of the planar domain.

Keywords Reaction-diffusion systems · Asymptotic speed of propagation · Diffusion heterogeneities · KPP systems · Different spatial dimensions

1 Introduction

Road-field models are systems of reaction-diffusion equations posed in different spatial dimensions that have been introduced in the context of mathematical biology in [2] in order to take into account the effect that a line of fast diffusion has on the propagation in a half-plane, where a logistic-type reaction takes place.

More precisely, in [2], the authors consider a density $v(x, y, t)$ that diffuses in the upper half-plane $\{(x, y) \in \mathbb{R}^2 : y > 0\}$, called the *field*, with diffusion coefficient $d > 0$, and reproduces according to a reaction term $f(v)$, which is assumed to be of Fisher-KPP type. On the so-called *road*, i.e. the boundary of the half-plane given by $\{(x, y) \in \mathbb{R}^2 : y = 0\}$, another density $u(x, t)$ diffuses with a possibly different coefficient $D > 0$. In addition, a symmetric exchange between the road and the field

A. Tellini (✉)
Universidad Politécnica de Madrid, ETSIDI, Departamento de Matemática Aplicada a la Ingeniería Industrial, Madrid, Spain
e-mail: andrea.tellini@upm.es

S. Dipierro (ed.), *Contemporary Research in Elliptic PDEs and Related Topics*, Springer INdAM Series 33, https://doi.org/10.1007/978-3-030-18921-1_12

is considered, with a fraction νv that passes from the field to the road and a fraction μu that, vice-versa, passes from the road to the field (μ, ν being positive constants). The corresponding reaction-diffusion system thus reads

$$\begin{cases} v_t - d\,\Delta v = f(v) & \text{for } (x,y,t) \in \mathbb{R}\times\mathbb{R}^+\times\mathbb{R}^+, \\ u_t - D\,u_{xx} = \nu\, v(x,0^+,t) - \mu\, u & \text{for } (x,t)\in\mathbb{R}\times\mathbb{R}^+, \\ -d\, v_y(x,0^+,t) = \mu\, u - \nu\, v(x,0^+,t) & \text{for } (x,t)\in\mathbb{R}\times\mathbb{R}^+, \end{cases} \tag{1}$$

where $\mathbb{R}^+$ denotes the set of positive numbers, $f : [0,\infty) \to \mathbb{R}$ is a Lipschitz function which is differentiable in 0 and satisfies

$$f(0) = f(1) = 0, \quad f > 0 \text{ in } (0,1), \quad f < 0 \text{ in } (1,\infty), \quad f(s) \leq f'(0)s \text{ for } s \in [0,\infty), \tag{KPP}$$

and $v(x,0^+,t) := \lim_{y\downarrow 0} v(x,y,t)$, $v_y(x,0^+,t) := \lim_{y\downarrow 0} v_y(x,y,t)$.

The study of such a system is motivated by many situations in nature in which some species or diseases spread faster along transportation networks (roads, rivers, railways) than in the surrounding environment. Some specific examples are the spreading of *Vespa velutina* in France (see [18]) or the early spread of HIV in the Democratic Republic of Congo (see [12]).

In [2], it has been proved that there exists a quantity, which will be denoted by c^*_{hp}, such that the solution of (1) starting from every continuous, compactly supported, nonnegative pairs $(u_0, v_0) \neq (0,0)$ (throughout all this work we will consider such a kind of initial data), converges to the unique positive steady-state of the system, $\left(\frac{\nu}{\mu}, 1\right)$, with an asymptotic speed of propagation in the direction of the road equal to c^*_{hp} (observe that the subindex refers to the domain, which is a half-plane).

By *asymptotic speed of propagation* in the direction of the road, i.e. the x direction, we mean that c^*_{hp} satisfies the following two properties:

(i) for all $c > c^*_{\text{hp}}$, $\lim_{t\to\infty} \sup_{\substack{|x|>ct \\ y\geq 0}} (u,v) = (0,0)$,

(ii) for all $a > 0$ and $c < c^*_{\text{hp}}$, $\lim_{t\to\infty} \sup_{\substack{|x|<ct \\ 0\leq y<a}} \left|(u,v) - \left(\frac{\nu}{\mu}, 1\right)\right| = 0$.

Such properties say that, asymptotically in time, the solution of the parabolic problem is close to the positive steady-state *inside* bounded rectangles expanding in the x direction at a speed *smaller* than c^*_{hp}, while it is still close to $(0,0)$ *outside* half-strips which are unbounded in y and expand in the in the x direction at a speed *larger* than c^*_{hp}.

The main result of [2] is a precise geometrical characterization of c^*_{hp} that, in particular, allows the authors to compare it with the speed of propagation of the Fisher-KPP equation, i.e. the first equation in (1), which is given by $c_{\text{KPP}} := 2\sqrt{df'(0)}$ (see, e.g., [1, 13, 15]). The results of [2] are summarized in the following theorem.

Theorem 1 ([2]) *Problem* (1) *admits an asymptotic speed of propagation in the x direction which will be denoted by c^*_{hp} and satisfies:*

(i) $c^*_{\mathrm{hp}} \geq c_{\mathrm{KPP}}$;
(ii) $c^*_{\mathrm{hp}} > c_{\mathrm{KPP}}$ *if and only if* $D > 2d$;
(iii) $\lim_{D\to\infty} c^*_{\mathrm{hp}}(D) = \infty$.

In particular, these results establish that the speed of propagation can never be smaller than the one of a homogeneous environment and that the road enhances such a speed if and only if the diffusion D on it is larger than a certain threshold given by $2d$. Finally, this enhancement can be made arbitrarily large, by taking a sufficiently large D.

Several works on road-field systems in a half-plane have been carried out afterwards, with the goal of ascertaining more features of these models: in [3] additional reaction and transport terms have been considered on the road, in [6] the asymptotic speed of propagation in every direction has been determined, in [7] the existence of traveling fronts has been investigated, in [4, 5] a nonlocal diffusion is taken into account on the road, in [16, 17] nonlocal exchange terms and the relation between such a model and (1) are considered, in [14] μ and ν are allowed to depend periodically on x. A work that treats more general fields, which nonetheless are still unbounded in every direction, is [11], where the case of asymptotically conic domains is studied.

Other works devoted to road-field systems are related to fields with bounded section in the y direction: in [19] the analogue of system (1) is studied in the case of a strip-shaped field bounded by two roads on which the diffusion is different with respect to the one in the field. Such a situation reads

$$\begin{cases} v_t - d\,\Delta v = f(v) & \text{for } (x,y,t) \in \mathbb{R}\times(-R,R)\times\mathbb{R}^+, \\ u_t - D\,u_{xx} = \nu\, v(x,\pm R^{\mp},t) - \mu\, u & \text{for } (x,t) \in \mathbb{R}\times\mathbb{R}^+, \\ \pm d\, v_y(x,\pm R^{\mp},t) = \mu\, u - \nu\, v(x,\pm R^{\mp},t) & \text{for } (x,t)\in\mathbb{R}\times\mathbb{R}^+. \end{cases} \tag{2}$$

In addition, in [19] the corresponding higher-dimensional case is considered, while in [20] the model with a strip bounded by only one road and homogeneous Dirichlet boundary conditions on the other line is handled; finally, [8–10] deal with (2) in the case of ignition-type reactions f.

The main result of [19] is the existence of an asymptotic speed of propagation, denoted by c^*_{st} (in order to refer to the strip-shaped field), in the x direction, which satisfies the properties summarized in the following theorem.

Theorem 2 ([19]) *Problem* (2) *admits an asymptotic speed of propagation c^*_{st} in the x direction which, in addition, satisfies:*

(i) $\lim_{R\downarrow 0} c^*_{\mathrm{st}}(R) = 0$;
(ii) $\lim_{R\to\infty} c^*_{\mathrm{st}}(R) = c^*_{\mathrm{hp}}$;
(iii) *if* $D \leq 2d$, *the function* $R \mapsto c^*_{\mathrm{st}}(R)$ *is continuous and increasing;*

(iv) if $D > 2d$, the function $R \mapsto c^*_{\text{st}}(R)$ is continuous, and it is increasing for $R \in (0, R_M)$ and decreasing for $R \in (R_M, \infty)$, where $R_M := \frac{\nu}{\mu} \frac{D}{D-2d}$. Moreover, in this case, there exist $R_{\text{hp}} \in (0, R_M)$ and $R_K \in (0, R_{\text{hp}})$ such that $c^*_{\text{st}}(R) > c^*_{\text{hp}}$ if and only if $R > R_{\text{hp}}$, and $c^*_{\text{st}}(R) > c_{\text{KPP}}$ if and only if $R > R_K$.

Observe that property (i) in Theorem 2 is new with respect to problem (1), whose speed of propagation is bounded away from 0, while (ii) can be seen as a continuous dependence result of the speed of propagation with respect to the domain. Indeed, one can think as one road in (2) to be fixed and, as $R \to \infty$, the other one lying further and further; thus the latter looses its effects on the propagation, and we recover problem (1).

Another similarity with Theorem 1 is the appearance of the same threshold $2d$ for the diffusion D, but now related to the monotonicity of c^*_{st} with respect to the size of the strip. As remarked in [19], the emergence of two types of monotonicity can be explained by the lack of reaction on the road: if $D \leq 2d$ it is more convenient for the population to propagate in the interior of the strip, where both the reaction and the diffusion are better than on the boundary; thus a larger strip makes the speed of propagation larger. On the contrary, if $D > 2d$, on the one hand it is better to have a larger field for the effect of the reaction to be greater, but, on the other hand, the roads are now more convenient for the diffusion and, by increasing R, they become further apart. The competition between these effects, entails the existence of an optimal distance of the roads which maximizes the speed of propagation.

By comparing Theorems 1 and 2, it is apparent that road-field systems may behave in an extremely different way according to whether the section of the field is bounded or not. In this work we pursue this study by analyzing the combined effect of a part of field with bounded width together with another one with unbounded width, the two parts being separated by two roads where the diffusion is different with respect to the one in the field.

With respect to (1) and (2), observe that we have to allow two-side exchanges; for this reason, we first generalize the analysis of [2] to the system

$$\begin{cases} v_t - d\,\Delta v = f(v) & \text{for } (x, y, t) \in \mathbb{R} \times \mathbb{R} \setminus \{0\} \times \mathbb{R}^+, \\ u_t - D\,u_{xx} = \nu\left[v(x, 0^+, t) + v(x, 0^-, t)\right] - 2\mu\,u & \text{for } (x, t) \in \mathbb{R} \times \mathbb{R}^+, \\ \mp d\,v_y(x, 0^\pm, t) = \mu\,u - \nu\,v(x, 0^\pm, t) & \text{for } (x, t) \in \mathbb{R} \times \mathbb{R}^+, \end{cases} \tag{3}$$

which corresponds to the case of a *plane* with a road of different diffusion. The factor 2 in the second equation of (3) takes into account the fact that the density u can pass to both sides of the surrounding field and gets positive contribution by the density v both from the upper and the lower part. These exchanges are compensated by the flux equations, i.e. the last relations in (3).

The main result that we provide for (3) is the following proposition, which, as it will be apparent in the proof of Proposition 7, will essentially be based on the study of the dependence of c^*_{hp} with respect to the exchange parameters μ and ν.

Proposition 1 *Problem* (3) *admits an asymptotic speed of propagation in the x direction, denoted by* c^*_{p1} *(referring to the planar field with one road), which satisfies:*

(i) $c^*_{\mathrm{p1}} \geq c_{\mathrm{KPP}}$*;*
(ii) $c^*_{\mathrm{p1}} > c_{\mathrm{KPP}}$ *if and only if* $D > 2d$*. In such a case,* $c^*_{\mathrm{p1}} < c^*_{\mathrm{hp}}$.

We observe that the section of the field in (3) is unbounded as in (1) and there is a lower bound on the asymptotic speed of propagation in the direction of the road, given again by the classical Fisher-KPP speed. Indeed, this is a general result which always holds true when the field has at least one component which is unbounded in every direction (see Lemma 1 below). Another point that (3) shares with (1) is that, when the diffusion in the field dominates—i.e. when $D \leq 2d$—the speed of propagation coincides with the one of the homogeneous case, while, when the diffusion on the road dominates, enhancement of the propagation speed takes place. Nevertheless, such an enhancement is reduced when the density is allowed to exchange on the two sides with respect to the case of one-side exchanges given by (1). This phenomenon is not a priori evident, since, despite the fact that in (3) the fraction of the density that leaves the line of fast diffusion is twice as much as in (1), also the contribution from the field doubles.

Finally, the last problem that we consider is

$$\begin{cases} v_t - d\,\Delta v = f(v) & \text{for } (x, y, t) \subset \mathbb{R} \times \mathbb{R} \setminus \{\pm R\} \times \mathbb{R}^+, \\ u_t - D\,u_{xx} = \nu\left[v(x, \pm R^+, t) + v(x, \pm R^-, t)\right] - 2\mu\,u & \text{for } (x, t) \in \mathbb{R} \times \mathbb{R}^+, \\ \mp d\,v_y(x, R^\pm, t) = \mu\,u - \nu\,v(x, R^\pm, t) & \text{for } (x, t) \in \mathbb{R} \times \mathbb{R}^+, \\ \mp d\,v_y(x, -R^\pm, t) = \mu\,u - \nu\,v(x, -R^\pm, t) & \text{for } (x, t) \in \mathbb{R} \times \mathbb{R}^+, \end{cases} \tag{4}$$

which describes a plane with *two* roads where the diffusion is different with respect to the one in the field, and for which the main result is the following.

Theorem 3 *Problem* (4) *admits an asymptotic speed of propagation in the x direction, denoted by* c^*_{p2} *(referring to the planar case with two roads), which satisfies:*

(i) $c^*_{\mathrm{p2}} \geq c_{\mathrm{KPP}}$*;*
(ii) $c^*_{\mathrm{p2}} > c_{\mathrm{KPP}}$ *if and only if* $D > 2d$*. In such a case,* $R \mapsto c^*_{\mathrm{p2}}(R)$ *is continuous, decreasing and satisfies*

$$\lim_{R \downarrow 0} c^*_{\mathrm{p2}}(R) = c^*_{\mathrm{hp}}, \qquad \lim_{R \uparrow \infty} c^*_{\mathrm{p2}}(R) = c^*_{\mathrm{p1}}. \tag{5}$$

In particular, $c^*_{\mathrm{p2}}(R) > c^*_{\mathrm{p1}}$ *for all R;*

(iii) if $D > 2d$, there exists $R^* \in (R_K, R_{\text{hp}})$, where R_K and R_{hp} are the ones of Theorem 2, such that $c^*_{\text{p2}}(R) > c^*_{\text{st}}(R)$ if and only if $R < R^*$.

Once again, we see that an unbounded field in every direction makes the asymptotic speed of propagation bounded from below by c_{KPP}, with the usual threshold of the diffusion on the road in order to have enhancement. The main novelty here is the fact that, contrarily to the case of a strip bounded by two roads of fast diffusion, when such enhancing roads are placed in the whole plane and the distance between them increases, the speed of propagation always decreases. This means that the densities take advantage of the reaction in the field, no matter how it is distributed, and separating the roads of fast diffusion reduces their effect on the enhancement. In addition, observe that, when the roads enhance the speed of propagation, having two of them gives a better enhancement than in the case with only one road, as it is natural to expect. Finally, relations (5) can be seen once more as a continuous dependence of the speed of propagation with respect to the domain: when the strip between the roads shrinks, the effect inside it becomes negligible, as if the exchanges where one-sided; while, if the distance between the roads tends to infinity, considering one of them to be fixed makes the effect of the other one disappear.

The results of Theorems 1–3 and of Proposition 1 are summarized in Fig. 1.

This chapter is distributed as follows: in Sect. 2 we recall some preliminary results, from basic features of road-field systems up to the general way to construct the asymptotic speed of propagation; in Sect. 3 we consider the problems with one road, i.e. (1) and (3), we recall the proof of Theorem 1 given in [2], and we prove Proposition 1; finally, in Sect. 4, we consider the remaining problems, those with two roads, recalling the proof of Theorem 2 given in [19] and providing the one of Theorem 3, which is the main new result of this chapter.

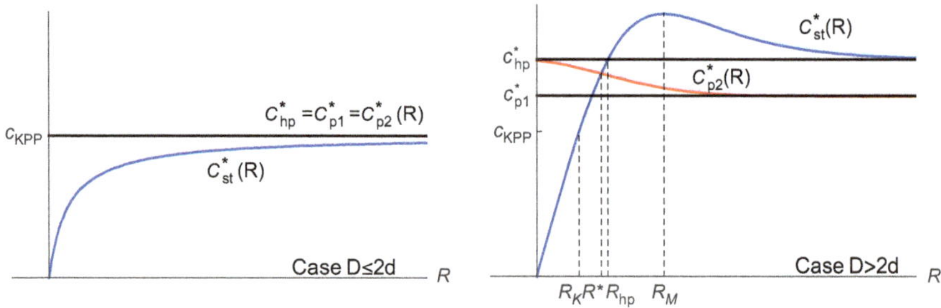

Fig. 1 Graphs of the asymptotic speed of propagation in the x direction for Problems (1), (2), (3) and (4), considered as a function of R: (left) the case $D \leq 2d$ and (right) the case $D > 2d$

2 Preliminary Results: Comparison Principles, Long-Time Behavior and Existence of the Asymptotic Speed of Propagation

In this section we recall some preliminary results that have been proved in [2] for system (1) and in [19] for system (2), and that can be easily adapted to the cases of systems (3) and (4). Without further mention, we stress that such results are valid for all the aforementioned systems, with the natural modifications due to the different domains in which they are posed. We begin with the following parabolic strong comparison principle.

Proposition 2 ([2]) *Let* $(\underline{u}, \underline{v})$ *and* $(\overline{u}, \overline{v})$ *be, respectively, a subsolution bounded from above and a supersolution bounded from below of* (1) *such that* $(\underline{u}, \underline{v}) \le (\overline{u}, \overline{v})$ *at* $t = 0$*, component-wise in their respective domains. Then,* $(\underline{u}, \underline{v}) < (\overline{u}, \overline{v})$ *for all* $t > 0$*, or there exists* $T > 0$ *such that* $(\underline{u}, \underline{v}) = (\overline{u}, \overline{v})$ *for all* $t < T$.

Then, we recall the well-posedness of the system, starting from nonnegative, bounded, continuous initial data (uniqueness, in particular, follows from Proposition 2).

Proposition 3 ([2]) *Let* (u_0, v_0) *be nonnegative, bounded and continuous. Then, there is a unique solution* (u, v) *satisfying* $\lim_{t\downarrow 0}(u, v) = (u_0, v_0)$.

The following is a comparison principle for a class of *generalized subsolutions*, that will be repeatedly used for the characterization of the asymptotic speed of propagation. Once again, we state it for system (1), although it is also valid, with the obvious modifications, for all the other systems.

Proposition 4 ([2]) *Let* (u_1, v_1) *be a subsolution of* (1) *bounded from above, and such that* u_1 *and* v_1 *vanish, respectively, on the boundary of an open set* E *of* $\mathbb{R}\times(0, +\infty)$*, and of an open set* F *of* $\{y > 0\}\times(0, +\infty)$ *(in the relative topologies). Assume in addition that*

$$v_1 \le 0 \qquad \text{in } \overline{E} \cap \{u_1 > 0\} \setminus \overline{F},$$
$$u_1 \le 0 \qquad \text{in } \overline{F} \cap \{v_1 > 0\} \setminus \overline{E}.$$

Then, setting

$$\underline{u} := \begin{cases} \max\{u_1, 0\} & \text{in } \overline{E}, \\ 0 & \text{otherwise}, \end{cases} \qquad \underline{v} := \begin{cases} \max\{v_1, 0\} & \text{in } \overline{F}, \\ 0 & \text{otherwise}, \end{cases}$$

for any supersolution $(\overline{u}, \overline{v})$ *of* (1) *bounded from below and such that* $(\underline{u}, \underline{v}) \le (\overline{u}, \overline{v})$ *at* $t = 0$*, we have* $(\underline{u}, \underline{v}) \le (\overline{u}, \overline{v})$ *for all* $t > 0$.

Next we present the result for the long-time behavior of the solutions. Although it is valid for all the systems (see [2, Section 4] for systems (1) and (3), and [19,

Section 3] for system (2)), we state it for system (4) and give a sketch of the proof, which is slightly different from the other cases, since it requires the combination of the pieces of field with bounded and unbounded section.

Theorem 4 *Let* (u, v) *the solution of* (4) *with a nonnegative, continuous compactly supported initial datum* $(u_0, v_0) \neq (0, 0)$. *Then*

$$\lim_{t\to\infty} (u, v) = \left(\frac{\nu}{\mu}, 1\right). \tag{6}$$

Proof We proceed in several steps.

Step 1. The pair $\left(\frac{\nu}{\mu}K, K\right)$, where K is a sufficiently large constant, is a stationary supersolution of (4) which lies above (u_0, v_0). Thus, the solution of (4) with this supersolution as an initial datum converges to a stationary solution of (4), denoted by (U_1, V_1), which, thanks to Proposition 2, satisfies

$$\limsup_{t\to\infty}(u, v) \leq (U_1, V_1). \tag{7}$$

In addition (U_1, V_1) is x independent and symmetric with respect to reflections about the x axis $\{y = 0\}$. Indeed, the solution of the parabolic problem and its limit as $t \to +\infty$ inherit the desired symmetries from the initial datum, from the fact that the Cauchy problems associated to (4) have a unique solution (thanks to Proposition 3), and that (4) is invariant by translations in x and by reflections about $\{y = 0\}$.

Step 2. By taking, for $\alpha, \beta, \varepsilon$ positive and small,

$$\underline{v} = \varepsilon \cos(\alpha x) \cos\left(\beta\left(y - R - 1 - \frac{\pi}{2\beta}\right)\right)$$

for $x \in \left(-\frac{\pi}{2\alpha}, \frac{\pi}{2\alpha}\right)$ and $y \in \left(R+1, R+1+\frac{\pi}{\beta}\right)$, together with its reflection about $\{y = 0\}$ and extending to 0 elsewhere, we obtain a stationary generalized subsolution of (4) which lies below (u, v), the latter considered at $t = 1$. Proposition 4 thus gives the existence of a stationary solution (U_2, V_2) of (4) which is symmetric about the x axis and such that

$$(U_2, V_2) \leq \liminf_{t\to\infty}(u, v). \tag{8}$$

Finally, a sliding argument as the one of [3, Lemma 2.3] allows us to obtain the independence on x of (U_2, V_2).

Step 3. We claim that the unique nonnegative, bounded stationary solution of (4) which is x independent and symmetric about $\{y = 0\}$ is $\left(\frac{\nu}{\mu}, 1\right)$. Thus, (7) and (8) allow us to obtain (6).

To prove the claim, consider a stationary solution $(U, V(y))$ with the mentioned symmetries. Thanks to the second equation in (4) for $y = R$, it satisfies

$$0 = \nu \left(V(R^+) + V(R^-) \right) - 2\mu U \tag{9}$$

and, thanks to the first one and the symmetry about $\{y = 0\}$,

$$\begin{cases} -dV''(y) = f(V(y)), & y \in (0, R), \\ V'(0) = 0. \end{cases}$$

We prove that $V(0) = 1$, which, thanks to (KPP), will entail that $V \equiv 1$ and, as a consequence from (9), $U = \frac{\nu}{\mu}$. If, by contradiction, $V(0) \in (0, 1)$, then (KPP) implies that V is concave and decreasing in $(0, R)$. By combining this with (9) and with the third equations in (4), we obtain

$$0 > dV'(R^-) = \mu U - \nu V(R^-) = \nu V(R^+) - \mu U = dV'(R^+),$$

thus V would be decreasing and concave for all $y > R$, which is impossible, since it is positive. Similarly, we can exclude that $V(0) > 1$, otherwise V would be convex and increasing for all $y \neq R$, thus unbounded. □

The following result, which relies on the comparison principles given in Propositions 2 and 4, will be used, together with the constructions performed in Sects. 3 and 4, to obtain the existence of the asymptotic speed of propagation. Once again, we state it for system (4) even if it is valid, with the obvious due modifications, for all the road-field systems considered in this work. Since it is one of the core results, we also provide a sketch of the proof (for the details we refer to [3, 19]).

Proposition 5 *Assume that there exists $c^* > 0$ such that:*

(i) for every $c \geq c^*$ there exist supersolutions of the linearized system around $(0, 0)$

$$\begin{cases} v_t - d\,\Delta v = f'(0)v & \text{for } (x, y, t) \in \mathbb{R} \times \mathbb{R} \setminus \{\pm R\} \times \mathbb{R}^+, \\ u_t - D\,u_{xx} = \nu \left[v(x, \pm R^+, t) + v(x, \pm R^-, t) \right] - 2\mu\, u & \text{for } (x, t) \in \mathbb{R} \times \mathbb{R}^+, \\ \mp d\, v_y(x, R^\pm, t) = \mu\, u - \nu\, v(x, R^\pm, t) & \text{for } (x, t) \in \mathbb{R} \times \mathbb{R}^+, \\ \mp d\, v_y(x, -R^\pm, t) = \mu\, u - \nu\, v(x, -R^\pm, t) & \text{for } (x, t) \in \mathbb{R} \times \mathbb{R}^+, \end{cases} \tag{10}$$

of the form

$$(\overline{u}, \overline{v}) = e^{\pm\alpha(x \pm ct)}(1, \phi(y)), \tag{11}$$

where α is a positive constant, and $\phi(y)$ is positive in the field;

(ii) *for all $c < c^*$, $c \sim c^*$ there exist arbitrarily small, nonnegative generalized stationary subsolutions $(\underline{u}, \underline{v})$ of*

$$\begin{cases} v_t - d\,\Delta v \pm c v_x = f(v) & \text{for } (x, y, t) \in \mathbb{R} \times \mathbb{R} \backslash \{\pm R\} \times \mathbb{R}^+, \\ u_t - D u_{xx} \pm c u_x = \nu\big[v(x, \pm R^+, t) + v(x, \pm R^-, t)\big] - 2\mu u & \text{for } (x, t) \in \mathbb{R} \times \mathbb{R}^+, \\ \mp d\, v_y(x, R^{\pm}, t) = \mu\, u - \nu\, v(x, R^{\pm}, t) & \text{for } (x, t) \in \mathbb{R} \times \mathbb{R}^+, \\ \mp d\, v_y(x, -R^{\pm}, t) = \mu\, u - \nu\, v(x, -R^{\pm}, t) & \text{for } (x, t) \in \mathbb{R} \times \mathbb{R}^+, \end{cases} \tag{12}$$

with $(\underline{u}, \underline{v})$ having compact support and being symmetric about $\{y = 0\}$.[1]

Then c^* is the asymptotic speed of propagation of problem (4).

Proof Take $k > 0$ sufficiently large so that $k(\overline{u}, \overline{v})$, where $(\overline{u}, \overline{v})$ is the supersolution given by assumption (i) for $c = c^*$ with the "−" sign, lies above (u_0, v_0). Observe that, since system (10) is linear, $k(\overline{u}, \overline{v})$ is still a supersolution of (10) and thus, thanks to (KPP), it is a supersolution to (4).

Consider $c > c^*$ and $x > ct$; then, thanks to Proposition 2,

$$(u, v) < k e^{-\alpha(x - c^* t)}(1, \phi(y)) < k e^{\alpha(c^* - c)t}(1, \phi(y)) \to 0$$

as $t \to \infty$, proving the first part of the definition of asymptotic speed of propagation for the propagation to the right. For the propagation to the left we reason similarly, by taking the supersolution in (i) with the "+" sign.

On the other hand, using the subsolutions given by assumption (ii), one can prove, following the same lines of Theorem 4, that, for $c < c^*$, with c arbitrarily close to c^*,

$$\lim_{t \to \infty} (u(x \pm ct, t), v(x \pm ct, y, t)) = \left(\frac{\nu}{\mu}, 1\right),$$

and then the second part of the definition of asymptotic speed of propagation follows by applying [3, Lemma 4.1] (see also [19, Lemma 4.4] for a proof of it). □

3 Characterization of the Asymptotic Speed of Propagation for Problems with One Road

This section is devoted to the construction and a geometric characterization of the asymptotic speed of propagation for the road-field systems considered in the introduction having one road of different diffusion, i.e. Problems (1) and (3).

[1] This symmetry condition is not required—and even meaningless—when the domain is the upper half-plane.

The following lemma, whose proof is based on [2, Lemma 6.2], constructs subsolutions with the characteristics of assumption (ii) of Proposition 5 (with (12) adequately replaced in each case by the corresponding parabolic problem with additional transport terms $\pm cv_x$, $\pm cu_x$), when $D \leq 2d$, $0 < c < c_{\mathrm{KPP}}$, and the field has at least one component whose section is unbounded in y, i.e. for systems (1), (3) and (4). Proposition 5 will thus entail that the speed of propagation for these three systems is larger than or equal to c_{KPP}.

Lemma 1 *Let $D \leq 2d$ and $0 < c < c_{\mathrm{KPP}}$. Then, there exist arbitrarily small, nonnegative generalized stationary subsolutions of* (12) *(and the analogous versions corresponding to systems* (1) *and* (3)*) with compact support and symmetric about* $\{y = 0\}$ *(when the domain has this symmetry too).*

Proof We look for subsolutions of the form

$$\underline{v} = \varepsilon\psi(x)\cos\left(\beta\left(y - R - 1 - \frac{\pi}{2\beta}\right)\right) \tag{13}$$

for β, ε positive and small, $y \in \left(R + 1, R + 1 + \frac{\pi}{\beta}\right)$, and $\psi(x)$ nonnegative with compact support to be determined. We also take the reflection of $\underline{v}$ about $\{y = 0\}$ and extend to 0 elsewhere in the field.

Observe that, if $\underline{v}$ is small enough (i.e., if ε is small enough), solves

$$-d\Delta v \pm cv_x = (f'(0) - \delta)v \tag{14}$$

for $\delta \in (0, f'(0))$, $\delta \sim 0$, and its support is contained in the field, then, by taking $\underline{u} = 0$, we obtain a subsolution to (12). For (13) to solve (14), $\psi(x)$ has to satisfy

$$d\psi'' \mp c\psi' + (f'(0) - \delta - d\beta^2)\psi = 0,$$

thus, since $0 < c < c_{\mathrm{KPP}}$, for $\delta, \beta \sim 0$ $\psi(x)$ is given by $e^{\lambda x}$, where $\lambda \in \mathbb{C} \setminus \mathbb{R}$ is a root of the associated characteristic polynomial. In order to obtain a real solution, we take the real part of ψ and, to have compact support in x, we take only one oscillation and extend to 0 elsewhere. □

The following proposition, which summarizes the content of [2, Sections 5–6], gives a geometrical characterization of c^*_{hp} and, combined with Proposition 5, allows us to prove Theorem 1. We recall the elements of its proof, since they will be used also in the rest of this chapter.

Proposition 6 ([2]) *Problem* (1) *admits an asymptotic speed of propagation c^*_{hp} in the x direction which, for $D \leq 2d$, satisfies $c^*_{\mathrm{hp}} = c_{\mathrm{KPP}}$, while, for $D > 2d$, it* *satisfies* $c^*_{\mathrm{hp}} > c_{\mathrm{KPP}}$ *and is the smallest value of* c *for which the curves*

$$\alpha^-_{d,\mathrm{hp}}(c,\beta) := \frac{c - \sqrt{c^2 - c^2_{\mathrm{KPP}} - 4d^2\beta^2}}{2d}, \quad \alpha^+_{D,\mathrm{hp}}(c,\beta) := \frac{c + \sqrt{c^2 + \frac{4\cdot\mu\cdot Dd\beta}{\nu + d\beta}}}{2D} \tag{15}$$

have real intersections.

Proof Let us begin with the case $D \leq 2d$. Thanks to Lemma 1, in order to apply Proposition 5, it is sufficient to construct, for every $c \geq c_{\mathrm{KPP}}$ supersolutions of the linearization of (1) around $(0, 0)$ of the form (11). To this end, we take $\phi(y) = \gamma e^{-\beta y}$, with $\beta \geq 0$, $\gamma > 0$ and, by plugging into the linearized system, we obtain that such a candidate is a solution (respectively, a supersolution) if and only if the following algebraic system, involving the unknowns α, β, γ and the parameter c,

$$\begin{cases} c\alpha - d\alpha^2 - d\beta^2 = f'(0) \\ c\alpha - D\alpha^2 = \nu\gamma - \mu \\ d\beta\gamma = \mu - \nu\gamma \end{cases} \iff \begin{cases} c\alpha - d\alpha^2 - d\beta^2 = f'(0) \\ c\alpha - D\alpha^2 = \frac{-\mu d\beta}{\nu + d\beta} \\ \gamma = \frac{\mu}{\nu + d\beta} \end{cases} \tag{16}$$

is satisfied (respectively, if and only if it is satisfied with the equality signs replaced by "$\geq$").

The first equation in (16) describes, for $c \geq c_{\mathrm{KPP}}$, a circle $\Sigma_d(c)$ in the (β, α) plane with center $\left(0, \frac{c}{2d}\right)$ and radius $\frac{\sqrt{c^2 - c^2_{\mathrm{KPP}}}}{2d}$. Observe that $\Sigma_d(c)$ degenerates to its center as $c \downarrow c_{\mathrm{KPP}}$.

When $D \leq 2d$ and $c \geq c_{\mathrm{KPP}}$, by taking $(\alpha, \beta, \gamma) = \left(\frac{c}{2d}, 0, \frac{\mu}{\nu}\right)$, which amounts to consider the center of Σ_d in the (β, α) plane, the relations in (16) are satisfied with "$\geq$", and we have constructed the desired supersolution.

To treat the case $D > 2d$, we explicitly write the curve given by the second relation of (16) as a function of β and the parameter c, obtaining the curve $\alpha^+_{D,\mathrm{hp}}(c, \beta)$ defined in (15)—we only consider the branch with the "+" in front of the square root, since this will be enough for the construction, as it will be apparent from the following discussion. We observe that such a curve intersects the α-axis in the point $\left(0, \frac{c}{D}\right)$; thus, since $D > 2d$, the circle Σ_d arises, for $c = c_{\mathrm{KPP}}$, above $\alpha^+_{D,\mathrm{hp}}$. In addition, the lower part of the circle, which is parameterized by the function $\alpha^-_{d,\mathrm{hp}}(c, \beta)$, introduced in (15) as well, is decreasing with respect to c and converges to 0 as $c \to \infty$, while $\alpha^+_{D,\mathrm{hp}}(c, \beta)$ is increasing in c and tends to ∞ as $c \to \infty$. Therefore, since these curves are regular, there exists a least value of c, denoted by c^*_{hp}, which is greater than c_{KPP} and for which they intersect for the first time, being tangent, and they intersect strictly for every greater c.

To conclude the proof, we show, thanks to Proposition 5, that c^*_{hp} is the asymptotic speed of propagation. By construction, there are solutions of (16) for every $c \geq c^*_{\mathrm{hp}}$, providing solutions of the linearized system.

To construct compactly supported subsolutions for $c < c^*_{\rm hp}$, $c \sim c^*_{\rm hp}$, consider the truncation of Problem (1) obtained by considering $0 < y < L$ and imposing $v(x, L, t) = 0$. Reasoning as above, i.e. studying the corresponding system for α, β, γ, it is possible to construct solutions of the linearized truncated system with penalization, i.e. with $f'(0)$ replaced by $f'(0)-\delta$, of type $e^{\pm\alpha(x\pm ct)}(1, \gamma \sinh(\beta(L-y)))$ for c greater than or equal to a certain value $c^*(L, \delta) < c^*_{\rm hp}$. Moreover, for c smaller than $c^*(L, \delta)$, arbitrarily close to it, it is possible to show by using Rouché's theorem (see [2, Lemma 6.1]) that the system for α, β, γ has complex solutions, giving complex solutions of the linearized truncated system. Taking the real part of such solutions, which oscillates in x, considering only one oscillation—as in the proof of Lemma 1—extending to 0 and taking small multiples, gives a compactly supported subsolution to the original problem. Since $\lim_{(L,\delta)\to(\infty,0)} c^*(L, \delta) = c^*_{\rm hp}$, this procedure allows us to construct subsolutions satisfying assumption (ii) of Proposition 5 for $c < c^*_{\rm hp}$, arbitrarily close to it, which concludes the proof. □

This geometric characterization allows us to prove almost immediately Theorem 1, for which we recall once more the elements of the proof given in [2].

Proof (of Theorem 1) The existence of $c^*_{\rm hp}$ and parts (i) and (ii) are contained in Proposition 6.

Passing to (iii), we observe that if, by contradiction, $c^*_{\rm hp}(D)$ was bounded, the second curve in (15) would converge locally uniformly to 0 as $D \to \infty$, thus it would not have any intersection with the first one, against the characterization of $c^*_{\rm hp}(D)$ given in Proposition 6. □

This completes the review of the results for the half-plane with one road and we pass now to construct the speed of propagation for the case of a plane with one road (3), which is the content of the following proposition. Then, we give the proof of Proposition 1.

Proposition 7 *Problem* (3) *admits an asymptotic speed of propagation* $c^*_{\rm p1}$ *in the* x *direction which, for* $D \le 2d$, *satisfies* $c^*_{\rm p1} = c_{\rm KPP}$, *while, for* $D > 2d$, *it satisfies* $c^*_{\rm p1} > c_{\rm KPP}$ *and is the smallest value of* c *for which the curves*

$$\alpha^-_{d,\rm p1}(c,\beta) := \frac{c - \sqrt{c^2 - c^2_{\rm KPP} - 4d^2\beta^2}}{2d}, \quad \alpha^+_{D,\rm p1}(c,\beta) := \frac{c + \sqrt{c^2 + \frac{4\cdot 2\mu\cdot Dd\beta}{\nu + d\beta}}}{2D} \tag{17}$$

have real intersections.

Proof The construction follows the same lines of the one of Proposition 6 and relies on Proposition 5. On the one hand, one looks for supersolutions of type $\overline{u} = e^{\pm\alpha(x\pm ct)}$, $\overline{v} = \gamma e^{\pm\alpha(x\pm ct)-\beta y}$ for $y > 0$ and $\overline{v} = \gamma e^{\pm\alpha(x\pm ct)+\beta y}$ for $y < 0$, and obtains the algebraic system

$$\begin{cases} c\alpha - d\alpha^2 - d\beta^2 = f'(0) \\ c\alpha - D\alpha^2 = 2\nu\gamma - 2\mu \\ d\beta\gamma = \mu - \nu\gamma \end{cases} \iff \begin{cases} c\alpha - d\alpha^2 - d\beta^2 = f'(0) \\ c\alpha - D\alpha^2 = \frac{-2\mu d\beta}{\nu+d\beta} \\ \gamma = \frac{\mu}{\nu+d\beta}, \end{cases} \tag{18}$$

which leads, when $D > 2d$, to the search for real intersections between the curves (17).

On the other hand, the construction of compactly supported subsolutions for $c < c^*_{\text{pl}}$, $c \sim c^*_{\text{pl}}$, follows, in the case $D \leq 2d$, from Lemma 1 and, in the case $D > 2d$, by truncating in y and using Rouché's theorem to obtain complex solutions, exactly as indicated in the proof of Proposition 6. □

Proof (of Proposition 1) The existence of c^*_{pl}, its lower bound and the threshold for enhancement with respect to c_{KPP} have already been proved in Proposition 7.

It only remains to show that $c^*_{\text{pl}} < c^*_{\text{hp}}$ when $D > 2d$, and, for this, it is sufficient to observe that $\alpha^-_{d,\text{pl}}(c, \beta)$ in (17) coincides with $\alpha^-_{d,\text{hp}}(c, \beta)$ in (15), while $\alpha^+_{D,\text{pl}}(c, \beta)$ is obtained from $\alpha^+_{D,\text{hp}}(c, \beta)$ by replacing μ with 2μ. As a consequence, thanks to the geometric characterization given in Propositions 6 and 7, if we explicitly point out the dependence of c^*_{hp} and c^*_{pl} with respect to the parameter μ, we have that, always for $D > 2d$, $c^*_{\text{pl}}(\mu) = c^*_{\text{hp}}(2\mu)$, and, in order to get the conclusion, it is sufficient to show that $\mu \mapsto c^*_{\text{hp}}(\mu)$ is decreasing. To this end, observe that the function $\mu \mapsto \alpha^+_{D,\text{hp}}(c, \beta, \mu)$ is strictly increasing and, by construction, the curves $\alpha^+_{D,\text{hp}}(c^*_{\text{hp}}(\mu), \beta, \mu)$ and $\alpha^-_{d,\text{hp}}(c^*_{\text{hp}}(\mu), \beta)$ are tangent for every μ. Thus, if $\mu_1 < \mu_2$, $\mu_1 \sim \mu_2$, the curves $\alpha^+_{D,\text{hp}}(c, \beta, \mu_2)$ and $\alpha^-_{d,\text{hp}}(c, \beta)$ are strictly secant for $c = c^*_{\text{hp}}(\mu_1)$ and, due to the monotonicities in c, this parameter has to be decreased in order to obtain the value for which they intersect for the first time, which, thanks again to Proposition 6, provides us with $c^*_{\text{hp}}(\mu_2)$. □

4 Characterization of the Asymptotic Speed of Propagation for Problems with Two Roads

In this section we construct the asymptotic speed of propagation for the two remaining problems (2) and (4), those with two roads. We preliminarily observe that such problems are symmetric with respect to reflections about $\{y = 0\}$. For this reason, we will construct the super- and subsolutions needed to apply Proposition 5 with the same symmetry, i.e. we will look for functions defined only on $\{y > 0\}$ and satisfying $v_y(0^+) = 0$, and then will consider their even extension on $\{y < 0\}$.

We begin with problem (2) for a strip-shaped field $\{y \in (-R, R)\}$. In this case, the construction of the speed of propagation is more complicated than in the cases presented in Sect. 3, since the eigenvalue problem $-\phi''(y) = \lambda\phi(y)$ has two types of positive eigenfunctions in $(-R, R)$ which satisfy $\phi'(0)=0$: $\cos(\sqrt{\lambda}y)$ for $\lambda \in (0, \frac{\pi}{2R})$, and $\cosh(\sqrt{-\lambda}y)$ for $\lambda < 0$. This entails that we have to consider two types of supersolutions of the form (11): the first type with $\overline{v}_1 = \gamma e^{\pm\alpha(x\pm ct)}\cos(\beta y)$, $0 < \beta < \frac{\pi}{2R}$, and the second one $\overline{v}_2 = \gamma e^{\pm\alpha(x\pm ct)}\cosh(\beta y)$. The geometric characterization of the asymptotic speed of propagation that we obtain in this case is the following one, and the proof we provide summarizes the results of [19, Section 4].

Proposition 8 ([19]) *Problem* (2) *admits an asymptotic speed of propagation c^*_{st} in the x direction which is the smallest value of the parameter c for which either the curves,*

$$\alpha^{\pm}_{d,\text{st},1} := \frac{c \pm \sqrt{c^2 - c^2_{\text{KPP}} + 4d^2\beta^2}}{2d}, \quad \alpha^{\pm}_{D,\text{st},1} := \frac{c \pm \sqrt{c^2 - \frac{4\mu D d\beta\sin(\beta R)}{\nu\cos(\beta R) - d\beta\sin(\beta R)}}}{2D}, \tag{19}$$

or

$$\alpha^{-}_{d,\text{st},2} := \frac{c - \sqrt{c^2 - c^2_{\text{KPP}} - 4d^2\beta^2}}{2d}, \quad \alpha^{+}_{D,\text{st},2} := \frac{c + \sqrt{c^2 + \frac{4\mu D d\beta\sinh(\beta R)}{\nu\cosh(\beta R) + d\beta\sinh(\beta R)}}}{2D} \tag{20}$$

have real intersections in the first quadrant of the (β, α) plane (in (19) *we consider $0 < \beta < \overline{\beta} < \frac{\pi}{2R}$, where $\overline{\beta}$ is the first zero of the denominator inside the square root of $\alpha^{\pm}_{D,\text{st},1}$).*

If intersection first occurs between the curves (19)*, then c^*_{st} is said to be of* type 1, *and will be denoted by $c^*_{\text{st},1}$, otherwise, if intersection first occurs between the curves in* (20)*, then we say that c^*_{st} is of* type 2 *and we denote it by $c^*_{\text{st},2}$.*

Proof The proof follows similar lines as the ones of Sect. 3: to construct the above-mentioned supersolutions of type 1, one has to find solutions of the following system

$$\begin{cases} c\alpha - d\alpha^2 + d\beta^2 = f'(0) \\ c\alpha - D\alpha^2 = \nu\gamma\cos(\beta R) - \mu \\ -d\beta\gamma\sin(\beta R) = \mu - \nu\gamma\cos(\beta R) \end{cases} \iff \begin{cases} c\alpha - d\alpha^2 + d\beta^2 = f'(0) \\ c\alpha - D\alpha^2 = \frac{\mu d\beta\sin(\beta R)}{\nu\cos(\beta R) - d\beta\sin(\beta R)} \\ \gamma = \frac{\mu}{\nu\cos(\beta R) - d\beta\sin(\beta R)} \end{cases} \tag{21}$$

(observe that, for $0 < \beta < \overline{\beta}$, $\gamma > 0$); while for supersolutions of type 2, one reduces to system

$$\begin{cases} c\alpha - d\alpha^2 - d\beta^2 = f'(0) \\ c\alpha - D\alpha^2 = \nu\gamma\cosh(\beta R) - \mu \\ d\beta\gamma\sinh(\beta R) = \mu - \nu\gamma\cosh(\beta R) \end{cases} \iff \begin{cases} c\alpha - d\alpha^2 - d\beta^2 = f'(0) \\ c\alpha - D\alpha^2 = \frac{-\mu d\beta\sinh(\beta R)}{\nu\cosh(\beta R) + d\beta\sinh(\beta R)} \\ \gamma = \frac{\mu}{\nu\cosh(\beta R) + d\beta\sinh(\beta R)}. \end{cases} \tag{22}$$

System (21) leads to find intersections between the curves in (19), while (22) between those in (20).

In order to conclude, assume that $c^*_{\rm st}$ is of type 1. Then, by using a generalized version of Rouché's theorem, whose proof can be found in [19], it is possible to reason as in the proof of Proposition 6 to show that for $c < c^*_{\rm st}$, $c \sim c^*_{\rm st}$, system (21) admits complex solutions which can be used to construct the desired compactly supported subsolutions. The same can be done by using system (22) when $c^*_{\rm st}$ is of type 2. We remark that no truncation is needed here to obtain a compact support in y, since y is already bounded. □

Remark 1 By studying the dependence on c of the curves (19) and (20), one can observe that in both cases they have real intersections for sufficiently large c, thus both provide us with supersolutions of the problem. As a consequence, one might think that two different values for $c^*_{\rm st}$ can be obtained, one for each pair of curves. Nonetheless, the analysis of [19] (see in particular Section 4 and Proposition 4.1) guarantees that the construction of compactly supported subsolutions only works in one of the two cases, entailing in particular that the definition of the type of $c^*_{\rm st}$ given in Proposition 8 is well posed.

Proof (of Theorem 2) The existence follows from Proposition 8. Moreover, it is possible to show (see [19, Section 4]) that

$$\text{if } D \leq 2d, \quad c^*_{\rm st} = c^*_{{\rm st},1} \text{ for all } R > 0, \tag{23}$$

$$\text{if } D > 2d, \quad c^*_{\rm st} = \begin{cases} c^*_{{\rm st},1} & \text{for } R \in (0, R_M), \\ c^*_{{\rm st},2} & \text{for } R > R_M, \end{cases} \tag{24}$$

where we use the notation introduced in Proposition 8. We are now ready to prove the qualitative properties of $c^*_{\rm st}$.

(i) For $R \sim 0$, (23) and (24) give that $c^*_{\rm st} = c^*_{{\rm st},1}$. In addition, as $R \downarrow 0$, the curve $\alpha^{\pm}_{D,{\rm st},1}$ converges to the horizontal lines $\alpha = 0$ and $\alpha = c/D$. Thus, for any fixed c, there are always intersections between such a curve and $\alpha^{\pm}_{d,{\rm st},1}$, which connects, in the first quadrant of the (β, α) plane, the points $\left(\sqrt{\frac{f'(0)}{d}}, 0\right)$ and (∞, ∞). Proposition 8 therefore gives that $\lim_{R\downarrow 0} c^*_{\rm st}(R) = 0$.

(ii) We distinguish the cases $D \leq 2d$ and $D > 2d$. In the former one, thanks to (23), we only have to consider system (21), and we observe that, when $c = c_{\rm KPP}$, if we take $(\alpha, \beta, \gamma) = \left(\frac{c_{\rm KPP}}{2d}, 0, \frac{\mu}{\nu}\right)$, the first and third relation of such a system hold true, while the second one holds true with the "$\geq$" sign. Thus, $c^*_{\rm st}(R) < c_{\rm KPP}$ for all R and $\limsup_{R\to\infty} c^*_{\rm st}(R) \leq c_{\rm KPP} = c^*_{\rm hp}$. On the other hand, observe that the construction of compactly supported subsolutions of Lemma 1 can be carried out for sufficiently large R, entailing that $\liminf_{R\to\infty} c^*_{\rm st}(R) \geq c_{\rm KPP}$, which concludes the proof in this case.
When $D > 2d$, according to (24), $c^*_{\rm st}$ is of type 2 for sufficiently large R. Now, the convergence of $c^*_{\rm st}(R)$ to $c^*_{\rm hp}$ as $R \to \infty$ follows from the geometric characterizations given in Propositions 6 and 8, observing that $\alpha^{+}_{D,{\rm st},2}$ converges, as $R \to \infty$, to $\alpha^{+}_{D,{\rm hp}}$ locally uniformly in β.

(iii) Once again, if $D \leq 2d$, (23) guarantees that $c^*_{\rm st} = c^*_{\rm st,1}$ for all R. One then proves that the curves $\alpha^{\pm}_{D,{\rm st},1}$ shrink continuously as R increases, while the curves $\alpha^{\pm}_{d,{\rm st},1}$ do not depend on R, entailing that $c^*_{\rm st,1}$ is continuous and increasing.

(iv) To prove that, if $D > 2d$, $c^*_{\rm st}(R)$ is increasing for $R \in (0, R_M)$, we use (24) and reason as in the previous point. Similarly, we use (24) and the fact that $\alpha^{+}_{D,{\rm st},2}$ increases, with respect to R, to obtain that $c^*_{\rm st} = c^*_{\rm st,2}$ is decreasing for $R > R_M$.
The continuity of the function $R \mapsto c^*_{\rm st}$ is obvious for $R \neq R_M$, since the curves in (19) and (20) depend continuously on R. For $R = R_M$, the conclusion is not direct, since a transition of type occurs. Nevertheless, one proves that $\alpha^{+}_{d,{\rm st},1}$ and $\alpha^{-}_{D,{\rm st},1}$ do not play a role for $R = R_M$, and observes that, for $\beta = 0$, the remaining curves in (19) and those in (20) match in a differentiable way, which allows us to obtain the continuity of $c^*_{\rm st}(R)$ also for $R = R_M$.
The existence and properties of $R_{\rm hp}$ and R_K now follow directly form the continuity and monotonicity properties of $c^*_{\rm st}$, together with properties (i) and (ii) of Theorem 2 and (ii) of Theorem 1. □

Finally, we pass to the case of the plane with two roads (4), for which, as already remarked, Lemma 1 entails that $c^*_{\rm p2} \geq c_{\rm KPP}$.

Differently from the case of the strip, here it is enough to consider only one type of supersolution. Indeed, the unique *positive* eigenfunctions of $-\phi''(y) = \lambda\phi(y)$ for $|y| > R$ are the ones of exponential type. The differential equation in the field being the same for $|y| < R$, this forces to take $\overline{v}(x, y, t) = \gamma_1 e^{\pm\alpha(x\pm ct)} \cosh(\beta y)$ in $(-R, R)$ (recall that, by symmetry, we consider functions satisfying $\overline{v}_y(x, 0^+, t) = 0$), excluding in this way the cosine. Moreover, in analogy with the constructions of Sect. 3, we take $\overline{u}(x, t) = e^{\pm\alpha(x\pm ct)}$ and $\overline{v}(x, y, t) = \gamma_2 e^{\pm\alpha(x\pm ct)-\beta(y-R)}$ for $y > R$. As usual, the constants $\alpha, \beta, \gamma_1, \gamma_2$ will be sought to be positive.

After these preliminaries, we show in the following proposition that these supersolutions suffice for the construction and characterization of the speed of propagation for this problem.

Proposition 9 *Problem* (4) *admits an asymptotic speed of propagation in the x direction* $c^*_{\rm p2}$, *which, for* $D \leq 2d$, *satisfies* $c^*_{\rm p2} = c_{\rm KPP}$, *while, for* $D > 2d$, *it satisfies* $c^*_{\rm p2} > c_{\rm KPP}$ *and is the smallest value of c for which the curves*

$$
\begin{aligned}
\alpha^{-}_{d,{\rm p2}}(c, \beta) &:= \frac{c - \sqrt{c^2 - c^2_{\rm KPP} - 4d^2\beta^2}}{2d}, \\
\alpha^{+}_{D,{\rm p2}}(c, \beta) &:= \frac{c + \sqrt{c^2 + 4\mu D\left(\frac{d\beta}{\nu+d\beta} + \frac{d\beta\sinh(\beta R)}{\nu\cosh(\beta R)+d\beta\sinh(\beta R)}\right)}}{2D}
\end{aligned}
\qquad (25)
$$

have real intersections.

Proof By plugging the above described candidate to supersolution into the linearization, the system that we obtain in this case reads

$$\begin{cases} c\alpha - d\alpha^2 - d\beta^2 = f'(0) \\ c\alpha - D\alpha^2 = \nu(\gamma_1 \cosh(\beta R) + \gamma_2) - 2\mu \\ d\beta\gamma_1 \sinh(\beta R) = \mu - \nu\gamma_1 \cosh(\beta R) \\ d\beta\gamma_2 = \mu - \nu\gamma_2 \end{cases} \iff$$

$$\iff \begin{cases} c\alpha - d\alpha^2 - d\beta^2 = f'(0) \\ c\alpha - D\alpha^2 = -\frac{\mu d\beta}{\nu+d\beta} - \frac{\mu d\beta \sinh(\beta R)}{\nu\cosh(\beta R)+d\beta\sinh(\beta R)} \\ \gamma_1 = \frac{\mu}{\nu\cosh(\beta R)+d\beta\sinh(\beta R)} \\ \gamma_2 = \frac{\mu}{\nu+d\beta}. \end{cases} \tag{26}$$

As in Sect. 3, if $D \leq 2d$, $(\alpha, \beta, \gamma_1, \gamma_2) = \left(\frac{c}{2d}, 0, \frac{\mu}{\nu}, \frac{\mu}{\nu}\right)$ provides us with supersolutions to (26) for $c \geq c_{\mathrm{KPP}}$. Thus, Proposition 5 and Lemma 1 allow us to conclude that $c^*_{\mathrm{p2}} = c_{\mathrm{KPP}}$ in this case. When $D > 2d$, instead, intersections between the curves in (25) provide us with solutions to (26) and, as a consequence, supersolutions to (4). Thanks to the monotonicity with respect to c, intersections between such curves exist for c greater than or equal to a certain value (greater than c_{KPP}) for which, as usual, the curves are tangent, and which will turn out to be c^*_{p2}.

Indeed, in order to apply Proposition 5, we only have to obtain compactly supported subsolutions in the case $D > 2d$ for $c < c^*_{\mathrm{p2}}$, $c \sim c^*_{\mathrm{p2}}$. To do so, we proceed as in the proof of Proposition 6: we consider the truncation at $y = L$, with $L \gg R$, by imposing $v(x, L, t) = 0$, and, by using Rouché's theorem, we prove that, for $c < c^*_{\mathrm{p2}}$, $c \sim c^*_{\mathrm{p2}}$, the associated linearized system with penalization admits complex solutions which allow us to obtain subsolutions whose support is compact also in the x variable. □

Proof (of Theorem 3) The existence part, (i) and the first part of (ii) are contained in Proposition 9. In (ii), it remains to prove the behavior with respect to R in the case $D > 2d$: the continuity and monotonicity follow since the map $R \mapsto \alpha^+_{D,\mathrm{p2}}$ is continuous and increasing, while $\alpha^-_{d,\mathrm{p2}}$ does not depend on R. Thus, since the curves in (25) are tangent for $c = c^*_{\mathrm{p2}}(R)$, if $R' > R$, $R' \sim R$, $\alpha^-_{d,\mathrm{p2}}(c, \beta)$ and $\alpha^+_{D,\mathrm{p2}}(c, \beta, R')$ are strictly secant for $c = c^*_{\mathrm{p2}}(R)$, and c has to be reduced in order to obtain the tangency situation.

The curve $\alpha^+_{D,\mathrm{p2}}$ converges locally uniformly to $\alpha^+_{D,\mathrm{hp}}$ as $R \downarrow 0$, and to $\alpha^+_{D,\mathrm{p1}}$ as $R \to \infty$. This proves the limits in (5). The fact that $c^*_{\mathrm{p2}}(R) > c^*_{\mathrm{p1}}$ follows from the monotonicity of $c^*_{\mathrm{p2}}(R)$ and the second limit in (5).

Passing to part (iii), by the continuity of $c^*_{\mathrm{st}}(R)$ and the properties of R_{hp}, we have $c^*_{\mathrm{st}}(R_{\mathrm{hp}}) = c^*_{\mathrm{hp}} > c^*_{\mathrm{p2}}(R_{\mathrm{hp}})$, where the last inequality follows from (ii) here. On the other hand, the properties of R_K, together with (ii) of Proposition 1 and (ii) here, give $c^*_{\mathrm{st}}(R_K) = c_{\mathrm{KPP}} < c^*_{\mathrm{p1}} < c^*_{\mathrm{p2}}(R_K)$. The existence of R^* and its properties now follow by the continuity and monotonicities of $c^*_{\mathrm{st}}(R)$ and $c^*_{\mathrm{p2}}(R)$. □

Acknowledgements This work has been supported by the Spanish Ministry of Economy, Industry and Competitiveness through contract Juan de la Cierva Incorporación IJCI-2015-25084 and project MTM2015-65899-P, and by the European Research Council under the European Union's Seventh Framework Programme (FP/2007-2013)/ERC Grant Agreement n.321186—ReaDi "Reaction-Diffusion Equations, Propagation and Modelling".

The author wishes to thank the anonymous referee for his/her comments which have improved the presentation of the results of this work.

References

1. D.G. Aronson, H.F. Weinberger, Multidimensional nonlinear diffusion arising in population genetics. Adv. Math. **30**, 33–76 (1978)
2. H. Berestycki, J.-M. Roquejoffre, L. Rossi, The influence of a line with fast diffusion on Fisher-KPP propagation. J. Math. Biol. **66**, 743–766 (2013)
3. H. Berestycki, J.-M. Roquejoffre, L. Rossi, Fisher-KPP propagation in the presence of a line: further effects. Nonlinearity **26**, 2623–2640 (2013)
4. H. Berestycki, A.-C. Coulon, J.-M. Roquejoffre, L. Rossi, Speed-up of reaction-diffusion fronts by a line of fast diffusion, in *Séminaire Laurent Schwartz—Équations aux Dérivées Partielles et Applications. Année 2013–2014*, Exp. No. XIX, pp. 25 (Ed. Éc. Polytech., Palaiseau, 2014)
5. H. Berestycki, A.-C. Coulon, J.-M. Roquejoffre, L. Rossi, The effect of a line with nonlocal diffusion on Fisher-KPP propagation. Math. Models Methods Appl. Sci. **25**, 2519–2562 (2015)
6. H. Berestycki, J.-M. Roquejoffre, L. Rossi, The shape of expansion induced by a line with fast diffusion in Fisher-KPP equations. Comm. Math. Phys. **343**, 207–232 (2016)
7. H. Berestycki, J.-M. Roquejoffre, L. Rossi, Travelling waves, spreading and extinction for Fisher-KPP propagation driven by a line with fast diffusion. Nonlinear Anal. **137**, 171–189 (2016)
8. L. Dietrich, Existence of travelling waves for a reaction–diffusion system with a line of fast diffusion. Appl. Math. Res. Express. **2015**(2), 204–252 (2015)
9. L. Dietrich, Velocity enhancement of reaction-diffusion fronts by a line of fast diffusion. Trans. Amer. Math. Soc. **369**, 3221–3252 (2017)
10. L. Dietrich, J.-M. Roquejoffre, Front propagation directed by a line of fast diffusion: large diffusion and large time asymptotics. J. Éc. Polytech. Math. **4**, 141–176 (2017)
11. R. Ducasse, Influence of the geometry on a field-road model: the case of a conical field. J. London Math. Soc. **97**, 441–469 (2018)
12. N.R. Faria et al., The early spread and epidemic ignition of HIV-1 in human populations. Science **346**, 56–61 (2014)
13. R.A. Fisher, The wave of advantage of advantageous genes. Ann. Eugen. **7**, 355–369 (1937)
14. T. Giletti, L. Monsaingeon, M. Zhou, A KPP road-field system with spatially periodic exchange terms. Nonlinear Anal. **128**, 273–302 (2015)
15. A.N. Kolmogorov, I.G. Petrovskii, N.S. Piskunov, Étude de l'équation de la diffusion avec croissance de la quantité de matière et son application àă un problème biologique. Bull. Univ. État Moscou, Sér. Intern. A **1**, 1–26 (1937)
16. A. Pauthier, Uniform dynamics for Fisher-KPP propagation driven by a line of fast diffusion under a singular limit. Nonlinearity **28**, 3891–3920 (2015)
17. A. Pauthier, The influence of nonlocal exchange terms on Fisher-KPP propagation driven by a line of fast diffusion. Commun. Math. Sci. **14**, 535–570 (2016)
18. C. Robinet, C. Suppo, E. Darrouzet, Rapid spread of the invasive yellow-legged hornet in France: the role of human-mediated dispersal and the effects of control measures. J. Appl. Ecol. **54**, 205–215 (2017)

19. L. Rossi, A. Tellini, E. Valdinoci, The effect on Fisher-KPP propagation in a cylinder with fast diffusion on the boundary. SIAM J. Math. Anal. **49**, 4595–4624 (2017)
20. A. Tellini, Propagation speed in a strip bounded by a line with different diffusion J. Differ. Equ. **260**, 5956–5986 (2016)

www.ingramcontent.com/pod-product-compliance
Ingram Content Group UK Ltd.
Pitfield, Milton Keynes, MK11 3LW, UK
UKHW021832270726
14058UKWH00001B/111